The Regional Geography of Canada

4.18 Annual Unemployment Rates by Province, 1986–2002   201
4.19 Average Earnings for All Earners, Canada, Provinces, and Territories,
        1980, 1990, and 2000   203
4.20 Major Phases for the Aboriginal Population in Canada   207
4.21 Population with French Mother Tongue, 1951–2001   212
4.22 Canada's Population by Mother Tongue, 1951–2001   213
4.23 Regional Population of Canada by Percentage, 1901–2001   214
5.1   Employment by Industrial Sector in Ontario, 2002   243
5.2   Automobile-Assembly Plants in Southern Ontario, 2002   255
5.3   Census Metropolitan Areas in Ontario, 1996–2001   262
5.4   Top 10 Countries by Birth of Immigrants to Toronto, 1991–2001   266
5.5   Top 10 Ethnic Origins in Ottawa–Hull, 2001   268
5.6   Urban Centres in Northern Ontario, 1996–2001   271
6.1   Timeline: Historical Milestones in New France   290
6.2   Timeline: Historical Milestones in the British Colony   293
6.3   Timeline: Historical Milestones in Confederation   294
6.4   Employment by Industrial Sector in Québec, 2002   301
6.5   Comparison of Ontario and Québec Industrial Structures, 1995 to 2002   302
6.6   Major Cities and Census Metropolitan Areas in Québec, 1996–2001   320
6.7   Population Change: Montréal and Toronto, 1951–2001   323
6.8   Population of Cities of Northern Québec, the North Shore,
        and the Gaspé Peninsula   325
7.1   Employment by Industrial Sector in British Columbia, 2002   354
7.2   Comparison of Ontario and British Columbia Industrial Structures,
        1995 to 2002   355
7.3   Manufacturing Shipments by Industry, British Columbia, 2001   356
7.4   Fishery Statistics, British Columbia, 2001   357
7.5   Mineral Production in British Columbia, 2001   360
7.6   Timber Harvest by Species, British Columbia, 2001   364
7.7   Forest Product Exports from British Columbia, 2001   365
7.8   Lumber Production by Provinces/Territories, 2002   372
7.9   Major Urban Centres in British Columbia, 1996–2001   373
7.10 Smaller Urban Centres in British Columbia, 1996–2001   376
8.1   Number of Farms in Western Canada, 1971–2001   400
8.2   Average Size of Farms in Western Canada, 1971–2001   400
8.3   Basic Statistics for Western Canada by Province, 2001   405
8.4   Employment by Industrial Sector in Western Canada, 2002   407
8.5   Shift in Acreage: Wheat, Durum, and Canola, 1986–7, 1995–6,
        and 2001–2   418
8.6   Shift in Price: Wheat, Durum, and Canola, 1986–7, 1994–5,
        and 2001–2   418
8.7   Mining Production by Value in Western Canada, 2002   421

8.8   Forest Area and Production in Western Canada, 1999   424
8.9   Major Urban Centres in Western Canada, 1996–2001   425
9.1   Basic Statistics for Atlantic Canada by Province, 2001   439
9.2   Population Change in Atlantic Canada, 1996–2001   450
9.3   Employment by Industrial Sector in Atlantic Canada, 2002   454
9.4   Atlantic Canada and Newfoundland Fisheries Landings, 1990–2001   462
9.5   Mineral Production in Atlantic Canada, 2002   466
9.6   Urban Centres in Atlantic Canada, 1996–2001   477
10.1 Comprehensive Land-Claim Agreements in the Territorial North   505
10.2 Population and Aboriginal Population, Territorial North, 2001   506
10.3 Components of Population Growth for the Territories, 1980
        and 2000–1   507
10.4 Net Number of Migrants and Migration Rates, 1986–2001   507
10.5 Estimated Employment by Industrial Sector in the Territorial
        North, 2003   512
10.6 Mineral Production in the Territorial North, 2002   525

Preface

The purpose of this book is to introduce university students to Canada's regional geography. In studying the regional geography of Canada, the student not only gains an appreciation for the country's amazing diversity but also learns how its regions interact with one another. By developing the central theme that Canada is a country of regions, this text presents a number of images of Canada, revealing its physical, cultural, and economic diversity as well as its regional complexity. A number of features such as photos, key terms, maps, vignettes, tables, graphs, and references and further readings are designed to facilitate and enrich student learning.

The Regional Geography of Canada divides Canada into six geographic regions: Ontario, Québec, British Columbia, Western Canada, Atlantic Canada, and the Territorial North. Each region has a particular regional geography, history, population, and a unique location. These factors have determined each region's character, set the direction for its development, and created a sense of place. In examining these themes, this book underscores the dynamic nature of Canada's regional geography. Part of the dynamism of Canada's regional geography is the changing relationship between Canada's six regions. Trade liberalization provides one element of change while regional tensions provide another element of change.

Following World War II, the liberalization of world trade and the creation of the North American trading bloc exposed Canada's economy to new challenges and dangers. The consequences of trade liberalization were sixfold:

- Canada's old spatial economic structure was transformed from one that was national to one that is continental.
- Canada's economy was integrated into the North American market.
- Canada's manufacturing sector was restructured.
- Rates of unemployment first increased and then declined.
- Canada's regional economies were reoriented to adjust to the larger North American market.
- Canada's economic growth and regional development have become tied more and more to trade with the United States.

This book explores both the national and regional implications of these economic changes and proposes scenarios for greater stability in the new global economic order.

Regional tensions have always existed in Canada. These tensions often place pressure on the existing social and political systems. This pressure for change reflects the 'potential' for a redress of the division of power within Canada's federation. From this perspective, regional tensions can lead to a shift of power within a region and may even alter relations between regions and/or the federal government. In this book, four tensions or faultlines are examined. These tensions occur between Aboriginal and non-Aboriginal Canadians, French and English Canadians, centralist and decentralist forces, and new immigrants ('newcomers') and those people born in Canada ('old-timers'). Each of these tensions exists in all parts of Canada but, because of spatial variations in Canada's human and/or physical geography, each appears more prominently in a particular region or regions. For instance, the Aboriginal/non-Aboriginal faultline is most deeply felt in the Territorial North, where a larger proportion of the population is Aboriginal. The French/English faultline manifests itself most commonly in Québec. The centralist/decentralist faultline is frequently played out between the federal government and Western Canada, often over resources. The impact of immigration is heavily concentrated in the three major cities of Toronto, Vancouver, and Montréal. This book explores the nature of these faultlines, the need to reach compromises, and the fact that these faultlines provide the country with its greatest strength—diversity. Ultimately, these faultlines are shown to be not divisive forces but forces of change that ensure Canada's existence as a country of regions.

Organization of the Text

This book consists of 11 chapters. Chapters 1 through 4 deal with general topics related to Canada's national and regional geographies— Canada's physical, historical, and human geography—thereby setting the stage for a discussion of the six main geographic regions of Canada. Chapters 5 through 10 focus on these six geographic regions. The core/periphery model provides a guide for the ordering of these six regions. The regional discussion begins with Ontario and Québec, which represent the traditional demographic, economic, and political core of Canada. The core regions are followed by British Columbia, Western Canada, Atlantic Canada, and the Territorial North. Chapter 11 provides a conclusion.

Chapter 1 discusses the nature of regions and regional geography, including the core/periphery model and its applications. Chapter 2 introduces the major physiographic regions of Canada and other elements of physical geography that affect Canada and its regions. Chapter 3 is devoted to Canada's historical geography, such as its territorial evolution and the emergence of regional tensions and regionalism. This discussion is followed, in Chapter 4, by an examination of the basic demographic, economic, and social factors that influence both Canada and its regions. This chapter explores the national and global forces that have shaped Canada's regions as well as the features of its population (size, urbanization, etc.). To sharpen our awareness of how economic forces affect local and regional developments, three major themes running throughout this text are introduced in these first four chapters. The primary theme is that Canada is a country of regions. Two secondary themes reflect the recent shift in economic circumstances and its effects on regional geography. One is the integration of the North American economy, and the second is the changing world economy. These two economic factors, described as continentalism and globalization, exert both positive and negative impacts on Canada and its regions. These economic forces are explored through the core/periphery model.

In Chapters 5 to 10, the text moves from a broad, national overview to a more regional focus. Each of these six chapters profiles one of Canada's large geographic regions: Ontario, Québec, British Columbia, Western Canada, Atlantic Canada, and the Territorial North. These regional chapters explore the physical and human characteristics that distinguish each region from the others and that give each region its special sense of place. To emphasize the economic specialty of each region, a predominant economic activity is identified and explored through a *Key Topic*. They are: the Automobile Industry in Ontario; Hydro-Québec in Québec; Forestry in British Columbia; (Agricultural) Transition in Western Canada; the Fishing Industry in Atlantic Canada; and Megaprojects in the Territorial North. From this presentation, the unique character of each region emerges. The book concludes with Chapter 11, which discusses the future of Canada as a country of regions.

Third Edition

The third edition of this text on the regional geography of Canada has involved four major revisions. First, I have added a fourth faultline that deals with immigration. While newcomers to Canada formed an essential part of its early historical development, these settlers came almost exclusively from one corner of the world—Europe. In the 1960s, a fundamental and far-reaching change in immigration policy took place and this change opened up the opportunity for people from Africa, Asia, and Latin America to come to Canada. The flow of these newcomers has been significant and now provides the main drive behind Canada's population growth. In turn, the ethnic composition and social fabric of Canada have become more diversified and this cultural and racial diversity is seen mainly in the great cities of Canada.

Second, I have brought the contemporary discussion of Canada's regional geography into the twenty-first century by looking at recent developments in the globalization phenomenon, by examining the sharper edge to Canada–US relations in the post-11 September 2000 era, and by using the vast data set found in the 2001 census. Many new tables and graphs are found in Chapter 4 ('Canada's Human Face') and in the six regional chapters.

Third, I have revamped each of the key focus topics in each of the regional chapters. In Ontario, for instance, the automobile industry faces new challenges—the safety net provided by the Auto Pact is gone, leaving Ontario's Big Three automotive assembly plants vulnerable to stiff competition from foreign producers, especially from Japan's leading automobile producers, Honda and Toyota. Over the last decade, both Japanese firms have increased their market share of automobile sales in Canada and Toyota is poised to overtake DaimlerChrysler as the third largest producer of automobiles in Canada. The key topics discussed in Atlantic Canada, British Columbia, and Western Canada have a common theme—hard times. Atlantic Canada's cod fishery remains in difficulty because of the demise of the northern cod stocks. By 2003, the cod stocks had not yet recovered and the impact on outport communities has been very hurtful. The forest industry in British Columbia has been particularly hard hit by the US government's imposition of duties on softwood lumber entering the United States and, more recently, the rising value of the Canadian dollar, which makes exports to the United States more expensive. Western Canada took a double blow—low grain prices and the collapse of beef exports due to the discovery of a single case of mad cow disease in Alberta. Grain farmers in Western Canada continue to face low prices, high farm subsidies in Europe and the United States, and recently imposed duties on wheat exports to the United States. The key topics for Québec (hydroelectricity) and the Territorial North (megaprojects) continue to flourish. As the demand and price for electricity inevitably increase, the great hydroelectric resources found in northern Québec take on an even more dominant place in that province's economy and in its strategy for industrial development. With the resource development agreements with the Cree and Inuit, the door is now open for additional hydroelectric projects in northern Québec. The key topic for the Territorial North is megaprojects. By far, the most outstanding megaproject is based on the fabulous diamond deposits in the Northwest Territories.

Fourth, I have given more attention to the effects of industrialization on the environment. The reader is directed to the section entitled 'Environmental Challenges' in Chapter 2 ('Canada's Physical Base') and to each regional chapter where a critical environmental issue is portrayed. Chapter 2 has an expanded discussion on the negative consequences of industry on our natural environment. Special attention is paid to acid rain and global warming.

Acknowledgements Past and Present

I have continued to rely on the help of colleagues across Canada. With each edition, I have benefited from the constructive comments of anonymous reviewers selected by Oxford University Press. I especially owe a debt of thanks to one of those anonymous reviewers who spiced his critical comments with words of encouragement that kept me going.

Professors at the University of Saskatchewan were particularly helpful—several provided me with their comments on specific matters that fall under their areas of specialization while some were pressed into reviewing draft versions of regional chapters. Most contributed photographs. These colleagues and friends include Alec Aitken, Bill Archibold, William Barr, Dirk de Boer, K.I. Fung, the late Walter Kupsch, Lawrence Martz, the late John McConnell, John Newell, Jim Randall, Jack Stabler, and Mike Wilson. Professors

from other institutions who played a role in shaping this book include Jim Miller, Robert MacKinnon, and Gilles Viaud from Cariboo College; Hugh Gayler and Dan McCarthy from Brock University; Ben Moffat from Medicine Hat College; and Keith Storey from Memorial University.

Thanks to a sabbatical leave granted to me by the University of Saskatchewan, I was able to spend the fall term of 1996 at the Département de géographie at the Université de Laval. My generous hosts not only put up with my limited command of French but, more importantly, contributed significantly to the design and content of my chapter on Québec. Among those contributing to my education were Louis-Edmond Hamelin, who must be considered the Dean of Canadian geographers; Jean-Jacques Simard, a well-known sociologist who has written extensively on Québec's north; and Benoit Robitaille, the Chair of the Département de géographie at Laval. Discussions of critical issues with Jacques Bernier, Marc St-Hilaire, and Eric Waddell proved extremely fruitful. For me, one of the highlights at Laval was the opportunity to present my ideas on Québec at one of the Friday afternoon seminars organized by Dean Louder. The constructive and stimulating responses from the cultural geographers attending that seminar were especially helpful to me.

Maps are a critical element in geography. I must thank Keith Bigelow, the cartographer in the Department of Geography at the University of Saskatchewan, for the help he provided on the maps. The library staff in Government Documents at the University of Saskatchewan was particularly helpful in my search for detailed information on Canada's regions.

The National Atlas of Canada and Statistics Canada have created important Web sites for geography students. These provide access to a wide range of geographic data and maps. The National Atlas of Canada has placed many of its printed maps on its Web site <www.atlas. gc.ca>. Statistics Canada has published key elements of its 2001 census data by provinces and territories on its Web site <www.statcan.ca>. Statistics Canada has also published these key statistics for all urban places on another Web site <ww2.statcan.ca>. Students can also access annually published statistical data and population estimates from the Statistics Canada site under the heading CANSIM (Canadian Socio-economic Information Management System). Joel Yan of Statistics Canada was particularly helpful in identifying suitable data from CAN-SIM used in this book.

The staff at Oxford University Press, but particularly Laura Macleod, Mark Piel, and Phyllis Wilson, made the preparation of the third edition a pleasant and rewarding task. Richard Tallman, who diligently and skillfully edited my manuscript into a polished finished product, deserves special mention.

Finally, a special note of appreciation to my wife, Karen.

Note To Students and Instructors

A package of Student's Resources to accompany this text is available at:
www.oup.com/ca/he/companion/bone
Login: Regional
Password: hudsonbay

Instructors, please contact your local Oxford University Press sales representative for further information regarding the instructor's manual and test item file.

Overview and Objectives

This chapter provides an introduction to the nature and purpose of regional geography, along with a rationale for examining Canada under a regional microscope. Four themes are examined: (1) the importance of history and physical geography in shaping Canada's regions; (2) the role of a sense of place within each region; (3) the emergence of tensions between regions and within Canadian society; and (4) the value of the core/periphery model. Canada has six geographic regions, each with its particular history and physical geography, sense of place, and position within the core/periphery model. Each region's major characteristics are presented in a series of regional profiles. To underline the fluid interaction among regions and with the outside world, a conceptual framework based on the core/periphery model is presented and then discussed. The implications for Canada and its geographic regions are then outlined.

- Define regional geography.
- Explore the regional perspective.
- Consider the emergence of tensions.
- Discuss the notion of a sense of place.
- Present the geographic regions of Canada.
- Describe the dynamic nature of regions.
- Examine the core/periphery model.
- Apply that conceptual framework to Canada.
- Outline a series of regional profiles.

The Regional Geography of Canada

Third Edition

Robert M. Bone

OXFORD
UNIVERSITY PRESS

OXFORD
UNIVERSITY PRESS

70 Wynford Drive, Don Mills, Ontario M3C 1J9
www.oup.com/ca

Oxford University Press is a department of the University of Oxford.
It furthers the University's objective of excellence in research, scholarship,
and education by publishing worldwide in

Oxford New York

Auckland Bangkok Buenos Aires Cape Town Chennai
Dar es Salaam Delhi Hong Kong Istanbul Karachi Kolkata
Kuala Lumpur Madrid Melbourne Mexico City Mumbai Nairobi
São Paulo Shanghai Taipei Tokyo Toronto

Oxford is a trade mark of Oxford University Press
in the UK and in certain other countries

Published in Canada
by Oxford University Press

Copyright © Oxford University Press Canada 2005

National Library of Canada Cataloguing in Publication

Bone, Robert M.
The regional geography of Canada / Robert M. Bone. — 3rd ed.

Includes bibliographical references and index.
ISBN 0-19-541933-2

1. Canada—Geography—Textbooks. I. Title.

FC76.B66 2004   917.1   C2004-903877-X

Cover and text design: Brett J. Miller

1 2 3 4 - 08 07 06 05
This book is printed on permanent (acid-free) paper ∞.
Printed in Canada

Contents Overview

Detailed Contents   vi
List of Figures   xii
List of Tables   xiv
Preface   xvii

Chapter 1   **Regions of Canada**   2
Chapter 2   **Canada's Physical Base**   42
Chapter 3   **Canada's Historical Geography**   86
Chapter 4   **Canada's Human Face**   156
Chapter 5   **Ontario**   228
Chapter 6   **Québec**   280
Chapter 7   **British Columbia**   336
Chapter 8   **Western Canada**   386
Chapter 9   **Atlantic Canada**   436
Chapter 10   **The Territorial North**   486
Chapter 11   **Canada: A Country of Regions**   536

             Appendix   550
             Additional Data on Exports, Income,
             Employment, Immigrants, Aboriginal
             Population, and Official Languages
             Index   565

Detailed Contents

List of Figures   xii
List of Tables   xiv
Preface   xvii

Chapter 1   **Regions of Canada**   2
             Overview and Objectives   2
             Introduction   3
             Geography as a Discipline   5
             The Nature of Regions   9
             The Nature of Regional Geography   10
             Canada's Geographic Regions   12
             The Dynamic Nature of Regions   15
             Economic Integration   16
             A Spatial Framework: The Core/Periphery Model   18
             Canada and the Core/Periphery Model   21
             Regional Profiles   24
             Understanding Canada's Regions   33
             Summary   34
             Notes   34
             Key Terms   37
             Bibliography   38
             Further Reading   40

Chapter 2   **Canada's Physical Base**   42
             Overview and Objectives   42
             Introduction   43
             Physical Variations within Canada   43
             The Nature of Landforms   45
             Physiographic Regions   46
             The Impact of Physiography on Human Activity   57
             Geographic Location   58
             Climate   59
             Climatic Zones   65
             Permafrost   69
             Major Drainage Basins   72

Environmental Challenges 75
Summary 79
Notes 79
Key Terms 81
Bibliography 83
Further Reading 84

Chapter 3   **Canada's Historical Geography** 86
Overview and Objectives 86
Introduction 87
The First People 88
The Second People 95
The Third People 99
The Territorial Evolution of Canada 101
Faultlines 108
The Centralist/Decentralist Faultline 109
The Aboriginal/Non-Aboriginal Faultline 121
The Immigration Faultline 128
The French/English Faultline 134
Summary 147
Notes 148
Key Terms 151
Bibliography 152
Further Reading 154

Chapter 4   **Canada's Human Face** 156
Overview and Objectives 156
Introduction 157
Canada's Population 161
Population by Geographic Region 165
Urban Population 168
Population Change 171
Age and Sex Structure 176
Population Trends and Canadian Society 178
The Economy 191
Canada's Economy and Labour Force 196
Population Trends and Demographic Faultlines 203
Summary 217
Notes 218

Key Terms		219
Bibliography		221
Further Reading		227

Chapter 5	**Ontario**	**228**
	Overview and Objectives	228
	Introduction	229
	Ontario within Canada	229
	Ontario's Physical Geography	230
	Environmental Challenges	232
	Ontario's Historical Geography	234
	Ontario Today	237
	Southern Ontario	245
	Key Topic: The Automobile Industry	250
	Northern Ontario	256
	Ontario's Urban Geography	261
	Ontario's Future	272
	Summary	273
	Notes	273
	Key Terms	274
	Bibliography	275
	Further Reading	278

Chapter 6	**Québec**	**280**
	Overview and Objectives	280
	Introduction	281
	Québec within Canada	283
	Québec's Physical Geography	284
	Environmental Challenges	287
	Québec's Historical Geography	287
	Québec Today	295
	Québec's Economy	299
	Southern Québec	302
	Key Topic: Hydro-Québec	307
	Northern Québec	312
	Québec's Urban Geography	321
	Québec's Future	325
	Summary	328
	Notes	328
	Key Terms	330

Bibliography 330
Further Reading 334

Chapter 7 **British Columbia** **336**
Overview and Objectives 336
Introduction 337
British Columbia within Canada 338
Environmental Challenges 340
Land Claims and the Aboriginal/Non-Aboriginal Faultline 341
British Columbia's Physical Geography 343
British Columbia's Historical Geography 346
British Columbia Today 352
British Columbia's Wealth 356
Key Topic: Forestry 364
British Columbia's Urban Geography 373
British Columbia's Future 378
Summary 379
Notes 379
Key Terms 380
Bibliography 381
Further Reading 384

Chapter 8 **Western Canada** **386**
Overview and Objectives 386
Introduction 387
Western Canada within Canada 387
Western Canada's Physical Geography 388
Environmental Challenges 392
Western Canada's Historical Geography 393
Western Canada Today 404
Key Topic: Transition 416
Western Canada's Resource Base 421
Western Canada's Urban Geography 425
Western Canada's Future 428
Summary 428
Notes 429
Key Terms 431
Bibliography 431
Further Reading 435

Chapter 9   **Atlantic Canada**   **436**
Overview and Objectives   436
Introduction   437
Atlantic Canada within Canada   437
Atlantic Canada's Physical Geography   439
Environmental Challenges   443
Atlantic Canada's Historical Geography   444
Atlantic Canada Today   448
Key Topic: The Fishing Industry   457
Atlantic Canada's Land Wealth   463
Atlantic Canada's Urban Geography   474
Atlantic Canada's Future   477
Summary   480
Notes   481
Key Terms   481
Bibliography   482
Further Reading   485

Chapter 10   **The Territorial North**   **486**
Overview and Objectives   486
Introduction   487
The Territorial North within Canada   487
The Territorial North's Physical Geography   490
Environmental Challenges   492
The Territorial North's Historical Geography   493
The Territorial North Today   502
Resource Development in the Territorial North   513
Key Topic: Megaprojects   519
The Territorial North's Future   527
Summary   528
Notes   528
Key Terms   530
Bibliography   531
Further Reading   534

Chapter 11   **Canada: A Country of Regions**   **536**
Overview and Objectives   536
Introduction   537
Regional Character   537
The Core/Periphery Model   539

Canada's Faultlines   543
The Last Century: An Overview   545
The Twenty-First Century   546
Notes   547
Bibliography   548
Further Reading   549

Appendix: Additional Data on Exports, Income,   550
Employment, Immigrants, Aboriginal Population,
and Official Languages

Index   565

List of Figures

1.1	'You've changed, Sam': The changing character of the United States	5
1.2	The six geographic regions of Canada	11
1.3	Canadian nordicity	13
1.4	Regional populations by percentage, 1871 and 2001	16
2.1	Physiographic regions and continental shelves in Canada	46
2.2	Maximum extent of ice, 18,000 BP	49
2.3	Seasonal temperatures in Celsius, January	61
2.4	Seasonal temperatures in Celsius, July	62
2.5	Annual precipitation in millimetres	63
2.6	Climatic zones of Canada	67
2.7	Natural vegetation zones	68
2.8	Soil zones	69
2.9	Permafrost zones	71
2.10	Drainage basins of Canada	72
3.1	Migration routes into North America	89
3.2	Culture regions of Aboriginal peoples	95
3.3	Aboriginal language families	96
3.4	Canada, 1867	101
3.5	Canada, 1873	103
3.6	Canada, 1905	105
3.7	Canada, 1927	106
3.8	Canada, 1999	108
3.9	The regions of Canada	116
3.10	Historic treaties	124
3.11	Modern treaties	126
4.1	The CN rail system	158
4.2	Canada's population zones	165
4.3	Capital cities	167
4.4	Annual number of immigrants admitted to Canada, 1901–2001	180
4.5	Proportion of immigrants born in Europe and Asia by period of immigration	181
4.6	Immigration: An increasingly important component of population growth in Canada	184
4.7	Unemployment rate for Canada, 1997 to March 2003	192
4.8	Annual profits, 1995–2002	193
4.9	Average earnings, 1980, 1990, and 2000	201
4.10	Earnings by education level	202
4.11	Comparison of male and female earnings	204

4.12 Share of immigrants in Montréal, Toronto, and Vancouver,
       1981, 1991, and 2001   206
4.13 Aboriginal population by ancestry, 1901–2001   208
4.14 Percentage of Aboriginal population by self-identification,
       provinces and territories, 2001   211
4.15 Net migrants by province, 1996–2001   215
5.1   Ontario, 2001   230
5.2   Central Canada   231
5.3   Automobile-assembly centres in Ontario   254
5.4   US lumber lobby wins again   260
5.5   Major urban centres in Central Canada   263
6.1   Québec, 2001   283
6.2   Referendum fatigue   298
6.3   Hydroelectric power in Central Canada   309
7.1   British Columbia, 2001   338
7.2   Vancouver wins Winter Olympics bid   339
7.3   Physiography of the Cordillera   344
7.4   Railways in British Columbia   348
7.5   Forest regions in British Columbia   366
7.6   Traditional US ball games   371
7.7   Major urban centres in British Columbia   374
8.1   Western Canada, 2001   388
8.2   Planning to implement Kyoto Accord   403
8.3   Major urban centres in Western Canada   406
8.4   Chernozemic soils in Western Canada   409
8.5   Agricultural regions in Western Canada   410
8.6   Western Canada's resource base, 2003   420
9.1   Atlantic Canada, 2001   438
9.2   Major fishing banks in Atlantic Canada   458
9.3   Natural resources in Atlantic Canada   464
9.4   Major urban centres in Atlantic Canada   475
10.1 The Territorial North, 2001   487
10.2 Land-claim agreements in the Territorial North   503
10.3 Major urban centres in the Territorial North   506
10.4 Value of fur production in the Northwest Territories   510
10.5 Resource development in the Territorial North   518

List of Tables

1.1 General Characteristics of the Six Canadian Regions, 2001   14
1.2 Key Steps Altering Canada's Economy   17
1.3 Basic Characteristics of a Core and a Periphery   19
1.4 Two Versions of the Core/Periphery Model   22
1.5 Social Characteristics of the Six Canadian Regions, 2001   25
2.1 Geological Time Chart   47
2.2 Latitude and Longitude of Selected Centres   59
2.3 Climatic Types   60
2.4 Air Masses Affecting Canada   64
2.5 Canadian Climatic Zones   68
2.6 Canada's Drainage Basins   73
3.1 Timeline: Old World Hunters to Contact with Europeans   90
3.2 Population of the Red River Colony, 1869   98
3.3 Canada's Population by Provinces and Territories, 1901 and 1921   100
3.4 Timeline: Territorial Evolution of Canada   104
3.5 Timeline: Evolution of Canada's Internal Boundaries   107
3.6 Members of the House of Commons by Geographic Regions   114
3.7 Population in Western Canada by Province, 1871–1911   130
3.8 Population and Ethnicity, Western Canada, 1916   132
3.9 Population by Colony or Province, 1851–1871   139
4.1 Population Change by Geographic Regions, 1951–2001   157
4.2 Population Density, 2001   162
4.3 Population Zones, 2001   166
4.4 Percentage of Urban Population by Region, 1901–2001   168
4.5 Population and Percentage Increase for Census Metropolitan Areas, 1996–2001   169
4.6 Population Increase, 1851–2001   173
4.7 Canada's Rate of Natural Increase, 1851–2002   174
4.8 Phases in the Demographic Transition Theory   175
4.9 Median Age, Percentage over 65, and Sex Ratio, by Region   176
4.10 Newcomers by Place of Birth   182
4.11 Immigration by Place of Birth   183
4.12 Demographic Impact of Immigration   185
4.13 Ethnic Origin of Canadians   186
4.14 Types of Economic Structure   194
4.15 Major Sectors of Canada's Labour Force   197
4.16 Changes in Percentage Distribution of Canada's Exports, 1995–2002   198
4.17 Labour Force Characteristics, 1998–2002   200

Chapter 1

Regions of Canada

■ Introduction

Geography helps us understand our world. Since Canada is a country of regions, Canada is best understood from a regional perspective. Each region, therefore, has a particular place within Canada and that place is shaped by many factors, but especially its history and physical geography. From that perspective, a strong sense of belonging emerges over time as people are fused with place. Finally, the place of Canada and its regions within North America and the rest of the global world are best understood from a spatial economic theory known as the core/periphery model.

Regional geography is dynamic. Since the late twentieth century, powerful economic, demographic, and social forces have been unleashed within Canada and these forces are transforming its very essence. In a recent article, Professors Bourne and Rose (2001) discussed and analyzed these forces of change. These forces often surface in a regional setting, creating local tensions that have national and international repercussions. Recent international events, but especially the fallout from the terrorist attacks on the United States of 11 September 2001, have magnified these tensions, giving them both a regional and a global significance. While the geological term 'faultlines' refers to cracks in the earth's crust, in this text it serves as a metaphor for stresses within Canadian society. Since Canada became a nation, four major faultlines have played a key role in its evolution. These faultlines often

flare up as regional tensions and reveal fundamental conflicts between English-speaking/French-speaking Canadians, centralists and decentralists, new and old Canadians, and Aboriginal/non-Aboriginal Canadians. These differences within Canadian society represent critical elements, and a failure to address them can result in flash points that threaten the regional cohesiveness of Canada. By seeking a balance between these tensions, Canada, over time, has become what John Ralston Saul (1997: 8–9) describes as a 'soft' country, meaning a society where conflicts are, more often than not, resolved through discussion and negotiations.[1] In this way, these tensions, rather than tearing at the very fabric of Canada, cement the country together.

As part of North America, Canada is closely involved with the United States. Yet, the geography of each country is different (Vignette 1.1). These differences often translate into political issues (the need for joint military security) and trade concerns (Canada's desire for access to the American market). While Canadian relations with the rest of the world are important, the long-term trend is clear—Canada's geopolitical interests, cultural development, and economic priorities are, for better or worse, linked to the United States. Canada's trade, for instance, is largely with the US. One advantage of sharing the North American continent with our powerful American neighbour is proximity to the American market, while another is the shared defence of the continent. Economic and military agreements like the North American Free Trade Agreement (NAFTA) and the North American Air Defence Agreement (NORAD) have, for the most part, worked well for Canada. Both of these agreements foster a closer co-operation between the two countries. While common interests have led Canada and the US to sign numerous agreements and to undertake joint ventures, Canada remains a distinct society (Adams, 2003).

Following the Al-Qaeda attacks of 11 September, the US took unprecedented measures to protect itself against future attacks. While its homeland security measures were designed to prevent terrorists from slipping into the United States, they have reshaped the architecture of Canada–US relations, transformed the 'unguarded' Canada–US border into a quasi-military checkpoint, and raised the issue of a North American security perimeter. Since Canada's economic well-being is dependent on American willingness to maintain an open border, the consequences of border restrictions are potentially crippling. A slowing of the flow of goods and services across the border would have a negative impact on Canada's eco-nomic well-being and would deal a devastating blow to Ontario's automobile industry, which depends so heavily on exports to the United States and parts from suppliers in the United States for its just-in-time automobile assembly operations. As Mark Steyn (2003: A14), columnist for the *National Post*, wrote:

> Within 48 hours of 9/11, it was clear that Canada had a choice: It could be inside a North American perimeter or outside a US perimeter. Given that the trucks were mostly backed up on the northern side of the border, the answer seemed obvious.

While this was clear to Steyn, a decision by Ottawa to join the American initiative for a North American security perimeter would mean a common position on immigration, military, and trade policies. In a friendly observation to Americans, Pierre Trudeau, in 1969, declared that 'living next to you is in some ways like sleeping with an elephant: No matter how friendly and even-tempered the beast, one is affected by every twitch and grunt' (Colombo, 1987: 54). Since the terrorist attacks on New York City and Washington,

---

Vignette 1.1   Two Different Physical Environments

Geography has played a role in shaping differences between Canada and the United States. While larger in geographic area than the United States, Canada contains much less agricultural land. As can be seen in Figure 1.2, over 90 per cent of Canada is classified as 'northern', meaning beyond the climatic limits of agriculture. The Canadian Shield protrudes southward to Lake Superior, thus separating the densely populated areas of southern Ontario from those of the Prairie provinces. No such physical barrier exists in the United States. Canada's vast space, its cold climate, and its physiography have made the creation of a transportation system to serve the Canadian population more challenging and certainly more expensive than in the United States. Geography has been kinder to the United States, making it a more temperate, arable, and manageable physical space than Canada's geographic space.

DC, the forces of continental integration have gained strength under the guise of North American security. Canada faces a difficult decision, namely, staying outside of the North American security perimeter and running the risk of greater delays at the US border, or subscribing to a North American security perimeter and harmonizing its border and immigration regulations with the United States. Dancing with the American elephant may facilitate commercial traffic crossing into the United States, but it may also diminish Canada's sovereignty.

Twitches and grunts from the American elephant often resonate more strongly in one or more regions of Canada. The explanation is simple: each region of Canada has a different set of trade relations with the United States. Exports to the United States reveal that primary and semi-processed products dominate trade from British Columbia, Western Canada, Atlantic Canada, and the Territorial North while manufactured goods are key exports from Ontario and Québec. For example, increased oil exports to the US market will have a positive impact on Western Canada, but will have no direct effect on the rest of Canada.

Geography as a Discipline

Geography has a profound effect on people living in a region. De Blij and Murphy (2001: 3) believe that 'Geography is destiny', meaning that for most people, place is the most power-

Figure 1.1 You've changed, Sam. Following Al-Qaeda suicide attacks on the World Trade Center in New York and the Pentagon in Washington, the United States declared an all-out war on terrorists and doubled its efforts to prevent future attacks by creating the Department of Homeland Security. This department is responsible for making America's borders more secure. One consequence has been to make the flow of goods and people from Canada to the United States more complicated and even unpleasant, especially for Canadian citizens born in the Middle East.

ful determinant of a lifetime's experiences. Not surprisingly, then, geographers are concerned about cultural and physical differences between places and how these differences affect local societies. Sharing of a common space is a critical factor in this spatial differentiation because it leads to a regional consciousness, and this consciousness is the cornerstone of regional geography. The concept of belonging to a place is not a new idea. The ancient Greeks recognized that the earth varied from place to place and that different peoples inhabited each place. Stimulated by the travels, writings, and map-making of scholars such as Herodotus (484–c. 425 BC), Aristotle (384–332 BC), Thales (c. 625–c.547 BC), Ptolemy (AD 90–168), and Eratosthenes (c. 276–c. 192 BC), the ancient Greeks coined the word 'geography' and mapped their known world. By considering both human and physical aspects of a region, geographers have developed an integrative approach to the study of our world. This approach, which is the essence of geogra-

phy, separates geography from other disciplines. The richness of geography is revealed in Vignette 1.2, drawn from *The Dictionary of Human Geography*, in which Peter Haggett's incisive definition of geography is provided.

Regional geography simplifies the fundamental goal of geography by dividing the complex world into smaller, more manageable spatial units called regions. A **region** is an area of the earth's surface that has distinctive characteristics. These characteristics vary according to the criteria geographers use.[2] A region may be defined in terms of its physical or human characteristics (Vignette 1.3). The Rocky Mountains constitute a physical region, while Acadia is a cultural region. Physical characteristics of a region include:

- geographic location (latitude and longitude);
- landforms (mountains, plateaux, and plains);
- climate;

Vignette 1.2   The Nature of Geography

[Geography is] the study of the earth's surface as the space within which the human population lives. The word comes from the Greek *geo*, the earth, and *graphein*, to write. Perhaps the best-known formal definition of the field was provided by the American geographer, Richard Hartshorne, in his *Perspective on the Nature of Geography* (1959): 'geography is concerned to provide accurate, orderly, and rational description and interpretation of the variable character of the Earth surface'. The last two terms in this definition need some elaboration. By 'variable character' geographers mean the spatial variation that can occur between the character of the earth's surface at one location and another. This variation may occur at all map scales, from the globe itself, say between continent and continent, down to a local level, say between one district and another within an urban area. By 'Earth surface' is meant that rather thin shell, only one-thousandth of the planet's circumference thick, that forms the habitat or environment within which the human population is able to survive.

Source: Peter Haggett, 'Geography', in R.J. Johnston, Derek Gregory, and David M. Smith, eds, *The Dictionary of Human Geography* (Oxford: Blackwell, 1986), 175. Reprinted by permission of Blackwell Publishers.

Vignette 1.3 Cultural Regions

Cultural regions are classified into three types: formal, functional, and vernacular. A formal cultural region is an area where people have one or more cultural traits in common. The Inuit, for example, inhabit the Canadian Arctic. They have a number of common cultural traits, including language (Inuktitut). A functional cultural region is distinguished by its political, social, or economic functions. Provinces and territories represent functional administrative regions. The Québec government has also engaged in a number of social functions designed to promote the French language and culture. Vernacular cultural regions are a third type. Such regions are held together by a sense of belonging. People living in vernacular regions have had a number of historical experiences that combine to provide them with a sense of regional identity. The Canadian Prairies, Central Canada, and the Maritimes are examples of vernacular regions.

- soil;
- natural vegetation.

Human or cultural characteristics include:

- culture (history, language, ethnicity, and religion);
- economy (resources, industries, and transportation);
- political identity (boundaries, political structure, and international relations);
- demographics (population and migration);
- urbanization;
- a sense of place.

A **sense of place** is a powerful psychological bond between people and a region. The physical nature and human characteristics of a region combine to form this psychological bond. Canada, with its vast spaces and magnificent diversity of physical settings, lends itself to strong regional bonds. The sea, for instance, exerts a profound influence on Maritimers who live along its coastline. The loss of the northern cod fishery in the 1990s brought great pain to the small fishing communities in Atlantic Canada but especially to the outports in Newfoundland (for more on this subject, see Chapter 9). The vast open expanses of the Prairie region also exert a powerful influence on its inhabitants. The North has a cold climate to which people have had to adapt in order to survive. Residents living and working in any area are exposed to contemporary issues and memories of past events that continue to shape their lives. Not all these memories are pleasant ones. For example, the Dust Bowl of the 1930s deeply marked Prairie farmers, forcing many to abandon their homesteads. While the Dust Bowl has faded into past collective memory, this semi-arid environment continues to influence life on the Prairies, making drought a real threat to Prairie agriculture. Indeed, two years of droughty growing conditions (summers of 2001 and 2002) have contributed to a significant drop in grain production and sales, leaving the Saskatchewan Wheat Pool deep in debt and near bankruptcy (for more on this topic, see Chapter 8).

The concept of a sense of place, then, recognizes that people living in a region have undergone a collective experience that leads to shared aspirations, concerns, goals, and values. Over time, such experiences develop into a social cohesiveness among those people living within a spatial unit. Often these experiences are coloured by extreme weather events in a region. The rainstorm and subsequent flooding in the Saguenay Valley of Québec in July 1996 was such an experience. Ice storms often have devastating effects by knocking out regional electrical systems. In early February 2003, New Brunswick was hit with a severe ice storm that dropped 60 millimetres of frozen rain, leaving some residents from Saint John to Moncton without power for several days while temperatures were hovering around -10°C.

A sense of place can also evolve from a region's history and geography. Early in its history, British Columbia was isolated from the rest of Canada by the Rocky Mountains, which form the largest mountain system in Canada, extending 1,200 km from the US–Canada border northward towards 60° N. While this formidable physical barrier, considered by many to be the backbone of Canada, has been surmounted by modern transportation systems, it remains an important physical and cultural feature unique to British Columbia. In *The Unknown Country*, Bruce Hutchison (1942: 266) described the Rocky Mountains' effect on people as he observed changes in the landscape from a passenger car on the Canadian Pacific train: 'Unlike the life of the Prairies or the East—so unlike it that, crossing the Rockies, you are in a new country, as if you had crossed a national frontier. Everyone feels it, even the stranger feels the change of outlook, tempo and attitude. What makes it so, I do not know.' Feelings of alienation run deep in

In 1811 Nova Scotia granted six families 800 acres of land, and the land was given the name Peggy's Cove. Originally a tiny fishing village, Peggy's Cove has become a major tourist attraction. (© Royalty-Free/Corbis/Magma)

British Columbia. This theme reverberates in Philip Resnick's *The Politics of Resentment: British Columbia, Regionalism and Canadian Unity* (2000). Resnick supports Hutchison's contention that BC is a distinct region, but he also stresses that a sense of alienation from Ottawa is widespread within the province. An equally powerful expression of place and alienation exists in Québec, where history and geography have had four centuries to nurture a strong sense of place and to give birth to a nationalist movement that has sought to separate Québec from the rest of Canada. New France was born and grew within the confines of the St Lawrence Lowlands. Since Québec's early days, the St Lawrence River has played a key role in its settlement and economic development. Prior to the British conquest of New France in 1760, this mighty river promoted the interests of this French colony by:

- providing a supply route to France, the mother country;
- facilitating internal movement within the colony, particularly to and from Québec City;
- encouraging each tenant farmer (*habitant*) to build his farmhouse near the river, which resulted in landholdings that took the form of long lots;
- allowing the fur traders (*coureurs du bois*) to penetrate far inland to trade with distant Indian tribes;
- giving French explorers access into the heart of North America, while the British settlers along the Atlantic Coast had no similar river system and had to contend with a formidable physical barrier, the Appalachian Mountains.

While Québec's territory greatly expanded after Confederation, the St Lawrence Lowlands region remains the core of the North American francophone community.

A region, then, is a synthesis of physical and human characteristics that, combined with its distinctiveness from surrounding regions, produce a regional character and a sense of place. People living and working in a region are conscious of belonging to that place and frequently demonstrate an attachment and commitment to their 'home' region. A sense of place, which falls somewhere between tribalism and nationalism, unites people on common issues and challenges confronting their region, and compels them to seek solutions to these issues and challenges. For Aboriginal Canadians, a sense of place is embodied in their concept of traditional ecological knowledge, while for new Canadians, a sense of place may seem elusive at first, but, with time, they too will develop such feelings. Indeed, the theme of this book is that Canada is a country of regions, each of which has a strong sense of place.

The Nature of Regions

Geographers conceptualize space in terms of regions. Regions are a critical intellectual concept and represent a framework for geographic studies. The geographer's challenge is to divide a large spatial unit like Canada into a series of 'like places'. To do so, a regional geographer selects the critical physical and human characteristics that logically divide a large spatial unit into a series of regions and that distinguish each region from adjacent ones. Towards the margins of a region, its core characteristics become less distinct and merge with those characteristics of a neighbouring region. In that sense, boundaries separating regions are best considered transition zones rather than finite limits.

Regions vary in many ways, one of which is scale. At the global scale, the continents and major cultural groupings constitute a dozen or so world regions. One popular world regional geography text divides the world into 12 world geographic realms (De Blij and Muller, 2001), while another has eight world regions (Clawson and Fisher, 1998). Each of these world divisions has a region (North America and Anglo America, respectively) that includes Canada.

At the scale of a nation-state, geographers organize space into much smaller regions. In this text, Canada is divided into six geographic regions such as Atlantic Canada and the Territorial North. Smaller regions can also be conceived. For example, urban geographers and planners conceive of a city as consisting of four basic parts: the central business district, the inner city, suburbs, and the urban countryside (Bunting and Filion, 2000: 246). In summary, geographic regions vary in size from **megaregions** (world regions) to **mesoregions** (segments of a nation-state) to **microregions** (areas within a locale). They may also vary from **natural regions** (for example, climatic regions) to **cultural regions** (for example, political regions).

Geographers sometimes divide regions further into subregions, which are simply a finer spatial presentation of a region. For example, the Cordillera is a major landform in North America. The Cordillera (as a region), however, contains a number of distinct landforms that form subregions. Three examples are the Rocky Mountains, the Interior Plateau, and the Coast Mountains. A more detailed account of Canada's landforms (including the Cordillera and its subregions) is presented in Chapter 2 under 'Physiographic Regions'. **Physiography** is the study of landforms, their underlying geology, and the processes that shape these landforms.

Regional models have their regions organ-

ized in a hierarchical fashion, i.e., subdividing a large geographic region into a set of smaller regions or establishing relationships between regions. Certainly, this is true of the core/periphery model, which exhibits both characteristics—it begins at a world scale but functions effectively at smaller scales, and is based on relationships between the core and its periphery (Table 1.4).

The Nature of Regional Geography

Once regions have been identified, the geographer's focus in describing and analyzing these can vary—this is the nature of regional geography. Often a regional geography approach can begin by presenting a region's geographic situation and physical nature, even though the main discussion focuses on a region's human geography, particularly its economic activities and social characteristics. A theoretical framework is sometimes employed to provide an abstract image of that space and the linkages between its various parts. This helps in understanding the main elements underlying a region's basic geographic structure and the processes operating within it. The nature of regional geography, however, goes well beyond theory and the host of facts associated with regional characteristics. It is concerned with a spatial synthesis that depicts a region's distinctive character.

There have been several approaches to regional geography. Traditionally, geographers perceive people and nature as interacting in a cause-and-effect interplay—an interplay between culture and the physical environment. Geographers may emphasize the influence of physical geography on human activities, that is, how each region's physical base provides a number of opportunities and limitations for the settlement of the land, including a set of resources that provides the basis of economic development. Geographers may also recognize that people, acting through their culture, have an impact on the physical environment. In this text, physical and historical geography continue to play a role, but more emphasis is given to contemporary economic, environmental, and social matters that, in today's world, are transforming relationships between Canada's regions and its place in the North American and world economies. The most critical events driving this transformation are new trading arrangements. NAFTA (the North American Free Trade Agreement) and

the World Trade Organization (formerly GATT, the General Agreement on Tariffs and Trade, which in the mid-1990s was transformed into the more formalized WTO) have unleashed powerful forces that are recasting Canada's economy into a continental and global context and that have implications for Canada's environment and society. Ironically, trade barriers between Canadian provinces are proving to be more difficult to remove.[3]

A new economic reality for Canada and its regions is emerging. As the trade barriers protecting the old national economic system are eliminated, the volume of trade with the United States and, to a lesser degree, Mexico is

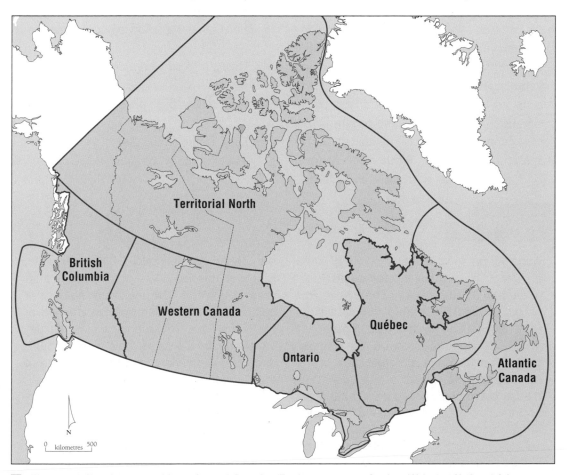

Figure 1.2 The six geographic regions of Canada. (Further resources: Student Web site, National Atlas section, Map 1. Web site instructions are found on p. xxi.)

increasing by leaps and bounds, thereby stimulating north-south trade patterns in North America. In fact, no other country in the world is so dependent on another country for its exports. From 1997 to 2002, Canada's exports to the United States increased from 80 per cent of its total exports to 85 per cent (Statistics Canada, 2003).

Canada's Geographic Regions

In this text, Canada consists of six large-scale geographic regions (Figure 1.2): Ontario, Québec, British Columbia, Western Canada, Atlantic Canada, and the Territorial North. Why have these six regions been selected as a basis of study? What is the logic underlying this selection? First of all, these six major regions represent the division of a huge country into manageable segments. Second, these regions are associated with identifiable physical features. For example, the terrain of the four Atlantic provinces is dominated by the Appalachian Uplands. While the Canadian Shield covers most of Ontario and Québec,

their most productive lands are found in the Great Lakes/St Lawrence Lowlands. The Rocky Mountains not only mark the beginning of the Cordillera but also provide a clear physical divide between Western Canada and British Columbia. Third, since provinces form the basic spatial unit for Statistics Canada, regional description and analysis is supported by census and other data. Finally, these six regions are commonly used by the media and scholars, suggesting that this regional division makes sense historically, politically, and geographically. Back as far as 1972, geographers used this regional structure for the preparation of a series of books on the regional geography of Canada.[4] In sum, then, these six geographic regions have the following advantages:

- They are readily understood by Canadians.
- They reflect the political nature of Canada.
- They facilitate the use of statistical data that are readily available at the provincial level.

Vignette 1.4   Nordicity

Nordicity presents a spatial sense of Canada from a particular perspective—the degree of 'northernness'. The five regions shown in Figure 1.2 are strikingly different from those used in this text (Figure 1.1). These differences illustrate the complex diversity of Canada and the challenge of defining regions. Nordicity is a quantitative measure of a place based on 10 natural and social variables. Each variable is assigned a range of values called polar units. Vancouver, for example, has 35 polar units, while Canada's northernmost centre, Eureka, has 857 polar units. The southern edge of northern Canada is defined as 200 polar units. South of this boundary lies the populated area of Canada. The maximum number of polar units (1,000) can only be attained at the North Pole. These polar units measure variations in Canada's cold environment, state of development, and its distance from the Canadian **ecumene**, that is, its densely inhabited core. The concept of nordicity has accomplished three goals: (1) it provides a numerical definition of each place in Canada; (2) it can accommodate change over time; and (3) it allows for a regionalization of Canada.

- They represent vernacular regions and therefore include a sense of both regional identity and place.
- They are associated with a core set of physical features.

As mentioned earlier, geographers classify regions according to a combination of physical and human characteristics. A geographer's selection of one or more of these characteristics depends on the nature of the investigation. Once selected, the characteristics serve as the criteria for defining the type and spatial extent of a particular region. The range in the type and extent of regions can be considerable. For instance, Canadian geographer Louis-Edmond Hamelin, in seeking to differentiate areas in Canada's North, conceived the notion of nordicity based on a composite of 10 natural and social factors (Vignette 1.4). The result was a vision of nordicity dividing Canada into five zones, two representing southern Canada and three, northern Canada (Figure 1.3). Compare Hamelin's regional division of Canada with the division used in this text (Figure 1.2). Hamelin's regions are employed in this text as subregions within Ontario, Québec, British Columbia, Western Canada, and

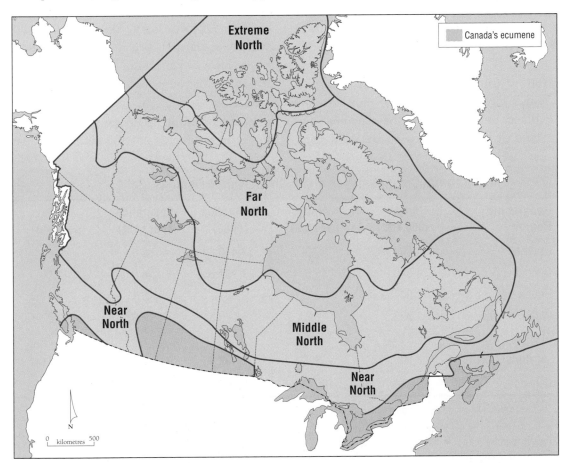

Canada's ecumene

Figure 1.3 Canadian nordicity. Nordicity, as expressed by the Near North, Middle North, Far North, and Extreme North, covers nearly 90 per cent of Canada.
Source: Adapted from Louis-Edmond Hamelin, *Canadian Nordicity: It's Your North Too*, translated by W. Barr (Montréal: Harvest House, 1979). Reprinted by permission of the author and translator.

Kimmirut (Lake Harbour) on Baffin Island in Nunavut was the hometown of David Panneok, a famous Inuk carver who passed away in 2002. Panneok used green soapstone and serpentine soapstone from a local quarry to carve his sculptures. In 2001, Kimmirut had a population of 433. (Barb and Ron Kroll)

the Near North and the Middle North, are identified (see Figures 1.3 and 5.2). This division reflects three factors: the hierarchical nature of regions; the flexibility of the concept of regionalization; and the interrelationship between existing regional systems, i.e., nordicity, core/periphery, and the six regions identified in this text.

Given the economic and demographic strength of Ontario and Québec, these two provinces continue to dominate Canada's economy and its political system. For that reason, these two provinces are classified as 'core' regions in this text, though their actual industrial core is found in southern Ontario and southern Québec (Figure 5.2). Ontario and Québec, along with British Columbia, combine to have 75 per cent of the Canadian population and account for over 75 per cent of Canada's gross domestic product (Table 1.1). Economic output is measured by **gross domestic product** (GDP), the monetary value

Atlantic Canada. For instance, Ontario is divided into two spatial parts—northern and southern Ontario. Within the economic hinterland of northern Ontario, two subregions,

Table 1.1	General Characteristics of the Six Canadian Regions, 2001				
Geographic Region	**Area* (000 km²)**	**Area (per cent)**	**Population (000s)**	**Population (per cent)**	**GDP (per cent)**
Ontario	1,068.6	10.8	11,410.8	38.0	40.6
Québec	1,540.7	15.5	7,237.5	24.0	21.1
British Columbia	947.8	9.5	3,907.7	13.1	12.0
Western Canada	1,963.5	19.8	5,073.3	16.9	20.2
Atlantic Canada	540.3	5.4	2,285.7	7.7	5.7
Territorial North	3,909.8	39.0	92.8	0.3	0.4
Canada	9,970.7	100.0	30,007.0	100.0	100.0

*Includes freshwater bodies such as the Canadian portion of the Great Lakes.

Note: Statistics Canada produces new population data on 1 July of each year and GDP figures at the end of each year. More recent figures can be obtained on the Statistics Canada Webpage: <http://www.statcan.ca/english/Pgdb/People/Population/demo02.htm>and <http://www.statcan.ca/english/Pgdb/Economy/Economic/econ15.htm>.

Sources: Statistics Canada, *2001 Census: Population and Dwelling Counts*, at <www12.statcan.ca/english/census01/products/standard/popdwell/Table-PR.cfm>, searched 28 January 2003; Statistics Canada, *Gross Domestic Product*, at <www.statcan.ca/english/Pgdb/econ15.htm>, searched 26 November 2002.

of all goods and services produced by an economy over a specified period. The remaining seven provinces, which account for roughly 24 per cent of Canada's population and economic output, are combined into two geographic regions. Atlantic Canada includes Newfoundland, New Brunswick, Nova Scotia, and Prince Edward Island, while Alberta, Saskatchewan, and Manitoba comprise Western Canada. By creating these two regions, the five geographic regions become more similar in territorial size, population levels, and economic output. The remaining geographic region, the Territorial North, extends over one-third of Canada's territory. However, few people live in this region, which accounts for less than 1 per cent of Canada's economic output. The Territorial North consists of Yukon, the Northwest Territories, and Nunavut.

The Dynamic Nature of Regions

Regions are not static entities. They and their characteristics change over time, as do the relationships between regions. Human beings, whose activities affect both the physical and human elements in a region, are the principal agents of regional change. This dynamism is revealed in a variety of ways. Some changes involve humans' alteration of the physical landscape. In the late nineteenth century, for example, the building of the Canadian Pacific Railway opened the West to settlement, thereby dramatically changing its land use. More recently, the construction of the Confederation Bridge from mainland Canada to Prince Edward Island established a direct highway link. Unfortunately, not all economic activities have had a singularly positive outcome. Human activities have resulted in serious damage to the physical environment and negative social implications for local inhabitants: clear-

cut logging practices in the mountainous areas of British Columbia; overfishing of the northern cod stocks on the Grand Banks off Newfoundland; and diversion of the Churchill River into the Nelson River in northern Manitoba. Similarly, over a century ago, the Klondike gold rush and its aftermath had an enormous impact on the social fabric and environment of Yukon.

Social changes, too, have had a profound impact on the state of Canadian regions. Social change is sometimes manifested in the courts or through elected governments' enactment of legislation. Without a doubt, the 1982 Canadian Charter of Rights and Freedoms has had a profound impact on Canadian society by supporting individual and minority rights.[5] Equally important, Aboriginal rights were recognized by the courts and governments, which in turn led to the first comprehensive land-claim agreement in 1984, followed by five more in the 1990s. Also, a new territory, Nunavut, was created. The net result is a new political geography in the Territorial North.[6]

Another example of social change is the empowerment of the French language in Québec. Between 1969 and 1977, three different provincial governments passed language laws in order to have the language of the French majority fully recognized as the province's primary language. The third of these laws, the so-called Charter of the French Language (Bill 101), was enacted in 1977. The purpose of Bill 101 was to make all of Québec's society function in French—in its various levels of government, in corporations both large and small, and in service industries. Even public advertisements were restricted to the French language. Within a few years, the impact of this language legislation on Québec society was profound, for it greatly strengthened the place of the French language within

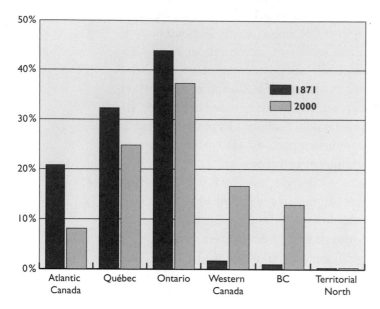

Figure 1.4 Regional populations by percentage, 1871 and 2001.
Sources: Adapted from Statistics Canada (1997c, 2002).

the business community and public institutions, and among the general public. While the English language was under attack, anglophones retained their own schools, as guaranteed in the British North America Act and, since 1982, as guaranteed by the Constitution Act of Canada.

Often, economic change also has a regional impact. The cause of change can originate either within or outside the country. For Canada's resource hinterland, sudden changes in world prices for its primary products can create a boom or a bust in the local economy. In such cases, economic change is highly concentrated in one region. For instance, the sharp rise in world oil prices in the 1970s suddenly transformed Alberta, Canada's principal oil-producing province, into an economically prosperous 'have' province. In other cases, economic change can affect more than one region. This is particularly true when Ottawa attempts to ameliorate regional disparities. In 1967 Ottawa initiated the fiscal equalization

payments program, which transfers wealth earned by 'have' provinces (Ontario, Alberta, and British Columbia) to economically disadvantaged 'have-not' provinces (Newfoundland, New Brunswick, Nova Scotia, Prince Edward Island, Manitoba, Saskatchewan, and Québec). Ottawa's goal is to ensure that all Canadians have access to a basic level of public services.

Over time, the balance of power between the regions changes, affecting the relationships between regions. In many ways, changes in the population size of regions provide a measure of power shifts. In 1871, 97 per cent of Canada's population lived in Atlantic Canada, Québec, and Ontario. One hundred thirty years later, the figure had dropped to 70 per cent (Figure 1.4). This remarkable demographic change illustrates the dynamic nature of regions and provides a key to understanding regional relations and conflicts.

Economic Integration

A particularly important aspect of change for Canada's regions has been, and continues to be, economic integration within a North American context. For a long time, Canada was a high-tariff country.[7] In 1879, John A. Macdonald introduced his economic policy of high tariffs known as the National Policy. His immediate goal was to stimulate the Dominion's faltering economy by encouraging Canadians to buy goods made by other Canadians. This required that tariffs be raised to keep out lower-priced American goods. One consequence of Macdonald's policy was the creation of an industrial core in southern Ontario and Québec. Another consequence was that the rest of the country remained a resource hinterland. Many would argue that this policy served Canada well, while others claim that it

Table 1.2	**Key Steps Altering Canada's Economy**
Date	**Event**
1947	The first GATT agreement was signed by Ottawa and 22 other governments. These countries agreed to reduce tariffs and eliminate import quotas among the members of GATT through multilateral negotiations and agreements.
1965	The Auto Pact between Canada and the United States created a continental market for automobiles. An important provision for Canada was that its share of automobile production be maintained. The three large American car manufacturers (General Motors, Ford, and Chrysler) could now expand their production to a continental scale. In this process, Canadian branch plants ceased to produce a wide variety of automobiles for the Canadian market and began to produce only a few models (but more of them) for the continental market.
1985	The Royal Commission on the Economic Union and Development Prospects for Canada examined the question of Canada–United States free trade and concluded that such an agreement would greatly benefit Canada.
1989	The Free Trade Agreement between Canada and the United States was approved. The aim of the FTA was to integrate the two economies. Unlike the Auto Pact, however, there were no assurances that Canada's share of economic activities would be maintained. Instead, the North American marketplace would determine the type and amount of economic activity in Canada.
1994	Superseding the FTA, the North American Free Trade Agreement (NAFTA) broadened the geographic area of the FTA to include Mexico. The North American economy now had a member with much lower wage rates, causing many labour-intensive industries to close their operations in Canada and reopen in Mexico.
1995	The final GATT treaty (the so-called Uruguay Round) was ratified by 124 governments on 1 January 1995. This agreement is the most far-reaching trade pact in world history and moves closer to the idea of world free trade. Unlike the arrangement under the seven previous GATT negotiating rounds, the World Trade Organization, a permanent institution with powers to enforce trade rules and assess penalties against members, was created.
1998	The proposed Multilateral Agreement on Investment would have augmented the concept of world free trade by requiring governments to treat foreign and domestic companies equally. Canada declared its willingness to participate in these negotiations, conducted by the Organization for Economic Co-operation and Development (OECD), but some Canadians expressed strong concern that such an agreement would preclude Ottawa from acting in the best interests of its own citizens. By 1998, strong opposition caused the OECD to abandon this proposal.
2000	In May a World Trade Organization appeal panel upheld an earlier WTO ruling that the Auto Pact offered favoured treatment to some countries over others and Canada moved to comply with the ruling and wind down the machinery of the Auto Pact by February 2001. The Canadian Auto Workers union had fought the decision because of the threat of non-unionized workers, and auto plants in Québec and Windsor, Ontario, have closed subsequent to the WTO ruling.
2001	Canada's open border with the United States was altered by the suicide mission of Al-Qaeda terrortorists who commandeered and then crashed two passenger planes into the World Trade Center in New York City and another one into the Pentagon in Arlington, Virginia. Almost immediately, Wash-

Table 1.2	Key Steps Altering Canada's Economy (cont'd)
Date	**Event**
	ington created a Department of Homeland Security, promoted a North American Security Perimeter, and announced the Enhanced Border Security and Visa Entry Reform Act that, together, had enormous implications for Canadian access to the United States. While the full impact has yet to be determined, every indication suggests much closer ties with the United States in areas of border security, defence of North America, and immigration. Such ties will allow continued access to the US market, which is the key to Canada's continued prosperity.
2002	Canada ratified the Kyoto Protocol. The *Kyoto Protocol* is a document signed by about 180 countries at Kyoto, Japan, in December 1997. This agreement commits 38 industrialized countries to cut their emissions of *greenhouse gases* between 2008 to 2012 to levels that are 5.2 per cent below 1990 levels. The United States government of George W. Bush refused to support the Kyoto Protocol. In December 2002 the Canadian Parliament ratified the Protocol, thereby committing Canada to reduce its annual greenhouse gas (GHG) emissions by 5.2 per cent, i.e., by 240 megatonnes (MT) below its 1990 level by 2012. Just how the cost of reducing greenhouse gas emissions will affect Canada's economy and its competitive position with the United States remains unclear, although with the announcement by Russia in 2003 that it, too, would not participate in the Kyoto agreement for economic reasons, the ultimate fate and moral impact of this environmental agreement on the international community remain in doubt.
2003	The United States is proceeding with a North American Security Perimeter. Canada and Mexico are still considering whether or not to participate, though Paul Martin, Canada's new Prime Minister, has indicated that Canada needs to be at the table on these discussions.

favoured Central Canada at the expense of the rest of Canada.

Whatever the truth, Canada has abandoned this protectionist policy, turning instead to the North American and global economies (Table 1.2). By unleashing such powerful economic forces, Canada's economy and society are changing. Canadians, now more outward looking, are seeking their place in the unfolding North American and world economies. Since the Canada–US Free Trade Agreement in 1989, Canadian firms have adjusted to the American marketplace with astonishing speed. Consequently, the volume of north/south trade now matches Canada's east/west trade. In this economic transformation, the Canadian National Railway has facilitated the strengthening of this north/south axis. Touted as North America's

railroad, CN, by purchasing the Illinois Central Railway, has quickly created a north/south railway system that fully integrates its Canadian operations into America. Since the terrorist attacks of 2001, rail integration and north/south traffic have become more complicated due to American concerns about internal security. Costly delays at the border are not in the interests of either country. Accordingly, Canada and the United States signed an agreement on 4 April 2003 to facilitate scrutiny of Canadian rail cargo to the United States.

A Spatial Framework: The Core/Periphery Model

A spatial framework provides an essential foundation for the study of regional geography. A spatial vision of the world enables a geography student to comprehend more readily the processes affecting different regions of our constantly changing world and to grasp the significance of numerous facts by associating them with a conceptual model. Certainly, the core/periphery model, sometimes called the heartland/hinterland model, provides a useful classification scheme for Canada's regions as well as a broad interpretation of relationships between those regions. By their very nature, however, abstract models simplify the real world. In doing so, when applied to actual regions, they often lose touch with the complexity and variety of economic, political, and social realities. For that reason, the spatial framework used in this book only serves as a backdrop to the regional geography of Canada. This text emphasizes the economic disparities between, and the interdependency among, regions, as well as the processes that shape and reshape regions. To assist the reader in comprehending this information, a conceptual framework based on the core/periphery model is employed.

The **core/periphery model** is an abstract theory devised and refined by scholars that explains how the capitalist economic system evolved into distinct spatial units. In the most simplistic terms, industrialization takes place in a few favoured areas known as cores. As these industrial cores expand, so does their need for raw materials, energy, and foodstuffs, so they must turn to their peripheries for sufficient supplies. In this way, the world is divided into two interrelated functional units: industrial cores and resource peripheries. The **core** controls the direction and rate of development of its **periphery**. A core region is also known as a **heartland**, and a periphery is sometimes referred to as a **hinterland**. As a core has virtually total economic and political power, a core and a periphery have a dominant/dependent relationship expressed by the characteristics shown in Table 1.3.

The core/periphery model describes an asymmetrical economic relationship between regions that function at different geographic levels. One classification identifies three levels, namely, global, regional, and local levels. These different levels are interrelated, but the highest-order core (the global level) has ultimate control of the system. Under these conditions, the core/periphery concept may be

Table 1.3	**Basic Characteristics of a Core and a Periphery**	
	Core	**Periphery**
Geographic:	Space and physical barriers overcome by a modern transportation system	Space and physical barriers continue to hinder economic development
Economic:	Industrial production emphasizes manufacturing	Primary production dominates
Cultural:	Attitudes, language, social customs, and values prevail	Forced to accept the core's culture
Political:	Controls the periphery	Is subservient to the core

Source: Adapted from Edward J. Malecki, *Technology and Economic Development* (New York: Wiley, 1991). Reprinted by permission of Pearson Education Limited.

regarded as a multiple system of 'nested' cores and peripheries, each operating at particular spatial levels but with all levels interconnected and dominated by the highest-order core. This notion of different geographic levels enriches our understanding of Canada's regional nature. Using Ontario as an example of a particular spatial level, we can argue that a core and a periphery exist within that province. Southern Ontario, where the vast majority of the province's population, industry, and political power are found, acts as the core. In sharp contrast, northern Ontario, with its tiny population and primary economy based on forestry, hunting, mining, and trapping, serves as a periphery.

In general, core regions tend to be relatively powerful and wealthy, while peripheral regions tend to be relatively weak and poor. In terms of trade, the core exchanges manufactured goods for primary products from its periphery. While this economic arrangement generates great wealth, the distribution of that wealth is often questioned. Some hold that this trade enriches the core and impoverishes the periphery. Others refute this claim, arguing that the market economy benefits all regions. Pushing the argument of the core/periphery model too strongly, however, leads to a philosophical cul-de-sac known as environmental determinism.[8] Geographers generally take the position that physical geography affects the range of developmental opportunities facing people living in a region. Nevertheless, there are several interpretations of the core/periphery model and its ability to distribute wealth. One interpretation is by neo-Marxist scholars, while another is by conventional economists who subscribe to free-market principles.

The neo-Marxist version of the core/periphery model substitutes regions for classes. The process driving this model is regional

exploitation rather than class exploitation. Sometimes this left-wing view takes the form of **dependency theory** (Frank, 1969). According to this theory, rich industrial countries or regions (the so-called core) extract the wealth generated by resource industries in developing countries or regions (the so-called periphery). This exploitive process eventually strips the developing country or region of its natural wealth, leaving it underdeveloped and impoverished. As a form of neo-colonial domination, cities within the peripheral region serve the interests of the core and thereby facilitate the development process unleashed by capitalist firms. Since branch operations of foreign companies are located in these cities, the primary role of these cities is to function as a node for the exploitation of the surrounding hinterland. From an ideological point of view, the functions of these centres are controlled by decisions made in the industrial core where the main headquarters of the foreign companies are located.

Conventional economists have interpreted the core/periphery model differently. While they agree that the core region is a dominant economic force, they believe that the core, by investing in the periphery, is not exploiting but developing that region. This process of regional development is often called 'modernization'. From the perspective of **modernization theory**, regional differences are only temporary and will disappear with time. The mechanism for such change lies in a competitive marketplace where regional inequalities are corrected by the mobility of the factors of production, particularly labour and capital. At the global level, capital and technology are likely to move to developing countries where the rate of return is higher because of lower labour costs. Within a nation-state, workers are attracted to areas of high wages, thereby creating a shortage of workers in the low-wage areas. At that point, wage levels supposedly rise in the low-wage area, reaching an equilibrium with wages in other regions.

Wallerstein's Global Version

Immanuel Wallerstein (1979) devised a world system analysis to study social change from a neo-Marxist perspective. He applied the notion of class exploitation to countries of the world by dividing the capitalist world economy into three spatial units. These are the core, semiperipheral, and peripheral units (Table 1.4). Wallerstein echoed the Marxist argument that the core appropriates the wealth generated by the periphery for its own enrichment. Since industrial cores operate within a global environment, a world system approach based on a core/periphery model is necessary to understand social and economic changes. Wallerstein argues that his world system analysis provides the key to understanding how the world core maintains its power by dominating its periphery.

Friedmann's Regional Version

In the 1960s, John Friedmann, an American scholar, was working as a regional planner in Venezuela where he devised a regional version of the core/periphery model (Table 1.4). He applied this model to regions within a single country. He viewed core regions as territorially organized subsystems of society that have a high capacity for generating and absorbing innovative change, and peripheral regions as subsystems whose development is determined chiefly by core region institutions on which they depend. From a modernization perspective, Friedmann conceived an industrial core and three varieties of peripheral regions. All

three are dominated in different ways by the industrial core region. The three peripheral regions are the upward transitional region, the downward transitional region, and a resource frontier. This division of the periphery into three subregions demonstrates the existence of important variations in peripheral resource bases and their economic relationship with the core.

Canada and the Core/Periphery Model

Because Friedmann's regional classification closely approximates the economic characteristics of the six geographic regions used in this book, his regional classification was adapted for this text. The application of Friedmann's version of the core/periphery model that underlies this text must be accompanied by three caveats:

1. This is only one of several well-known models designed by geographers to conceptualize spatial organization (see the discussion of the staples thesis, below) and it is in no way definitive.
2. Dramatic shifts in interregional relationships can take place over time, prompted by any number of factors, both internal and external.
3. State intervention in the marketplace can either redistribute national wealth (thus counteracting the tendency of market forces to concentrate wealth in a few favoured regions) or it can accelerate this concentration of wealth by using political or military power (thus protecting the wealthy regions).

Despite these caveats, Friedmann's core/periphery model establishes a reasonable con-

Table 1.4	Two Versions of the Core/Periphery Model
Type of Spatial Unit	**Global Version**
Core Regions	Highly industrialized regions that dominate trade, control the most advanced technologies, and have high levels of productivity within diversified economies. Western Europe, North America, and Japan form the principal core regions at the world scale.
Semiperipheral Regions	Falling between core and peripheral regions, semiperipheral regions are able Regions to exploit nearby peripheral areas, but are themselves exploited and dominated by the core regions. They are countries that were once peripheral, indicating that peripheral (and possibly core) status is not necessarily permanent. Brazil, India, and South Korea, having developed an industrial base, have moved into the semiperipheral category.
Peripheral Regions	Peripheral regions have undeveloped or very specialized economies characterized by old technologies and low levels of productivity. Egypt, Indonesia, and Zambia fall into the peripheral group.
Type of Spatial Unit	**Global Version**
Core Region	The core region is the focus of economic, political, and social activity. Most people live in the core, which is highly urbanized and industrialized. The core has a high capacity for innovation and economic change. Innovations and economic advances are disseminated downward through the national urban hierarchical system to the periphery. Central Canada is Canada's core region.
Upward Transitional Region	The upward transitional region's economy and population are growing as both capital and labour flow into this rapidly developing area. While initial development occurred in the primary sector, there is now a greater emphasis on manufacturing and service activities. British Columbia is an example of an upward transitional region.
Downward Transitional Region	In a downward transitional region, the economy is declining, unemployment is rising, and out-migration is occurring. Often this is an 'old' region that is dependent on resource development for its economic growth. Now that these resources have passed their prime or have been exhausted, the regional economy has stalled. Atlantic Canada is an example of a downward transitional region.
Resource Frontier	Located far from the core region, few people live in this frontier and little development has taken place. Resource companies are just beginning to penetrate into this remote area. As energy and mineral deposits are discovered, the prospects for economic growth are enhanced. The Territorial North represents Canada's last resource frontier.

Sources: John Friedmann, *Regional Development Policy: A Case Study of Venezuela* (Cambridge, Mass.: MIT Press, 1966); Immanuel Wallerstein, *The Capitalist World-Economy* (Cambridge: Cambridge University Press, 1979).

text to analyze the striking regional economic disparities evident across Canada. By combining the abstract notion of the core/periphery model with geographic reality, the model attempts to shed some light on the power imbalances—economic, political, and social— that have existed between the major regions of Canada since Confederation and the effect of differences in resources on the balance of power. As mentioned earlier, in Canada the centre of power has resided in Central Canada, especially in the Windsor–Québec City axis. From the perspective of the core/periphery model, all the other regions of Canada function as the periphery (hinterlands); they supply the natural resources (raw materials and foodstuffs) that fuel manufacturing activity, employment, and urban growth in the dominant core region.

For the first 122 years of Confederation, the natural advantages enjoyed by Central Canada were reinforced by high tariffs. In this way, Confederation favoured Central Canada, making it dominant and the peripheral areas dependent. Since 1967, Ottawa has directed federal transfers (also known as equalization payments) to economically disadvantaged regions of the country. Ottawa's capacity to redress regional disparities has diminished with economic integration with the United States. In the 1970s, Ottawa eagerly intervened in regional economic matters. The primary federal agency was the Department of Regional Economic Expansion, commonly referred to as DREE (Savoie, 1986). From 1967 to 1982, DREE served as a super ministry of regional development. Since 1983, Ottawa has continued its policy of regional subsidies. Eventually the programs were centralized under the federal Department of Human Resources.

In terms of application to Canada's regional geography, three major attractions of the core/periphery model are:

- Canada's regional disparities lend themselves to such a spatial framework.
- Relations between Canada's regions, whether economic or political, reflect the core/periphery model.

- The model is related to Harold Innis's **staples thesis**, which served to reinterpret Canada's early economic history.[9]

How closely does the theoretical model fit the realities of Canada? The answer to this question changes with time. From the days of Macdonald's 1879 National Policy to the Auto Pact of 1965, Canada's economy was closed, with high tariffs preventing foreign goods from reaching the Canadian domestic market. At that time, the combination of a nationalist economic policy and the regional nature of Canada produced a near mirror image of the core component of the model. Highly industrialized areas in southern Ontario and southern Québec functioned as the industrial core, supplying manufactured products to the rest of Canada. The role of the Canadian resource hinterland, however, was linked much more closely to global markets, so its image as a periphery was somewhat blurred. Why did Canada's resource hinterland sell so much of its primary products at world prices to external markets and yet buy Canadian manufactured goods at artificially high domestic prices? This peculiar situation was created by Ottawa's trade policy. The federal government placed high tariffs on imported foreign manufactured goods, but rarely took similar action to ensure high prices for raw materials and foodstuffs within Canada. Given the small size of the domestic market, many **primary products**, such as grain, lumber, and minerals, became necessarily dependent on world markets.

With the introduction of the Auto Pact in 1964, the economic advantage of a single market and production system for automobiles in Canada and the United States soon became apparent. The next step towards a continental economy took place in 1989 when the Free Trade Agreement was signed. In

1994, this agreement was superseded by the North American Free Trade Agreement (NAFTA), which included Mexico (Table 1.2). The FTA and NAFTA forced manufacturing firms based in Canada to compete within the much larger continental market. During this restructuring process, some manufacturing companies survived while others failed. Not surprisingly, this economic integration caused many American-owned branch plants to close their Canadian plants, thereby allowing their larger and more efficient plants in the United States to serve the entire North American market. Canadian-based firms, including a few branch plants, turned to a more specialized production line suited to the North American market. Such an approach allowed these firms to achieve **economies of scale**. That is, by increasing their production, these firms reduce the average cost of their product. During this period, three economic events took place within the Canadian hinterland:

- The price of manufactured goods available in the Canadian hinterland declined because less expensive foreign imports became available, thereby reducing costs to consumers in the hinterland.
- The removal of trade barriers permitted more Canadian products into the American domestic market, thereby stimulating the hinterland economy. In several cases, the Americans found ways to curtail these Canadian products from entering the United States. American restrictions on Canadian softwood lumber represent the most serious of these trade disputes.
- The elimination of the rail transportation subsidy for grain in 1995 made the export of bulky products like spring wheat more expensive for Prairie farmers, forcing them to diversify by seeking alternative agricul-

tural crops that are either sold locally (canola) or have a high enough value to permit shipment by rail (durum wheat).

Therefore, the shift towards freer trade has lowered the cost of production, widened the range of products exported, and encouraged the diversification of the economies of Canada's hinterland regions. In spite of these benefits, a few sectors, including the softwood lumber and beef industries, have been hurt by US trade barriers.

In the next section, brief profiles of each of Canada's six geographic regions provide a preview of each region's predominant characteristics. The order of these regional profiles reflects the core/periphery model used in this text. Each profile also touches on some of the major themes to be developed in later chapters and defines the Key Topics in Chapters 5 to 10.

Regional Profiles

The statistical data on the area, population, and economic strength of Canada's regions (Table 1.1) and their social characteristics (Table 1.5) underscore the following key factors in studying Canada's regional geography:

- Of the over 30 million Canadians, more than 60 per cent live and work in Ontario and Québec. This demographic fact illustrates the dominant position of these provinces in the nation's economic and political matters and supports the notion of a core region.
- With over 80 per cent of the Québec population declaring French as their mother tongue, that province's position as the centre of French language and culture in Canada is indisputable. From a linguistic perspective, Québec is a distinct part of

Canada; from a federalist point of view, Québec's geopolitical situation makes it an indispensable part of Canada; but from a separatist perspective, this cultural/linguistic difference provides the raison d'être for an independent country.

- While Aboriginal Canadians reside in all provinces, they form the largest proportion of the population in the Territorial North. In 2001, Aboriginal Canadians comprised just over half of the territorial population. In Nunavut, the figure is around 85 per cent. Since 1999, the government of Nunavut has provided the Inuit with political expression of their culture. With this demographic power, Aboriginal northerners—especially the Inuit—have a more secure position in the affairs of the Territorial North than in the five southern geographic regions.

Ontario

Ontario remains the economic powerhouse of Canada. Its historic development goes back to the eighteenth century. Shortly after the American War of Independence began in 1775, British settlers fled from New England and resettled in the remaining British colonies, including in the area now known as southern Ontario. Although Ontario was settled much later than Atlantic Canada and Québec, it soon became the dominant economic and political power in Canada. Much of the credit for its industrial development goes to the protective trade barriers first imposed under the National Policy in 1879, which continued unabated until 1989 when the FTA was signed, and the favourable physical features of the Great Lakes–St Lawrence Lowlands. The fertile soils and mild climate of the region attracted many

Table 1.5	Social Characteristics of the Six Canadian Regions, 2001			
Geographic Region	French* (000s)	French (per cent by region)	Aboriginal Peoples**	Aboriginal Peoples (per cent by region)
Ontario	307,295	2.7	188,315	1.6
Québec	5,918,390	83.4	79,400	1.1
British Columbia	16,905	0.4	170,025	4.4
Western Canada	46,375	0.9	436,450	8.6
Atlantic Canada	241,375	10.6	37,785	2.4
Territorial North	1,040	1.1	45,865	51.7
Canada	6,531,375	22.0	976,310	3.3

*French includes those who speak French most often at home.

**In 2001 the census asked two questions to determine the number of Aboriginal peoples. The 2001 Aboriginal population, as determined by a question about identity, was 976,310. The three census categories for Aboriginal population were North American Indian (608,850), Métis (292,310), and Inuit (45,070). The 2001 Aboriginal population, as determined by a question based on ethnic origin, was 1,319,890. The explanation for the difference in the number of Aboriginal peoples may be one of 'multiple identity'.

Sources: Statistics Canada, Aboriginal Peoples of Canada: A Demographic Profile (2003), at: <www12.statcan.ca/english/census01/products/analytic/companion/abor/contents.cfm>; Statistics Canada, Census of Population: Language, Mobility and Migration (2003), at: <www.statcan.ca/Daily/English/021210/d021210a.htm>

settlers. As the number of settlers increased, a large domestic market emerged. In addition, this area's close proximity to the American manufacturing belt enabled it to develop close ties with American manufacturing companies. By 2001, over 11.4 million residents claimed Ontario as their home, making this region the largest in Canada. Toronto, the country's largest city, is situated on the shores of Lake Ontario in southern Ontario. As the financial capital of Canada, this megacity houses most of the corporate headquarters of major Canadian and foreign firms. Altogether, southern Ontario, along with southern Québec, forms a powerful economic axis within Canada and, since the Free Trade Agreement, within North America. Northern Ontario, on the other hand, is a sparsely populated area that functions as a resource hinterland. Ontario is noted for seven other important characteristics:

- Three physiographic regions exist in Ontario: the Hudson Bay Lowland, the Canadian Shield, and the Great Lakes–St Lawrence Lowlands.
- The population is heavily concentrated in the Great Lakes–St Lawrence Lowlands, with a secondary cluster in the Canadian Shield. Few people live in the Hudson Bay Lowland.
- Ontario can be divided into two subregions. Southern Ontario is an industrial core while northern Ontario is a resource hinterland. In turn, northern Ontario can be divided into the Near North and the Middle North (Figure 5.2).
- Manufacturing is strong, particularly the automobile assembly and parts sector.
- Its economy and population are growing with both capital and people coming to Ontario. Ontario, but especially Toronto,

In this aerial view of downtown Ottawa, Ontario, Parliament Hill is shown in the foreground while the Library and Archives Canada building is situated between the Alexandra and the Macdonald-Cartier bridges. These two bridges span the Ottawa River and link Ottawa with Hull, Québec. (Don Loveridge/Valan Photos)

is the most attractive destination for immigrants.

- Outside of Québec, the largest number of francophones reside in Ontario. In 2001, French-speaking Ontarians comprised 4.5 per cent of the total population, or 509,265 people, up by nearly 2 per cent from 1996.

- Key Topic: Given the effect of the Auto Pact on Ontario's economy, the automobile industry is examined in more detail in Chapter 5. The struggle for market share between the Big Three firms (General Motors, Ford, and DaimlerChrysler) and foreign firms, but especially Toyota and Honda, holds a central place in this discussion.

Québec

Québec is both an economic power and a cultural homeland. Like Atlantic Canada, settlement in Québec began in the seventeenth century, thus giving it a long and rich history. The fertile lowlands of the St Lawrence Valley represent the heartland of Québec and it serves as the cultural homeland of francophones across North America. In the nineteenth century, arable land in the St Lawrence Valley was occupied and Québec's settlers spread southward into the Eastern Townships of the Appalachian Uplands and northward into the Canadian Shield, but limited agricultural land kept their population numbers low. With the acquisition of new territories in 1898 and 1912, Québec extended its territorial reach to the Arctic lands that border Hudson Strait. Currently, these northern lands remain a resource frontier while agricultural and manufacturing activities are concentrated in the St Lawrence Lowlands. Montréal, the principal city of Québec, is situated on an island in the

Montréal, Québec, is a leading centre of art and music in Canada. Lea Vivot's bronze sculpture, *Mother and Child*, can be found on Montréal's Sherbrooke Street West. (Robert Bone)

St Lawrence River. Because of its language, civil laws, and history, Québec has a culture that is strikingly different from that of other regions of Canada, yet because of its geographic position and economic linkages with the rest of Canada and the United States, Québec is an integral part of both the Canadian and the larger North American economy. According to the Friedmann classification, Québec consists of an industrial core surrounded by a resource hinterland. Québec is noted for many cultural, economic, and physical features:

- Québec is by far the largest province by geographic size and ranks second in population size.
- Québec straddles four physiographic units: the Hudson Bay Lowland, the Canadian Shield, Appalachia, and the St Lawrence Lowlands.
- With a varied agricultural and resource base, Québec has a highly diversified economy. Agriculture is mostly dairy, fruit, and vegetable production, while the resource base includes forests, minerals, and hydroelectricity.
- Most manufacturing is concentrated in the Montréal region. Manufacturing is changing with labour-intensive firms declining and high-technology firms increasing.
- Québec's population reached 7.2 million by 2001. However, slow population growth is weakening the province's political position within Canada. From 1871 to 2001, Québec's share of Canada's population fell from 34.2 per cent to 24.0 per cent.
- Most Quebecers subscribe to the 'two founding nations' perception of Canada that views the creation of Canada as a compact between the French and English.
- Key Topic: The development of hydroelectricity in northern Québec to supply southern consumers and factories demonstrates the heartland/hinterland structure of this province. These key elements are discussed at greater length in Chapter 6, including the recent agreements between the provincial government and the Cree and Inuit that have opened the North to more hydroelectric development and other forms of resource exploitation.

British Columbia

With its mild climate, vast forests, and varied mineral wealth, British Columbia is one of the

Coal Harbour in Vancouver, British Columbia, is the starting point for the Vancouver–Alaska cruise, one of the world's most popular cruises. Every year more than one million passengers, on more than 300 sailings, pass through the Port of Vancouver's Canada Place and Ballantyne cruise ship facilities. (Vancouver Sun)

most richly endowed geographic regions. Like Western Canada, British Columbia is an upward transitional region within Canada's periphery. If its rapid economic and population growth continues, this Pacific region is destined to become a core region early in the twenty-first century. While staple production played a key role in its early economic development, service activities based on high technology and tourist-oriented businesses have now become its strongest economic sector. Even so, forestry, natural gas, and, to a lesser degree, mining and fishing remain crucial economic activities. Since 1998, British Columbia's economic growth has slowed because of a drop in Asian customers' demand for primary products and the high tariffs imposed by the US government on softwood products shipped to the United States. Nevertheless, the province's population continued to increase during this difficult economic period. By 2001, its population had reached 3.9 million, which represents an increase of nearly 5 per cent over the 1996 figure. British Columbia is distinguished by six additional characteristics:

- Settlement of this British colony began in the mid-nineteenth century, shortly before 1871 when British Columbia joined Canada.
- In terms of physiography, British Columbia lies within the Cordillera, except for its northeastern sector where the Interior Plains are found.
- Because of the Pacific Ocean's influence, British Columbia's coastal region has the mildest climate in Canada.
- Most of the nearly 4 million British Columbians reside along the Pacific coast.
- Because of its robust economy and mild climate, British Columbia attracts many newcomers from other provinces and for-

eign countries, particularly the Far East.
- Key Topic: The extensive forests of British Columbia are a renewable resource that once seemed inexhaustible but now has reached its natural limit. As well, US trade barriers have hurt the BC softwood lumber industry. These two unresolved issues that are harrying the forest industry are examined in Chapter 7.

Western Canada

Thanks to the remarkable economic growth in Alberta, Western Canada is an upward transitional region within the Canadian periphery. Still, the region remains heavily dependent on agriculture and resource development. Internal differences exist, as revealed by the spatial split between a vibrant Alberta economy and sluggish economies found in both Manitoba and Saskatchewan. Another dichotomy exists between rural and urban places. While rural areas are suffering from economic decline and depopulation, the major cities of Calgary, Edmonton, Winnipeg, Saskatoon, and Regina are enjoying both economic and population growth. Today, some 5 million Canadians live in this western hinterland. While agriculture remains a key to the region's prosperity, the western economy has diversified, with most growth occurring in its service and resource industries. This trend towards diversification is most noticeable in Alberta, where oil and gas developments propelled the province's economy into high gear, ensuring its position as a 'have' province. Saskatchewan and Manitoba, with fewer energy resources, still depend heavily on their agricultural sectors. When drought struck Western Canada in the summers of 2001 and 2002, both grain and livestock industries suffered great losses, one from crop failure and the other from the collapse of graz-

Pincher Creek, Alberta, is famous for its livestock industry. More recently, Pincher Creek has become the site of wind energy production, helping to make Alberta the leading province for wind-produced electricity in Canada. Powerful chinook winds from the Rocky Mountains make Pincher Creek a particularly attractive site for wind turbines. (Barrett and MacKay Photography, Inc.)

ing land. While the farmers and ranchers were in the front line of this natural disaster, the ripple effects nearly forced the Saskatchewan Wheat Pool into bankruptcy. Other characteristics of Western Canada include:

- The Interior Plains and the Canadian Shield are the main physiographic regions, though the eastern edge of the Rocky Mountains and part of the Hudson Bay Lowlands fall within Western Canada.
- Most of the population is in the Interior Plains.
- Agriculture is limited by a dry continental climate that can cause crop failures.
- Distance from ports remains a limiting factor for agricultural and industrial exports.

- Aboriginal peoples form 8.6 per cent of the region's population while francophones make up 3.1 per cent of the population.
- Key Topic: Agriculture is in transition but remains troubled by dry growing conditions and by low world prices, thus making moisture and global prices critical elements in the uneven economic growth .of Western Canada. Together, they form the focus for this region's agricultural economy.

Atlantic Canada

Atlantic Canada is the smallest geographic region in Canada. Except for the Territorial North, it has the smallest population (2.3 million in 2001). From a geographic and historical perspective, Atlantic Canada consists of two parts: the Maritimes and Newfoundland/Labrador. Unlike other regions, the sea has played a strong, even dominant, role in the lives of Atlantic Canada's inhabitants. During the twentieth century Atlantic Canada did not fare well in Confederation, perhaps the result of its small resource base and a geographic location that limited the region's access to its natural market in New England through high tariffs, and to Central Canada because of high shipping costs. As well, the region's fish, forests, and minerals have been exploited for a long time, which has sometimes resulted in depreciation of the resource base, notably the mineral resources and cod stocks. Consequently, the economy and population of Atlantic Canada grew more slowly than those of other regions. Atlantic Canada is a disad-

vantaged region of Canada. Its heavy dependency on equalization payments, unemployment insurance, and other transfer payments is a measure of its weak economic position within Confederation. Other indicators are high unemployment rates and a steady outmigration of its residents. In terms of Friedmann's core/periphery model, Atlantic Canada most closely approximates his downward transitional region. Offshore oil and gas developments have slowed this trend, and they might prove to be the catalyst to reverse it. Six other characteristics of Atlantic Canada are:

- Atlantic Canada has a long and rich history. In 1604, Port Royal (now Annapolis Royal) was founded in present-day Nova Scotia.
- The region comprises part of the Appalachian physiographic region and

Halifax, Nova Scotia, is the largest city in Atlantic Canada. Founded in 1749 when Governor Edward Cornwallis and 2,500 settlers created Canada's first permanent British settlement, Halifax has grown into a modern metropolitan city. (Barrett and MacKay Photography, Inc.)

includes a vast continental shelf.

- Halifax is the largest metropolitan city in Atlantic Canada.
- Offshore energy developments have sparked hopes for an economic revival in Newfoundland and Nova Scotia.
- The nearly 300,000 Acadians living in New Brunswick form the largest francophone population cluster outside of Québec.
- Key Topic: Given the importance of the fisheries to Atlantic Canada, this topic is explored in more detail in Chapter 9 from three perspectives: the failure of the cod stocks to recover, the shift to other forms of fishing, and the inclusion of the Mi'kmaq in the lobster fishery.

The Territorial North

At 3.9 million km², the Territorial North is the largest geographic region in Canada. Stretching from 60° N to the North Pole, the Territorial North includes nearly 40 per cent of the territory of Canada and four physiographic regions. The four physiographic regions are the Cordillera, the Interior Plains, the Canadian Shield, and the Arctic Lands. Settlement is concentrated in the southern Cordillera and Interior Plains. Even so, with just under 100,000 residents, the Territorial North has a population smaller than that of Prince Edward Island (135,294 in 2001). Aboriginal peoples form just over half of this population. Development in this resource hinterland has been

Inuit cultural performers Aqsarniit (which means Northern Lights) perform in the Hall of Honour on Parliament Hill in Ottawa for the installation of the new Governor-General, Adrienne Clarkson, in 1999. (CP/Tom Hanson)

hampered by remoteness from world markets and by a cold environment. Energy and mineral resources dominate the northern economy, while renewable resource development based on agriculture and forestry is limited. Wildlife is the major renewable resource and many Aboriginal families depend on it for much of their meat and fish. The Territorial North is a resource frontier in Friedmann's classification system. Here are some other characteristics:

- The Territorial North has an extremely cold environment with much of the ground permanently frozen.
- Nearly half of the people live in Whitehorse, Yellowknife, and Iqaluit, while the rest live in small towns and villages.
- Although the resource economy is the driving force behind the northern economy, most inhabitants are employed in the public sector. The public sector is heavily dependent on Ottawa for most of its funds.
- By 2003, five Aboriginal groups had signed comprehensive land-claim agreements that provide both cash and land. The cash enables them to enter the market economy, and the land enables them to maintain their land-based economy.
- A new territory, Nunavut, hived off from the eastern part of the Northwest Territories, became a reality in 1999.
- The emergence of a new political geography that recognizes Aboriginal peoples is a bold step that can only strengthen the social fabric and regional structure of Canada.
- Key Topic: Large-scale resource projects, which play a critical role in the northern economy, are covered in Chapter 10. Special attention is paid to the diamond industry, which has had a significant

impact on the economy of the Northwest Territories and may have similar results for Nunavut.

Understanding Canada's Regions

As a prelude to understanding the many inter-related physical and human dimensions of Canada's geographic regions, the next three chapters focus on Canada's physical geography, history, and human geography. Chapter 2 ('Canada's Physical Base') explores the wide range of natural and geomorphic processes that account for wide variations in landforms across Canada. Seven physiographic regions are described with brief accounts of their geology, soils, natural vegetation, and climate. A major theme is the physical environment's influence on human occupancy of the land and the effects of human occupancy on the physical environment. Chapter 2, therefore, is more than a mere backdrop for the remainder of this text; rather, it illuminates how human activities and human decision-making both shape and are shaped by the physical environment in which they occur, and it introduces the extent of environmental problems resulting from industrial development, as described in each of the regional chapters.

Chapter 3 ('Canada's Historical Geography') focuses on the arrival of Canada's First Peoples, early French and British settlements, and the territorial evolution of Canada. Four faultlines affect Canada and its six geographic regions differently both in time and location. These tensions that often lie dormant only to flare up unexpectedly are between English-speaking/French-speaking Canadians, centralists and decentralists, new and old Canadians, and Aboriginal/non-Aboriginal Canadians. These tensions provide a background for the subject of the next chapter and for subsequent regional chapters. Chapter 4 ('Canada's Human Face') is directed to contemporary issues dealing with demography (baby boom and bust), Canada's pluralistic society (the impact of a more open immigration policy), and a continental economy (the effect of the Free Trade Agreement). This chapter also looks at the population trends associated with Canada's four faultlines.

Summary

The world is a complex place. To simplify the complexities of space, regional geographers divide the world and countries into regions, which vary by scale but are often interrelated in a hierarchical order. A regional geographer selects critical physical, historic, and human characteristics that logically divide a large spatial unit into a series of regions. Towards the margins of a region, its main characteristics become less distinct and merge with those of a neighbouring region. For that reason, boundaries are best considered as transition zones. Canada is a country of regions. Shaped by its history and physical geography, Canada

is distinguished by six geographic regions: Ontario, Québec, British Columbia, Western Canada, Atlantic Canada, and the Territorial North. Within Canada and its regions, four key tensions exists that, through their interaction, demonstrate the very essence of Canada as a 'soft' nation, where conflicts are usually resolved or ameliorated through compromise rather than by political or military power.

The essential foundation for studying regional geography is to conceptualize places and regions as components of a constantly changing global system. The core/periphery model provides an abstract spatial framework for understanding the general workings of the modern capitalist system. It consists of an interlocking set of industrial cores and resource peripheries. This model can function at different geographic scales and serves as an economic framework for interpreting Canada's regional nature. In addition to the core region, three types of regions devised by Friedmann extend our appreciation of the diversity of the Canadian periphery. They are: (1) an upward transitional region, (2) a downward transitional region, and (3) a resource frontier.

Notes

1. While Canadians and Americans occupy the same continent, historic events and geographic differences laid the foundation for the emergence of two *different* societies within one continent. These differences are found in many aspects of the two societies. Each country has a different approach to gun control legislation, multiculturalism, and a national medical system. As a result of these and other differences, some scholars believe that Canadians are more trusting of their governments and more tolerant of social diversity than are Americans (Hartz,

1955; Lipset, 1990; Lemon, 1996; Saul, 1997).

How has history contributed to such beliefs? Governance provides one key. Each country has followed a different path in developing its political system of governance. This difference began with the birth of each country. The United States fought a War of Independence to break away from Great Britain. Following the defeat of the British forces, Americans established a presidential style of government whereby each eligible voter has a vote to determine the selection of the President, although, as the 2000 American election demonstrated, the Electoral College in the US means that these votes do not directly elect the President. In that election, Al Gore, the Democratic candidate, won the popular vote by half a million over George W. Bush, but Gore lost the election because his opponent secured more Electoral College votes. In sharp contrast, the colonies that eventually formed Canada did not gain a measure of political independence from Britain until 1867. Then, too, the American Constitution of 1776 calls for 'life, liberty, and the pursuit of happiness' while the British North America Act of 1867 (now known as the Constitution Act, 1867) stresses 'peace, order, and good government'. The American approach to governance reflects American revolutionary idealism that stresses individualism, while Canadians were forced to balance individual and collective rights through a parliamentary system. The vote in a Canadian federal election is not cast for a candidate for Prime Minister but for a local candidate representing a political party. Generally, the leader of the party with the most local (riding or constituency) candidates elected to Parliament then becomes Prime Minister. Originally, the large French-speaking population in Lower Canada (Québec) necessitated a search for compromise with the English-speaking population in Upper Canada (Ontario) and the Maritimes. Consequently, the Fathers of Confederation agreed to a federal state with certain powers assigned to the provinces. The provinces in Canada have more autonomous political and economic power than do the states in the US. This balancing act in Canada is much more complicated today because others, such as visible minorities and Aboriginal peoples, have joined French-speaking Quebecers in seeking a distinct place within Canada. Canada, unlike the United States, was forged from evolutionary political pragmatism that requires both compromise on key political issues and a balance between individual and collective rights (New, 1998; Saul, 1997). For this reason, Saul (1997: 8–9) describes Canada as a 'soft' country and the United States as a 'hard' one.

2. Geographers have interpreted Canada's regions differently. Thirty-six years ago, Kenneth Hare (1968: 3) attempted to capture the richness of the country by describing Canada as 'a great geographical system, whereby a vast area of land, much of it then empty, has been organized and made productive'. From his perspective, the organization of Canada (into regions) was a direct reflection of its physical geography. J. Lewis Robinson had the same approach. In 1982, however, Larry McCann de-emphasized the role of physical geography in regional geography and increased the role of history within a theoretical framework known as the core/periphery model. In doing so, McCann made a bold departure from the more traditional approach to regional geography.

3. Even with the 1995 Agreement on Internal Trade (AIT), major impediments to interprovincial trade remain in such areas of provincial jurisdiction as agriculture and trucking standards. The reason is simple—provinces use these barriers to promote their economies. While interprovincial trade contributes less proportionally to Canada's gross domestic product than before the Free Trade Agreement, it remains important. In 2001, interprovincial trade formed about 20 per cent of Canada's gross domestic product.

4. In 1972, Canadian geographers used this same regional structure to prepare a series of six regional books on Canada for the 22nd International Geographical Congress. The editors of this regional series were the well-known geographers of the 1970s: L. Gentilcore, F. Grenier, A.G. Macpherson, P.J. Smith, J.L. Robinson, and W.C. Wonders.

5. The Charter ensures that every Canadian has the right to vote for candidates who will represent them at various levels of government. However, Aboriginal governments can select their leaders by following traditional practices, which often means that only chiefs can vote for provincial and national Indian leaders.

6. As a result of the federal government's recognition of Aboriginal rights, land-claim settlements have begun to modify the cultural, economic, and political landscape of the Territorial North and the northern areas of many provinces. Already the James Bay Cree, Inuit, and Innu of Québec, the Inuvialuit, Gwich'in, and Sahtu Dene of the Northwest Territories and Yukon, the Inuit of Nunavut, and the Nisga'a of BC have obtained land-claim settlements and received land and cash in exchange for surrendering their Aboriginal rights to vast areas of land. In 1999 and 2000 respectively, the Labrador Inuit and the Dogribs of the NWT reached agreements-in-principle. The James Bay Cree also achieved a regional form of self-government in northern Québec. While the North remains a resource hinterland, land-claim settlements ensure that those Aboriginal peoples will have more control over resource development and will therefore be better able to protect the wildlife and their hunting lifestyle.

7. Prior to Confederation, British colonies were part of the imperial colonial system. During the 1840s, Britain abandoned the mercantile system of preferential trade with its colonies in favour of free trade. Britain believed that free trade would reduce the cost of imported raw materials and foodstuffs, but it also meant lower prices for primary products produced in British North America. The British North American colonies sought other trade arrangements. In the late 1840s, the Province of Canada used a tariff to protect its manufacturing firms. In 1854, the Reciprocity Treaty with the United States was put into place. Under this treaty, staple goods could move freely between both countries. Trade flourished, particularly between the Maritimes and New England. In 1866, the United States cancelled this treaty, forcing the British North American colonies to find an alternative, which was to trade among themselves. From this perspective, Confederation was intended to promote trade between the four founding provinces.

8. Environmental determinism is based on the principle that the physical environment determines human affairs. Geographers reject that position, although they recognize that the environment does exert a strong influence on the nature of human activities in various regions of the world. Students can find a more complete discussion of environmental determinism and other philosophical options in geography, including possibilism, positivism, humanism, and Marxism, in William Norton, *Human Geography*, 5th edn (Toronto: Oxford University Press, 2004). The writings of most geographers reflect either one of these philosophical positions or some combination. The core/periphery model, for example, sprang from Marxist scholarship. Because of the model's powerful spatial implications, other scholars holding different philosophical positions have modified this theory by removing its economically deterministic character. Instead, they accept that external forces (such as global institutions like the WTO) and internal forces (such as federal-provincial agreements like equalization payments) can modify the impact of the physical environment on regional development. Hinterlands, therefore, are not locked into a single outcome because of their physical geography.

9. The staples thesis, devised by Harold Innis (1930), is a Canadian variant of the core/periphery model. Within this theoreti-

cal context, Innis conceived of Canada's early history in terms of its regional geography but with the country functioning as a resource economy driven by foreign demands. From this core/periphery perspective, Innis projected a new vision of Canada's economic history, asserting that the export of Canada's natural resources to Great Britain, the United States, and other advanced industrial countries affected the national and regional economies of Canada and their social and political systems. Innis insisted that Canada is a nation because of its geography; that is, Canada's history was triggered by the exploitation of a major resource (staple) found in a region and shaped by its trade relation with external markets. In some cases resource development led to economic diversification; in other cases it did not. In several books, Innis provided detailed historical accounts of staples development in several regions of Canada. For instance, the fur trade in early colonial Canada did not result in the diversification of the northern economy but, rather, in the enrichment of England and France (Innis, 1930). The wheat trade, on the other hand, led to the diversification of the economy of Western Canada. The staples thesis, then, provided Innis with a Canadian version of the core/periphery model (Barnes, 1993; Hayter and Barnes, 2001).

Key Terms

core

An abstract area or real place where economic power, population, and wealth are concentrated; sometimes described as an industrial core, heartland, or metropolitan centre.

core/periphery model

A theoretical concept based on a dual spatial structure of the capitalist world and a mutually beneficial relationship between its two parts, which are known as the core and the periphery. While both parts are dependent on each other, the core (industrial heartland) dominates the economic relationship with its periphery (resource hinterland) and thereby benefits the most from this relationship. The core/periphery model can be applied at several geographical levels, including international, national, and regional.

cultural regions

Geographic areas defined by one or more cultural features such as religion.

dependency theory

A neo-Marxist interpretation of the dual spatial structure of the capitalist world that is based on an exploitive relationship between its two parts, known as the core and periphery. As a consequence of this exploitive economic relationship, the periphery is eventually stripped of its natural resources and left in a state of underdevelopment. Neo-Marxist scholars see underdevelopment as resulting from capitalist exploitation.

economies of scale

A reduction in unit costs of production that results from an increase in output.

ecumene

The portion of the land that is settled.

gross domestic product

An estimate of the total value of all materials, foodstuffs, goods, and services produced by a country or province in a particular year.

heartland

A geographic area in which a nation's industry, population, and political power are concentrated; also known as a core.

hinterland

A geographic area based on resource development that supplies the heartland with many of its primary products; also known as a periphery.

megaregions

Extremely large geographic units such as Canada.

mesoregions

Medium-size geographic units such as the Cordillera.

microregions

Small geographic units such as Cape Breton Island.

modernization theory

Economic theory that explains how economic growth occurs and how wealth generated by this growth is distributed throughout the world. The core/periphery model, for example, illustrates how the periphery is 'developed' by the core.

natural regions

Geographic areas defined by one or more physical features such as a soil type.

periphery

The weakly developed area surrounding an industrial core; also known as a hinterland.

physiography

A study of landforms, their underlying geology, and the processes that shape these landforms; geomorphology.

primary products

Goods derived from agriculture, fishing, logging, mining, and trapping; non-processed products.

region

An area of the earth's surface that is defined by its distinctive human or natural characteristics. Boundaries between regions are often transition zones where the main characteristics of one region merge into those of a neighbouring region. Geographers use the concept of regions to study parts of the world.

regional geography

The study of the geography of regions and the interplay between physical and human geography, which results in an understanding of human society, its physical geographical underpinnings, and a sense of place.

sense of place

The special and often intense feelings that people have for the region in which they live. These feelings are derived from a variety of experiences; some are due to natural factors such as climate, while others are from cultural factors such as language. Whatever its origin, a sense of place is a powerful psychological bond between people and their region.

staples thesis

The idea that the history of Canada, especially its regional economic and institutional development, was linked to the discovery, utilization, and export of key resources in Canada's vast frontier. Harold Innis devised this thesis in the early 1930s, and his ideas continue to influence Canadian scholars.

Bibliography

Adams, Michael. 2003. *Fire and Ice: The United States, Canada and the Myth of Converging Values*. Toronto: Penguin.

Alexandroff, Alan S., and Don Guy. 2003. 'What Canadians have to say about relations with the United States', *Border Papers No 73*, C.D. Howe Institute.

Allan, John, Doreen Massey, and Alan Cochrane. 1998. *Rethinking the Region*. London: Routledge.

Barnes, Trevor, ed. 1993. 'Focus: A Geographical Appreciation of Harold A. Innis', *The Canadian Geographer* 37, 4: 352–64.

Bunting, Trudi, and Pierre Filion, eds. 2000. *Canadian Cities in Transition: The Twenty-First Century*, 2nd edn. Toronto: Oxford University Press.

Clawson, David L., and James S. Fisher. 1998. *World Regional Geography: A Development Approach*. Toronto: Prentice-Hall.

Columbo, John Robert, ed. 1987. *New Canadian Quotations*. Edmonton: Hurtig Publishers.

De Blij, J.H., and Peter O. Muller. 1994. *Geography: Realms, Regions and Concepts*. Toronto: John Wiley.

———— and Alexander B. Murphy. 1998. *Human Geography: Culture, Society, and Space*, 6th edn. Toronto: John Wiley.

d'Haenens, Leen, ed. 1998. *Images of Canadianness: Visions on Canada's Politics, Culture, Economics*. Ottawa: University of Ottawa Press.

Frank, A.G. 1969. *Capitalism and Underdevelopment in Latin America*. New York: Monthly Review Press.

French, H.M., and O. Slaymaker, eds. 1993. *Canada's Cold Environments*. Montréal and Kingston: McGill-Queen's University Press.

Friedmann, John. 1966. *Regional Development Policy: A Case Study of Venezuela*. Cambridge, Mass.: MIT Press.

Getis, Arthur, and Judith Getis. 1995. *The United States and Canada: The Land and the People*. Boston: William C. Brown.

Hamelin, Louis-Edmond. 1979. *Canadian Nordicity: It's Your North Too*, trans. W. Barr. Montréal: Harvest House.

Hartz, Louis. 1995. *The Liberal Tradition in America*. New York: Harcourt Brace Jovanovich.

Hayter, Roger, and Trevor J. Barnes, 2001. 'Canada's Resource Economy', *The Canadian Geographer* 45, 1: 36–41.

Hutchison, Bruce. 1942. *The Unknown Country: Canada and Her People*. Toronto: Longmans, Green and Company.

Innis, Harold. 1930. *The Fur Trade in Canada: An Introduction to Canadian Economic History*. New Haven: Yale University Press.

Lemon, James T. 1996. *Liberal Dreams and Nature's Limits: Great Cities of North America Since 1600*. Toronto: Oxford University Press.

Lipset, Seymour M. 1990. *Continental Divide: The Values and Institutions of the United States and Canada*. New York: Routledge.

McCann, L.D., ed. 1982. *A Geography of Canada: Heartland and Hinterland*. Scarborough, Ont.: Prentice-Hall.

Malecki, Edward J. 1991. *Technology and Economic Development*. New York: Wiley.

Marsh, James H., ed. 1988. *The Canadian Encylopedia*, 2nd edn. Edmonton: Hurtig.

Matthews, Ralph. 1983. *The Creation of Regional Dependency*. Toronto: University of Toronto Press.

Norton, William. 2004. *Human Geography*, 5th edn. Toronto: Oxford University Press.

Resnick, Philip. 2000. *The Politics of Resentment: British Columbia Regionalism and Canadian Unity*. Vancouver: University of British Columbia Press.

Robinson, J. Lewis. 1989. *Concepts and Themes in the Regional Geography of Canada*, 2nd edn. Vancouver: Talonbooks.

Saul, John Ralston. 1997. *Reflections of a Siamese Twin: Canada at the End of the Twentieth Century*. Toronto: Viking.

Savoie, Donald J. 1986. *The Canadian Economy: A Regional Perspective*. Toronto: Methuen.

Scott, Allen J. 1998. *The Coming Shape of Global Production, Competition, and Political Order*. Oxford: Oxford University Press.

Simpson, Jeffrey. 2000. *Star-Spangled Canadians*. Toronto: HarperCollins Canada.

Shidelar, Janet. 1999. *The Geography of Canada Bibliography Series: Vol. 7, Canada as a Whole*. Plattsburgh, NY: Plattsburgh State University, Center for the Study of Canada.

Stanford, Quentin H., ed. 1998. *Canadian Oxford World Atlas*, 4th edn. Toronto: Oxford University Press.

Statistics Canada. 1993. *Population Estimates by First Official Language Spoken 1991*. 1991 Census of Canada. Catalogue no. 94–320. Ottawa: Industry, Science and Technology.

————. 1994. *Canada's Aboriginal Population by Census Subdivision and Census Metropolitan Area*. 1991 Census of Canada. Catalogue no. 94–326. Ottawa: Industry, Science and Technology.

————. 1997a. 1996 Census: Nation Tables—Population by Mother Tongue, Showing Age Groups, for Canada, Provinces and Territories, 1996 Census—20% Sample Data, 2

December 1997 [on-line database], Ottawa. Searched 15 July 1998: <http://www.statcan.ca/english/census/>.

———. 1997b. *The Daily—1996 Census: Mother Tongue, Home Language and Knowledge of Languages*, 2 December 1997 [on-line database], Ottawa. Searched 14 July 1998: <http://www.statcan.ca/Daily/English/>.

———. 1997c. *A National Overview: Population and Dwelling Counts*. 1996 Census of Canada. Catalogue no. 93–357–XPB. Ottawa: Industry Canada.

———. 1998a. *The Daily—1996 Census: Aboriginal Data*, 13 January 1998 [on-line database], Ottawa. Searched 14 July 1998: <http://www.statcan.ca/Daily/English/>.

———. 1998b. *The Daily—1996 Census: Ethnic Origin, Visible Minorities*, 17 February 1998 [on-line database], Ottawa. Searched 16 July 1998: <http://www.statcan.ca/Daily/English/>.

———. 1998c. *Canadian Economic Observer*. Catalogue no. 11–010–XPB. Ottawa: Statistics Canada.

———. 2002. Census of Canada 2001—Census Geography. Highlights and Analysis: Canada's 2001 population [on-line database], Ottawa. Searched 8 March 2003: <http://www12.statcan.ca/English/census01/>.

———. 2003. Canadian Statistics: Imports and Exports of Goods on a Balance-of-payments Basis [on-line database], Ottawa. Searched 8 March 2003: <http://www.statcan.ca/english/Pgdb/gblec02a.htm>2.

Steyn, Mark. 2003. 'Join America? They don't want us', *National Post*, 20 Jan., A14.

Taras, David, and Beverly Rasporich, eds. 1997. *A Passion for Identity: An Introduction to Canadian Studies*, 3rd edn. Toronto: Nelson.

Telford, Hamish, and Harvey Lazar. 2003. *Canada: The State of the Federation 2001: Canadian Political Culture(s) in Transition*. Montréal and Kingston: McGill-Queen's University Press.

Thomas, David M., ed. 2000. *Canada and the United States: Differences that Count*, 2nd edn. Peterborough, Ont.: Broadview Press.

Wallerstein, Immanuel. 1979. *The Capitalist World Economy*. Cambridge: Cambridge University Press.

Warkentin, John. 1997. *Canada: A Regional Geography*. Scarborough, Ont.: Prentice-Hall.

Further Reading

Hare, Kenneth F. 1968. 'Canada', in John Warkinton, ed., *Canada: A Geographical Interpretation*. Toronto: Methuen, 3–12.

Professor Hare had an illustrious career as a geographer, climatologist, and a senior university administrator. In 1968, he wrote the lead chapter in a major geography book on Canada, in which he described a Canada that had just emerged from 20 years of rapid economic expansion and population growth. The Auto Pact, signed in 1965, had already had an impact on Ontario's economic growth and its success foretold more economic gains from trade with the United States. The baby boom had just peaked, marking the end of Canada's high rate of natural increase. Yet, Canada was about to change, and change radically, from the impact of three critical social events: the emergence of a powerful separatist party, the Parti Québécois, which formed the government of Québec in 1976; the increasing assertion of Aboriginal rights following the Liberal government's 1969 White Paper on Aboriginal peoples; and the impact of the 1967 immigration policy, which began to shift the main flow of new Canadians from Europe to the entire world. For students, Professor Hare's version of Canada in the 1960s serves as a useful historical picture just prior to these critical changes.

Bourne, Larry S. and Damaris Rose. 2001. 'The Changing Face of Canada: The uneven geographies of population and social change', *The Canadian Geographer* 45, 1: 105–19.

Over the last 50 years, Canada and its regional geography have seen dramatic changes, as noted above. Larry Bourne and Damaris Rose, two prominent Canadian geographers, have identified the demographic and social forces that caused these changes. Bourne and Rose argue that these forces are not yet spent. In fact, they predict that these demographic and social forces will continue to shape the country's future more so than economic and political forces. However, their impact is not evenly spread across Canada, but, in a synergetic manner, they come together in particular places at particular times to reshape the regional and local rural/urban landscape. In their analysis,

Bourne and Rose identify processes of change that, in combination, have recast Canadian society into a different format. To prove their case, the authors focus their attention on four critical forces:

- population and spatial demography, including the changing components of population growth;
- lifestyles, families, and living arrangements, including changes in family structure, domestic relations, and household composition;
- social diversity in the urban system caused by immigration and migration;
- the labour market nexus, including shifts in the linkages between the domestic or household sphere, the sphere of work and production, and the changing nature of the state and civil society.

Climate and physiography have a powerful influence on human occupation of any land, creating the natural conditions within which settlement and industrial development occur. The link to the core/periphery model is obvious: areas with advantageous locations, climates, and natural resources attract the majority of settlers and gradually form population and economic cores, leaving less advantageous areas with fewer people and weaker economies.

Different physiographic regions and climatic zones have different soil, natural vegetation, and wildlife zones. This chapter provides a basic introduction, emphasizing the interrelationships between the influence of physical geography on human occupation and humans' impact on the land.

- ▣ Show how physical geography has shaped the regional nature of Canada.
- ▣ Examine the geological structure, origins, and characteristics of Canada's physical base and seven physiographic regions.
- ▣ Describe the main global factors that influence climate.
- ▣ Explore the interrelationships among environmental factors.
- ▣ Discuss how physical conditions influence human occupation of the land, with reference to the core/periphery model.
- ▣ Examine how human activities and the physical environment are interrelated.
- ▣ Consider the impact of human activities on Canada's natural environment and the potential threat of global warming.

Chapter 2 Canada's Physical Base

■ Introduction

The earth provides a wide variety of natural settings for human beings. For that reason, physical geography helps us understand the regional nature of our world. The basic question posed in this chapter is: Why is Canada's physical geography so essential to an understanding of its regional geography? The answer lies in the regional character of Canada's physical geography and in the interrelationships between physical geography and human settlement and activity. Physical geography is an underlying factor in shaping Canada's national and regional character, and it provides a fundamental explanation for the distribution of population within Canada.[1] In fact, population differences between Canada and the United States can be, in part, attributed to physical geography (Vignette 2.1). In this text, physical geography also provides the raison d'être for the physical basis of the core/periphery model. The argument is a simple one: regions with a more favourable physical base are more likely to develop into core regions, while regions with less favourable physical geographic conditions remain peripheral and dependent on the core.

Physical Variations within Canada

Geographers recognize that the physical nature of Canada varies in a number of ways.

Vignette 2.1	Two Different Geographies

Canada and the United States occupy the northern and central parts of North America, yet each nation has strikingly different geographies. Canada, while much larger in geographic area, has a much smaller area suitable for agriculture and settlement. Much of Canada lies in high latitudes where polar climates and permafrost place these lands far beyond the limits of commercial agriculture and settlement. Most Canadians therefore live in a narrow zone close to the border with the United States (see Figure 4.3). Here, more temperate climates prevail. Geography, therefore, has been kinder to the United States, giving it more suitable physical space for settlement and allowing its population to reach 285 million in 2001, compared to 30 million for Canada. Geography, therefore, is partly responsible for the much smaller population in Canada. Canada, for instance, has a population density of 3 people per square kilometre compared to 29 in the United States. This physical reality, best described as Canada's northern handicap, limits the areas suitable for settlement. Immigrants to Canada have recognize this geographic fact and most take up residence in one of Canada's three largest cities.

For example, climate varies from place to place. The Maritimes have a mild, wet climate, while the Arctic has a cold, dry climate. Climate also affects the shape of landforms (mountains, plateaux, and lowlands) through a variety of weathering and erosional processes. Major landforms illustrate the regional distinctiveness of Canada's physical geography. For instance, the Canadian Prairies have totally different features compared to those of the Canadian Shield. The Prairies have a flat to gently rolling landscape, while the adjacent Canadian Shield consists of rugged, rocky, hilly terrain.

Geographers perceive an interaction between people and the physical world. Natural features and processes (such as landforms and climate) affect human life in many ways. In turn, human activities often have an impact on the natural environment. This interactive two-way relationship is a fundamental component of regional geography. Favourable physical conditions can make a region more attractive for human settlement. The combination of a mild climate and fertile soils in the Great Lakes–St Lawrence Lowlands encourages agricultural activities, while the St Lawrence River and the Great Lakes provide low-cost water transportation to local and world markets. The favourable physical features of this region have allowed it to become Canada's industrial heartland.

As scientists who study the spatial aspects of nature and the processes that shape nature, physical geographers are concerned with all aspects of the physical world: physiography (landforms), bodies of water, climate, soils, and natural vegetation. Regional geographers, however, are more interested in how physical geography varies and subsequently influences human settlement of the land. The Rocky Mountains, for instance, offer few opportunities for agricultural settlement, but the spectacular scenery has led to the emergence of an economy based on tourism. Regional geographers are also concerned about the effect of human activities on the natural environment. In most cases, humans have a negative impact on the environment. For example, within the Bow Valley of the Rocky Mountains, extensive land developments have reduced the size of the natural habitat of wild animals such as bears and elk. Ironically, if more land is converted into golf courses, resort facilities, and housing developments, the animals that make this wilderness region so unique and attractive to tourists may no longer be able to survive. Another example is urban sprawl, which has gobbled up some of Canada's best farm land in the Niagara Peninsula, the Fraser Valley, and the Okanagan Valley. In our contemporary world, therefore, humans are the most active and, some would say, the most dangerous agents of environmental change.

The discussion of physical geography in this chapter and in the six regional chapters is designed to provide basic information about the natural environment and its essential role in the regional geography of Canada. To that end, the following points are emphasized:

- While physical geography varies across Canada, it has distinct and unique regional patterns.
- Landforms are one aspect of this physical diversity.
- Climate, soils, and natural vegetation are another aspect and provide the basis for biodiversity.
- The impact of human activity is changing the natural environment and, in the case of industrial pollution, there are long-term negative implications for all life forms.
- Physical geography has a powerful impact

on Canadians by making certain areas more attractive for settlement and urban/industrial development. This relationship between the natural environment and the human world forms the basis of the core/periphery model.

We begin our discussion of physical geography by examining the nature and origin of landforms.

The Nature of Landforms

The earth's surface features a variety of landforms: mountains, plateaux, and lowlands. These landforms are subject to change by various physical processes. Some processes create new landforms while others reduce them. The earth, then, is a dynamic planet whose surface is actively shaped and reshaped over time. For instance, the process known as weathering breaks down the earth's surface. Weathering consists of both chemical and physical processes that break down rocks into smaller particles. Geomorphic processes (such as moving water, ice, and wind) transport these smaller particles to other locations where they are deposited. This deposition eventually results in the formation of new landforms. Geomorphic processes also erode the land and in doing so reshape the earth's surface.

The earth's crust, which forms less than 0.01 per cent of the earth and is the thin solidified shell of the earth, consists of three types of rocks: igneous, sedimentary, and metamorphic. When the earth's crust cooled about 3.5 billion years ago, **igneous rocks** were formed from molten rock known as magma. Nearly 3 billion years later, **sedimentary rocks** were formed from particles derived from previously existing rock. Through weathering, rocks are broken down and transported by water, wind,

or ice and then deposited in a lake or sea. At the bottom of a water body, these sediments form a soft substance or mud. In geological time, they harden into rocks. Hardening occurs because of the pressure exerted by the weight of additional layers of sediments and because of chemical action that cements the particles together. Since only sedimentary rocks are formed in layers (called **strata**), this feature is unique to this type of rock. **Metamorphic rocks** are distinguished from the other two types of rock by their origin: they are igneous or sedimentary rocks that have been transformed into metamorphic rocks by the tremendous pressures and high temperatures beneath the earth's surface. Metamorphic rocks are often produced when the earth's crust is subjected to folding and faulting. **Faulting** is a process that fractures the earth's crust, while **folding** bends and deforms the earth's crust.

The earth's crust is broken into at least 14 huge slabs or plates, each moving in response to the currents of molten material just below

Basalt is a hard, black volcanic rock that, when cooled, can form various shapes, including tabular columns. These weathered basalt columns are found on Axel Heiberg Island, Nunavut. (David Nunuk/Science Photo Library)

the crust. This motion, originally described as **continental drift**, is now known as **plate tectonics**. It can result in the folding and faulting of the earth's crust.[2] For example, as these huge plates drift, they may collide with one another, thus causing earthquakes. Over sufficient geological time, these plates have compressed parts of the earth's crust into mountain chains.

Physiographic Regions

The earth's surface can be classified into a series of physiographic regions. A **physiographic** region is a large area of the earth's crust that has distinct characteristics. There are three key characteristics of a physiographic region:

- It extends over a large, contiguous area with similar relief features.
- Its landform has been shaped by a common set of geomorphic processes.
- It possesses a common geological structure and history.

Canada has seven physiographic regions (Figure 2.1). The Canadian Shield is by far the

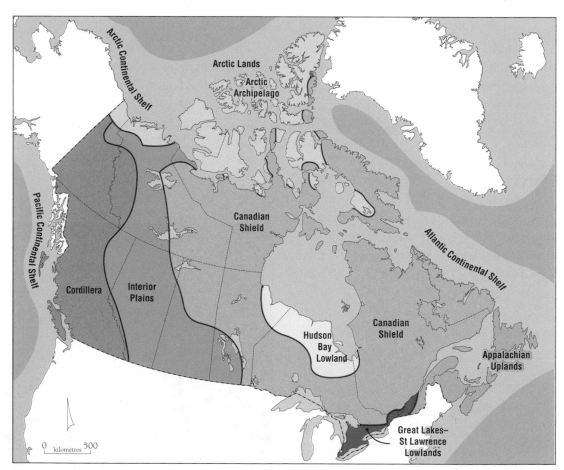

Figure 2.1 Physiographic regions and continental shelves in Canada. Arctic Lands consists of a dozen large islands and numerous small islands that together are known as the Arctic Archipelago. The Canadian Shield is the largest physiographic region and extends beneath the Interior Plains, the Hudson Bay Lowland, and the Great Lakes–St Lawrence Lowlands. (Further resources: Student Web site, National Atlas section, Maps 2 and 3. Web site instructions are found on p. xxi.)

largest region, while the Great Lakes–St Lawrence Lowlands is the smallest. Perhaps the most spectacular and varied topography (i.e., the landforms of the earth's surface) occurs in the Cordillera, while the Hudson Bay Lowland has the most uniform relief. The remaining three regions are the Interior Plains, Arctic Lands, and Appalachian Uplands.

Each physiographic region has a different geological structure. These structural differences have produced a particular set of mineral resources in each physiographic region. For example, formed from the solidification of the earth's crust about 3.5 billion years ago, the Precambrian crystalline rock that makes up the Canadian Shield contains deposits of copper, diamonds, gold, nickel, iron, and uranium. The famous Sudbury nickel mines are in the Canadian Shield. Other physiographic regions were formed much later, as shown in the geological time chart (Table 2.1). The formation of the Interior Plains began about 500 million years ago when rivers deposited sediment in a shallow sea that existed in this area. Over a period of about 300 million years, more and more material was deposited into this inland sea, including massive amounts of vegetation and the remains of dinosaurs and other creatures. Eventually, these deposits were solidified into layers of sedimentary rocks 1 to 3 km thick. As a result, the Interior Plains have a sedimentary structure that contains oil and gas deposits. Such variation in the geological structure of each physiographic region has produced unique mineral resources for the human occupants of these regions to extract. Furthermore, as these regions developed their various resources, differences in regional economies began to take shape.

The Canadian Shield

As noted previously, the Canadian Shield is the largest physiographic region in Canada. It extends over nearly half of the country's land mass. The Canadian Shield forms the ancient geological core of North America. More than 3 billion years ago, molten rock solidified into the Canadian Shield (Table 2.1). Today, these ancient Precambrian rocks are not only exposed at the surface of the Shield but also underlie many of Canada's other physiographic regions.

During the last ice advance, the surfaces of

Table 2.1	Geological Time Chart	
Geological Era	**Geological Time (millions of years ago)**	**Physiographic Region(s) Formed**
Precambrian	600 to 3,500	Canadian Shield
Paleozoic	250 to 600	Appalachian Uplands, Interior Plains, and Arctic Lands
Mesozoic	100 to 250	Interior Plains
Cenozoic Quaternary	0 to 100	Cordillera The Quaternary Period is divided into the Pleistocene Epoch (ice ages) and the Holocene Epoch (the post-glacial period).

the Canadian Shield and those of other physiographic regions were subjected to glacial erosion and deposition (Vignette 2.2). **Glacial erosion** and **deposition** are caused by giant ice sheets slowly grinding over the earth's surface. As the ice sheet moved over the surface of the Canadian Shield, the ice scraped, scoured, and scratched its massive rock surface. During the movement of ice sheets, a variety of loose materials such as sand, gravel, and boulders were trapped within the ice sheet. As the ice sheet reached its maximum extent, the edge of the ice sheet melted, depositing rocks, soil, and other debris. This debris is called till. Towards the end of the ice age, these ice sheets melted in situ, depositing whatever debris they contained. Sometimes the water from the melting ice was blocked from reaching the sea by the

Vignette 2.2   The Last Ice Age

During the earth's geological history, climatic cooling has produced a number of ice ages. The last **ice age** took place during the Pleistocene Epoch, which began nearly 2 million years ago (Table 2.1). Over the last 10,000 years, world temperatures increased and this geological time period is called the Holocene Epoch. Climatic cooling is not fully understood, but scientists have two theories. One is that climate cooling occurs when the distance between the earth and the sun is at its maximum every 21,000 years, due to variations in the earth's orbit. The other theory postulates that an increase in the amount of dust in the atmosphere from erupting volcanoes causes climatic cooling. Both theories offer explanations for a reduction in the amount of sunshine (solar energy) reaching the surface of the earth, which would cool the planet enough to trigger an ice age. While the average world temperature has increased over the last few decades—thus supporting the notion of global warming, which is attributed to the burning of fossil fuels—geologists believe that we are living in an interglacial period and that, within the next 100,000 years, the climate will again cool, resulting in another ice age.

At least 20 times during the last two million years (Pleistocene Epoch), huge sheets of ice, perhaps over 5 km thick, spread over Canada and the northern edge of the United States and then quickly retreated. The most recent glacial advance is called the late Wisconsin, reaching as far south as the state of Wisconsin (hence the name of this particular glacial advance). The late Wisconsin ice sheet reached its maximum extent 18,000 years ago.

The late Wisconsin ice advance consisted of two major ice sheets, the Laurentide and the Cordillera. The Laurentide ice sheet was centred in the Hudson Bay area. As its mass increased, the sheer weight of the ice sheet caused it to move, eventually covering much of Canada east of the Rocky Mountains. In the Cordillera, a series of alpine glaciers coalesced into the Cordillera ice sheet, which spread westward into the continental shelf off the Pacific coast and eastward, eventually merging with the Laurentide ice sheet.

Roughly 15,000 years ago, the climate began to warm, causing these ice sheets to retreat. Seven thousand years ago, the last main remnants of these ice sheets were in the Rocky Mountains and in the uplands of the Canadian Shield in northern Québec-Labrador and on Baffin Island. Today, the largest glaciers in Canada are in the mountains of Ellesmere Island. These ancient ice sheets and alpine glaciers have had a lasting impact on Canada's landforms.

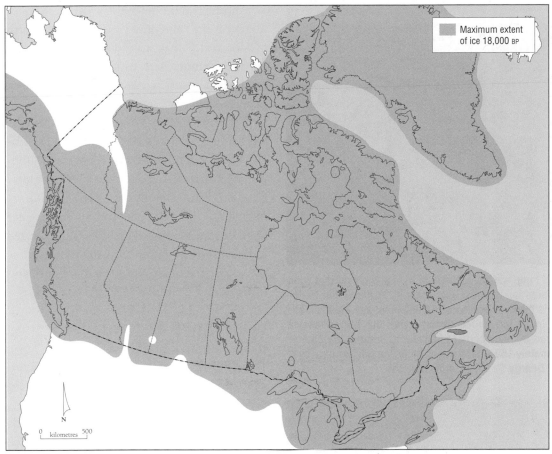

■ **Figure 2.2 Maximum extent of ice, 18,000 BP**. The last advance of the Wisconsin ice sheet covered almost all of Canada and extended into the northern part of the United States. Geologists believe that the present 'warm' climate is an interlude before the next ice advance. (Further resources: Student Web site, National Atlas section, Map 4. Web site instructions are found on p. xxi.)

retreating ice sheet. These waters then formed temporary lakes. Once this ice was removed, these waters surged towards the sea.

Evidence of the impact of these processes on the surface of the Canadian Shield is widespread. Drumlins and eskers, both depositional landforms, are common to this region. **Drumlins** are long, low hills composed of till (material deposited and shaped by the movement of an ice sheet), while **eskers** are long, narrow mounds of sand and gravel deposited by melt water streams found under a glacier. There are also **glacial striations**, which are scratches in the rock surface caused by large

rocks embedded in the slowly moving ice sheet.

The Canadian Shield consists mainly of a rugged, rolling upland. Shaped like an inverted saucer, the region's lowest elevations are along the shoreline of Hudson Bay, while its highest elevations occur in Labrador and Baffin Island, where the most rugged and scenic landforms in the Canadian Shield are found. The Torngat Mountains, for instance, provide spectacular scenery with a coastline of fjords. These mountains reach elevations of 1,600 m, making them the highest land east of the Rocky Mountains. They also form the boundary between northern Québec and Labrador.

Eskers take the form of long, narrow mounds of sand and gravel that wind their way across glaciated landscapes. This esker is located in the Northwest Territories and was likely formed some 13,000 years ago when a meltwater stream flowed in an ice-walled tunnel inside the Laurentide Ice Sheet. Eskers are found in many parts of Canada. In Ontario, the Boulter esker extends approximately 160 km from near North Bay (Mattawa) to near Orillia (Washago). (© George D. Lepp/Corbis/Magma)

The stunning St Elias Mountain Range in Yukon Territory was subjected to intense erosion from alpine glaciers during the last ice age. Its sharp peaks and arêtes are a result of the powerful impact of mountain glaciers. (Al Harvey/The Slide Farm)

Another area of the Canadian Shield, known as the Laurentides, is located just north of Montréal. It has many lakes and hills that are the summer and winter playground for local residents as well as for tourists from Ontario and New England.

Other areas of the Canadian Shield are dotted by large communities that operate as single-industry mining towns, such as the iron-mining town of Labrador City in Labrador/Newfoundland, the nickel centre of Sudbury in Ontario, and the copper mining and smelter town of Flin Flon in Manitoba. Often located in remote places, resource towns such as these are vulnerable to closure if the mining operation ceases.

The Cordillera

The Cordillera, a complex region of mountains, plateaux, and valleys, occupies over 16 per cent of Canada's territory. With its north/south alignment, the Cordillera extends from southern British Columbia to Yukon. The Cordillera, classified as a young geological structure, was formed about 40 to 80 million years ago (Table 2.1) when the North American tectonic plate slowly moved westward, eventually colliding with the Pacific plate. The collision compressed sedimentary rocks into a series of mountains and plateaux now known as the Cordillera. These ancient sedimentary rock strata can be seen on exposed mountain sides in the Rockies. Along the Pacific coast, tectonic plate movement continues, making the coast of British Columbia vulnerable to both earthquakes and volcanic activity. With the vast majority of the population of this region clustered along the coast in the cities of Vancouver, Victoria, New Westminster, and Nanaimo, the potential damage and loss of life from a major earthquake (measuring 7.0 or

Vignette 2.3   Alpine Glaciation

While glaciers still exist in the Rocky Mountains, they are slowly melting and retreating. During the late Wisconsin ice advance about 18,000 years ago, these glaciers grew in size and eventually covered the entire Cordillera. At that time, alpine glaciers advanced down slopes, carving out hollows called **cirques**. As the glaciers increased in size, they spread downward into the main valley, creating **arêtes**, steep-sided ridges formed between two cirques. As it moved through the valley, the glacier eroded the sides of the river valleys, creating distinctive U-shaped glacial valleys known as **glacial troughs**. The Bow Valley is one of Canada's most famous glacial troughs. Cutting through the Rocky Mountains, the Bow Valley now serves as a major transportation corridor. It has also developed into an international tourist area. The centre of this tourist trade is the world-famous ski resort of Banff.

Located along the Continental Divide between British Columbia and Alberta, the Athabasca Glacier forms part of the massive Columbia Icefield. Known as the 'mother of rivers', the meltwaters from the Columbia Icefield nourish four river systems (the Saskatchewan, Columbia, Athabasca, and Fraser river systems) whose waters empty into three oceans—the Atlantic, Arctic, and Pacific oceans. (Barrett and Mackay Photography, Inc.)

greater on the Richter scale) might be the worst natural disaster to strike Canada. The strongest earthquake ever recorded in Canada shook the sparsely populated Queen Charlotte Islands in August 1949. This earthquake measured 8.1 on the Richter scale.

In more recent geological times, the Cordillera ice sheet altered the landforms of the region. Over the last 20,000 years, alpine glaciation has sharpened the features of the mountain ranges in the Cordillera and broadened its many river valleys (Vignette 2.3). The Rocky Mountains are the best known of these mountain ranges. Most have elevations between 3,000 and 4,000 m. Their sharp, jagged peaks create some of the most striking landscape in North America. The highest mountain in Canada—at nearly 6,000 m—is

Mount Logan, part of the St Elias Mountain Range in southwest Yukon.

The Interior Plains

The Interior Plains region is a vast sedimentary plain that covers nearly 20 per cent of Canada's land mass. The Interior Plains are wedged between the Canadian Shield and the Cordillera, extending from the Canada–US border to the Arctic Ocean. Within the Interior Plains, most of the population lives in the southern area where a longer growing season permits grain farming.

Millions of years ago, a huge shallow inland sea occupied the Interior Plains. Over the course of time, sediments were deposited into this sea. Eventually, the sheer weight of these deposits produced sufficient heat and pressure to transform these sediments into sedimentary rocks. The oldest sedimentary rocks were formed during the Paleozoic era, about 500 million years ago (Table 2.1). Since then, other sedimentary deposits have settled on top of them, including those associated with the Mesozoic era when dinosaurs roamed the earth.

Tectonic forces have had little effect on the geology of this region. For that reason, the Interior Plains are described as a stable geological region. For example, sedimentary rocks formed millions of years ago remain as a series of flat rock layers within the earth's crust. Geologists have used such sedimentary structures as geological time charts. In Alberta and Saskatchewan, rivers have cut deeply into these soft rocks, exposing Mesozoic strata. The Alberta Badlands provide an example of this rough terrain. Archaeologists have discovered many dinosaur fossils within these Mesozoic rocks in southern Alberta and Saskatchewan.

Beneath the Interior Plains, valuable deposits of oil and gas are in sedimentary structures called **basins**. Known as fossil fuels, oil and gas deposits are the result of the capture of the sun's energy by plants and animals in earlier geologic time. The storage of this energy in the form of hydrocarbon compounds takes place in sedimentary basins. The Western Sedimentary Basin is the largest such basin. Most oil and gas production in Alberta comes from this basin. Fossil fuels are non-renewable resources, meaning that they cannot regenerate

The Alberta Badlands were formed at the end of the last ice age when vast amounts of meltwater flowed through the Red Deer River and its tributaries. These quick-moving waters easily cut through soft sedimentary rocks to reach rock layers that date back to the late Cretaceous Period. The Dinosaur Trail that explores these badlands and the Royal Tyrrell Museum of Paleontology are located near Drumheller, Alberta. (© Royalty Free/Corbis/Magma)

themselves. Renewable resources, such as trees, can reproduce themselves.

As the Laurentide ice sheet melted and began to retreat from the Interior Plains about 12,000 years ago, the surface of the region was covered with as much as 300 m of debris deposited by the ice sheet. Huge glacial lakes were formed in a few places. Later, the melt water from these lakes drained to the sea, leaving behind an exposed lakebed. Lake Agassiz, for example, was once the largest glacial lake in North America and covered much of Manitoba, northwestern Ontario, and eastern Saskatchewan—its lakebed is now flat and fertile land that provides some of the best farmland in Manitoba. When glacial waters escaped into the existing drainage system, they cut deeply into the glacial till and sedimentary rocks, creating huge river valleys known as **glacial spillways**. The geological history of the Interior Plains accounts for the great variety of landforms found within the area.

Just north of Edmonton, the Interior Plains slope towards the Arctic Ocean. The Athabasca River marks this northward course as it eventually enters the Mackenzie River. Edmonton lies on the banks of the North Saskatchewan River, which flows eastward into Lake Winnipeg. These waters then enter the Nelson River, which empties into Hudson Bay. Across this section of the Interior Plains, the land slopes towards Hudson Bay. Elevations decline from 1,200 m just west of the Rocky Mountains to about 200 m near Lake Winnipeg. These changes in elevation create three subregions within the Prairies: the Manitoba Lowland, the Saskatchewan Plain, and the Alberta Plateau. Typical elevations are 250 m in the Manitoba Lowland, 550 m in the Saskatchewan Plain, and 900 m in the Alberta Plateau. The Cypress Hills, however, provide a sharp contrast to the flat or rolling terrain of the Canadian Prairies (Vignette 2.4), as do the deeply incised river valleys of the Peace River country.

Vignette 2.4   Cypress Hills

The Cypress Hills, a subregion of the Interior Plains, consist of a rolling plateau-like upland that is deeply incised by fast-flowing streams. Situated in southern Alberta and Saskatchewan, this subregion is the highest point in Canada between the Rocky Mountains and Labrador. These hills are an erosion-produced remnant of an ancient higher-level plain formed in the Cenozoic era (see Table 2.1). With an elevation of over 1,400 m, these hills rise 600 m above the surrounding plain formed about 50 million years ago from materials borne eastward by rivers originating in the Rocky Mountains. During the maximum extent of the Laurentide ice sheet about 18,000 years ago, the higher parts of the Cypress Hills remained above the Laurentide ice sheet. Known as **nunataks**, these areas served as refuge for animals and plants. As the alpine glacier melted, streams flowing from the Rocky Mountains deposited a layer of gravel up to 100 m thick on these hills.

Today, the Cypress Hills area is a humid 'island' surrounded by a semi-arid environment and has an entirely different natural vegetation compared to the area surrounding it. Unlike the grasslands, the Cypress Hills have a mixed forest of lodgepole pine, white spruce, balsam poplar, and aspen. The Cypress Hills also contain many varieties of plants and animals found in the Rocky Mountains.

The Hudson Bay Lowland

The Hudson Bay Lowland region was formed when the Laurentide ice sheet no longer blocked the Atlantic Ocean from entering what is now Hudson Bay. This inland extension of the Atlantic Ocean has been termed the Tyrrell Sea. This inland saltwater sea reached its maximum extent about 7,000 years ago, extending over much of the lowlands surrounding Hudson and James bays. With the huge weight of the ice sheet removed, the earth's crust began to rise, forcing the Tyrrell Sea to retreat. This process is called **isostatic rebound** (Vignette 2.5). Slowly the isostatic rebound caused the seabed of the Tyrrell Sea to rise above sea level, thus exposing a low, poorly drained coastal plain (most of which is called the Hudson Bay Lowland). This process of isostatic rebound continues today, making the Hudson Bay Lowland the youngest of the physiographic regions in Canada (Table 2.1).

The Hudson Bay Lowland comprises about 3.5 per cent of the area of Canada. It lies mainly in northern Ontario, though a small portion stretches into Manitoba. This region extends from James Bay along the west coast of Hudson Bay to just north of the Churchill River.

Much of the ground's surface consists of wet peatland known as **muskeg**. Low ridges of sand and gravel are interspersed between these extensive areas of muskeg. These ridges are the remnants of former beaches of the Tyrrell Sea. Because of its almost level surface, much of the land is poorly drained. Underneath the peatland are recently deposited marine sediments mixed with glacial till. With few resources to support human activities, the region has only a handful of tiny settlements. From this perspective, the Hudson Bay Lowland is one of the least favourably endowed physiographic regions of Canada. Moosonee (at the mouth of the Moose River in northern Ontario) and Churchill (at the mouth of the Churchill River in northern Manitoba) are the largest settlements in the region. Each has a population of just over 1,000 people. These two settlements, formerly fur-trading posts, are now the termini of two northern railways (the Ontario Northland Railway and the Hudson Bay Railway, respectively).

Arctic Lands

The Arctic Lands region stretches over 25 per cent of the area of Canada. Centred in the

| Vignette 2.5 | **Isostatic Rebound** |

At its maximum extent about 18,000 years ago, the weight of the huge Laurentide ice sheet caused a depression in the earth's crust. When the ice sheet covering northern Canada melted, this enormous weight was removed, and the elastic nature of the earth's crust allowed it to return to its original shape. This process, known as **isostatic rebound** or uplift, follows a specific cycle. As the ice mass slowly diminishes, the isostatic recovery begins. This phase is called a **restrained rebound**. Once the ice mass is gone, the rate of uplift reaches a maximum. This phase is called a **postglacial uplift**. It is followed by a period of final adjustment called the **residual uplift**. Eventually the earth's crust reaches an equilibrium point and this isostatic process ceases. In the Canadian North, this process began about 11,000 years ago and has not yet completed its cycle.

Canadian Arctic Archipelago, this region lies north of the Arctic Circle. It is a complex composite of coastal plains, plateaux, and mountains. The Arctic Platform, the Arctic Coastal Plain, and the Innuitian Mountain Complex are the three principal physiographic subregions. The Arctic Platform consists of a series of plateaux composed of sedimentary rocks. This subregion is in the western half of the Arctic Archipelago around Victoria Island. The Arctic Coastal Plain extends from the Yukon coast and the adjacent area of the Northwest Territories into the islands located in the western part of the Beaufort Sea. The third subregion, the Innuitian Mountain Complex, is located in the eastern half of the Arctic Archipelago. It is composed of ancient sedimentary rocks. Like the Rocky Mountains, its sedimentary rocks were folded and faulted. However, unlike the Rocky Mountains, the plateaux and mountains in the Innuitian subregion were formed in the early Paleozoic era (Table 2.1). At 2,616 m, Mount Barbeau on Ellesmere Island is the highest point in the Arctic Lands region.

Across these lands, the ground is permanently frozen to great depths, never thawing even in the short summer. This cold thermal condition is called **permafrost**. Physical weathering, consisting mainly of differential heating and frost action, shatters bedrock and produces various forms of patterned ground. **Patterned ground** consists of rocks arranged in polygonal forms by minute movements of the ground caused by repeated freezing and thawing. Patterned ground and **pingos** (ice-cored mounds or hills) give the Arctic Lands a unique landscape.

The climate in this region is cold and dry. In the mountainous zone of Ellesmere Island, glaciers are still active. That is, as these alpine glaciers advance from the land into the sea, the

Tundra polygons are a type of patterned-ground found in Arctic Canada. Tundra polygons are formed by the repeated freezing of water in cracks in the ground and take centuries to form. (Peter Dunwiddle/Visuals Unlimited)

Some 400 icebergs reach the coast of Newfoundland and Labrador each spring and summer. Almost all come from the Greenland Ice Sheet. Icebergs consist of a floating mass of freshwater ice and are a hazard to ships and offshore oil rigs. (Search4Stock Inc.)

ice is 'calved' or broken from the glacier, forming icebergs. On the plains and plateaux, it is a polar desert environment. The term 'polar desert' describes barren areas of bare rock, shattered bedrock, and sterile gravel. Except for primitive plants known as lichens, no vegetation grows. Aside from frost action, there are no other geomorphic processes, such as water erosion, to disturb the patterned ground.

Most people live in the coastal plain in the western part of this physiographic region. The three largest settlements are situated at the mouth of the Mackenzie River. Inuvik has a population of almost 3,000, while Aklavik and Tuktoyaktuk are smaller communities.

The Appalachian Uplands

The Appalachian Uplands region represents only about 2 per cent of Canada's land mass. Sometimes known as Appalachia, this physiographic region consists of the northern section of the Appalachian Mountains (stretching south to the eastern United States), though few mountains are found in the Canadian section. With the exception of Prince Edward Island (Vignette 2.6), its terrain is a mosaic of rounded uplands and narrow river valleys. These weathered uplands are the remnants of ancient mountains that underwent a variety of ero-

sional processes almost 500 million years ago. A combination of weathering and geomorphic processes (including water, wind, and ice) has worn down these mountains, creating a much subdued mountain landscape. The highest elevations are in the Gaspé Peninsula. Here, Mount Jacques Cartier, at an elevation of 1,268 m, is found. The coastal area has been slightly submerged; consequently, ocean waters have invaded the lower valleys, creating bays or estuaries. The result is a number of excellent small harbours and a few large ones, such as Halifax harbour. The island of Newfoundland consists of a rocky upland with only pockets of soil found in valleys. Like the Maritimes, it has an indented coastline where small harbours abound. The nature of this physiographic region has favoured settlement along the heavily indented coastline where there is easy access to the fishing banks.

The Great Lakes–St Lawrence Lowlands

The Great Lakes–St Lawrence Lowlands physiographic region is small but important. Extending from the St Lawrence River near Québec City to Windsor, this narrow strip of land is wedged between the Appalachians, the Canadian Shield, and the Great Lakes.[3] Near

Vignette 2.6   Prince Edward Island

Unlike other areas of Appalachia, Prince Edward Island has a flat to rolling landscape. While the island is underlain by sedimentary strata, these rocks are a relatively soft, red-coloured sandstone that is quickly broken down by weathering and erosional processes. Occasionally, outcrops of this sandstone are exposed, but for the most part the surface is covered by reddish soil that contains a large amount of sand and clay. The heavy concentrations of iron oxides in the rock and soil give the island its distinctive reddish-brown hue. Prince Edward Island, unlike the other provinces in this region, has an abundance of arable land.

Vignette 2.7	**Champlain Sea**

About 12,000 years ago, vast quantities of glacial water from the melting ice sheets around the world drained into the world's oceans. Sea levels rose because of this additional water, causing the Atlantic Ocean to surge into the St Lawrence and Ottawa valleys, perhaps as far west as the edge of Lake Ontario. Known as the Champlain Sea, this body of water occupied the depressed land between Québec City and Cornwall and extended up the Ottawa River Valley to Pembroke. These lands had been depressed earlier by the weight of the Laurentide ice sheet. About 10,000 years ago, the earth's crust rebounded sufficiently to cause the Champlain Sea to retreat. However, the sea left behind marine deposits, which today form the basis of the fertile soils in the St Lawrence Lowlands.

the eastern end of Lake Ontario, the Canadian Shield extends across this region into the United States where it forms the Adirondack Mountains in New York state. Known as the Frontenac Axis, this part of the Canadian Shield divides the Great Lakes–St Lawrence Lowlands into two distinct subregions.

As the smallest physiographic region in Canada, the Great Lakes–St Lawrence Lowlands comprises less than 2 per cent of the area of Canada. As its name suggests, the landscape is flat to rolling. This topography reflects the underlying sedimentary strata and its thin cover of glacial deposits. In the Great Lakes subregion, flat sedimentary rocks are found just below the surface. This slightly tilted sedimentary rock, which consists of limestone, is exposed at the surface in southern Ontario, forming the Niagara Escarpment. A thin layer of glacial and lacustrine (i.e., lake) material, deposited after the melting of the Laurentide ice sheet in this area about 12,000 years ago, forms the surface, covering the sedimentary rocks.

In the St Lawrence subregion, the landscape was shaped by the Champlain Sea, which occupied this area for about 2,000 years. It retreated about 10,000 years ago and left marine materials, which are now broad terraces that slope gently towards the river (Vignette 2.7). The sandy to clay surface materials are a mixture of recently deposited sea, river, or glacial materials. For the most part, this subregion's soils are fertile, which, when combined with a long growing season, allow agricultural activities to flourish.

The physiographic region lies well south of 49° parallel, which forms the US–Canada border west of Ontario. The Great Lakes subregion extends from 42° N to 45° N, while the St Lawrence subregion lies somewhat further north, reaching towards 47° N. As a result of its southerly location, its proximity to the industrial heartland of the United States, and its favourable physical setting, the Great Lakes–St Lawrence region is home to Canada's main ecumene and manufacturing core.

The Impact of Physiography on Human Activity

Not only do physiographic regions provide a basic understanding of the physical shape and geological structure of Canada, they have also exerted a powerful influence over the geographic pattern of early settlers' land selection. In some instances, settlers were attracted to certain types of land while they avoided other types. Two examples illustrate this point. In the seventeenth century, the St Lawrence Lowlands was an attractive area for the establishment of a French colony because of its agricultural lands and its accessibility by water to France. Few settlers ventured beyond this favoured area. To the north was the rocky Canadian Shield, while to the south was the Appalachian Uplands. Neither of these surrounding physiographic regions offered attractive land for farming. Instead, Indians occupied these lands where they hunted game and trapped furs in order to barter with French traders for European goods.

Physical features can also create barriers to settlement. The Rocky Mountains were such a barrier in the nineteenth century. In 1867, when the Dominion of Canada was formed, the Rocky Mountains isolated the small British colony on the southern tip of Vancouver Island from the settled area of Canada. At that time, communications and trade with adjacent American settlements along the Pacific coast proved much easier and quicker than the overland route used by fur traders to reach Montréal. Until the completion of the CPR in 1885, the Rocky Mountains were such an imposing physical barrier that many residents of Vancouver Island favoured joining the United States rather than the Dominion of Canada.

In our contemporary world, such physical barriers are no longer the obstacles they once were. Technological advances in transportation and communications have greatly reduced the **friction of distance**, the term used to describe how interactions between two points decrease as the distance between them increases. The obstacles presented by physical barriers and distance have been greatly diminished.

Geographic Location

Canada's location in the northern half of North America is a critical factor in its physical geography. For instance, Canada has a much cooler climate than the country to its south, the United States of America (Vignette 2.1). As well, Canada is surrounded by the Arctic, Atlantic, and Pacific oceans, making it a marine nation (Vignette 2.8). The Arctic Ocean is covered by a slow-moving permanent ice pack, making ocean navigation impractical. A measure of geographic location on the earth's surface is provided by latitude and longitude. Because the earth is a spherical body, this measure is given in degrees. By **latitude**, we mean the measure of distance north

Vignette 2.8   Facts about Canada as a Maritime Nation

- At 243,792 km, Canada has the longest coastline in the world, forming 25 per cent of the world's coastline.
- With an offshore economic zone stretching seaward some 200 nautical miles and comprising 3.7 million km², Canada has the largest offshore zone in the world.
- Canada, with two million lakes and rivers covering 755,000 km² or 7.6 per cent of the country's landmass, has the largest freshwater system in the world.

- From the Gulf of St Lawrence to Lake Superior, Canada's inland waterway extends over 3,700 km, making it the longest in the world.
- The Arctic Archipelago covers 1.4 million km², making it the largest archipelago in the world.
- Approximately 7 million Canadians live along its coastal literal.

Adapted from Fisheries and Oceans Canada: <www.dfo-mpo.gc.ca/communic/facts-info/facts-info_e.html>.

Table 2.2	Latitude and Longitude of Selected Centres	
Centre	**Latitude**	**Longitude**
Windsor, Ontario	42°18' N	83°01' W
Alert, Northwest Territories	83°63' N	60°05' W
St John's, Newfoundland	47°34' N	52°43' W
Victoria, British Columbia	48°26' N	123°20' W
Whitehorse, Yukon	60°41' N	135°08' W

and south of the equator. For example, Ottawa is 45 degrees 24 minutes north of the equator. Degrees and minutes are expressed as ° and ', respectively. Since the distance between each degree of latitude is about 110 km, Ottawa is about 5,000 km north of the equator. By **longitude**, we mean the distance east or west of the prime meridian. As the equator represents zero latitude, the prime meridian represents zero longitude. It is an imaginary line that runs from the North Pole to the South Pole and passes through the Royal Observatory at Greenwich, England. Canada lies entirely in the area of west longitude. Ottawa, for example, is 75°28' west of the prime meridian. Within Canada, latitude and longitude vary enormously. The variation in latitude has considerable implications for climate, which in turn affects the types of soils, natural vegetation, and wildlife found in each climatic zone. Examples of the range of latitudes and longitudes found in Canada are shown in Table 2.2.

Climate

Our physical world encompasses more than just landforms, physiographic regions, and geographic location. Climate, for instance, plays a key role in our physical world. **Climate** describes average weather conditions for a specific place or region over a very long period of time while weather refers to current state of the atmosphere with a focus of weather conditions that affect people living in a particular place. In short, climate is what we can expect while weather is what we get. Extreme weather events—such as blizzards, droughts, and ice storms—are also part of climate and often have very powerful impacts on humans. Extreme weather events have brought people together to combat natural disasters, and, as indicated in Chapter 1, have contributed to their sense of belonging to a region.

Floods provide such an example. Often, they reoccur. As de Loë (2000: 357) explains, 'Floods are considered *hazards* only in cases where human beings occupy floodplains and shoreland.' What weather conditions provide conditions for flooding of such landforms? Often heavy rainfall combined with a sudden rise in temperatures that causes rapid snowmelt triggers catastrophic floods. An excellent example is found in the flat Manitoba Lowland where the normally benign Red River winds its way from North Dakota in the United States northward to Lake Winnipeg. Since 1950, residents of Winnipeg and other communities along the Red River have suffered through five spring floods, in 1950, 1979, 1996, 1997, and 2001. In 1950, the Red River flood drove over 100,000 people from their homes. Following that disaster, the Red River Floodway, a wide channel nearly 50 km long, was constructed. Its purpose was to divert the flood waters around the city of Winnipeg. However, small communities in the Red River Basin remained vulnerable to flooding. In 1997, the largest flood in the twentieth century occurred (Rasid et al., 2000). While the Red River Floodway saved Winnipeg, the towns of Emerson, Morris, Ste Agathe, and St Adolphe and the surrounding farm buildings and lands were less fortunate.

Since climate is relatively stable over a long period of time, it plays a key role in the formation of soils and natural vegetation. One outcome is the emergence of global patterns of soils and natural vegetation. Climatic conditions vary around the world and within Canada. In the Köppen climatic classification scheme for the world, for example, there are 25 climate types that reflect different temperature patterns and seasonal precipitation patterns. Seven of Köppen's climatic types are found in Canada and the equivalent Canadian climatic type is shown in Table 2.3.

Table 2.3   Climatic Types

Köppen Classification	Canadian Climatic Zone	General Characteristics
Marine West Coast	Pacific	1. Warm to cool summers; mild winters 2. Precipitation throughout the year with a maximum in winter
Highland	Cordillera	1. Cooler temperature at similar latitudes because of higher elevations
Steppe	Prairies	1. Cool summers and cold winters 2. Low annual precipitation
Humid continental	Great Lakes–St Lawrence Lowlands	1. Warm, humid summers and short, cold winters 2. Moderate annual precipitation with little seasonal variation
Humid continental, cool	Atlantic Canada	1. Cool, humid summers and short, cool winters
Subarctic	Subarctic	1. Short, cool summers; long, cold winters 2. Low annual precipitation
Tundra	Arctic	1. Extremely cool and very short summers; long, cold winters 2. Very low annual precipitation

Sources: Adapted from Robert W. Christopherson, *Geosystems: An Introduction to Physical Geography*, 3rd edn (Upper Saddle River, NJ: Prentice-Hall, 1998); F. Kenneth Hare and Morley K. Thomas, *Climate Canada* (Toronto: Wiley, 1974).

Climatic Controls

These are the three dominant climatic controls that affect Canada's weather and climate and that are related to the global atmospheric and oceanic circulation system:

- Variations in the amount of solar energy reaching different parts of the earth's surface correspond with latitude and temperature. That is, lower latitudes receive more solar energy and therefore have higher temperatures than higher latitudes.

- The global circulation of air masses causes a westerly flow of air across Canada, although invasions of air masses from the Arctic and the Gulf of Mexico can temporarily disrupt this general pattern of air circulation.

- Distance from oceans plays an important role in temperature and precipitation. That is, as distance from oceans increases, the annual temperature range increases and the annual amount of precipitation decreases.

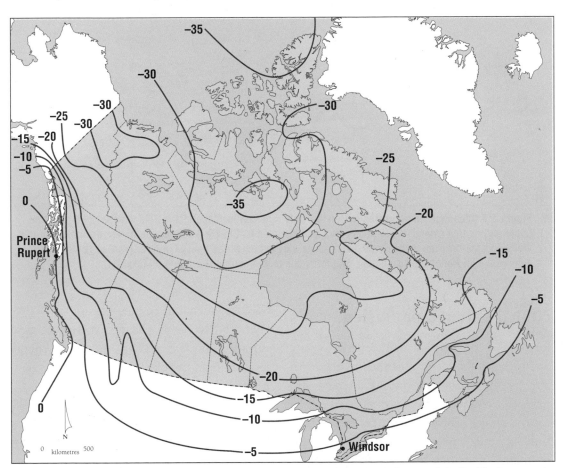

Figure 2.3 Seasonal temperatures in Celsius, January. The moderating influence of the Pacific Ocean and its warm air masses is readily apparent in the 0 to −5° C January isotherm. For example, Prince Rupert, located near 55° N, has a warmer January average temperature (0° C) than Windsor (-2° C), which is located near 42° N. (Further resources: Student Web site, National Atlas section, Map 5. Web site instructions are found on p. xxi.)

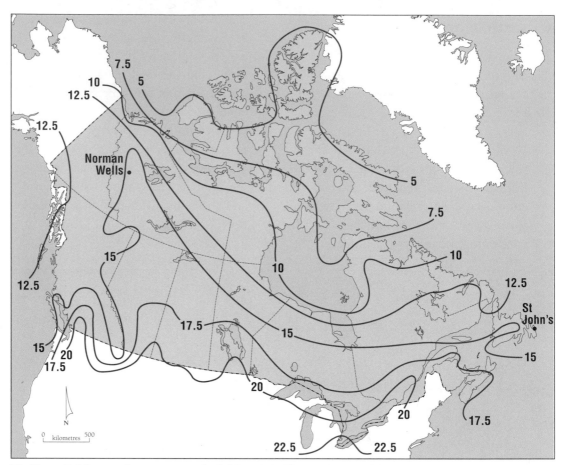

Figure 2.4 Seasonal temperatures in Celsius, July. The continental effect results in very warm summer temperatures that extend into high latitudes, as illustrated by the 15° C July isotherm. For example, Norman Wells, located near the Arctic Circle, has warmer July temperatures than St John's. (Further resources: Student Web site, National Atlas section, Map 5. Web site instructions are found on p.xxi.)

Global Circulation System

Regional climates are controlled by the amount of solar energy absorbed by the earth and its atmosphere and then converted into heat. The amount of energy received at the earth's surface varies by latitude. In low latitudes around the equator, there is a net surplus of energy (and therefore high temperatures), but in high latitudes around the North and South poles, more energy is lost through re-radiation than is received, and therefore annual average temperatures are extremely low.

Canada, where settlements extend from 42° N (Windsor) to 83° N (Alert), experiences great variation in the amount of solar energy received (and therefore great variation in temperatures).

The **global circulation system** redistributes this energy (i.e., energy transfers) from low latitudes to high latitudes through the circulation system in the atmosphere (system of winds and air masses) and the oceans (system of ocean currents). For example, the Japan Current warms the Pacific Ocean, bringing milder weather to British Columbia. On Cana-

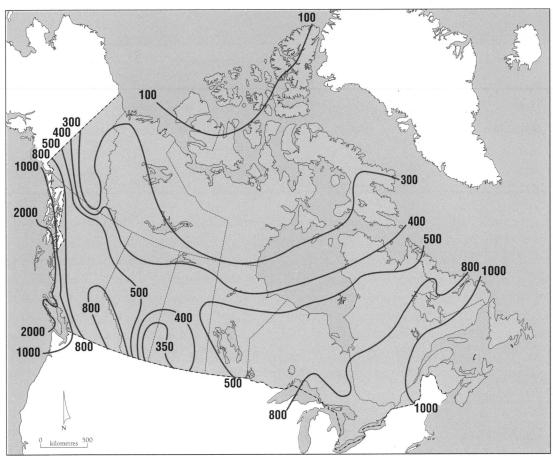

Figure 2.5 Annual precipitation in millimetres. The lowest average annual precipitation occurs in the Territorial North, indicating the dry nature of the Arctic air masses that originate over the ice-covered Arctic Ocean. The highest average annual precipitation occurs along the coast of British Columbia due to the moist marine air masses and the coastal mountains. (Further resources: Student Web site, National Atlas section, Map 6. Web site instructions are found on p. xxi.)

da's east coast, the opposite process occurs as the Labrador Current brings Arctic waters to Atlantic Canada. While Halifax, at 44°40' N, lies about 500 km closer to the equator than Victoria, at 48°26' N, Halifax's winter temperatures, on average, are much lower than those experienced in Victoria.

The atmospheric circulation system travels in a west-to-east direction in the higher latitudes of the northern hemisphere, causing air masses that develop over large water bodies to bring mild and moist weather to adjacent land masses. Such air masses are known as **marine**

air masses. In this way, energy transfers ultimately determine regional patterns of global weather and climate (see Tables 2.3 and 2.4). Air masses originating over large land masses are known as **continental air masses**. These air masses are normally very dry and vary in temperature depending on the season. In the winter continental air masses are cold, while in the summer they are associated with hot weather.

Canada experiences warmer and moister weather in its lower latitudes and colder and drier conditions in its higher latitudes. However, Canada's coastal areas (particularly its

Table 2.4	Air Masses Affecting Canada		
Air Mass	Type	Characteristics	Season
Pacific	Marine	Mild and wet	All
Atlantic	Marine	Cool and wet	All
Gulf of Mexico	Marine	Hot and wet	Summer
Southwest US	Continental	Hot and dry	Summer
Arctic	Continental	Cold and dry	Winter

Pacific coast) experience smaller ranges of seasonal temperatures and more annual precipitation than do inland or continental areas at the same latitude (Figures 2.3, 2.4, and 2.5). Winnipeg, for example, experiences a much greater daily and annual range in temperature than does Vancouver, even though both lie near 49° N. The principal reason is Vancouver's greater proximity to the ameliorating effects of the Pacific Ocean.

Air Masses

Air masses are large bodies of air with similar temperature and humidity characteristics. They form over large areas that have uniform surface features and relatively consistent temperatures. Such areas are known as source regions. The Pacific Ocean is a marine source region, while the interior of North America is a continental source region. During a period of about a week or so, an air mass may form over a source region, taking on the temperature and humidity characteristics of that source region. Canada's weather is affected by five air masses associated with the northern hemisphere. For example, Pacific air masses bring mild, wet weather to British Columbia's coast for most of the year. These air masses are much stronger in the winter, so British Columbia normally experiences greater precipitation in winter than in

summer. In some years, British Columbia can have a relatively dry summer. The general characteristics of the five major air masses affecting Canada's weather are shown in Table 2.4.

Air masses bring moisture from oceans to land bodies. Across Canada, precipitation is unevenly distributed (Figure 2.5). The lowest average annual precipitation occurs in the Territorial North, indicating the dry nature of the Arctic air masses that originate over the ice-covered Arctic Ocean. The highest average annual precipitation takes place along the coast of British Columbia. Here, much falls as frontal and orographic rainfall due to two factors: (1) the warm Pacific Ocean serves as a source for eastward-moving Pacific air masses that often contain large quantities of water vapour, and (2) along the British Columbia coast, precipitation occurs either as the warm Pacific air mass rises over a colder one or as this same air mass must rise over the coastal mountain ranges. In both cases, the water vapour condenses and falls as rain or, at higher elevations, as snow. The three principal types of precipitation are discussed in Vignette 2.9.

Climate, Soils, and Natural Vegetation

As noted earlier, climate affects the development of soils and the growth of natural vegeta-

Vignette 2.9   Types of Precipitation

As air masses rise, their temperature drops. This cooling process triggers condensation of water vapour contained in the air masses. With sufficient cooling, water droplets are formed. When these droplets reach a sufficient size, precipitation begins. Precipitation refers to rainfall, snow, and hail. There are three types of precipitation. **Convectional precipitation** results when moist air is forced to rise because the ground has become particularly warm. Often this form of precipitation is associated with thunderstorms. **Frontal precipitation** occurs when warm air masses are forced to rise over colder (and denser) air masses. **Orographic precipitation** results when air masses are forced to rise over high mountains. However, as those same air masses descend along the leeward slopes of those mountains (that is, the slopes that lie on the east side of the mountains), their temperature rises and precipitation is less likely to occur. This phenomenon is known as the **rain shadow effect**.

tion. In fact, the interdependency of climate, soils, and natural vegetation is so strong that physical geographers have identified an orderly and interrelated global pattern of climatic, soil, and natural vegetation zones. This relationship is revealed in the three maps indicating climatic, natural vegetation, and soil zones (Figures 2.6, 2.7, and 2.8). As shown in Table 2.5, climate determines to a large extent the **soil order** and native vegetation that exist in a given region and hence influences land use, such as crop cultivation, forestry, or grazing. Together with topography, climate determines the land's suitability for human settlement.

Climate has a direct impact on many economic activities. Long, cold winters cause people to use more energy; Canadians are among the highest consumers of energy in the world. Variations in precipitation affect economic activities. Several years of below-normal precipitation often have a negative effect on agriculture, forestry, and hydroelectric production. For example, the 1988 drought in Canada cost the national economy approximately $1.8 billion in decreased agricultural and hydroelectric output, increased costs of fighting forest fires, and a loss of commercial timber and wildlife

habitat (Shabbar et al., 1997: 3016). Certainly, the drought of 2001 and 2002 reaped havoc on farmers and ranchers in Alberta and Saskatchewan. The lack of rainfall made pastures next to useless and grain farmers faced crop failure. The bountiful hay harvest in Ontario, Québec, and the Maritimes demonstrated the regional nature of this recent Prairie drought and, with the voluntary shipment of hay from these provinces to Alberta and Saskatchewan in the fall of 2002, indicated the cohesion of Canada's rural society.

Climatic Zones

The earth's atmosphere is a perpetually moving global system of air circulation that works to adjust the differences in pressure and temperature over different parts of the globe. This global circulation system, together with ocean bodies and major topographic features, affects Canada's weather. There is a climatic order within our complex and dynamic atmosphere. This order is expressed in several ways, including climatic zones. A **climatic zone** is an area of the earth's surface where similar weather conditions occur. Long-term data describing annual, seasonal, and daily temperatures and precipitation are used to define the extent of a climatic zone. Similar weather conditions occur in a particular area for complex reasons. For example, land near large water bodies usually receives more precipitation than land far from large water bodies. As well, coastal settlements have a small range of seasonal temperatures due to the sea's cooling effect in the summer and its warming effect in the winter. Land-locked places, however, do not benefit from the influence of the sea and have a much wider range of annual, seasonal, and daily temperatures.

Canada lies in the northern half of North America. This geographical location has several consequences for Canadians:

- Canada, located in the middle and high latitudes, receives much less solar energy than the continental United States and Mexico. It therefore has shorter summers and longer winters.
- Canada is noted for its long, cold winters, which affect Canadians in many ways. 'Coldness', wrote French and Slaymaker, 'is a pervasive Canadian characteristic, part of the nation's culture and history' (1993: i). They go on to state that winter's effects include not only low absolute temperatures but also exposure to wind chill, snow, ice, and permafrost.
- Continental climates are widespread in Canada's interior. These climates, but notably the Subarctic climate, are formed over large areas of the interior of Canada and are characterized by cold, dry winters and warm, dry summers. Except for the Pacific and Atlantic climates, Canadians live in areas with continental-type climates.
- Marine climates are limited in their geo-

graphic extent to the Pacific coast of British Columbia and to Atlantic Canada.

Canada has seven climatic zones (Figure 2.6): the Pacific, Cordillera, Prairies, Great Lakes–St Lawrence, Atlantic, Subarctic, and Arctic. The Subarctic zone is the largest climatic zone. It extends over much of the interior of Canada and is found in each geographic region. Though the Arctic climate exists along the Labrador coast and in the extreme northern reaches of Québec, the Subarctic climate prevails in the northern areas of Atlantic Canada, Québec, Ontario, and Western Canada, and it is present in northeast British Columbia. As well, the Subarctic climate is found in the Territorial North and is the principal climate in the Northwest Territories. The Subarctic climatic zone extends into much higher latitudes in northwest Canada than in northeast Canada because of warmer temperatures in the northwest. In northwest Canada, the average July temperature often reaches or exceeds 10° C, thus permitting the growth of trees. In similar latitudes of northeast Canada, summer temperatures are much lower. In the extreme north of Québec, for example, the average July temperature is below 10°C, thus resulting in tundra rather than a tree vegetation cover. The Subarctic climatic zone therefore has a southeast to northwest alignment (Figure 2.6). This alignment, somewhat modified in Québec, is caused by three factors:

- A continental effect means that land heats up more rapidly than water in the summer, resulting in higher summer temperatures in continental areas, such as Yukon and the Mackenzie Valley, compared to coastal areas, such as the coast of Hudson Bay and Baffin Island, at the same latitude. Also, oceans warm up more slowly than land because oceans reflect more solar energy and because solar energy is distributed throughout the water body.

- The snow cover in the western section of the Subarctic is much thinner than in the eastern half. Pacific air masses dominate the weather pattern in this area in the spring and bring warmer weather. A thinner snow cover and warmer spring temperatures cause snow to disappear more quickly in the western Subarctic. Once the snow is gone, temperatures rise sharply.
- The Atlantic Ocean (including Hudson Bay) cools northern Québec and Labrador

(thus breaking the pronounced southeast to northwest alignment of the Subarctic climatic zone). Part of that cooling effect is due to the Labrador Current, which brings Arctic waters to the middle latitudes of Atlantic Canada, and to the marine air masses that originate over the Atlantic Ocean. Combined with the deeper snow pack, this keeps spring temperatures low in the eastern subregion.

Each climatic zone has a particular natural vegetation type and soil (Table 2.5; Figures 2.7 and 2.8). The Subarctic climate, for instance, is

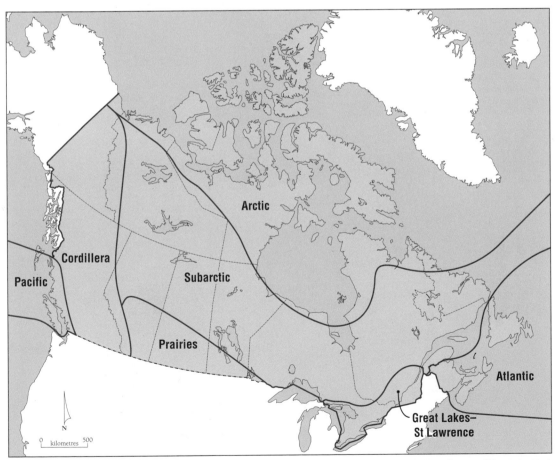

Figure 2.6 Climatic zones of Canada. Each climatic zone represents average climatic conditions in that area. Canada's most extensive climatic zone, the Subarctic, is associated with the boreal forest and podzolic soils. (Further resources: Student Web site, National Atlas section, Map 7. Web site instructions are found on p. xxi)

Table 2.5	Canadian Climatic Zones		
Canadian Climatic Zone	**Natural Vegetation Type**	**Soil Order**
Pacific	Coastal rainforest	Podzolic
Cordillera	Montane and boreal forests	Mountain complex
Prairies	Grassland and parkland	Chernozemic
Great Lakes–St Lawrence	Broadleaf and mixed forests	Podzolic
Atlantic	Mixed and boreal forests	Podzolic
Subarctic	Boreal forest	Podzolic
Arctic	Tundra and poplar desert	Cryosolic

Note: See Figures 2.6, 2.7, and 2.8. Also see Key Terms at end of chapter for definitions of soil orders.

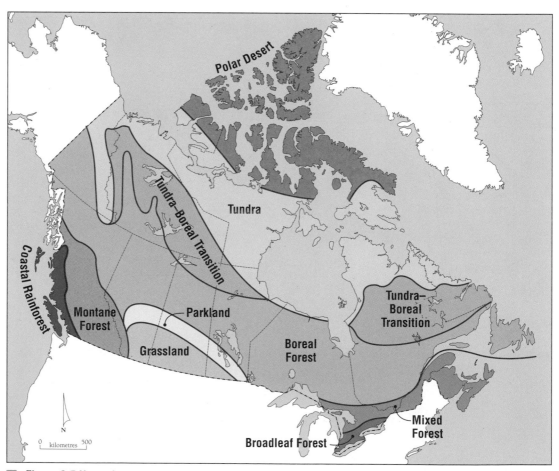

■ **Figure 2.7 Natural vegetation zones**. These natural vegetation zones have 'core' characteristics, which diminish towards their edges. (Further resources: Student Web site, National Atlas section, Map 8. Web site instructions are found on p.xxi) Transitions exist between natural vegetation zones. Two major transition zones shown here are the Tundra-Boreal Transition and the Parkland.

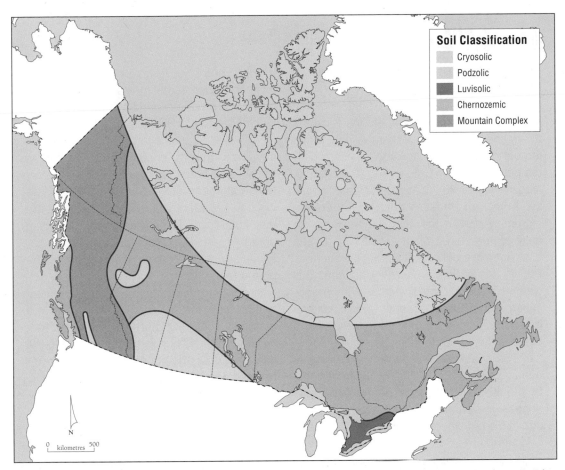

Soil Classification
- Cryosolic
- Podzolic
- Luvisolic
- Chernozemic
- Mountain Complex

Figure 2.8 Soil zones. Most agricultural land is in luvisolic and chernozemic soil zones that together comprise about 5 per cent of Canada's land base. (Further resources: Student Web site, National Atlas section, Map 9. Web site instructions are found on p. xxi)

associated with the boreal forest and podzolic soils (Table 2.5 and Vignette 2.10). The core characteristics of each of these climatic zones are presented in the appropriate regional chapter, e.g., the Pacific and Cordillera climatic zones are presented in the chapter on British Columbia.

Permafrost

A particularly distinctive feature of Canada's physical geography is permafrost. As noted earlier in this chapter, permafrost is permanently frozen ground with temperatures at or below zero for at least two years. The vast extent of permafrost in Canada provides a measure of the size of the country's cold environment (Figure 2.9). Permafrost exists in the Arctic and Subarctic climatic zones and occurs at higher elevations in the Cordillera zone. Overall, permafrost is found in just over two-thirds of Canada's land mass.

Permafrost extends deeply into the ground. North of the Arctic Circle, permafrost may extend more than several hundred metres into the ground. Further south, permafrost is less frequent and where it occurs, it rarely penetrates more than 10 m into the ground. Per-

Vignette 2.10   Subarctic Climate Type

The Subarctic climatic type has the greatest seasonal variation in temperatures of all the climatic types in Canada, with long, cold winters and short, warm summers. As is typical of continental climates, extremely cold winter temperatures occur. January minimum daily temperatures often drop to -40° C and sometimes even to -50° C. Winters are influenced by Arctic air masses and are therefore extremely dry. In the short summer period, daily temperatures often exceed 20° C and occasionally reach 30° C. During the summer, Pacific air masses usually dominate this weather pattern, providing most of the precipitation in the Subarctic zone. Under these air mass conditions, the annual temperature range is quite broad, perhaps reaching 80° C.

There are also important variations in annual precipitation within the Subarctic zone. In the western subzone of the Subarctic, annual precipitation is low—about 40 cm—due to the rain shadow effect of the Cordillera.

In the eastern subzone, annual precipitation is much higher, sometimes exceeding 80 cm, most of which is provided by the Atlantic and Gulf of Mexico air masses.

The warm but short summers provide adequate growing conditions for coniferous trees. For example, average monthly summer temperatures exceed 10° C, thereby promoting tree growth. Black and white spruce are the most common species in the Canadian boreal forest. Birch and poplar also occur, especially along the southern edge of the boreal forest. Stands of Jack pine trees indicate an area that is recovering from a forest fire. Beneath this coniferous forest, there are podzolic soils. Wetlands are widespread: much of the land is poorly drained due to the disrupted drainage pattern caused by glaciation and permafrost. Wetlands contain numerous lakes, peat bogs, and marshes. Canadians often refer to this type of poorly drained land as muskeg.

Tundra vegetation, such as lichens, mosses, and sedges, grow in sheltered areas within the Canadian Arctic in order to minimize exposure to arctic winds. Arctic plants have a short reproductive cycle that can be completed in the brief summer. This red flowering plant is purple saxifrage (*Saxifraga oppositifolia*). (Al Harvey/The Slide Farm)

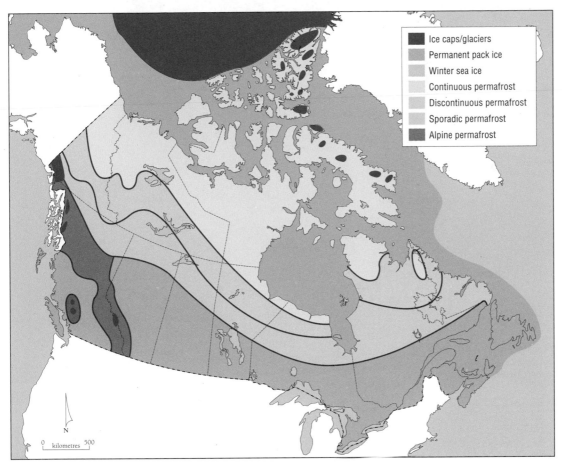

Figure 2.9 Permafrost zones. Canada's cold environment is demonstrated by the permanently frozen ground that extends over two-thirds of the country. (Further resources: Student Web site, National Atlas section, Map 10. Web site instructions are found on p. xxi.)

mafrost is found in all six of Canada's geographic regions and reaches its most southerly position along 50° N in central Québec. Along the southern edge of permafrost, there is a transition zone where small pockets of frozen ground have a depth of less than 1 m. Further south, these pockets of permafrost disappear.

Permafrost is divided into four types. **Alpine permafrost** is found in mountainous areas and takes on a vertical pattern as elevations of a mountain increase. Over most of Canada, however, permafrost follows a zonal pattern, which does not correspond to latitude but rather to the annual mean temperatures

that fall below zero.[4] The zonal pattern has a northwest to southeast alignment, that is, from Yukon to central Québec (see Figure 2.9).

As the mean annual temperature varies, the type of permafrost also changes. **Continuous permafrost** occurs in the higher latitudes of the Arctic climatic zone, where at least 80 per cent of the ground is permanently frozen, although it also extends into northern Québec. Continuous permafrost is associated with very low mean annual air temperatures of -15° C or less. **Discontinuous permafrost** occurs when 30 to 80 per cent of the ground is permanently frozen. It is found in the Subarctic climatic

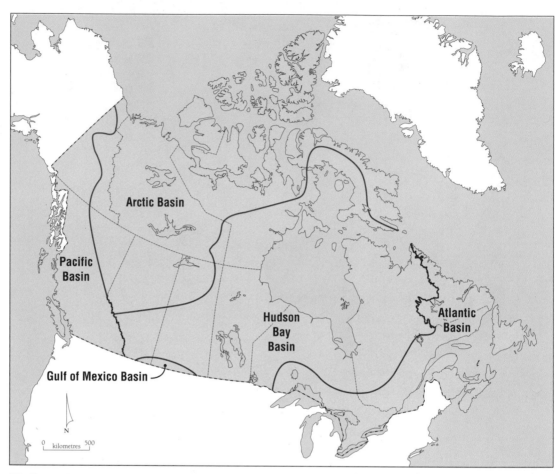

Figure 2.10 Drainage basins of Canada. The Hudson Bay drainage basin is by far the largest of the five basins in Canada. It also serves as a boundary between southern Alberta and British Columbia, and between northern Québec and Labrador. (Further resources: Student Web site, National Atlas section, Map 11. Web site instructions are found on p. xxi.)

zone where mean annual air temperature ranges from -5° C in the south to -15° C in the north. Sporadic permafrost is found mainly in the northern parts of the provinces, where less than 30 per cent of the area is permanently frozen. Sporadic permafrost is associated with mean annual temperatures of zero to -5° C.

Major Drainage Basins

Facing three oceans, Canada is clearly a maritime nation (Vignette 2.10). Canada has four major drainage basins (Figure 2.10 and Table

2.6): the Atlantic Basin, the Hudson Bay Basin, the Arctic Basin, and the Pacific Basin. In addition, a small portion of southern Alberta and Saskatchewan forms part of the Mississippi River system, which drains into the Gulf of Mexico. A **drainage basin** is land that slopes towards the sea and is separated from other lands by topographic ridges. These ridges form drainage divides. The continental divide of the Rocky Mountains, for example, separates those streams flowing to the Pacific Ocean from those flowing to the Arctic and Atlantic oceans. While each stream has a drainage basin, those streams flowing to the same ocean combine their basins to form a major drainage basin.

Historically, the rivers in each basin have played major roles in the development of the country. For example, Aboriginal peoples and Europeans both used the St Lawrence and Mackenzie rivers as transportation routes during the fur trade. These rivers remain important waterways today.

Atlantic Basin

The Atlantic Basin is centred on the Great Lakes and the St Lawrence River and its tributaries, but the basin also includes Labrador. The Atlantic Basin has the third largest drainage area and also the third greatest stream-flow. The largest hydroelectric development in this drainage basin is located at Churchill Falls in Labrador. Development of the lower Churchill River has reached the discussion stage between the governments of Newfoundland/Labrador and Québec and three Aboriginal groups, the Labrador Inuit, the Innu First Nation, and the Labrador Métis. Earlier hydroelectric developments took place

along the St Lawrence River in southern Québec and along its tributary rivers that flow out of the Laurentide Upland of the Canadian Shield. Rivers such as the Manicouagan River originate in the higher elevation of the Laurentide Upland. Here, the combination of abundant precipitation, natural lakes, and a sharp increase in elevation provides ideal natural conditions for the generation of hydroelectric power. Because there is a large market for electrical power in the St Lawrence Lowlands, virtually all potential sites in the Laurentides have been developed.

The Hudson Bay Basin

The Hudson Bay Basin is the largest drainage basin in Canada (Table 2.6). It occupies about 3.8 million km². In the West, its rivers originate in the Rocky Mountains and flow to Hudson Bay. In the East, the headwaters of its rivers in the uplands of northern Québec flow westward into James Bay. In northern Ontario and Manitoba, rivers drain into James and Hudson bays.

The combination of large rivers and sudden drops in elevation that occur in the Cana-

Table 2.6	Canada's Drainage Basins	
Drainage Basin	**Area** **(million km²)**	**Stream-flow** **(m³/second)**
Hudson Bay	3.8	30,594
Arctic	3.6	20,491
Atlantic	1.6	21,890
Pacific	1.0	24,951
Gulf of Mexico	<0.1	—
Total	10.0	105,135

Sources: A.H. Laycock, 'The Amount of Canadian Water and Its Distribution', in M.C. Healey and R.R. Wallace, eds, *Canadian Aquatic Resources* (Ottawa: Department of Fisheries and Oceans, 1987), 32; Philip Dearden and Bruce Mitchell, *Environmental Change and Challenge: A Canadian Perspective* (Toronto: Oxford University Press, 1997), 142.

dian Shield makes this part of the basin ideal for developing hydroelectric power stations. In fact, most of Canada's hydroelectric power is generated in the Canadian Shield area of the Hudson Bay Basin—the largest installations are on La Grande Rivière in northern Québec and on the Nelson River in northern Manitoba. La Grande Rivière's hydroelectric developments are the first stage in the James Bay Project. The Great Whale River Project was to follow the completion of the hydroelectric projects on La Grande Rivière, but a variety of circumstances (low energy demand, low prices in New England, and strong opposition from environmental groups and the Cree Indians of northern Québec) stalled its development.

The Arctic Basin

The Arctic Basin is Canada's second-largest drainage basin. The Mackenzie River dominates the drainage system in this basin. Along with its major tributaries (the Athabasca, Liard, and Peace rivers), the Mackenzie River is the second-longest river in North America. However, because of low precipitation in the Arctic, this basin has only the fourth-largest stream-flow. There are few hydroelectric projects in the Arctic Basin because of the long distance to markets, with the exception of the hydroelectric development on the Peace River in British Columbia. Here, power from the Gordon M. Shrum generating facility is trans-

Niagara Falls, one of the most spectacular waterfalls in the world, lies along the border with the United States. On the Canadian side, the Horseshoe Falls are 54 metres high and 675 metres wide. Niagara Falls was formed some 10,000 years ago when an ice front, which separated lakes Erie and Ontario, melted. Water now flows from Lake Erie northward across the Niagara Escarpment into Lake Ontario. (© Royalty Free/Corbis/Magma)

mitted to the population centres in southern British Columbia and to the United States, primarily to the states of Washington, Oregon, and California.

The Pacific Basin

The Pacific Basin is the smallest basin. However, it has the second-highest volume of water draining into the sea. Heavy precipitation along the coastal mountains of British Columbia accounts for this unusually high streamflow. As a result, the Pacific Basin is the site of one of Canada's largest hydroelectric projects. Located at Kemano, this facility is owned and operated by Alcan, which uses the electrical generating station to supply power to its aluminum smelter at Kitimat. The ice-free, deepwater harbour at Kitimat and low-cost electric power generated at Kemano make Kitimat an ideal location for an aluminum smelter.

Environmental Challenges

Canadians once believed that their resources were infinite so that human activities could never harm the environment. While Canadians have traditionally exploited the country's natural wealth for their benefit, the current extent and pace of human impacts on the environment have reached and in some locations exceeded the danger point. Many serious environmental problems facing Canada today are the consequence of such human actions (Dearden and Mitchell, 1997: ch. 1) and geographers are calling for more protected areas and parks (Slocombe and Dearden, 2002: ch. 12). The rush to develop our resources has placed greater and greater demands on the natural environment. The discharge of toxic chemicals and raw sewage into our oceans, rivers, and lakes has fouled the waters, some-

times with disastrous consequences; car exhaust is a major source of air pollution in cities and a leading contributor of carbon dioxide to the atmosphere; the loss of forest habitat has reduced the habitat for many plants and animals. Canada's air, land, and water are subject to pollution from a variety of sources. Air pollution, for instance concerns geographers at three levels:

- *Global scale.* Air pollution contributes to global warming and, through the process of the global atmospheric system, adds minute toxic particles to the Canadian Arctic.
- *Regional scale.* Air pollution has damaged the forests and lakes of Ontario, Québec, and Atlantic Canada through acid rain.
- *Local scale.* Air pollution known as smog in Canada's major cities now poses a health hazard.

The environment that we want is slipping away as pollution of our air, land, and water continues. One sign of environmental degradation is the threat to Canada's biodiversity by a host of human-induced factors. Hydroelectric projects, often touted as a perfect means of harnessing rivers to produce a valuable product, can have serious environmental consequences for the local region. Sometimes, these consequences take the form of resource conflicts. The Pacific Basin provides such an example. This basin forms an integral part of the natural biological cycle for salmon (a renewable resource), which live most of their lives in the Pacific Ocean. When they reach maturity, however, they return to their spawning grounds in the headwaters of the various tributaries that flow into the major rivers of the Pacific Basin, including the Nechako River, which is an important tributary of the Fraser

River. Alcan Canada's construction of a hydro-electric dam on this river flooded prime salmon spawning grounds and thus created a resource conflict.

Another sign, indeed one with much broader implications for the habitat in the Arctic, results from the dependency of our industrial world on obtaining energy from fossil fuels. The release of carbon dioxide into the atmosphere may be the primary cause of the recent warming of the atmosphere. Over the past decade, for example, average temperatures in the Arctic have increased with the result that the ice cover on the Arctic Ocean has thinned and the length of time that land-fast ice in Hudson Bay exists has decreased. Both these events have had a negative impact on the time available for seal hunting and on the size of the habitat for polar bears.

In the remainder of this chapter, two important results of industrial pollution—acid rain and global warming—are examined. In each regional chapter, a local environmental challenge is examined: for Ontario, the pollution of the Great Lakes; for Québec, toxic wastes in the St Lawrence; for British Columbia, clear-cut logging; for Western Canada, the threat to groundwater by wastes from large-scale hog barns; for Atlantic Canada, the toxic wastes at Cape Breton; and for the Territorial North, the conflict between placer gold mining and fish habitat.

Acid Rain

Acid rain—precipitation that has an unusually acidic chemical composition—affects the forests and lakes of Ontario, Québec, and Atlantic Canada. Acidity levels are described in terms of the pH factor, which measures the acidity or alkalinity of a substance on a scale of 0 to 14, with 0 being extremely acidic and 14

being extremely alkaline. The average pH of normal rainfall is 5.6, making it slightly acidic. Acid rain, however, has a pH of 2.4 or less. At this pH level, acid rain is a serious environmental problem in many parts of the world, including eastern Canada.

Most acid rain results from the chemicals derived from the burning of fossil fuels in industrial plants and from automobile exhaust. Sulphur dioxide and nitrogen oxides are emitted in the smoke from coal-burning plants, while automobile engines are particularly prolific producers of nitrogen oxides. Acid rain is created when oxides of sulphur and nitrogen change chemically as they dissolve in water vapour in the atmosphere and then return to earth as tiny droplets with high concentrations of sulphuric acid and nitric acid.

The most visible impact of acid rain is in urban areas where it slowly corrodes limestone buildings and defaces marble sculptures. Less visible but just as destructive are the effects of acid rain on forests, lakes, and soils. In eastern Canada, acid rain has damaged many forests and caused a sharp decline in fish stocks in many lakes. The principal areas in Canada affected by acid rain are located in Ontario, Québec, and the Maritime provinces. The problem of acid rain is compounded by the fact that pollutants emitted from coal-burning electrical power plants in the Great Lakes area of the United States fall as acid rain in Ontario and Québec.

Global Warming

Global warming could significantly alter Canada's environment (Bouchard, 2001). If world temperatures rose sufficiently, Canada's climatic zones, followed by its natural vegetation zones, would shift northward. Under this scenario, the geographic extent of the Arctic would be greatly reduced because the greatest increases in annual temperatures would take place in the Arctic due to the loss of snow cover (the so-called **albedo effect**). With a reduced period of snow cover, solar energy would be able to warm the ground more effectively and thus cause temperatures in the atmosphere to increase.

The economic consequences of global warming for Canada would be far-reaching. Some consequences would be favourable; many others would not. Global warming would affect the Arctic and Surbarctic more than the temperate areas of Canada. For example, the ice-free shipping season in Hudson Bay and the Arctic Ocean would be greatly extended but, with a rising sea level, at a cost of flooding coastal towns and cities. Then, too, global warming could thaw the permanently frozen ground known as permafrost. Melting of the ice in permafrost could cause massive ground subsidence, resulting in an irregular relief referred to by physical geographers as 'thermokarst topography'. This would disrupt transportation and pipeline systems, and play havoc with foundations for buildings, bridges, and other human-made structures (Bone et al., 1997: 265–74). In southern Canada, the impacts would see agricultural activities take place further north, but, on the negative side, grain agriculture in the Canadian Prairies might be subject to greater risk of drought. Water transportation would greatly benefit from longer navigation seasons. The Great Lakes and the St Lawrence River, for instance, would be navigable year-round.

But what is the basis for this theory of global warming? Two factors are at play: (1) the adding of more and more greenhouse gases to the atmosphere; and (2) the mechanics of the greenhouse effect. Only a few gases (called greenhouse gases) in our atmosphere can absorb solar energy. These gases (including carbon dioxide, methane, and water vapour) form less than 0.1 per cent of the atmosphere. Carbon dioxide is the principal greenhouse gas. It is transparent to short-wave radiation from the sun but opaque to the long-wave radiation emitted by the earth. In this way, carbon dioxide and other greenhouse gases heat the atmosphere, causing air temperatures to rise. This process is known as the greenhouse effect. The chief source of greenhouse gases comes from the burning of fossil fuels.

Theoretically, those who argue the case for global warming have a very strong case, based on the greenhouse effect hypothesis (as the percentage of greenhouse gases in the atmosphere increases, the capacity of the atmosphere to absorb solar energy also increases, thus causing air temperatures to rise); based on predictions made by computer simulations of the atmosphere called general circulation models (GCMs); and based on recorded increases in world temperatures over the past three decades. Some climatologists call for a major change in our climate within 50 years due to global warming. Cohen (1997: 1) predicted an increase in the mean world temperature at a minimum of 2° C and a maximum of 5° C by 2050. Some believe that these changes have already begun. Unlike past climatic changes (Vignette 2.11), global warming is caused by human actions. Hence, we have the power to prevent it. Although no one can predict the precise impacts of global warming on Canada, climatic change would translate into longer, hotter summers and shorter, milder winters.[5]

Is global warming under way? Nobody knows for certain, but as more carbon dioxide and other greenhouse gases are added to the atmosphere, the greenhouse effect should cause air temperatures to rise. Certainly the

Vignette 2.11   The Little Ice Age

Both major and minor climatic changes have occurred during the earth's history. A major cooling of the world's climate occurred in the Pleistocene Epoch of the Cenozoic era (Table 2.1). During the geological period, some 20 major ice advances occurred, the last being the Wisconsin.

However, minor cooling of the world's climate has also occurred. Minor cooling refers to relatively short periods of time when the global climate is slightly cooler than normal. A minor cooling, known as the Little Ice Age, took place between 1450 and 1850. The Little Ice Age had a dramatic impact on human beings living in the Arctic. During the Little Ice Age, the global climate was much cooler than it is today. This period was characterized by lower temperatures and longer winters in higher latitudes. In northern Canada, the ice cover over the Arctic Ocean was more extensive, which prevented bowhead whales from entering these waters. The consequences for the Thule inhabitants, who had developed a hunting economy based on the bowhead whale, were devastating. They were forced to hunt smaller game—seals and caribou. The results were twofold: the new hunting system could not support as many people and it required smaller, more mobile hunting groups. Archaeologists believe that the Inuit are the descendants of the Thule people.

mean world air temperature has increased over the last 250 years, but is it due to the burning of fossil fuels or something else such as a natural solar cycle? For instance, the world's temperature declined during the Little Ice Age (Vignette 2.11).

This brings us to the question, what scientific evidence supports global warming? A prominent scientific body, the Intergovernmental Panel on Climatic Change (IPCC), has played a lead role in assessing global climate change. The IPCC, established in 1988 by the World Meteorological Organization and the United Nations Environment Program, has produced a comprehensive assessment on climate change every five years beginning in 1991. In 2001, the authors of the Third Assessment Report of the IPCC concluded that new and stronger evidence indicates that most of the warming observed over the last 50 years is attributable to human activities. They also announced that the global average surface temperature has increased by 0.6 C° since the late nineteenth century, and that the 1990s was the warmest decade and 1998 the warmest year since 1861 (Houghton et al., 2001: 26).

The same research group predicts rapid increases in world temperatures over the next 100 years, with increases in the range of 1.7° C to 4.2° C (ibid., 70). However, temperature increases have so far fallen within the so-called normal temperature variation associated with the earth's climate, that is, within plus or minus 1 per cent of the world mean air surface temperature. Until world temperatures exceed this limit, global warming has not 'officially' begun.

While the physics of global warming in the greenhouse model are elementary, the actual process of climate change is extremely complex and remains unclear. Global warming seems inevitable, but other events might diminish the amount of solar energy reaching the earth's surface and thereby reverse the temperature trend. For example, volcanic eruption, by releasing huge amounts of dust into the air, could reflect significant amounts of incoming solar energy to outer space and thereby chill the world's climate. But can we count on such an event occurring to reverse the current warming trend?

Summary

Physical geography varies across Canada. This variation is critical in understanding Canada's regional character. At the macro level, physiographic regions represent large areas with similar landforms and geological structures. Climate creates a similar zonal arrangement of soils and natural vegetation. Climate therefore determines to a large extent the type of soil and native vegetation in a given region and hence influences the utilization of the land. Together with topography, climate partly determines the land's ability to support a population. This link between the physical and human worlds identifies those regions having a more favourable mix of physical characteristics for economic development, and translates the abstract core/periphery model into a geographic reality. The Great Lakes–St Lawrence Lowlands region is the most favoured physical region in Canada. Physical barriers, such as the Rocky Mountains, and extreme climatic conditions, such as the very long and cold winters in northern Canada, have also affected the historical settlement of the country and continue to influence contemporary economic activities.

Canada has several physiographic regions and climatic zones. The seven physiographic

regions are: the Canadian Shield, the Cordillera, the Interior Plains, the Hudson Bay Lowland, the Arctic Lands, the Appalachian Uplands, and the Great Lakes–St Lawrence Lowlands. The seven climatic zones are: the Pacific, Cordillera, Prairie, Great Lakes–St Lawrence, Atlantic, Subarctic, and Arctic.

Because Canada lies in the northern latitudes, permafrost is common in the northern areas of provinces (Newfoundland, Québec, Ontario, Manitoba, Saskatchewan, Alberta, and British Columbia) and in the Territorial North. Within Canada, rivers flow into four major drainage basins: the Atlantic, Hudson Bay, Arctic, and Pacific. Only a tiny area of Alberta and Saskatchewan, where waters originating in Canada drain into the Gulf of Mexico, belongs to the Mississippi Basin.

In our contemporary world, humans are the most active and dangerous agents of environmental change. The cultivation of the land, exploitation of its renewable and non-renewable resources, and processing of its primary products have forever changed our natural environment in many ways. The construction of the Confederation Bridge, which joins Prince Edward Island to the mainland of Canada, is one example. Often, however, industrial activities have damaged the environment by causing air pollution, acid rain, and global warming. Both air pollution and acid rain represent serious environmental problems today, while global warming may prove to be the environmental challenge of the twenty-first century.

Notes

1. Canadians have various visions of themselves, their region, and their country. For the most part, these visions are rooted in the physical nature and historical experiences

that affected Canada and its regions. For example, people see Canada as a northern country because of its location in the North American continent and because of its climates, which are often noted for long, cold winters. Hamelin's concept of nordicity (as discussed in Chapter 1) exemplifies this northern perspective. Artists, too, have found this northern theme appealing. Gilles Vigneault, one of Québec's best-known chansonniers, wrote the song 'Mon pays'. Though referring to Québec, the opening line, *Mon pays ce n'est pas un pays, c'est l'hiver* (My country is not a country, it is winter), resonates equally well in all regions of Canada.

2. Tectonic forces press, push, and drag portions of the earth's crust in a slow but steady movement. This process is the basis for the theory of continental drift. In 1912, Alfred Wegener suggested that long ago all the earth's continents formed one huge land mass. Tectonic action caused the breakup of this land mass into huge slabs. These slabs of the earth's crust drifted slowly on the molten mass (magma) beneath the earth's crust.

3. The Champlain Sea covered Anticosti Island and the northern tip of the island of Newfoundland. For the purposes of this text, the eastern extent of this physiographic region ends just east of Québec City.

4. The mean annual temperature of a location on the earth's surface is a measure of the energy balance at that point. Solar energy is the source of heat for the earth, and this energy is returned to the atmosphere in a variety of ways. Therefore, a global energy balance exists. However, there are regional energy surpluses and deficits in different parts of the world. For example, the Arctic has an energy deficit, while the tropics have a surplus. These energy differences drive the global atmospheric and oceanic circulation systems. When the mean annual temperature is below zero Celsius, it indicates that an energy deficit exists.

5. Slaymaker and French discuss the effects of global warming and climatic change on Canada's North. They maintain that climatic change caused by the greenhouse effect would alter Canada's cold environments more dramatically than the country's other environments. The authors describe possible changes to sea ice, permafrost, snow cover, sea level, and natural vegetation. There are four key factors for such a remarkable climatic change in northern Canada:

    • The percentage of carbon dioxide in the atmosphere over the Territorial North is much greater than that over southern Canada.
    • The thinning of the ozone layer over the Territorial North permits more solar radiation to enter the atmosphere.
    • The release of methane gases from the muskeg in northern Canada will add more greenhouse gases to the northern atmosphere.
    • The reduction in the duration of snow cover will expose the northern lands to solar radiation for a longer time.

In the same book, Ledrew presents the nature of climatic change, while Smith describes the impact of climatic change on permafrost. See chapters 11, 12, and 13 in *Canada's Cold Environments*, edited by Hugh M. French and Olav Slaymaker (Montréal and Kingston: McGill-Queen's University Press, 1993).

Key Terms

albedo effect
Proportion of solar radiation reflected from the earth's surface.

alpine permafrost
Permanently frozen ground that is found at high elevations.

arêtes
Sharp mountain ridges that are formed between two cirques.

basins
Structural depressions in sedimentary rock that are caused by a bending of sedimentary strata into huge bowl-like shapes. Petroleum may accumulate in sedimentary basins.

chernozemic
A soil order identified by a well-drained soil that is often dark brown to black in colour; associated with the grassland and parkland natural vegetation types and located in the Prairies climatic zone.

cirques
Large, shallow depressions found in mountains caused by the plucking action of alpine glaciers.

climate
An average condition of weather over a very long period of time.

climatic zone
A geographic area where similar types of weather occur.

continental air masses
Homogeneous bodies of air that have taken on moisture and temperature characteristics of the land mass of their origin. Continental air masses are normally dry and cold in the winter and dry and hot in the summer.

continental drift
The movement of the earth's crust. This concept is also known as plate tectonics.

continuous permafrost
Extensive areas of permanently frozen ground in the Arctic, where at least 80 per cent of the ground is permanently frozen.

convectional precipitation
An upward movement of moist air causes the air to cool, resulting in condensation and then precipitation.

cryosolic
A soil order associated with permafrost and poorly drained land; soil is either lacking or extremely thin; associated with the tundra and polar desert vegetation types and located in the Arctic climatic zone.

deposition
The deposit of material on the earth's surface by various processes such as ice, water, and wind.

discontinuous permafrost
Permanently frozen ground mixed with unfrozen ground in the Subarctic. At its northern boundary about 80 per cent of the ground is permanently frozen, while at its southern boundary about 30 per cent of the ground is permanently frozen.

drainage basin
Land sloping towards the sea; an area drained by rivers and their tributaries.

drumlins
Landforms created by the deposit of glacial till and shaped by the movement of the ice sheet.

eskers
Long, sinuous mounds of sand and gravel that were deposited on the bottom of a stream flowing under a glacier.

faulting
The breaking of the earth's crust.

folding
The bending of the earth's crust.

friction of distance
The effect of distance on spatial interaction; that is, as distance increases, the number of spatial interactions (such as telephone calls) diminishes.

frontal precipitation
When a warm air mass is forced to rise over a colder air mass, condensation and then pre-

cipitation occur.

glacial erosion

The scraping and plucking action of moving ice on the surface of the land.

glacial spillway

A deep and wide valley formed by the flow of massive amounts of water originating from a melting ice sheet or from water escaping from glacial lakes.

glacial striations

Scratches or grooves in the bedrock caused by rocks embedded in the bottom of a moving ice sheet or glacier.

glacial trough

A U-shaped valley carved by an alpine glacier.

global circulation system

The movement of ocean currents and wind systems that redistribute energy around the world.

greenhouse effect

The absorption of long-wave radiation from the earth's surface by the atmosphere.

Holocene Epoch

The current geological division of the Geological Time Chart. It began some 10,000 years ago and is associated with the warm climate following the last ice age.

ice age

A long cold period accompanied by the appearance of continental ice sheets. The most recent ice age is called the Pleistocene Ice Age, which began some 2 million years ago.

igneous rocks

Rock formed when the earth's surface first cooled or when magma or lava reached the earth's surface.

isostatic rebound

The uplifting process of the earth's crust following the removal of an ice sheet that, because of its weight, depressed the earth's crust. Also known as postglacial uplift.

latitude

An imaginary line parallel to the equator that encircles the globe.

longitude

An imaginary line that runs through both the North and South poles.

luvisolic

A soil order identified by a well-drained soil that is often grey-brown in colour; associated with the broadleaf and mixed forest natural vegetation types in the Great Lakes–St Lawrence climatic zone.

marine air masses

Large homogeneous bodies of air with moisture and temperature characteristics similar to the ocean where they originated. Marine air masses are normally moist and mild in both winter and summer.

metamorphic rocks

Formed from igneous and sedimentary rocks by means of heat and pressure.

muskeg

A wet, marshy area found in areas of poor drainage, such as the Hudson Bay Lowland. Muskeg contains peat deposits.

nunataks

An unglaciated area of a mountain that stood above the surrounding ice sheet.

orographic precipitation

Rain or snow created when air is forced up the side of a mountain, thereby cooling the air and causing condensation followed by precipitation.

patterned ground

The arrangement of stones and pebbles in polygonal shapes. Patterned ground occurs in the Arctic where continuous permafrost exists and where frost shattering is the principal erosion process.

permafrost

Permanently frozen ground.

physiographic region

A large geographic area where a single landform, such as the Interior Plains, is found.

pingos

Hills or mounds that have an ice core and that are found in areas of permafrost.

Pleistocene Epoch

A minor division of the Geological Time Chart beginning nearly 2 million years ago. It forms part of the Quaternary Period and is associated with some 20 ice ages.

podzolic

A soil order identified by a poorly drained soil that is often grey in colour; associated with the boreal forest and the coastal rainforest and with climates that have large amounts of precipitation, such as the Pacific, Atlantic, and Subarctic climatic zones.

postglacial uplift

The slow rising of the earth's crust following the removal of an ice sheet that, because of its weight, depressed the earth's crust. Also known as isostatic rebound.

Quaternary Period

This geological period consisted of the Pleistocene and Holocene Epochs.

rain shadow effect

Results in a dry area on the lee side of mountains where air masses descend, causing them to become warmer and drier.

residual uplift

The final stages of isostatic rebound.

restrained rebound

The first stage of isostatic rebound.

sedimentary rocks

Rocks formed from the accumulation, in a layered sequence, of sediment deposited in the bottom of an ocean.

soil order

Classes of soil based on observable soil properties and soil-forming processes. In Canada there are nine soil orders, including chernozemic, cryosolic, and podzolic.

sporadic permafrost

Pockets of permanently frozen ground mixed with large areas of unfrozen ground. Sporadic permafrost ranges from a trace of permanently frozen ground to an area having up to 30 per cent of its ground permanently frozen.

strata

Layers of sedimentary rock.

subsidence

A downward movement of the ground. In areas of permafrost, subsidence occurs when large blocks of ice within the ground melt, causing the material above to sink or collapse.

till

Unsorted glacial deposits.

Bibliography

Archibold, O.W. 1995. *Ecology of World Vegetation*. London: Chapman & Hall.

Bone, R.M. 2003. *The Geography of the Canadian North: Issues and Challenges*, 2nd edn. Toronto: Oxford University Press.

———, Shane Long, and Peter McPherson. 1997. 'Settlements in the Mackenzie Basin: Now and in the Future 2050', in Cohen (1997: 265–74).

Bouchard, Mireillle. 2001. 'Un défi environnemental complexe du XXIe siècle au Canada: l'identification et la compréhension de la réponse des environnements face aux changements climatiques globaux', *The Canadian Geographer* 45, 1: 54–70.

Briggs, David, Peter Smithson, and Timothy Ball. 1993. *Fundamentals of Physical Geography*. Toronto: Copp Clark Pitman.

Canada. 1997. *Canada Year Book 1998*. Ottawa: Minister of Supply and Services.

Christopher, Robert W. 1998. *Geosystems: An Introduction to Physical Geography*, 3rd edn. Upper Saddle River, NJ: Prentice-Hall.

Cohen, Stewart J. 1990. 'Bringing Global Warming Issue Closer to Home: The Challenge of Regional Impact Studies', *American Meteorological Society* 71, 4: 520–6.

———, ed. 1997. *The Final Report of the Mackenzie Basin Impact Study*. Downsview: Environment Canada.

De Loë, R. 2000. 'Floodplain Management in Canada: Overview and Prospects', *The Canadian Geographer* 44, 4: 354–68.

Dyke, Arthur S., and Victor K. Prest. 1987. 'Late Wisconsinan and Holocene History of the Laurentide Ice Sheet', *Géographie physique et quaternaire* 41, 2: 237–64.

Fisheries and Oceans Canada. 2003. *Fast Facts*. Available at: <www.dfo-mpo.gc.ca/communic/facts-info/facts-info_c.htm>.

French, H.M., and O. Slaymaker, eds. 1993. *Canada's Cold Environments*. Montréal and Kingston: McGill-Queen's University Press.

Fulton, Robert J., and Victor K. Prest. 1987. 'The Laurentide Ice Sheet and Its Significance', *Géographie physique et quaternaire* 41, 2: 181–6.

Hare, F. Kenneth, and Morley K. Thomas. 1974. *Climate Canada*. Toronto: Wiley.

Houghton, J.T., Y. Ding, D.J. Griggs, M. Noguer, P.J. van der Linden, X. Dai, K. Maskell, and C.A. Johnson. 2001. *Climate Change 2001: The Scientific Basis*. Contribution of Working Group I to the Third Assessment Report of the Intergovernmental Panel on Climate Change. Cambridge: Cambridge University Press. See <http://www.ipcc.ch/pub/reports.htm>.

Laycock, A.H. 1987. 'The Amount of Canadian Water and Its Distribution', in M.C. Healey and R.R. Wallace, eds, *Canadian Arctic Resources*. Ottawa: Department of Fisheries and Oceans, 13–42.

Marsh, James H., ed. 1988. *The Canadian Encyclopedia*, 2nd edn. Edmonton: Hurtig.

Mitchell, Bruce, ed. 1995. *Resource and Environmental Management in Canada: Addressing Conflict and Uncertainty*, 2nd edn. Toronto: Oxford University Press.

Norton, William. 2001. *Human Geography*, 4th edn. Toronto: Oxford University Press.

Nuttall, Mark, and Terry V. Callaghan, eds. 2000. *The Arctic: Environment, People, Policy*. Amsterdam: Harwood Academic Publishers.

Rasid, Harun, Wolfgang Haider, and Len Hunt. 2000. 'Post-flood Assessment of Emergency Evacuation Policies in the Red River Basin, Southern Manitoba', *The Canadian Geographer* 44, 4: 369–86.

Shabbar, Amir, Barrie Bonsal, and Madhav Khandekar. 1997. 'Canadian Precipitation Patterns Associated with Southern Oscillation', *Journal of Climate* 10: 3016–27.

Slocombe, D. Scott, and Phillip Dearden. 2002. 'Protected Areas and Ecosystem-Based Management', in Dearden and Rick Rollins, eds, *Parks and Protected Areas in Canada: Planning and Management*, 2nd edn. Toronto: Oxford University Press, 295–320.

Trenhaile, Alan S. 1998. *Geomorphology: A Canadian Perspective*. Toronto: Oxford University Press.

Young, Steven B. 1989. *To the Arctic: An Introduction to the Far Northern World*. New York: Wiley.

Further Reading

Dearden, Philip, and Bruce Mitchell, 1997. *Environmental Change and Challenge: A Canadian Perspective*. Toronto: Oxford University Press.

Canada's vastness is often equated with unspoiled wilderness and unlimited resources. Perhaps there was some validity to these images in the distant past when our numbers were smaller and our technology more limiting, but not now. False and mis-

leading images are firmly put to rest in *Environmental Change and Challenge*. Our growing population and technological advances have placed our fragile environment at risk. Already, we have stepped over the line by decimating northern cod stocks and adding toxic chemicals to the Arctic food chain so that the Inuit must limit their consumption of seals and other sea mammals. After exposing a multitude of environmental problems caused by human activities, Dearden and Mitchell then turn to the matter of environment and resource management.

The scope of this book is very broad, ranging from how ecosystems function to how humans can conserve its energy, lands, and waters. For geography students, the case studies provide a rich and varied record of how our physical environment has been abused and how Canadians are just learning how to manage the environment. The authors begin by discussing groundwater and soil contamination in southwestern Ontario. Once contaminated by agricultural/industrial chemicals or toxic wastes, groundwater and soil are no longer suitable for human use for some considerable time. Unknowingly, the authors were foretelling the Walkerton, Ontario, disaster. In May 2000, seven people died and 2,300 became ill from E. coli poisoning after drinking from Walkerton's contaminated public water system. Seven months later, Walkerton's water was declared suitable for human consumption. The Walkerton disaster is only the tip of the iceberg. Disposal of urban and rural waste threatens Canada's groundwater. In urban Canada, cities are looking to rural sites to dispose of their mounting piles of garbage. In rural Canada, the trend towards larger and larger cattle and hog operations generates more and more animal waste. Disposal in local lagoons or settling tanks poses a threat to the groundwater and eventually to the drinking water of nearby farmsteads, rural villages, and towns.

Forests are another concern. Chapter 21 details the struggle to preserve one of the world's last great temperate rainforests. Clayoquot Sound in British Columbia saw a collision of two forces—those who want to harvest the old-growth trees of the Coastal Temperate Rainforest and those who want to preserve them. Faced with a fierce public outcry and opposition from the traditional owners of these lands, the Nuu-Chah-Nulth tribes, the British Columbia government decided to leave one-third of the land as a wilderness area and to allow selective cutting in the remaining area. Dearden and Mitchell conclude that the Clayoquot Sound experience confirms that resource and environmental issues are often about the management of conflict and disputes.

Overview and Objectives

St Boniface in Winnipeg, Manitoba
(Kennon Cooke/Valan Photos)

Canada's historical geography began when the first humans arrived in North America, perhaps 30,000 years ago. Since then, history and geography have combined to forge Canada's regional geography. In this chapter we look at three migrations: (1) from Asia to North America; (2) from France and Britain to the four original colonies; and (3) from Central Europe and Tsarist Russia to the Canadian West. By the end of World War II, these three groups had made Canada a multicultural and pluralistic society. They have also led to tensions or faultlines based on fundamental differences between English- and French-speakers, centralists and decentralists, 'new' and 'old' Canadians, and Aboriginal/non-Aboriginal Canadians. The search for solutions to these tensions is a dominant feature of Canadian society.

- Describe the arrival of Canada's first people.
- Examine the colonization of Canada by the French and the British.
- Account for the settlement of Canada's West by peoples from Central Europe and Tsarist Russia.
- Present the territorial evolution of Canada.
- Introduce the four faultlines.
- Discuss the notion of 'One Country, Two Visions'.
- Suggest 'solutions' to complex problems.

Canada's Historical Geography

■ Introduction

In one sense, Canada is a young country. Its formal history began in 1867 with the passing of the British North America Act by the British Parliament. In another sense, Canada is an old country whose history goes back much further than 1867. As an old country, its history has followed many twists and turns, but three events stand out because they continue to have a profound impact on the nature of Canadian society. These events are the arrival of the first people in North America, the colonization of North America by France and England, and the coming of people from Central Europe and Tsarist Russia. After 1867, Canada expanded in size until it became the second largest country in the world. This expansion continued and only ceased in 1949, when Newfoundland joined Canada. As the country was spread over such a huge territory, Canada developed a distinct regionalized character shaped by history and physical geography.

Within the federal system of government, political reality favours Ontario and Québec partly because the governing party is dependent on holding a majority of seats from the combined ridings of those two provinces. Provincial governments, while having defined powers, have insufficient tax revenue to meet all their responsibilities, such as health care and education.[1] Consequently, an ongoing battle between provincial governments and Ottawa takes place annually over transfer payments from the federal government to the provinces. With no federal body representing provincial/territorial interests, regionally based political parties, such as the Reform Party of Canada (which became the Canadian Alliance, which became the Conservative Party by amalgamating with the Progressive Conservative Party) and the Bloc Québécois, have emerged. Both add to the country's diverse character. Therefore, much of this great political experiment called Canada involves the difficult task of forging a unity among its various political members.

While Canada is admired by most peoples of the world, seeds of regional discontent exist in each province and territory. Regional discontent, often related to economic disparities and squabbling over natural resources, has generated ongoing tensions between the federal government and the provinces, as well as tensions among the provinces over issues such as language laws and constitutional amendments. To use *Globe and Mail* columnist Jeffrey Simpson's metaphor, these tensions can be likened to faultlines—large cracks in the earth's crust caused by tectonic forces. Like geological faultlines, Canada's economic, social, and political faultlines divide regions and people, and threaten to destabilize Canada's integrity as a nation. For long periods of time, these cracks in Canadian society remain dormant, but they can shift at any time, dividing the country into wrangling factions. A recent example was the Kyoto Protocol, ratified by the federal government in December 2002, which caused the political temperature between Alberta and Ottawa to rise sharply.

Four key tensions are examined in this chapter: (1) English-speaking/French-speaking Canadians; (2) centralists and decentralists; (3) 'new' and 'old' Canadians; and (4) Aboriginal/non-Aboriginal Canadians. In each case, differences and disagreements occur between a core (for example, Central Canada) and a periphery (for example, the rest of Canada). Central Canada is often the symbol of the political power held by Ottawa, and at other times of the economic power held by Ontario and Québec. Although the periphery refers in general to the rest of Canada, it can also represent a particular geographic region or even a group of people with a common complaint against the core.

Of the four faultlines, the French/English rift is by far the most threatening to Canada. For example, if another Québec referendum at some time in the future points conclusively to separation from the rest of Canada, it would have a devastating impact by separating Atlantic Canada from the rest of Canada. Yet at the same time, the historic and contemporary interaction between French- and English-speaking Canadians is the very process that makes Canada so unique. This give and take between the two so-called founding peoples often spills over to affect the Aboriginal, immigration, and regional issues that test Canada's unity. Oddly enough, because of the interaction between Aboriginal and non-Aboriginal Canadians, new and old Canadians, centralists and decentralists, and English-speaking and French-speaking Canadians, events affecting one faultline often have a positive or negative effect on the other three.

Chapter 3 begins by tracing the historical roots of Canada that result in geographic real-

ities. One aspect of these realities is the emergence of tensions between regions and/or groups that live in those regions. Often these disagreements pit Ottawa against the provinces. Frequently, these disagreements fall into three categories: (1) the issue of national interests overriding provincial/territorial ones; (2) the issue of federal funding for provincial programs; and (3) the issue of federal responsibilities for Aboriginal peoples, have-not provinces, and the three territorial governments. Examples for each category would be (1) the Kyoto Accord, (2) health care, (3) First Nations and the territorial governments.

The First People

Around 30,000 years ago, the first people to set foot on North American soil were Old World hunters who crossed a land bridge (known as Beringia) into Alaska and Yukon.

Beringia is now well below the surface of the Pacific Ocean, but at the time of the last ice advance, vast amounts of water were converted into ice, causing the sea level to drop by at least 100 metres and thus creating a land bridge between Asia and North America. However, these Old World hunters were blocked from proceeding south by ice that was perhaps 4 km thick. When an ice-free corridor split the Wisconsin ice sheet into the Laurentide and Cordillera ice sheets, they were able to migrate southward into the heart of North America (Figure 3.1). Geologists have estimated that this ice-free corridor, which was just east of the Rocky Mountains, appeared about 12,000 to 13,000 years ago. Just when the first people travelled along this narrow corridor is unknown, but it probably occurred after tundra vegetation had taken hold, supplying food for the animals that also migrated into the heart of North America. At the same time, sufficient

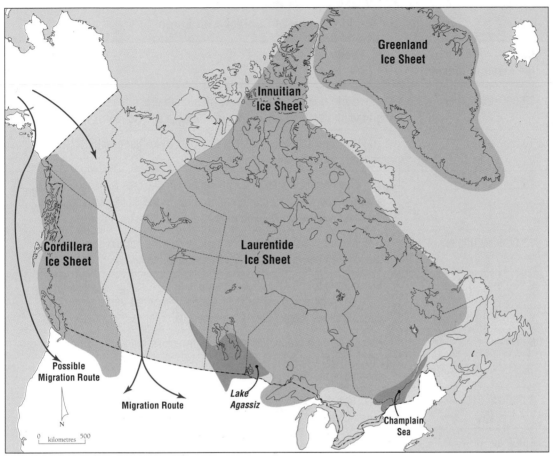

■ **Figure 3.1 Migration routes into North America.** The Wisconsin ice sheet had retreated by 12,000 BP, leaving an ice-free corridor between the two remaining parts of the Wisconsin ice sheet (the Cordillera and the Laurentide ice sheets). In places where the meltwater could not flow to the sea, water collected in low-lying areas and created glacial lakes such as Lake Agassiz. As explained in Vignette 2.7, however, the Champlain Sea was not a glacial lake but an extension of the Atlantic Ocean.

ice had melted to cause the ocean level to rise, thus submerging the Beringia land bridge, which made further migration from Asia to North America virtually impossible until the marine technology of the Paleo-Eskimos emerged about 5,000 years ago (Table 3.1).

Some archaeologists even speculate that Old World hunters may have penetrated into the lands south of the ice sheet much earlier. According to one theory, these early hunters made a coastal migration along the edge of the Cordillera ice sheet that covered all the mountains and valleys of British Columbia. By slow-

ly moving southward along the unglaciated islands adjacent to the ice-covered British Columbia coast, these early people could have circumvented the Cordillera ice sheet by island-hopping, thereby reaching the unglaciated area of North America much earlier than 12,000 years ago. However, archaeological evidence of such a route is lacking because these ancient campsites, if they existed, are now well below the current sea level.

Although the question of when, precisely, humans first reached North America is still unsettled, archaeological evidence indicates

Table 3.1	Timeline: Old World Hunters to Contact with Europeans
Date (BP = before present)	Event
30,000–25,000 BP	Old World hunters of the woolly mammoth cross the Beringia land bridge into the unglaciated areas of Alaska and southern Yukon.
18,000 BP	The Wisconsin ice sheet reaches its maximum geographic extent, covering virtually all of Canada.
15,000 BP	The Wisconsin ice sheet retreats rapidly in western Canada, exposing a narrow ice-free area along the foothills of the Rocky Mountains. Ice persists in northern Québec until about 6,000 years ago.
13,000–12,000 BP	Descendants of the Old World hunters migrate southwards through the ice-free corridor east of the Rocky Mountains.
9,000 BP	Mammoths and mastodons become extinct, forcing early inhabitants of North America to adjust their hunting practices and thereby become more mobile and less numerous.
5,000 BP	Paleo-Eskimos (known as the Denbigh) cross the Bering Strait to the Arctic Coast of Alaska. Later they move eastward along the Canadian section of the Arctic coast.
3,000 BP	The Dorset people, who were descendants of the Denbigh, develop a more advanced technology suited for an Arctic environment.
1,000 BP	A second wave of marine hunters, known as the Thule, migrate across the Arctic, eventually reaching the coast of Labrador. At the same time Vikings reach Greenland and North America, where they establish a settlement on the north coast of Newfoundland (L'Anse aux Meadows).
1497	John Cabot lands on the east coast (Newfoundland or Nova Scotia).
1534	Jacques Cartier plants the flag of France near Baie de Chaleur.

that the Old World hunters had reached the southern United States by at least 11,000 years ago. Archaeologists have assigned the name Paleo-Indian to these first people of North America because they shared a common hunting culture, which was characterized by its uniquely designed fluted-point stone spearhead. In time, the Paleo-Indians developed more distinct hunting cultures that Thomas (1999:10) divides into three groups:

- Clovis culture from 9500 to 8500 BC;
- Folsom culture from 9000 to 8200 BC;
- Plano culture from 8000 to 6000 BC.

Paleo-Indians

How did these Old World hunting cultures evolve into a New World hunting system? The Paleo-Indians, the people who devised the fluted points, were descendants of the Old

World hunters. The oldest fluted points found in North America are about 11,500 years old. These spearheads, along with the bones of woolly mammoths, have been discovered in the southern part of the Canadian Prairies. By 9,000 years ago, many of the large species, such as the woolly mammoths and the mastodons, had become extinct, possibly as a result of excessive hunting and/or climatic change. After the extinction of the woolly mammoths and mastodons, later Paleo-Indian cultures, which archaeologists refer to collectively as Folsom and Plano, developed a variety of unfluted stone points with stems for attachment to spear shafts, which made weapons more suitable for hunting buffalo and caribou. About 8,000 years ago, these hunters in the grasslands of the interior of Canada hunted buffalo, while those in the tundra and forest lands of northern and eastern Canada depended on caribou for most of their food. However, these smaller prey species could not support large numbers of people, so the Paleo-Indians had to develop new survival strategies. These strategies involved remaining in one area (and presumably keeping other peoples out of that area), developing effective hunting techniques for the local game, and making extensive use of fish and plants to supplement their principal diet of game.

This link between geographic territory and hunting societies marked the development of Paleo-Indian **culture areas**, which were geographic regions with the following two characteristics: (1) a common set of natural conditions that resulted in similar plants and animals, and (2) inhabitants who used a common set of hunting, fishing, and food-gathering techniques and tools. Under these conditions, Paleo-Indians formed more enduring social units that became the forerunners of the numerous North American Indian tribes at the time of contact with Europeans.

Indians

Climatic conditions vary across North America, ranging from a tropical climate in Mexico to an arctic climate just north of the Great Lakes–St Lawrence Lowlands. At that time, a great ice sheet still occupied most of Ontario and Québec. These varied conditions provided different opportunities for Indians, at least some of whom were descendants of the Paleo-Indians. Most archaeologists support the idea that Algonquians are direct descendants of Paleo-Indians, but they are less certain about Athapaskans, whose ancestors may have arrived in North America some time after the Paleo-Indian culture emerged about 11,500 years ago.

About 5,000 years ago, Indians living in the tropical climate of Mexico began to domesticate plants and animals. This agricultural system and its people gradually spread northward into areas with more restrictive growing conditions. These climatic differences required Indians to adapt their agricultural system accordingly. About 3,000 years ago, Indians in what is today the eastern United States planted corn, beans, and squash, which supplemented their diet of game and fish.

Agriculture was not possible north of the Great Lakes–St Lawrence Lowlands because of its shorter growing season for corn and other crops. Algonquian-speaking Indians who lived north of the Great Lakes had to hunt big-game animals, particularly caribou, for sustenance. They also traded with those more sedentary Indians, such as the Huron and Iroquois, who practised a form of slash and burn agriculture in the Great Lakes–St Lawrence Lowlands and in the Ohio Valley. By the sixteenth century, the Huron tribe controlled the agriculture lands between Lake Simcoe and the southeastern corner of Georgian Bay, where about 7,000 acres were under cultivation and where Indian villages, with populations as large as 5,000 persons, were commonplace (Dickason, 2002: 52). In western North America, agriculture spread northward along the valleys of the Mississippi River and its major tributary, the Missouri River. Tribes in the Canadian Prairies continued to hunt buffalo but did engage in trade for agriculture products with tribes along the upper reaches of the Missouri River. In northwest Canada, Athapaskan-speaking Indians, whose ancestors probably came from Asia much later (perhaps between 7,000 and 10,000 years ago), continued to practise a nomadic lifestyle of hunting and gathering. They moved about in the forest lands stalking big-game animals and made summer hunting trips to the tundra where the caribou had their calving grounds.

Arctic Migration

Arctic Canada was settled much later than the forested lands of the Subarctic. Before people could occupy the Arctic Lands, two developments were necessary. The first was the melting of the ice sheets that covered Arctic Canada. About 8,000 years ago, only small remnants of the great Laurentide ice sheet remained in northeastern Canada, leaving the western Arctic free of the ice sheet. The second development was the emergence of a hunting technique that would enable people to live in an Arctic environment. About 5,000 years ago, the Paleo-Eskimos, who were living along the northeast coast of Asia and in Alaska, had developed an Arctic sea-based hunting technique, and they began to move eastwards along the coast of Alaska and then along the Arctic coast of Canada. This hunting culture of Paleo-Eskimos is known as the Denbigh. Unlike previous marine hunting societies, these people invented a harpoon and other tools that enabled them to hunt seals and other marine mammals along the Arctic coast, though they also relied heavily on terrestrial game such as the caribou. About 3,000 years ago, a second migration from Alaska took place. Known as the Dorset culture, this culture replaced the Denbigh one. The third and final Arctic migration took place roughly 1,000 years ago, when the Thule people spread eastward from Alaska and gradually succeeded their predecessors. The Thule, the ancestors of the Inuit, hunted the bowhead whale and the walrus. However, with the cooling of the climate (known as the Little Ice Age), whales no longer entered the Arctic

A tent ring representing the remains of a Thule house is located near Igloolik, Nunavut. Tent rings represent a site where a Thule family had located their tents. The stone rings mark the edges of the tent skin walls that were weighted down with rocks. (Alec Aitken)

Ocean in large numbers because this ocean was covered by ice for most of the year, so the Thule turned to smaller game such as the seal and the caribou.

Initial Contact

By the time of European contact, the descendants of Old World hunters occupied all of North and South America. North American Indian and Inuit tribes met the European explorers searching for a trade route to the Orient. While the population of these tribes can only be estimated, many scholars now believe there may have been as many as 500,000 Indians and Inuit living in Canada at the time of contact. The greatest concentrations were found along the Pacific coast, where marine resources provide abundant food, and in the Great Lakes/St Lawrence Lowlands, where a primitive form of agriculture supported relatively large sedentary populations. Following contact, their numbers dropped sharply, perhaps declining to 100,000. Loss of hunting grounds to European settlers and the spread of new diseases by explorers, fur traders, and missionaries greatly contributed to this depopulation. In the early seventeenth century, the sudden collapse of Huronia, the most powerful Indian group in the region, took place shortly after contact with the French. Its numbers dropped from 21,000 to less than half this number within a decade. This demographic catastrophe was not unique to Huronia, but it does provide one example of the deadly impact of European diseases and colonial alliances on the Aboriginal peoples. French missionaries brought diseases to the Huron villages, while the Iroquois, who opposed the French/Huron alliance, attacked the Huron villages and eventually destroyed the Huron Confederacy in 1649. In little more

than a generation, Huron villages were abandoned, the cornfields of Huronia reverted to forest, and its people were greatly reduced in numbers and scattered across the land, some finding their way to the north shore of the St Lawrence in Québec, others being captured and assimilated by the Iroquois, and another remnant joining related tribes in what are today Michigan and Ohio.

John Cabot, the first European explorer, reached the east coast in 1497. He was followed by others, including Jacques Cartier and Martin Frobisher. In 1534 Cartier made contact with two Indian tribes along the Gaspé

Relationships between explorers and Aboriginal peoples were not always peaceful. In this painting by John White, the artist records a fight between Frobisher's crew and Baffin Island Inuit. (*Courtesy © British Museum 202756*)

Vignette 3.1   Contact between Cartier and Donnacona

In 1534, Jacques Cartier made contact at the Gaspé coast with Donnacona, the leader of the St Lawrence Iroquois. The Iroquois, whose village (Stadacona) was located at the site of Québec City, came to the Gaspé to fish for mackerel. The next year Cartier returned. After reaching the Iroquois village of Hochelaga, which later became the site of Montréal, Cartier realized that the St Lawrence did not provide a sea passage to China. Like the Spanish in Mexico and Peru who discovered gold and silver, Cartier hoped to find wealth in the New World. In 1541, on his third voyage to Stadacona, Cartier hoped to find diamonds and gold. He left France with several hundred colonists. By then, Donnacona and all but one of the Iroquois captives that Cartier had taken with him when he returned to France in 1536 were dead. The French authorities, who wished to conceal this fact, decided not to allow the surviving Indian woman to return to Stadacona with Cartier. The French arrived at Cap Rouge where they established a settlement just west of Stadacona. When the St Lawrence Iroquoians realized that Cartier was not going to return their chief and the other Iroquois captives, relationships quickly soured. Over the winter, the Iroquois killed at least 35 of the French settlers. Cartier and his surviving party left for France as soon as the river ice melted. He took along rocks that he believed contained gold and diamonds, but, like the ore mined on Baffin Island by Frobisher's men, Cartier brought back fool's gold and quartz. As there seemed to be no prospect for mineral wealth, the king of France lost interest in the New World. Sixty years went by before the French made another attempt at establishing a settlement at Stadacona. In 1608, Samuel de Champlain founded Québec City.

Source: Adapted from Ramsay Cook, *The Voyages of Jacques Cartier* (Toronto: University of Toronto Press, 1993).

coast, and in 1576 Frobisher encountered an Inuit encampment along the Arctic coast of southern Baffin Island. Both explorers were searching for a route to Asia, but instead discovered a new continent and peoples. In both instances, contact with North Americans ended badly. Lives were lost on both sides, and the ore that the explorers took back to Paris and London respectively proved worthless (Vignette 3.1). Instead of gold, they had found fool's gold (iron pyrites).

Culture Regions

At the time of contact with Europeans, Aboriginal peoples in Canada occupied specific territories where most obtained their food from hunting, fishing, and gathering activities. Within each area, these people developed survival techniques suitable for the local environment. Such areas are called culture regions. Aboriginal peoples adapted to seven culture regions: Eastern Woodlands, Eastern Subarctic, Western Subarctic, Arctic, Plains, Plateau, and Northwest Coast (Figure 3.2). The Inuit occupied the Arctic cultural region. In the Eastern Subarctic, the Cree were the principal Algonquian tribe. The Cree in this region had developed a technology—snowshoes—to hunt moose in deep snow. In the Western Subarctic region, the Athapaskans, including a number of Dene tribes, hunted caribou and

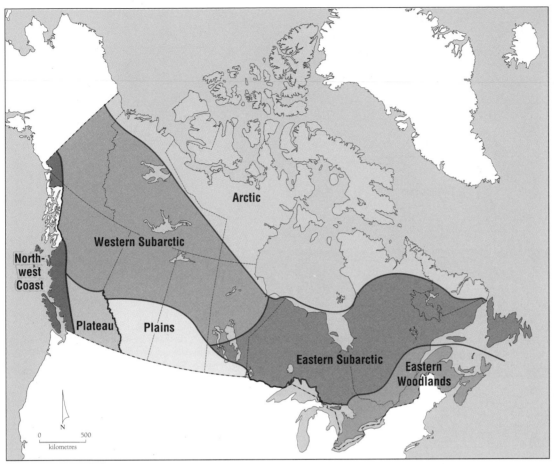

Figure 3.2 Culture regions of Aboriginal peoples. Common resources and natural conditions in certain geographic areas of North America were associated with similar ways of hunting wildlife and organizing economic and social activities.

other big-game animals. Northwest Coast Indians harvested the rich marine life found along the Pacific coast. Tribes such as the Haida, Nootka, and Salish comprised the Northwest Coast cultural region. In the southern interior of British Columbia, the Plateau Indians—including the Carrier, Lillooet, Okanagan, and Shuswap—occupied the valleys of the Cordillera, forming the Plateau cultural region. Across the grasslands of the Canadian West, Plains Indians such as the Assiniboine, Blackfoot, Sarcee, and Plains Cree hunted bison. The Iroquois and Huron lived in the Eastern Woodlands of southern Ontario

and Québec. Both groups combined agriculture with hunting. In the Maritimes, the Mi'kmaq and Maliseet also occupied the Eastern Woodlands where they hunted and fished. The complexity and diversity of Aboriginal peoples can be gleaned from the spatial arrangement of their languages (Figure 3.3).

The Second People

The colonization of North America by the French and the British was the second major development in Canada's early history. France and England established colonies in North

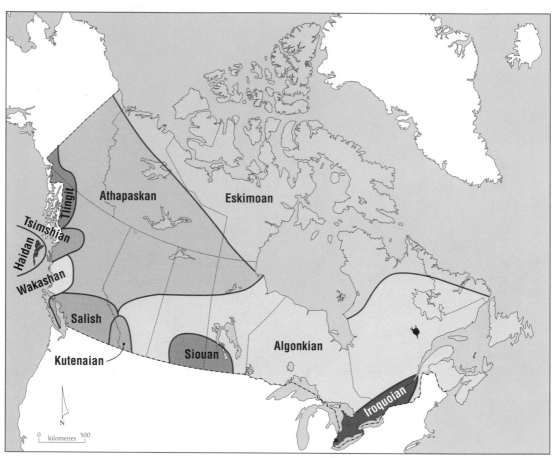

Figure 3.3 Aboriginal language families. Aboriginal peoples in Canada form a very diverse population. At the time of contact, there were over 50 distinct Aboriginal languages spoken. These languages formed 11 language families. Five of these language families are found in one natural cultural region, the Northwest Coast. Following contact, language loss was swift. By the end of the twentieth century, only two Aboriginal languages, Cree and Inuktitut, had over 10,000 speakers. (Further resources: Student Web site, National Atlas section, Map 12. Web site instructions are found on p.xxi.)

America in the seventeenth century. Québec City, founded in 1608 by Samuel de Champlain, was the first permanent settlement in Canada. By 1663, the French population in New France was about 3,000 compared to a population of about 10,000 Indians (mainly Huron and Iroquois), who lived in the same area of the St Lawrence Valley, the Great Lakes, and the Ohio Basin. By 1750, the French Canadians constituted most of the population in New France, where they numbered about 60,000, while the Indian population had dropped sharply because of disease and warfare. Following the British conquest of New France in 1760, the flow of French colonists ceased and British immigrants began to move to what used to be New France. In doing so, they increased the number of British settlers. Meanwhile, the French-Canadian population now had to depend entirely on natural population increase to expand their numbers.

The first large wave of British immigrants consisted of refugees from the United States. These Loyalists had supported Britain during

In 1759, British forces, under the command of General James Wolfe, defeated the French troops led by Louis-Joseph, Marquis de Montcalm, on the Plains of Abraham, which lay just west of the walled capital of New France, Québec City. (*Courtesy National Archives of Canada 128079*)

The American Revolution (1775-83) divided the residents of the Thirteen British Colonies. With the defeat of British forces, those British subjects who did not support the rebel cause were forced to leave, losing their property and sometimes their lives. Considered traitors by Americans, Loyalists were often subjected to mob violence. (North Wind Picture Archives)

the American War of Independence (1775–83). After the defeat of the British army, they sought refuge in other parts of the British Empire, including its North American possessions. In North America, most Loyalists settled in Nova Scotia, while a smaller number moved to the Eastern Townships of Québec to the east and south of Montréal. Others settled in what is now southern Ontario.

The second wave of immigrants came from the British Isles. From 1790 to 1860, almost a million people migrated from the British Isles to British North America, mostly because of deteriorating economic conditions in Great Britain and because of job opportunities in the merchant cod fishery. In the 1840s, the potato famines in Ireland added to the problems in the Old World, causing terrible hardships for the Irish people. Thousands fled the countryside and many left for the New World to settle in the towns and cities of British North America.

These two waves of migration greatly changed Canada. At the time of Confederation in 1867, the population of British North America had reached 3 million. Approximately 80 per cent of these British subjects lived in the Great Lakes–St Lawrence Lowlands, while about 20 per cent inhabited Atlantic Canada.

Across the rest of British North America, Aboriginal peoples made up most of the population. In the Red River Colony, a new Aboriginal people, the Métis, who were of Native and European descent, had emerged. By 1869, the Métis, who were split between French- and English-speaking, greatly outnumbered white settlers and fur traders in the Red River Colony. The colony had a population of nearly 12,000 and the Métis formed over 80 per cent this population (Table 3.2).

But more important, the old balance of demographic power between the French and English had been reversed. British migration to Canada had, over the course of 100 years, changed the demographic balance between French-speaking and English-speaking Canadians. Approximately 60 per cent of the European population was English-speaking. The new demographics also resulted in large English-speaking populations in the principal cities of Québec. Migration had also changed the ethnic composition of the English-speaking population from almost entirely English to a mixture of English, Irish, Scottish, and Welsh. Moreover, by the 1860s, Canada's ethnic character varied by region. In Atlantic Canada, the Scottish and the Irish outnum-

Table 3.2	**Population of the Red River Colony, 1869**	
Ethnic Group	**Population Size**	**Population Percentage**
Whites born in Canada	294	2.5
Whites born in Britain or a foreign country	524	4.4
Indians	558	4.7
Whites born in Red River	747	6.2
English-speaking Métis	4,083	34.1
French-speaking Métis	5,757	48.1
Total Population	11,963	100.0

Source: Lower (1983: 96).

bered the English. In Québec, the English and Irish formed a sizable minority in the towns and cities, though rural Québec remained solidly French-speaking except in the Eastern Townships. Ontario, like Atlantic Canada, was decidedly British.

The separation of the British and French in different geographic parts of British North America ensured the continued existence of two visions of Canada. The greatest threat to Canadian unity stems from these two irreconcilable visions (a further discussion is found in the subsection 'The French/English Faultline'). In 1867, 92 per cent of the population was either British or French. Each linguistic group developed its own vision of Canada: French Canadians saw Canada as two founding peoples, while English-speaking Canadians favoured the notion of equality among provinces. In English Canada, the notion of equal provinces grew out of the following factors:

- The nature of Confederation was such that provincial powers were shared equally.

- The British formed the majority of the population in three of the four provinces, thereby dominating the political affairs of those provinces.
- The British, while a minority within Québec, were the dominant business group.

The Third People

The hegemony of the British and French began to weaken in the early part of the twentieth century when the fertile lands of Western Canada were settled in large part by neither English-Canadian or French-Canadian farmers. In 1870, Ottawa obtained the vast land of the Hudson's Bay Company and was faced with the question of settling the land. Little progress was made until after the completion of the CPR railway in November 1885. Now the time was right for massive immigration to settle these lands. For Ottawa, there were two key advantages in occupying these lands. First, the threat of American settlers moving into the

| Vignette 3.2 | The Land Survey System and the Settling of the Praires |

In 1872, the federal government passed the Dominion Lands Act. This legislation established a system of survey that divided the land into square townships made up of 36 sections, each measuring 1 mile by 1 mile, with allowances for roads. Each section was further subdivided into four quarters, each quarter section measuring $\frac{1}{2}$ mile by $\frac{1}{2}$ mile and comprising 160 acres. This survey system gave the Canadian Prairies a distinctive 'checkerboard' pattern.

Ottawa sought to populate the arable lands of Western Canada by offering 'free' land to farmers. The free land took the form of a homestead (a quarter section or 160 acres). A free grant of one quarter section was available to persons 21 years or older with the payment of a $10 fee. Upon fulfilling cultivation and residency requirements within three years of acquiring the property, the homesteader would receive title to the land. Since Ottawa had made substantial land grants to the Canadian Pacific Railway and to the Hudson's Bay Company, not all the land was free. Both the Canadian Pacific Railway and the Hudson's Bay Company sold their land at market prices.

Table 3.3	Canada's Population by Provinces and Territories, 1901 and 1921			
Political Unit	**Population 1901**	**Per cent**	**Population 1921**	**Per cent**
Ontario	2,182,947	40.6	2,933,662	33.4
Québec	1,648,898	30.7	2,360,510	26.9
Nova Scotia	459,574	8.6	523,837	6.0
New Brunswick	331,120	6.2	387,876	4.4
Manitoba	255,211	4.8	610,118	6.9
Northwest Territories*	21,064	0.4	8,143	0.1
Prince Edward Island	103,259	1.9	88,615	1.0
British Columbia	178,657	3.3	524,582	6.0
Yukon	27,219	0.5	4,147	>0.1
Saskatchewan	91,279	1.7	757,510	8.6
Alberta	73,022	1.3	588,454	6.7
Canada	5,371,315	100.0	8,787,949**	100.0

*Saskatchewan and Alberta did not become provinces until 1905 and were officially included in the population of the Northwest Territories.
**Includes 485 members of the armed forces.
Source: Statistics Canada (2003).

Canadian West and annexing these lands would be diminished. Second, the creation of a grain economy would provide freight for the Canadian Pacific Railway, thereby helping turn it into a viable operation. But where to find the people? Some came from Ontario, Québec, and Atlantic Canada to claim their 160 acres as homesteaders while others came from Britain and the United States (Vignette 3.2).

Still, much of Western Canada was unoccupied. Clifford Sifton, the Minister of the Interior, accepted the challenge to settle the West. By the beginning of the twentieth century, Sifton had launched an aggressive and innovative advertising campaign to lure people from Britain and the United States to 'The Last Best West', but this effort failed to bring sufficient immigrants. At that point, he recognized the need to go beyond these two countries. In a break with past immigration policy, Sifton turned his attention to the people of Central

Europe, Scandinavia, and Tsarist Russia. Land-hungry peasants from the Ukraine formed the largest single group of immigrants but Doukhobors and Mennonites also came, giving Western Canada a distinct and different mix of ethnic groups and landholdings from the rest of the homesteaders, with communal rather than individual landholdings the norm among some groups. Sifton recognized that peasants from Tsarist Russia were well suited for settling the Canadian Prairies, and Sifton was willing to accommodate their religious concerns and social customs, including communal farming practices. From 1901 to 1921, Western Canada's population increased from 400,000 to 2 million (Table 3.3). As these ethnic groups and individuals spread across the Prairies, their impact was enormous on a landscape that only recently was populated largely by vast herds of buffalo and semi-nomadic Indian tribes (see Kerr and Holdsworth, 1990: Plate 17).

By opening the door for immigration from European countries without a French or British background, Sifton's immigration policy changed the face of Canada. His goal of settling the West was accomplished, and a new dimension had been added to Canada's social fabric—people with neither a French nor a British background. While the majority of these immigrants were homesteaders, some took jobs in Canada's boom resource industry while others settled in Canada's major cities, especial Montréal and Toronto. While most soon learned English, this cultural/linguistic difference from the two founding peoples surfaced later in the twenti-eth century as a powerful force for multicultur-alism and pluralism in English-speaking Canada. The immigration to the Canadian Prairies from 1896 to 1914 forms the topic for more detailed discussion under the subheading 'The Immigration Faultline'.

The Territorial Evolution of Canada

A brief presentation of the territorial evolution of Canada marks the start of our account of Canada's formal history as a nation. The British North America Act of 1867 united the colonies

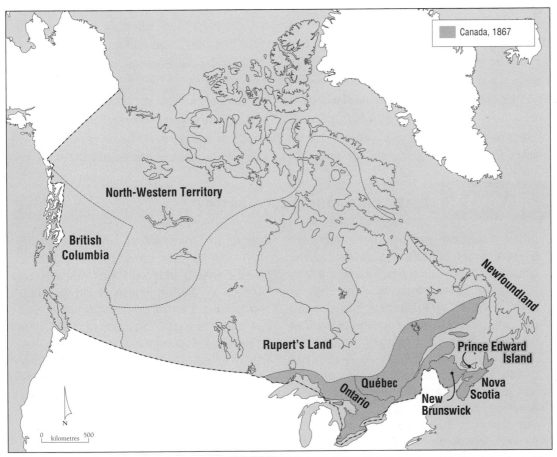

■ **Figure 3.4 Canada, 1867**. At Confederation, Canada was only a fraction of its current territorial extent. The Hudson's Bay Company controlled most of British North America, including Rupert's Land and the North-West-ern Territory. (Further resources: Student Web site, National Atlas section, Map 13. Web site instructions are found on p. xxi.)

of New Brunswick, Nova Scotia, and the Province of Canada (formerly Upper Canada and Lower Canada) into the Dominion of Canada.[2] Thus, Canada began as a small country, consisting of what is now known as southern Ontario, southern Québec, New Brunswick, and Nova Scotia (Figure 3.4).

The British government was a strong advocate of uniting its colonies in North America. For Britain, the union of its North American colonies had three advantages: (1) a better chance for their political survival against the growing economic and military strength of the United States; (2) an improved environment for British investment, especially for the proposed trans-Canada railway; and (3) a reduction in British expenditures for the defence of its North American colonies. The British colonies perceived unification differently. The Province of Canada, led by John A. Macdonald, pushed hard for a united British North America because it would have a larger domestic market for its growing manufacturing indus-

tries and a stronger defensive position against a feared American invasion. The Atlantic colonies showed little interest in such a union. As they were part of the British Empire, the attraction of joining the Province of Canada had little appeal. Furthermore, unlike the Canadians, Maritimers continued to base their prosperity on a flourishing transatlantic trading economy with Caribbean countries and Great Britain. From 1840 to 1870, the backbone of the Maritime economy was the construction of wooden sailing-ships. Shipbuilding was so important that this period was known as the 'Golden Age of Sail' in the Maritimes. Even diplomatic pressure from Britain to join Confederation had little effect on Maritime politicians. But the Fenian raid into New Brunswick in 1866 and the termination of the Reciprocity Treaty with the United States quickly changed public opinion in the Maritimes (Vignette 3.3).[3] Shortly after the Fenian raids, the legislatures of both New Brunswick and Nova Scotia voted to join Confederation.

Vignette 3.3   America's Manifest Destiny

In the nineteenth century, many Americans believed in the doctrine of Manifest Destiny, which was based on the belief that the United States would eventually expand to all parts of North America, thus incorporating Canada and Mexico into the American republic. From its beginnings along the Atlantic seaboard, the United States had greatly increased its territory by a combination of force, negotiation, and purchase. To Americans, such an expansion was an expression of their right to North America. As well, it would rid North America of the much-hated European colonial powers.

Not surprisingly, the Fathers of Confederation were concerned about American

designs on British North America. First, in 1866, the Fenians raided Upper Canada, Lower Canada, and New Brunswick with the intention of seizing British North America and holding it for ransom until Ireland was free of British rule. Second, in 1867, the United States purchased Alaska from the Tsar of Russia, leaving British Columbia wedged between American territory to its north and south. When the United States purchased Alaska from the Russians in 1867, the exact boundary along the coastline south of 60° N was uncertain. Canada and the United States settled this final border dispute in 1903.

Within a decade, the territorial extent of Canada expanded from four British colonies to the northern half of North America. The new Dominion grew in size with the addition of other British colonies and territories. Manitoba and the North-West Territories, which included the Hudson's Bay Company's Rupert's Land and the North-Western Territory in the Far Northwest (1870), British Columbia (1871), and Prince Edward Island (1873) joined the new Dominion (Figure 3.5), while the British government transferred its claim to the Arctic Archipelago to Canada in 1880. Negotiations between Ottawa and each British colony took place and treaties were signed with Indian tribes, but no such negotiations took place with the Red River Métis. The result was the Red River Rebellion of 1869–70, the formation of the Métis Provisional Government, and then negotiations with Ottawa that led to the entry of Manitoba into Confederation.

By 1880, Canada stretched from the Atlantic to the Pacific and north to the Arctic. While still part of the British Empire, Canada

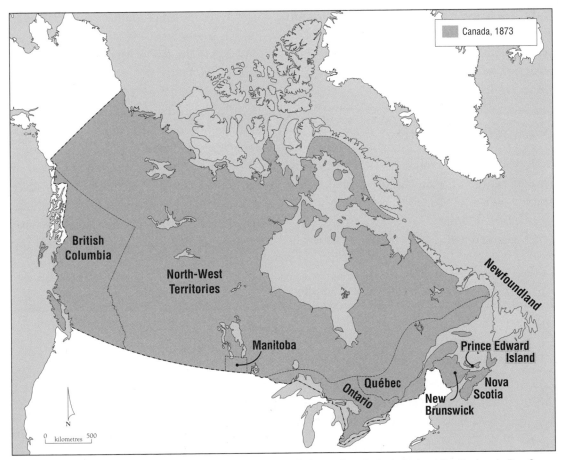

■ **Figure 3.5 Canada, 1873**. Within seven years of Confederation, Canada had obtained the Hudson's Company's lands (including the Red River Colony), and two British colonies (British Columbia and Prince Edward Island) had joined the new Dominion. For the North-Western Territory and Rupert's Land, the Crown paid the HBC £300,000, granted the Company one-twentieth of the lands in the Canadian Prairies, and allowed it to keep its 120 trading posts and adjoining land. In 1870, these lands were renamed the North-West Territories. (Further resources: Student Web site, National Atlas section, Map 13. Web site instructions are found on p. xxi)

had begun the slow journey to independence and nationhood. In 1905 the provinces of Alberta and Saskatchewan were created, and much later, in 1949, Newfoundland joined Canada, completing the union of British North America into a single political entity. (The territorial evolution of Canada is illustrated in Figures 3.4 to 3.8 and Table 3.4.)

Within a decade after Confederation, Canada's territorial extent had burgeoned, making it the second largest country in the world, which posed a serious problem for the nation's leaders, who had to somehow transform this vast territory into a nation. Sir John A. Macdonald, Canada's first Prime Minister, resolved this in part by authorizing the construction of a transcontinental railway, the Canadian Pacific Railway.

National Boundaries

Well before Confederation in 1867, wars and treaties between Britain and the United States shaped many of Canada's boundaries. The southern boundary of New Brunswick, Québec, and Ontario was settled in 1783 when

Britain and the United States signed the Treaty of Paris. Under this treaty, the United States gained control of the Indian lands of the Ohio Basin. Earlier, Britain had formally recognized the rights of Indians to these lands in the Royal Proclamation of 1763, which provided the constitutional framework for negotiating treaties with Aboriginal peoples. This recognition was the basis of Aboriginal rights in Canada (see 'Aboriginal Rights' in this chapter). Based on the fur trade route to the western interior, the boundary of 1783 passed through the Great Lakes to the Lake of the Woods.

In 1818, Canada's southern boundary was adjusted; it was set at 49° N from the Lake of the Woods to the Rocky Mountains. As for the northwestern boundary, there was some question as to where British territory ended and Russian territory began. British and Russians had come into contact through the fur trade. In the eighteenth century, Russian fur traders established trading posts along the Alaskan coast. Indians travelled along the Yukon River through the interior of Alaska to trade at these Russian forts. By the early nineteenth century, the Hudson's Bay Company had reached the

Table 3.4	Timeline: Territorial Evolution of Canada
Date	**Event**
1867	Ontario, Québec, New Brunswick, and Nova Scotia unite to form the Dominion of Canada.
1870	The Hudson's Bay Company's lands are transferred by Britain to Canada. The Red River Colony enters Confederation as the province of Manitoba.
1871	British Columbia joins Canada.
1873	Prince Edward Island becomes the seventh province of Canada.
1880	Great Britain transfers its claim to the Arctic Archipelago to Canada.
1949	Newfoundland and Labrador joins Canada to become the tenth province.

tributaries of the Yukon River. In 1825, Britain and Russia set the northern boundary at 141° W (the Treaty of St Petersburg). Russia also agreed to relinquish to Britain her claims to the coastal regions south of 54°40' N to 42° N. In Atlantic Canada, the boundary between Maine and New Brunswick had not been precisely defined in 1783. Minor adjustments took place in 1842 when Britain and the United States concluded the Webster-Ashburton Treaty.

The last major territorial dispute between Britain and the United States took place over the Oregon Territory. The Oregon Territory stretched along the Pacific Coast from Alaska to Mexico—Mexico at that time extended northward to 42° N. In the early part of the nineteenth century, the Hudson's Bay Company established a fur-trading post at the mouth of the Columbia River, where it conducted trade with the local Indian tribes. In the 1830s, American settlers crossed the Rocky Mountains into the Columbia Valley, where they proceeded to cultivate the fertile soils of the Willamette Valley. Because both Britain and the United States had a foothold in the Oregon Territory, sovereignty was uncertain. Great Britain's claim was based on two factors: that (1) the Hudson's Bay Company had estab-

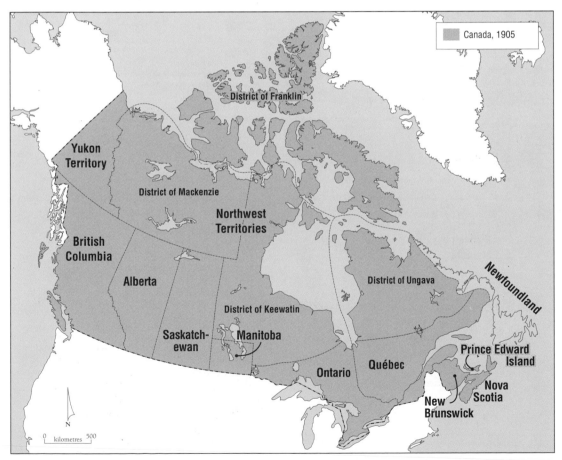

Figure 3.6 Canada, 1905. By 1905, two new provinces (Alberta and Saskatchewan) and one territory (Yukon) were created out of the North-West Territories. In 1905, what remained was renamed the Northwest Territories. Ontario, Québec, and Manitoba expanded their boundaries. (Further resources: Student Web site, National Atlas section, Map 13. Web site instructions are found on p. xxi.)

lished the first settlement (Fort Vancouver) in the region, and (2) the Hudson's Bay Company exerted control over the land where the Indians trapped fur-bearing animals. The American claim was centred on the fact that the vast majority of settlers were Americans. In the final outcome, there was no doubt that occupancy was a more powerful claim to disputed lands than claims based on exploration and the presence of a fur-trading economy. The boundary between British and American territory was set at 49° N with the exception of Vancouver Island, which extended south of this parallel.

Internal Boundaries

Since Confederation, the internal boundaries of Canada have changed (Figures 3.4 to 3.8). These changes have created new provinces (Alberta and Saskatchewan) and expanded the area of existing provinces (Manitoba, Ontario, and Québec). In all cases, these political changes took land away from the Northwest Territories.

In 1870, the boundary of Manitoba formed a tiny rectangle comprising little more than the Red River Colony. The province's western boundary, while extended in 1881 and 1884,

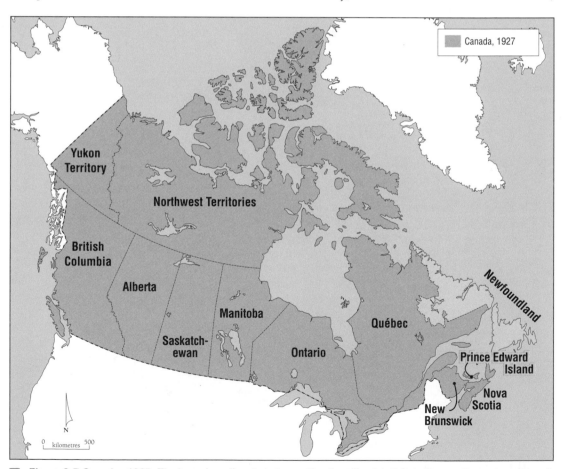

Figure 3.7 Canada, 1927. The boundary dispute between Newfoundland and Canada over the eastern boundary of Québec was resolved in Newfoundland's favour. Ontario, Québec, and Manitoba expanded again and attained their present boundaries. (Further resources: Student Web site, National Atlas section, Map 13. Web site instructions are found on p. xxi.)

did not reach its present limit until 1912. By this time, Manitoba spread north to the boundary with the Northwest Territory (now Nunavut) and east to the Lake of the Woods.

Québec, too, received northern territories. In 1898, its boundary was extended northward to the Eastmain River and then eastward to Labrador. In 1912, Ottawa assigned Québec more territory that extended its lands to Hudson Strait. Canada also believed that the province of Québec should extend to the narrow coastal strip along the Labrador coast, while the colony of Newfoundland contended that Newfoundland owned all the land draining into the Atlantic Ocean. In 1927, this dispute between two British dominions (Canada and Newfoundland) went to London. The British government decided that the boundary between Québec and Labrador was not the narrow coastal strip proposed by Canada but rather the watershed of those rivers flowing into the Atlantic Ocean. While the Québec government has never formally accepted this

ruling, Québec has conducted its affairs with Newfoundland as if Labrador were part of Newfoundland. For example, Hydro-Québec has made arrangements to purchase hydroelectric power produced in Labrador and the two provincial governments have discussed plans to produce more hydroelectric power in Labrador, sell it to Hydro-Québec, and then transmit the power through Québec to markets in New England.

In the years following Confederation, Ontario gained two large areas. In 1899, Ontario had its western boundary set at the Lake of the Woods (previously this area belonged to Manitoba), while its northern boundary was extended to the Albany River and James Bay. In 1912, Ontario obtained its vast northern lands, which stretch to Hudson Bay.

In 1905, Canada formed two new provinces, Alberta and Saskatchewan. The final adjustment to Canada's internal boundaries occurred in 1999 with the establishment

Table 3.5	Timeline: Evolution of Canada's Internal Boundaries
Date	**Event**
1881	Ottawa enlarges the boundaries of Manitoba.
1898	Ottawa approves of extending Québec's northern limit to the Eastmain River.
1899	Ottawa decides to set Ontario's western boundary at the Lake of the Woods and extend its northern boundary to the Albany River and James Bay.
1905	Ottawa announces the creation of two new provinces, Alberta and Saskatchewan.
1912	Ottawa redefines the boundaries of Manitoba, Ontario, and Québec, which extend their provincial boundaries to their present position.
1927	Great Britain sets the boundary between Québec and Labrador. Québec has never accepted this decision.
1999	A new territory, Nunavut, is hived off from the Northwest Territories in the Eastern Arctic.

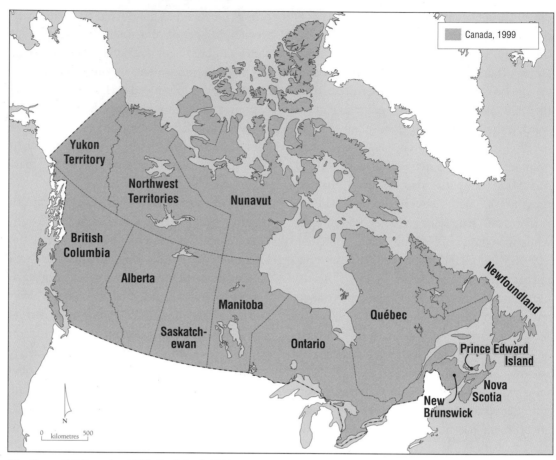

Figure 3.8 Canada, 1999. On 1 April 1999, Nunavut became a territory.

of the government of the territory of Nunavut (Figure 3.8 and Table 3.5).

Faultlines

Canada's regional geography is partly defined by its faultlines—tensions within Canadian society that often begin as a regional issue but soon turn into a national issue. Often these disagreements are between one or more regions and Ottawa. The federal government, because it is charged with establishing national policies and programs, has the power to settle such issues. However, sometimes federal policies and programs seem to favour the two largest provinces, Ontario and Québec.

In the Kyoto agreement signed by Ottawa in December 2002, for example, the federal government saw the reduction of greenhouse gases as beneficial to all Canadians but Alberta saw this agreement as placing it in a disadvantageous position because of its potentially negative impact on Alberta's heavy oil sands plants that discharge great amounts of carbon dioxide into the atmosphere. In their minds, this perception was supported by the exemption of the automobile industry from reducing its greenhouse gas emissions. Other provinces, but particularly Alberta with its large petroleum industry, would have to the bear most of the costs of reducing greenhouse gas emissions, and therefore interpreted

Ottawa's actions as favouring Ontario.

Ultimately, the challenge facing the federal government is to seek a balance between the different factions, but such efforts rarely satisfy all parts of Canada. Nevertheless, the process of searching for solutions to these tensions informs Canadians of different views and, in doing so, lessens the tension—perhaps not initially but usually over time. Two examples are found in the French/English and the centralist/decentralist faultlines. The French/English faultline pits Québec against the rest of Canada. For example, Canada's linguistic geography has the population of one region speaking mainly French while the rest are mainly English-speaking populations. Language is only one element of the French/English faultline, but it is often the catalyst that ignites the flames of separatism and enrages the rest of Canada. In 1969, the federal government made bilingualism an official policy of the Canadian government. Bilingualism was a reaction to French-speaking Canadians' dissatisfaction with their place within Canada. Bilingualism was designed to make French-speaking Canadians feel more at home in different parts of the country and to appease Québec and quell the flames of separatism. Initially, westerners resented bilingualism and a few claimed that 'French was being shoved down their throats'. Some 30 years later, bilingualism is accepted across the country and French-speaking Canadians outside of Québec have access to federal services in their native language. As well, the Acadians in New Brunswick now live in a bilingual province.

The centralist/decentralist faultline takes many forms. In recent years, besides the division over the Kyoto Protocol, the federal gun registration program has created territorial/federal tensions. Ottawa sought to control firearms and, in that way, assist the police to combat crime. In the Territorial North, the three governments oppose gun registration because the vast majority of their residents who hunt for food strongly resent the registration system and some refuse to participate. In addition, the gun control legislation has created divisiveness between Central Canada and Western Canada, as well as between urban and rural Canada.

The Centralist/Decentralist Faultline

Geography poses powerful challenges to Canada's national unity. These challenges stem from Canada's vast space, economic competition between regions, regional trade patterns with the United States, and internal political differences between the provinces and Ottawa. In combination, these challenges to national unity are manifest as regional self-interest that often results in regional tensions. Sometimes these internal tensions are expressed by a group of individuals, such as fishers in Atlantic Canada, or through common interests among provinces, such as the oil-producing provinces. Whatever their form or geographic extent, these internal forces produce regional pressures that can divide the country.

Regional self-interest and resulting tensions (whether cultural, economic, or political) are a natural outcome of Canada's physical and human geography. For better or worse, regionalism is a fact of life in Canada and it may well be the most telling characteristic of Canada's national character. The regional challenge to Canada's east-west alignment, and therefore to Canada's national unity, may be summed up in four ways, namely, those centrifugal forces that:

• separate Canadian regions from each other by great distances, making trade and commerce between those regions more difficult (geographic forces);

- lead regions to compete economically with each other (forces derived from regional self-interest);
- cause political competition over funding between the provinces and Ottawa (forces derived from the division of political powers in the Canadian Constitution);
- draw Canadian regions into the economic orbit of the United States (continental economic forces).

The powerful pull of trade has drawn Canada's regions into the American economy. As a consequence, the natural markets of North America have a north–south orientation. The Maritimes are geographically, historically, and economically tied to New England, while southern Ontario is linked to Michigan, Ohio, Pennsylvania, and New York. British Columbia is part of the Pacific Northwest and indeed has felt part of this international region referred to as Cascadia for many years.

Regional Tensions

The challenge of geography, therefore, has forced Ottawa to seek solutions to the problems bred by regional tensions, which are partly exacerbated by the regional implications of trade with the United States. Political solutions over the years have attempted to overcome geography by binding the new territories to the core of the country and by balancing regional economic disparities.

The first challenge facing the government of Sir John A. Macdonald was clear: a railway must span the country from the Atlantic Ocean to the Pacific Ocean. Without a more substantial presence, Macdonald's Conservative government feared that the US might annex the unoccupied parts of Western Canada (Vignette 3.3). From Ottawa's perspective, the railway would exert effective sovereignty over its western territories. Though the magnitude of this project was nearly beyond the capacity of the country, Ottawa proceeded. Its goals for this transcontinental railway were:

- to link the West with the rest of Canada;
- to settle the Canadian Prairies;
- to provide an export route for Prairie grain;
- to create a market in the West for eastern industries.

At the same time as Ottawa secured its western borders, it launched a new economic initiative in 1879, the National Policy, which reinforced the centralist/decentralist faultline. The National Policy and other federal policies achieved economic growth, but this growth took place at different rates across Canada. Rightly or wrongly, Canadians living outside Central Canada believe that the concentration of political power in Central Canada has had an unfair influence over national policies and therefore favoured economic development in Central Canada over developments in the rest of the country. In the following pages, these two themes—economic and political power struggles—are explored from the perspective of Central Canada (the core) and the rest of Canada (the periphery). In the case of economic power, Central Canada refers to the industrial core in southern Ontario and southern Québec. In the case of political power, Central Canada refers to Ottawa and indirectly to the influence of Ontario and Québec in Ottawa. Together these two power struggles are the basis of the centralist/decentralist conflict or faultline.

Centralists advocate a strong central government, national policies that exert a political dominance over provinces, and a strong national economy (which means an expanding industrial core). Decentralists seek to strengthen the powers allocated to provinces. In particular, decentralists call for a devolution of federal powers to the provincial governments and the expansion and diversification of regional economies. Economic diversification is deemed necessary to generate more jobs and thereby encourage population growth. Eventually, it is thought that economic diversification will lead to larger populations in hinterland regions and thus more political representation, and therefore power, in Ottawa.

Economic Power Struggle

In 1879, following the creation of the National Policy, Canada's manufacturing industries were protected from foreign goods by high tariffs and bolstered by lower customs duties on raw materials. The goal was to strengthen Canada's economy by creating a national industrial base. In fact, this economic policy fostered the development of an economic core and a resource hinterland. For Central Canada, the benefits from the policy were considerable as it ensured the region's role as Canada's industrial heartland. It also resulted in the concentration of financial power in Montréal and Toronto. Other regions were not so fortunate. The hinterland had no such advantage as a result of the federal economic policy. Instead, it was compelled to sell its raw materials and foodstuffs on the world market and buy its manufactured goods in the domestic market. For the hinterland, this arrangement translated into 'selling their products cheap' and 'buying expensive goods from Central Canada'. Feelings ran high because people in the hin-

terland regarded the National Policy as an arrangement that benefited Central Canada at the expense of the rest of the country. They became suspicious of other national programs announced by Ottawa because they appeared to be designed to favour Central Canada and thereby to exploit the other regions.

The National Policy accentuated the economic differences in various parts of the country. In doing so, it increased regional tensions, which resulted in the centralist/decentralist (or heartland/hinterland) divide. While each hinterland region faced a different set of economic relations with the industrial heartland of Canada, the basic economic relationship was the same—Central Canada produced the manufactured goods, while hinterlands produced raw materials and foodstuffs. But what prevented hinterlands from developing a manufacturing base under the National Policy?

Geography prevented the manufacturing companies in the Maritimes from reaching markets in Central Canada and the growing market in the West. While the steel mill in Cape Breton did supply some of the steel rails for the construction of the Canadian Pacific Railway, most Maritime firms could not overcome the distance-cost barrier to reach Central Canada and the West. Maritime industrialists were also no longer able to sell their products in the nearby market of New England because of high American tariffs. The combination of a small local market, great distance to the continental markets in Canada, and high American tariffs stalled economic development in the Maritimes. Before 1867, most Maritimers were uncertain about the benefit of Confederation; afterwards, many became convinced that union with Canada was a 'bad deal' for their region.

At the same time, unrest in Western Canada was widespread in rural areas. Under the national economic policy, western farmers had to sell their grain on the world market where prices were low, but they had to purchase their farm machinery from manufacturers in Central Canada where prices were high. By the 1920s, this unrest took the form of a political protest when the Progressive Party argued for free trade to allow farmers access to lower-priced American farm machinery. Geography compounded the problems facing grain farmers, who were located in the heart of North America. The cost of transporting their grain to foreign markets was very high, thereby making their returns even lower. How could farmers overcome the great distance to their grain markets, particularly those in Europe? They thought that the solution to this transportation problem lay in the construction of a railway to the nearest tidewater point—Hudson Bay! The Hudson Bay Railway, completed in 1929, never fulfilled the high expectations of western farmers because the savings derived from the shorter rail distance were offset by higher marine insurance costs due to the danger of icebergs. Farmers, however, believed the higher insurance charges were simply another trick by Central Canada to hold onto the grain trade at the expense of westerners.

British Columbia, though linked to the rest of Canada by the Canadian Pacific Railway, remained beyond the economic pull of Central Canada. British Columbia's economic order was driven by its geographic position on the west coast. The region's natural markets were overseas. At first, its raw materials, such as fish, forests, and mineral products, were shipped to markets in the western United States, the British Isles, and the Far East. As well, because of British Columbia's geographic proximity to Western Canada, many of the Prairies' products were transported to the port of Vancouver for transshipment overseas. In 1914 the opening of the Panama Canal had a great impact on the forest industry in British Columbia because it provided a 'shorter' shipping route to markets in the British Isles, Western Europe, and the east coast of the United States. Like Canadians in Western Canada, British Columbians resented having to pay for higher-priced manufactured goods from Central Canada compared to those available just across the border.

The Territorial North suffered a different fate. Until World War II, Ottawa simply ignored it. Regarded by the federal government as a remote wilderness inhabited by Indians and Eskimos (later known as Inuit), the Territorial North remained a fur economy and did not participate in the events transforming Canada's emerging industrial society until well into the twentieth century. Ottawa's laissez-faire approach limited federal expenditures to Aboriginal peoples in the Territorial North. The few northern Royal North-West Mounted Police posts (later the Royal Canadian Mounted Police) served as Ottawa's main expression of administration. Only when resource developments occurred, such as the Klondike gold rush, did Ottawa hurry to ensure a broader federal presence. After World War II, the federal government's laissez-faire policy was replaced by state involvement in northern development. This change was sparked by Prime Minister John Diefenbaker's 'northern vision', which translated into investments in northern infrastructure, especially highways leading to resources.

Political Power Struggle

While the core/periphery model emphasizes the core's economic dominance over the periphery, the core also exerts other forms of dominance, including political dominance.

Within the context of Canada, Ottawa not only represents the core but, by favouring Central Canada (Ontario and Québec), extends the advantages of this political dominance to a particular area of Canada. This leads to a sense of alienation and frustration among the remaining regions. For example, in the late 1970s, provincial-federal relations soured during debates over constitutional reform. The western provinces (Alberta, British Columbia, Manitoba, and Saskatchewan), supported by Nova Scotia and Newfoundland, had developed their own agenda for constitutional reform to obtain more powers for the provinces, including a reform of the Senate and the Supreme Court, to reflect regional interests. Québec Premier René Lévesque did not participate in these discussions, leaving it to the anglophone provinces to 'dismantle' Confederation. Ottawa's most consistent ally was Ontario, which confirmed hinterland charges that Ontario was the chief beneficiary of old federalism.

But does Ottawa really favour Central Canada or is it just a problem of balancing national against regional interests? The reality lies in the division of powers between the two levels of government (see Vignette 3.4). Can such a power imbalance exist in a federal state? How much power does a province need? How much power can the central government give up? These questions go to the heart of a federal state and the fragile relations between the central and regional governments. This tug of war is also played by the separatist movement in Québec, but for an end (independence) that is different from that sought by other provinces and premiers (more powers). In 1968, Jacques Parizeau likened the Canadian federation to a chicken being plucked by provincial governments led, of course, by Québec. When the last feather is plucked, the chicken will perish. Voilà, Québec has achieved its goal of independence (Nemni, 1994: 175).

Just how are regions represented in Parliament? Canada's political geography is shaped by the election of representatives from various political parties to the House of Commons.[4] In the 2000 federal election, the number of representatives by provinces and territories was

Vignette 3.4   Federal/Provincial Powers

Under Canada's federal system, the powers of government are shared between the federal government and the 10 provincial governments. All provinces have the same powers. At the time of Confederation, the powers of the provinces were carefully enumerated (and thus limited), while those of the federal government were not limited. Provincial governments are responsible for education, health and welfare, highways, civil law (property and civil rights), local government, and natural resources, while Ottawa has a much wider mandate, including authority over defence and external affairs, criminal law, money and banking, trade, transportation, citizenship, and Indian affairs. The two levels of government were assigned joint jurisdiction over agriculture, immigration, and taxation. Territorial governments are assigned their powers from Ottawa, while municipal governments obtain theirs from the provincial governments. First Nations are seeking self-government. While First Nations' status is not yet defined by the federal government, Indian leaders often refer to a 'third level of government', meaning a political level somewhere between municipal and provincial.

determined by the distribution of population according to the 1991 Census of Canada (redistribution of seats will take place when the 2001 census figures are used to determine the new electoral map of Canada). In the 2000 federal election, the total number of seats was 301 (Table 3.6). The number elected and their electoral districts are determined by the size of Canada's population, while the assignment of electoral districts across the country is determined by Canada's population distribution. There are two exceptions. First, provinces must have at least the same number of seats in the House of Commons as in the Senate. Prince Edward Island has four seats in the Senate and therefore has four seats in the House of Commons, though its population only warrants one or possibly two seats. Second, each territory is allocated one seat irrespective of its population. When the Northwest Territories was split, it was fortunate that of the two elected representatives, one resided in the new NWT and the other in Nunavut. Based on the 1991 census, the electoral district of Nunavut had the smallest population with 21,242 while Calgary Centre had the largest at 117,418. Prince

Edward Island, with four seats, had an average population per seat of 32,441. Since Ontario and Québec have the largest populations, they also have the largest number of seats, and therefore hold the political balance of power. The 2000 federal election results support this proposition, with the majority of seats in the House of Commons held by Ontario with 103 and Québec with 75 for a total of 178 or 59.1 per cent of the total number of seats. This concentration of electoral power is further intensified by a concentration of political power within the governing party, which is naturally concerned with national issues and priorities, not regional ones. From the standpoint of those outside Central Canada, national policies appear to favour Ontario and Québec. The Senate is supposed to provide a regional balance in Canada's political system by giving a political voice to 'regions'. However, since its members are appointed by the Prime Minister largely on the basis of long-time service to the political party currently in power, senators are more concerned about party loyalty than regional interests. In short, the Senate fails to provide a regional counterweight to the House

Table 3.6   Members of the House of Commons by Geographic Regions

Geographic Region	Number of members (seats)	Population 2001	Seats per 1,000 population
Territorial North	3	92,779	1:30
Atlantic Canada	32	2,285,729	1:71
Western Canada	54	5,073,323	1:94
Québec	75	7,237,479	1:97
Ontario	103	11,410,046	1:111
British Columbia	34	3,907,738	1:115

Source: Adapted from Canada (2003).

The confluence of the Red and Assiniboine rivers is known as the Forks. Today, the Forks lies in the heart of Winnipeg. In times past, the strategic location of the Forks provided Plains Indians with ready access by canoe to the lands south of the 49th Parallel and to the vast western interior. In 1738, the French explorer La Vérendrye established Fort Route at the Forks. With the founding of the Red River Colony in 1812, the Forks became the focal point of the Red River Colony. (Ron Garnett/AirScapes)

of Commons. Instead, this role falls to the provincial governments.

The struggle over political power takes place within the six geographic regions of Canada (Ontario, Québec, British Columbia, Western Canada, Atlantic Canada, and the Territorial North), which are far from equal. Ontario and Québec have most of Canada's population, its manufacturing industries, and national corporate headquarters. Of the six geographic regions, Ontario and Québec (Central Canada) are at the top of the economic and political hierarchy. In sharp contrast, the Territorial North is at the bottom. It has no industrial base; it has the smallest population of any

geographic region; and it is far from the decision-making centres in Canada. Using population size and economic power (as measured by gross domestic product) as two indicators of the strength of regional power, the remaining regions fall between Ontario and the Territorial North (Table 1.2).

The distribution of political power takes other forms—indicators of regional power include the richness and sustainability of a resource base, and access to the national, American, and international markets. Perhaps one of the more notable indicators of power is tax rates. The British North America Act gave Ottawa unlimited taxing powers, while provin-

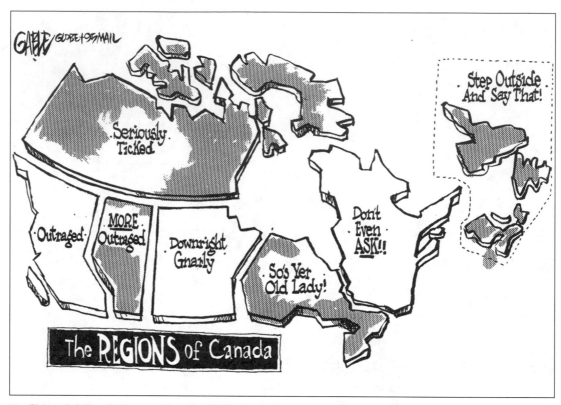

Figure 3.9 The Regions of Canada. Political cartoonist Brian Gable has aptly captured the regionalized tensions that are manifested in Canada.

Source: Globe and Mail, 11 December 1995: A14. Reprinted with permission from The Globe and Mail.

cial governments' ability to raise revenue through taxation is limited. Given that the provinces' levels of economic development vary, the ability to raise revenue through taxation is also uneven. For that reason, the rate of personal taxation varies from province to province. The 'have' provinces—Alberta, British Columbia, and Ontario—have the lowest personal tax rates, while the 'have-not' provinces have much higher rates. Newfoundland/Labrador, for example, has the highest personal income tax rate in Canada, while Alberta has the lowest. The same geographic pattern holds true for provincial sales tax rates: Newfoundland/Labrador has the highest rate, while Alberta has no sales tax.

These spatial variations in political and eco-nomic power make Canada a 'troublesome country to govern'. Each Prime Minister from Sir John A. Macdonald to Paul Martin has had to balance national economic interests against regional ones. Federal interventions in the marketplace are attempts to protect national interests, but they sometimes result in regional alienation.

Western Alienation

Western alienation represents a serious problem for Ottawa. It takes many forms but it is based on western provinces' lack of power, whether real or perceived, to control their destiny. The intensity of this regional alienation increases from Manitoba to British Columbia.

In the federal election of 27 November 2000, this anger was captured by the remarks of Liberal incumbent Raymond Chan, who lost to the Canadian Alliance candidate in the BC riding of Richmond. Chan blamed the electoral system for fuelling western frustration with Ottawa. Simply stated, BC's political power does not match its demographic and economic power. Consequently, Ottawa does not recognize the aspirations and needs of the western provinces. Chan, the former Minister for Asia-Pacific Affairs in the Liberal government, was quoted in the *National Post* (Fife and Bellavance, 2000: A6):

> We have to pay closer attention to the aspirations of the West, but the way I see it, the political structure has a bias toward the East. If you compare the Atlantic provinces, where the population is much smaller, they have 32 seats and in B.C., we have three [four] million people and economically we are very important for the country and yet we only have 34 seats.

Western alienation is a historical fact. At first, western anger was directed at the CPR. Later it was focused on Ottawa. The federal election of 2000 saw few Liberals elected in the four western provinces, indicating that western dissatisfaction with the federal government remained high, but not quite matching the dark days of 1980 when Ottawa introduced the National Energy Program.

National Energy Program

Resource development falls under provincial powers, yet the federal government made two initiatives that affected Alberta's petroleum industry. The first took place in the 1960s when the Diefenbaker government attempted to stimulate oil and gas developments in oil-producing provinces by allocating the Ontario market to the West. The second initiative, the National Energy Program, developed under the government of Pierre Trudeau, was Ottawa's attempt to achieve three goals: energy security, greater Canadian ownership of the oil industry, and a greater amount of wealth produced by the oil industry for Ottawa. The National Energy Program lasted only four years (1980–4), but its chief legacy was the western oil-producing provinces' distrust of the federal government. Why was this so?

Without a manufacturing base, the Canadian hinterland has had to depend on its natural resources as an extremely important source of regional development and provincial tax revenue. Thanks to its geology, Alberta has abundant petroleum deposits. This fact, plus the sudden rise in oil prices in the 1970s, turned Alberta from a have-not province to a have province. Since natural resources fall under the jurisdiction of the provincial governments, the federal government has no authority to 'tax' natural resources, yet with its national energy policy, Ottawa was able to force oil-producing provinces to share their rapidly increasing oil revenue with the federal government. Ottawa acted in this way, knowing full well that it would lead to western alienation, because it believed the National Energy Program was in the national interest.

Before the National Energy Program, the report of the Royal Commission on Energy (the Borden Commission) in 1959 provided the basis for a more favourable federal policy on energy for Alberta. Under the Diefenbaker government, Ottawa created a dual market for oil to promote oil development in Western Canada. The western market, extending from southern Ontario (including the Toronto area) to the Pacific coast, was reserved for oil produced from Canadian wells (primarily from Alberta). The eastern market consisted of eastern Ontario (including the Ottawa area), Québec, and Atlantic Canada. Lower-cost oil from the Middle East and Venezuela supplied the eastern market. The purpose of the dual market was to create a larger market for western oil even though its price in Toronto was 10 to 15 per cent higher than that of offshore oil refined in Montréal. Perspectives on this policy varied: Ottawa saw a decreased dependency on foreign oil; Alberta saw an expanded oil industry; Ontario saw slightly higher oil costs; and Québec, with its large oil refinery complex, saw its natural market area reduced (southern Ontario). Naturally, Québec was not pleased with this federal policy because its Montréal-based petroleum refineries could not sell their products in the Toronto market. As a result, oil refining expanded in Alberta (Edmonton) and Ontario (Sarnia) instead of Québec (Montréal).

In the 1970s, the price of world oil rose rapidly because of the Organization of Petroleum Exporting Countries (OPEC) strategy: by curtailing their supply of oil, the world price would rise. From 1972 to 1980, the world price of oil increased from US $2 a barrel to over US $20 a barrel. From 1973 to 1978, agreements were reached between the oil-producing provinces and the federal government to match the domestic price with the world price. Rising oil prices generated a huge profit for the oil companies and greatly increased oil revenues for Alberta and, to a much lesser degree, Saskatchewan and British Columbia. In 1979, however, Ottawa refused to match domestic oil prices with world prices, thereby creating a lower domestic price. For Central Canada, such a national policy gave its industrial firms a decided advantage compared to those across the US border. With world prices continuing to rise, subsidization of oil refineries in Atlantic Canada and Québec was necessary to meet the lower domestic price. The federal government needed a new energy strategy that addressed two key concerns. How could Ottawa pay for this price differential? Should not part of this 'windfall' that companies and provincial governments receive be redirected to the federal government?

In 1980, the federal government announced the National Energy Program, which enabled Ottawa to obtain a larger share of oil revenues through new taxes on the oil industry and other measures. Alberta objected vigorously along with the other two oil-producing provinces, Saskatchewan and British Columbia. Albertans believed that Ottawa forced oil-producing provinces to accept this program in order to ensure supplies of low-cost oil to Ontario and Québec. In 1982, Peter Smith, a professor of geography at the University of Alberta, presented the Albertan position by denouncing the policy as 'one more manifestation of the customary heartland outlook; what is good for Ontario and Québec has to be best for the rest.' Smith went on to state that:

> To Canada's misfortune, the issue brings the negative side of regionalism to the fore by causing regional sentiment to be focused on a siege mentality rather than on the sense of organized community responsibility a functional region should possess if it is to assume a political identity. The individualism of political regions is not just an Albertan reality; it is a Canadian one. And no misreading of the heartland-hinterland relationship should be allowed to obscure that historical and geographical fact. (Smith, 1982: 301–2)

Current Struggles

The nature of Canadian federalism demands a high level of co-operation between the federal and provincial governments because of shared or overlapping responsibilities, authority, and funding in many areas of public policy. National purposes can often only be achieved with provincial co-operation, yet tensions frequently arise between Ottawa and provincial governments because of shared powers, interdependency, and conflicting goals. Added to this situation, provincial governments are demanding more and more powers that now fall under the federal government's jurisdiction. All of this complicates federal-provincial relations, sometimes leading to bitter political confrontations. Negotiations between the two levels of government have come to an impasse more than once.

At these times, politicians have accused those on the other side of bad faith.

A major bone of contention between Ottawa and the provinces has been funding for a series of 'national' social programs. Social programs are the responsibility of provincial governments, but because these programs are so costly, only the have provinces can afford them. In some cases, Ottawa is directly involved and delivers the program, for example, Employment Insurance and the Canada Pension Plan. In other cases, Ottawa provides financial support to the less affluent provinces and territories through transfer payments, notably equalization payments and block cash transfers for health and post-secondary education under the Established Programs Financing Act. In 2002–3, transfer payments, including equalization payments, totalled $47.6

Vignette 3.5   Equalization Payments

Regional equity is an important theme in Canada. In 1957, the federal and provincial governments recognized the importance of providing similar public services in each of the provinces. Equalization payments would provide the funds necessary to achieve this goal. This program has ensured that per capita revenues of all provinces from shared taxes (personal income taxes, corporate income taxes, and succession duties) matched those of the wealthiest provinces. Until 1995, $200 billion had been transferred from Ottawa to the have-not provinces. About half of these payments had gone to Québec, 35 per cent to Atlantic Canada, and 15 per cent to Western Canada. The territorial governments also received funds ($1.3 billion in 2002–3) from the federal government, but these transfers fell under a separate program (Territorial Formula Financing).

In 2002–3, equalization payments amounted to $10.3 billion. Québec obtained $45.2 per cent, Atlantic Canada 35.8 per cent, Western Canada 14.3 per cent, and British Columbia 4.7 per cent.

Equalization payments and Territorial Formula Financing form a significant portion of the gross revenues for the five eastern provinces and almost the entire revenues for the three territories. While Québec receives the largest amount of equalization payments, these federal transfers amount to about 30 per cent of Québec's gross general revenues. For the four Atlantic provinces, however, these funds form about half of their gross general revenues. Fiscal dependency is even greater in the territories, where transfer payments amount to about 90 per cent of territorial revenues.

billion. Such financial support has enabled all provinces and territories to provide similar public services to their citizens, but the level of fiscal dependence is very high for certain regions. The Territorial North, Atlantic Canada, and Québec (Vignette 3.5) are heavily dependent on transfer payments from the federal government.

Federal funding for programs operating within provinces began in the 1950s with agreements for universities (1951) and national hospital insurance (1958), followed in the 1960s by the Canada Assistance Plan (1966), and universal health-care insurance (1968). The federal government has created an elaborate social safety net through the implementation of these social programs. Canada did not come by this social safety net easily because the division of powers between the federal and provincial governments often stood in the way. The provinces have the sole power to establish social programs, but only the federal government has the ability to pay for them. This seemingly insurmountable political conundrum was not solved until after World War II. As a consequence of extraordinary powers the federal government used for mobilizing and co-ordinating the war effort, after the war Ottawa wielded unprecedented political and financial power over the provinces. By offering to pay for part of these programs through grants, Ottawa was able to 'induce' the provinces to deliver programs with 'national standards'. Such financial dependence forms the basis of the current tension between Ottawa and the provinces.

By the early 1990s, Ottawa could no longer avoid dealing with its massive debt. While previous federal governments had begun the process of reducing this debt by raising taxes and reducing services, the Liberal government elected in 1993 began the job in earnest. Much of the federal debt was reduced by a cut in transfer payments that forced provinces to pay a larger share of their social expenditures, including health-care costs. Ottawa's 1997–8 budget for transfer payments, including equalization payments, was reduced by $5 billion (from $25 billion to $20 billion). Excluding equalization payments, federal transfers now fall under the Canada Health and Social Transfer Act. These transfers fell from $18.5 billion in 1996–7 to $12.5 billion in 1998–9. By 2001, transfers were to stabilize at $11.1 billion. Before the 1997–8 budget year had ended, the severity of these cuts was felt by Canadians supported by social welfare programs, Canadians seeking medical services, and Canadians enrolled in post-secondary education programs.

Provincial governments were furious with these 'unilateral' changes in transfer payments because they now had to make up the shortfall or reduce their services. Provinces chose to reduce services rather than increase taxes, so their health-care programs, social services, and universities suffered. At the end of its 1997–8 budget year, the federal government had eliminated its operating debt and began to produce a substantial surplus. By 2002, Ottawa had increased transfer payments to provincial governments, but provinces and even city governments are demanding that Ottawa hand over more funds. In the 2003 federal budget, Ottawa increased its funding for health care with a promise to increase spending by $30.9 billion over the next five years, and it created a 10-year infrastructure program for cities. The basic problem remains unresolved, however, because provinces are responsible for delivering social programs but do not have the capacity to generate sufficient tax revenue. Similarly, cities must provide an expensive infrastructure to function well but have an even smaller capacity to raise funds.

The Aboriginal/Non-Aboriginal Faultline

In 1867, the British North America Act made Ottawa responsible for the Indian tribes. Later Ottawa's responsibility was extended to all Aboriginal peoples to include the Inuit and the Métis. For that reason, the Aboriginal/non-Aboriginal faultline is cast into an Ottawa versus Aboriginal peoples framework. In this model, Ottawa acts like a core, while Aboriginal Canadians living on the edge of Canadian society serve as the periphery. Originally, Ottawa's objective was the assimilation of Indian peoples into Canadian society. Instead, Aboriginal peoples were marginalized. Since the 1970s, Ottawa has adopted a more enlightened policy, recognizing two facts: Aboriginal rights, and the failure of the federal government's assimilation policy. Aboriginal peoples' struggle for power is rooted in two questions: Who are the Aboriginal peoples of Canada? What are Aboriginal rights?

Aboriginal Peoples

The Canadian Constitution Act of 1982 refers to Indians, Métis, and Inuit under the umbrella term **Aboriginal peoples**, that is, those now living in Canada who can trace their ancestry to the original inhabitants who were in North America before the time of contact with Europeans in the fifteenth century. Indians are further distinguished between status, non-status, and treaty Indians. People legally defined as **status (or registered) Indians** are registered or entitled to be recorded as Indians, according to the Indian Act as amended in June 1985, and have certain rights acknowledged by the federal government, such as tax exemption for income generated on a reserve. In 1980, 317,000 Canadians were status (or reg-

istered) Indians. By 2002, the number of status Indians had grown to almost 700,000. **Non-status Indians** are people of Indian ancestry who are not registered as Indians and therefore have no rights under the Indian Act. **Treaty Indians** are status or registered Indians who are members of (or can prove descent from) a band that signed a treaty. They have a legal right to live on a reserve and participate in band affairs. Less than half live on reserves. The **Métis** are people of European and North American Indian ancestry. The **Inuit** are Aboriginal people located mainly in the North. In 2001, 1,319,890 people reported Aboriginal ancestry but only 976,305 claimed to have an Aboriginal identity. This latter figure was composed of 608,850 Indians, 292,310 Métis, and 45,070 Inuit (Statistics Canada, 2003).

The Indians, Inuit, and Métis are a highly diversified population. One indication of their cultural diversity is linguistic classification. There were approximately 55 distinct Aboriginal languages (of 11 language families) spoken in Canada at the time of original contact. Five of the language families were spoken along the Pacific coast, while only two were spoken east of Manitoba and one (Inuktitut) in the Arctic (Figure 3.3). The largest language family is Algonkian. There are 15 distinct Algonkian-based languages, the most common of which are Cree and Ojibwa. Inuktitut, the Inuit language, has regional dialects and is spoken across the Canadian Arctic.

Another measure of Aboriginal diversity is self-identification. Many Indians prefer to identify themselves with the name of their tribal group, such as Cree or Iroquois, while others use the name of their First Nation (band) for a more precise identification. For example, the Cree occupy a vast territory that stretches from northern Québec to Alberta. There are many Cree tribes within that territory. A Cree living in northern Saskatchewan might identify himself or herself as a member of a Cree band, such as the Lac la Ronge Indian band.

In 2002, the largest First Nation was southern Ontario's Six Nations of the Grand River with a population of 21,618. The next five largest were the Mohawk of Akwesasne (9,771) at St Regis on the Ontario–Quebec border near Cornwall, the Blood (9,358) in southern Alberta, Kahnawake (9,902) near Montréal, Saddle Lake (7,941), Cree reserve outside Edmonton, and the Lac La Ronge Cree in northern Saskatchewan (7,459). The majority do not live on their band's reserve, but live off-reserve in urban centres.

Aboriginal peoples are reclaiming their identity and place names. Some bands are relinquishing the names given to them by Europeans in favour of their original names, such as Anishinabe (for Ojibwa) and Gwich'in (for Kutchin). The landscape is also being reclaimed. For example, the Arctic community of Frobisher Bay, named after the English explorer, Martin Frobisher, is now Iqaluit, which means 'the place where the fish are', the capital city of Nunavut ('our land').

Aboriginal Rights

Aboriginal rights are group or collective rights that stem from Aboriginal peoples' occupation of the land before contact. Aboriginal peoples' traditional attitudes and values towards land and wildlife are strikingly different from those held by most non-Aboriginal Canadians. For Canada's First Peoples, the land has not only economic value but also cultural, political, and spiritual value. Aboriginal attitudes and values are based on their former economic and social systems, that is, the subsistence hunting system that was in place before contact with Europeans.

Land rights are the most fundamental Aboriginal rights. Indeed, it is from land rights that most other collective rights flow, such as self-determination and self-government. Aboriginal peoples first began to secure land rights before Confederation and this process continues today. Some Aboriginal peoples, including the Métis, are still negotiating with Ottawa.[5] Treaties set aside land for Indian bands, land that is called a **reserve** and is held collectively by and for the benefit of the band.

The reasons for signing treaties varied depending on the historical context. Authorities acting on behalf of the Crown often signed treaties to secure Aboriginal peoples as allies during times of turmoil, such as the War of 1812, or simply to procure land for the growing numbers of settlers coming to Canada. During the expansion to the West, treaties were signed to provide a place for Indian tribes so that Indian wars common south of the border would not erupt in Canada. For Aboriginal peoples, treaties often promised reserves that would not be available to settlers and support during the transition from semi-nomadic hunting to sedentary farming. The numbered treaties for the Plains Indians therefore offered protection from the anticipated flood of settlers and some guarantee that the federal government would care for them now that their principal source of food, the buffalo, was gone. However, treaty assurances of federal assistance were often not met (see Carter, 2004; Brownlie, 2003).

The terms of each treaty varied, although they generally included cash gratuities and presents at the signing of the treaty, annual payments in perpetuity, the promise of educational and agricultural assistance, the right to hunt and fish on Crown land until such land was required for other purposes, as well as land reserves to be owned by the Crown in trust for the Indians. In Treaty No. 6, for example, the amount of land assigned to each tribe was determined by its population size, i.e., each family of five received one square mile. Reserves are collectively owned by First Nations bands, though legally they are owned by the Crown in trust for them.

Conflicting ideas as to the significance of treaties between the signing parties largely shaped Aboriginal and non-Aboriginal relations in Canada during the twentieth century. When treaties were signed, Crown authorities viewed them as vehicles for extinguishing Aboriginal rights and titles to land and thus for opening the land to agricultural settlement. Aboriginal peoples, however, viewed them as agreements between 'sovereign' powers to share land and resources. With such diverse perceptions of treaties, disagreements were inevitable. Added to the issue of perception, some Aboriginal groups never signed treaties, while other treaties were never fulfilled. The latter part of the twentieth century witnessed various movements to repair injustices related to Aboriginal land rights (for example, through land claims) and to recognize Aboriginal peoples' inherent right to self-determination (for example, through constitutional reform). Through these attempts and through the signing of modern treaties, Aboriginal peoples are seeking a new place in Canadian society (Vignette 3.6) in the hope of participating in the larger economy while still retaining their culture.

From Hunting Rights to Modern Treaties

History sometimes makes strange allies. Shortly after Pontiac, chief of the Odawa, led a successful uprising against the British in 1763, Britain decided to form an alliance with him and other Indian leaders.[6] For strategic reasons, George III issued the Royal Proclamation of 1763, which identified a part of British territory west of the Appalachian Mountains as Indian lands. At that time, Britain believed that it could claim 'uninhabited land', which the British defined as land without permanent occupation (that is, no cultivated land or permanent settlements). However, the British also believed that Indians had a limited ownership over the lands they inhabited, and that therefore such lands must be purchased from the owners. This somewhat ambiguous concept remains the basis of land claims by Canadian Aboriginal peoples (Figure 3.10).

The legal meaning of Aboriginal title to land has evolved over time. Until the 1970s, Ottawa recognized two forms of land rights. Reserve lands were one type of right or ownership, which the Canadian government held for Indian people. The second type was a usufructuary right to use Crown land for hunting and trapping. At that time, Crown lands (both provincial and federal) included most of Canada's unsettled areas. Indian, Inuit, and Métis families lived on Crown lands, continuing to hunt, trap, and fish. However, federal and provincial governments could sell such lands to individuals and corporations or grant them a lease to use the land for a specific purpose, such as mineral exploration or logging, without compensating the Aboriginal users of those lands. By the 1960s, many Aboriginal groups still did not have treaties with the Canadian government. Atlantic Canada, Québec, the Territorial North, and British Columbia contained huge areas where treaties had not yet been concluded. As a consequence, Aboriginal peoples had no control over developments on these lands.

A combination of events radically changed this situation. One factor was the emergence of

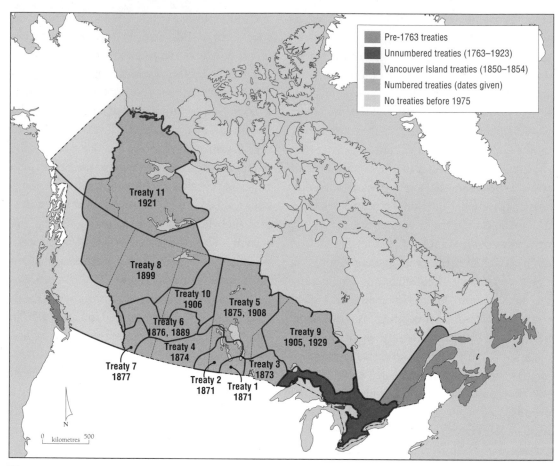

Figure 3.10 Historic treaties. The first treaties, made between the British government and Indian tribes, were 'friendship' agreements. In Upper Canada the Robinson (or unnumbered) treaties set aside reserve lands in exchange for the title to the remaining lands. With the settlement of lands in the Canadian West, Indians became concerned about their future, so many numbered treaties included provisions for agricultural supplies. By 1921, when the last numbered treaty was signed, many Aboriginal peoples in Atlantic Canada, Québec, and British Columbia were without treaties. (Further resources: Student Web site, National Atlas section, Map 14. Web site instructions are found on p. xxi.)

Native leaders who understood the political and legal systems. They used the courts to force the federal and provincial governments to address the issue of Aboriginal rights and land claims. The first major event took place in the late 1960s. In 1969, proposed reforms of the Indian Act (known as the White Paper) galvanized treaty Indians into action. The federal government White Paper proposed to treat all Canadians equally. For Indians, it meant the abolition of their treaty rights and the reserve land system. About the same time, the Nisga'a in northern British Columbia took their land claim, known as the *Calder* case, to court. In 1973, the Supreme Court of Canada narrowly ruled against (by a vote of four to three) the Nisga'a argument that the tribe still had a land claim to territory in northern British Columbia. However, in their ruling, six of the seven judges agreed that Aboriginal title to the land had existed in British Columbia at the time of Confederation. Furthermore, three judges said that Aboriginal title still existed in British Columbia because it had not been extinguished by the British Columbia government, while three other judges stated that Aboriginal title had been extinguished by the various laws passed by the British Columbia government since 1871. The seventh judge ruled against the Nisga'a claim on a legal technicality. The Supreme Court's narrow verdict and the legal opinion of three judges that Aboriginal title still existed changed the course of Aboriginal land claims in Canada. Now the federal government agreed that Aboriginal peoples who have not signed a treaty may very well have a legal claim to Crown lands.

The James Bay Project in northern Quebec and the proposed Mackenzie Valley Pipeline Project in the Northwest Territories added fuel to the political fire over Aboriginal rights. The possible impact of these industrial projects on Aboriginal peoples was made clear through the Mackenzie Valley Pipeline Inquiry of 1974–7 (the Berger Inquiry) into possible environmental and socio-economic impacts and in the media. It was obvious that Aboriginal organizations were prepared to take action to defend their land claims. Their position in the 1970s was 'no development without land-claims settlements'. All these events changed both the public's views of Aboriginal rights and the government's position. At first grudgingly and then more willingly, governments, corporations, and Canadian society recognized the validity of Aboriginal land claims. The James Bay and Northern Québec Agreement in 1975 was the first modern land-claim agreement in

---

Vignette 3.6   Modern Treaties

Modern treaties began in 1975 with the signing of the James Bay and Northern Québec Agreement (Figure 3.11). Since then, all modern treaties are either specific or comprehensive agreements between an Aboriginal group and the federal government. **Specific land-claim agreements** attempt to rectify shortcomings in the original treaty agreement with a band or seek to redress failure on the part of the federal government to meet the terms of the treaty. By 1990, over 500 specific claims had been filed with Ottawa. Most claims involved relatively small amounts of land. However, in 1992, a major agreement was signed between 26 First Nations in Saskatchewan and Ottawa. Called the Settlement of Treaty Land Entitlement in Saskatchewan, it involved a payment of $445 million over 12 years to allow these First Nations to purchase land that they should have received at the time of treaty or that they lost through Ottawa's mis-

handling of their affairs.

A **comprehensive land-claim agreement** is sought when a group of Aboriginal people, who have not yet signed a treaty, can demonstrate a claim to land through past occupancy. In 1984, the Inuvialuit of the Western Arctic became the first Aboriginal people to settle a comprehensive land claim with the federal government. In exchange for surrendering their Aboriginal claim to all this land, they received 91,000 km², $45 million (in 1977 dollars) in financial compensation, and guaranteed rights over resource management. Since 1984, comprehensive agreements have been signed between the government and the Gwich'in (1992), Sahtu (1993), Inuit of the Nunavut Settlement Area (1993), and the Yukon First Nations (1993). A comprehensive agreement with the Nisga'a (2000) marked the first modern land-claim agreement in British Columbia.

Canada (Figure 3.11). Since then, seven comprehensive claims have been finalized in northern Canada: (1) the Inuvialuit Final Agreement (1984); (2) the Gwich'in Final Agreement (1992); (3) the Sahtu Final Agreement (1993); (4) the Nunavut Final Agreement (1993); (5) the Yukon Umbrella Final Agreement (1993); (6) the Nisga'a Final Agreement (2000); (7) the Dogrib (Tlicho) Final Agreement (2003). The Labrador Inuit Agreement is expected to be ratified in 2005.

Many claims remain unsettled, including claims in British Columbia and Labrador. Until

they are concluded, relations between those pursuing agreements and the federal government will remain strained. For example, virtually the entire province of British Columbia, except for Vancouver Island, is claimed by First Nations. In British Columbia, progress has been slow. Until 1992, the provincial government believed that British occupancy had extinguished Aboriginal title. However, in 1992, the British Columbia government accepted the principle of Aboriginal land claims. The following year, Ottawa and Victoria agreed to a formula for paying outstanding

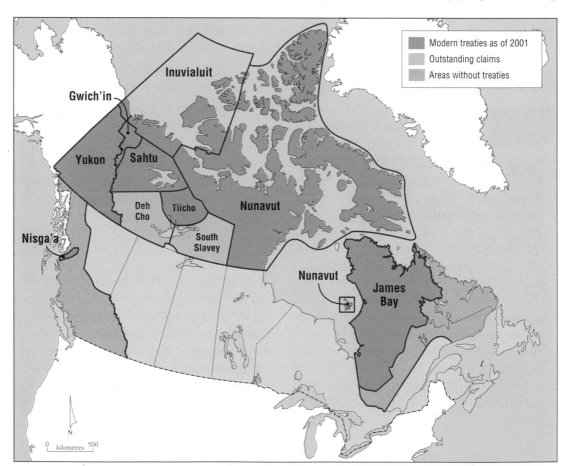

Figure 3.11 Modern treaties. The first modern treaty, the James Bay and Northern Quebec Agreement, was signed in 1975. Since then, modern treaties have fallen into two categories: comprehensive and specific. By 2001, the main areas without treaties were much of British Columbia, Labrador, and lands in Québec. (The original inhabitants of Newfoundland, the Beothuk, had perished from disease, encroachment, and slaughter by the early nineteenth century.)

land claims. The federal government would pay 90 per cent of the money needed to settle outstanding claims and the province would provide the land. In 2000, the Nisga'a agreement was finalized, while other First Nations are negotiating with the British Columbia Treaty Commission. In a backward step, the new BC Liberal government called for a referendum in 2001 on the issue of Indian land claims. By 2003, however, the same British Columbia government announced its desire to settle outstanding claims.

Those Aboriginal groups who have concluded modern treaties are moving forward. They are able to focus on economic and cultural developments rather than expending their energies on land-claim negotiations. In 1993, the Nunavut agreement broke new ground by effectively establishing self-government over an entire territory. Since then, modern land-claim agreements, such as the Nisg'a Agreement in 2000, have included arrangements for self-government. As a result, a gap is emerging within the Aboriginal community between those who have a modern treaty and those who do not. Also, as with countries and with regions, some Aboriginal groups reside on lands that are rich in natural resources, resource developments, and development potential (e.g., oil and gas deposits, oil sands, pipelines, prime timber land) that provide a base for economic growth and considerable wealth, while many other groups live in areas with little resource potential where even subsistence from the land is marginal if not impossible.

Bridging the Aboriginal/Non-Aboriginal Faultline

Aboriginal peoples are taking the control of their affairs away from Ottawa. Some Indian and Inuit peoples have made substantial advances in economic development, while others have moved into the area of self-government. Unfortunately, some Aboriginal peoples, including the Métis, have not yet begun this process and remain on the margins of Canadian society. For most, fortunately, the process of change has started.

For historic reasons, while some Aboriginal peoples have made treaties, others have not. This difference is significant because a treaty, particularly a modern treaty (a comprehensive or specific land-claim agreement), provides the land, capital, and an administrative organization necessary to initiate this process of economic, social, and political change. In 1996, the report of the Royal Commission on Aboriginal Peoples identified two major goals: Aboriginal economic development and self-government. The gap between Aboriginal and non-Aboriginal societies will not be bridged until these goals are achieved. The principal factor is transferring power from Ottawa (the political power core) to the various Aboriginal communities (the politically weak periphery).

How well each Aboriginal community will manage its affairs is unknown, but breaking the dependency on the federal government will be an important step. Such a step will result in a new and more positive relationship between Aboriginal and non-Aboriginal Canadians. That a new relationship is unfolding is demonstrated by the new power to co-assess with the federal government industrial impacts on the land and people in the settlement areas of comprehensive land claims. The Labrador Inuit provide another example when they gained a share of the royalties from the Voisey's Bay nickel mine within their land-claim agreement.

The Immigration Faultline

The history of immigration to Canada is a complex and sometimes controversial topic. In the historic past, immigration was often an instrument of colonial powers. After the British conquest of New France, for example, the British government set the immigration policy and the French-speaking majority in Canada did not have a say in the shaping of this policy. This power lay exclusively with the colonial power. The British government's objective was to offset the large French-speaking population by encouraging large-scale immigration from the British Isles and outlawing immigration from France. In the case of the Acadians, the British colonial powers, beginning in 1755, deported many of these people to England and other English colonies and many others fled to Québec and to the Louisiana Territory. At the same time, the British sought to resettle the area with British subjects. Colonial-style immigration, therefore, not only generated tensions between the existing population and the newcomers, but it also imposed a way of life and a set of institutions on the existing population and often marginalized these people.

While the existing population historically has viewed newcomers by asking how the new immigrants will benefit them and their society, the colonial power has taken the opposite position, i.e., how will the existing population benefit us? The economic, military, and social relationship between New France and the Huron Confederacy illustrates this point. In 1609, the Huron chiefs met with Samuel de Champlain to discuss both trade and a military alliance. The Huron had three objectives: (1) to gain access to European goods, including firearms, by supplying the French with beaver pelts; (2) to improve their material well-being with the trade goods and, in turn, trade these goods to more distant Indian tribes for profit; and (3) to strengthen their military position against their traditional enemies, the Iroquois, who were allied with the Dutch and later the English traders based in New York. The French had two objectives: (1) to secure a supply of furs; and (2) to convert the Huron to Christianity. At the height of the fur trade in the seventeenth century, New France greatly prospered and the Huron accounted for around half of the furs shipped to France (Dickason, 2002: 104). Trade was so important to the Huron tribes that when the French insisted that the Huron allow Jesuit missionaries to live among them as a condition for continued trade, the Huron reluctantly agreed. Unfortunately, the missionaries brought with them smallpox and other diseases that quickly swept through the Huron tribes, causing a sharp decline in their population. Another factor leading to the demise of Huronia was the French reluctance to trade firearms for furs. Since the English traders had no such hesitation, the Iroquois, though smaller in numbers, quickly became more powerful than the Huron and, with that military superiority, were able to harass the Huron fur brigades travelling to New France and later to attack and destroy Huron villages.

In this chapter, our attention is focused on the impact of immigration on Manitoba following the purchase of Hudson's Bay lands by Ottawa, and then on the subsequent settling of the Canadian Prairies by many people who were not of British ancestry. This historic period stretches from 1870 to 1914. During this time, although the face of colonialism had changed it had not softened, so immigrants had to conform to Canadian society. The experiences of the Métis of the Red River Colony and the Doukhobors—very clearly not British immigrants—exemplify this demand to conform to the majority society.

Manitoba: The Métis against the Newcomers

With the transfer of the vast lands administered by the Hudson's Bay Company, Canada changed from a small territory to a truly continental country. While the boundary between Western Canada and the United States was determined by 1874, the survey of lands for agricultural settlement took place in the 1880s. The survey system, based on a township and range model, stamped a rectangular-shaped grid on the cultural landscape, thus determining the shape and placement of farms, roads, and towns (Vignette 3.2). As Moffat (2002: 204) points out, this survey system 'enabled the division of western lands among the HBC, the Canadian Pacific Railway (CPR) and homesteaders, and set aside two sections in each township for the future of local education.' However, Ottawa failed to recognize the landholdings of the Métis and relegated Indians to reserves. Ottawa did not inform the residents of the Red River Colony of its plans for the Red River Colony nor did the government recognize local landholdings. Events quickly spun out of control, resulting in the Red River Rebellion.

The Red River Rebellion pitted the existing population of the Red River Colony against the newcomers whose agricultural system posed a deadly threat to the existing Métis hunting economy. Even before the arrival of settlers, surveyors sent by Ottawa ignored the long lot holdings of the Métis along the Red and Assiniboine rivers. In 1869, the Red River Colony was the only settled area of any size in the North-West Territories, with a population of nearly 12,000 evenly divided between French- and English-speaking residents (Table 3.2). Most consisted of mixed-blood people born of French and British fur traders and Indian mothers who had settled in long lots along the banks of the two major rivers, and whose economy was based on the buffalo hunt and subsistence farming. By early 1869, news of the pending transfer of Hudson's Bay Company lands to Ottawa had reached the colony, and the arrival of land surveyors resulted in open hostility. When Canadian land surveyors began to survey lands occupied by the Métis, the Métis feared for their rights to those lands and even their place in the new society. Matters came to a boil when, in October 1869, Louis Riel put his foot on the surveyor's chain and told them to leave. Thus, the Red River Rebellion began.

Two months later, the Métis under Riel formed a provisional government and began to negotiate with Canada over the terms of entry into Confederation. During these negotiations, the concerns of the Métis were recognized. The Manitoba Act that led to the admission of Manitoba into Confederation as Canada's fifth province ensured the use of English and French languages within the government of the new province and established a dual system of Protestant and Roman Catholic schools.

Within a decade, the population balance was broken due to an influx of immigrants from Ontario, many of whom either belonged to or supported the views of the Orange Order, a Protestant fraternal organization with strong anti-Catholic beliefs. Some newcomers saw no place for the Métis and Indians in the emerging society, thus creating tensions between the existing population and the newcomers. From 1871 to 1881, Manitoba's population increased from 25,228 to 62,260, with most immigrants coming from Ontario, the British Isles, and the United States (Table 3.8). At this time, those of British ancestry formed 54 per cent of the population, other Europeans made up 17 per cent, Métis, 17 per cent, and Indians, 11 per cent (Census of Canada, 1882: Table III). The newly formed English-speaking majority focused their attention on the dual school system. By 1891, Manitoba's population exceeded 150,000 (Table 3.7). In an example of the tyranny of the majority, the English-speaking population argued that with so few French-speaking students funding for the Catholic school was not warranted. In 1890, the government of Manitoba abolished public funding for Catholic schools. This decision took on national significance by becoming a critical issue between Québec and the rest of the country (see below).

What caused this influx? One reason was that Ontario no longer had a surplus of agricultural land and sons of farmers looked to the unsettled lands in the Great Plains of the United States and to Manitoba. A second reason was that the promise of a railway would make farming in Manitoba more viable. However, the completion of the Canadian Pacific Railway from Fort William on Lake Superior to

Selkirk, Manitoba, was delayed and was only completed in 1882. By that time grain could be transported by rail and ship to eastern Canada and Great Britain rather than by the more circuitous steamship route to St Paul, Minnesota, and then by rail to New York. Wheat farming in Manitoba had become a profitable business because of advances in agricultural machinery, farming techniques, and rising prices for grain. Equally important, new strains of wheat, first Red Fife and then Marquis, both of which ripened more quickly than previous varieties of wheat, thus lessened the danger of crop loss due to frost. Marquis wheat, which matured seven days earlier than Red Fife, allowed wheat cultivation to take place in the parkland belt of Saskatchewan and Alberta where the frost-free period was shorter than in southern Manitoba.

The Doukhobors

The settling of Western Canada represents both a major migration effort and the formation of a new cultural landscape. By 1871, the newly founded country of Canada had three

Table 3.7	Population in Western Canada by Province, 1871–1911		
Year	Manitoba*	Saskatchewan*	Alberta*
1871	25,228		
1881	62,260	21,652	9,875
1891	152,506	40,206	26,593
1901	255,211	91,279	73,022
1911	461,394	492,432	374,295

*The boundaries of Manitoba did not reach their present limits until 1881. Since Saskatchewan and Alberta became provinces in 1905, their populations for 1881, 1891, and 1901 were calculated from the censuses of Canada for 1881 and 1891.

Sources: Census of Canada 1880–81, vol. 1: 93–6; Census of Canada 1890–91, vol. 1: 112–13; Statistics Canada (2003).

distinct rural landscapes—Ontario, with its small but contiguous farms concentrated in the fertile lowlands bordering Lake Erie and Lake Ontario and stretching north to the Canadian Shield; the Maritimes, with small pockets of farmland found in the river valleys of the hilly Appalachian Mountains, as well as the more extensive agricultural areas in Prince Edward Island and the Annapolis Valley; and Québec, with its long lots along the St Lawrence River. The emerging cultural landscape of Western Canada took two forms—the rectangular appearance of its rural landholdings and the symbols of ethnic diversity as expressed by farm buildings and churches.

Many immigrants came from non-English-speaking European countries, marking a break from the traditional sources of immigration, the British Isles and the United States. Others came from Ontario and, to a lesser degree, from other provinces. Within two decades of entering Confederation, Manitoba's population had increased by just over 600 per cent (Table 3.7). Most were of British stock, but substantial numbers of Mennonities and Icelanders had also come to Manitoba. At the same time, few settlers had reached Saskatchewan and Alberta, though the Métis had relocated in Saskatchewan, primarily around the settlement of Batoche on the South Saskatchewan River just north of Saskatoon. By 1895, Western Canada had a predominantly British population that had established political and social institutions.

In the next decade, the volume of immigrants from Central Europe, Scandinavia, and Tsarist Russia increased substantially. As peasants, they were prepared for the harsh physical conditions associated with breaking the virgin prairie land and were willing to deal with the psychological stress of living on isolated farmsteads in a foreign country where

their native tongue was not accepted. As their numbers grew, the anglophone majority became concerned about these newcomers and their possible effect on the existing social structure. The demographic impact of the non-British migration to Western Canada is shown in the 1916 census (Table 3.8).

This wave of Central Europeans had tremendous implications for Western Canada. While most newcomers assimilated into the larger society, a few did not. Often these ethnic groups settled in one area where they were somewhat insulated from the larger society and where they attempted to maintain their traditional customs, language, and religion. The federal government, by providing land reserves for ethnic groups such as the Mennonites and Doukhobors, reinforced this tendency.

While they were successful farmers, the cultural differences between the more conservative Doukhobors and Canadian society were too great for the majority society to accept. While some Doukhobors were able to integrate into local society, the Community Doukhobors were not able to adapt. The Community Doukhobors remained faithful to their religious beliefs that emphasized communal living and that minimized dealings with the state, including informing the state about births and deaths of their members.

These people were deeply religious Russian peasants who rejected both the practices and beliefs of the Russian Orthodox Church and the secular authority of the Tsar. They were communalists and pacifists, and by refusing to serve in the army of Tsarist Russia they were considered by the state as outlaws and therefore needed to be punished. Consequently, they were persecuted by both the church and the state. Seeking to be left alone, the Doukhobors sought a place far from the forces of authority where they could practise their religion and communal lifestyle. In choosing to settle in Canada they were granted blocks of land and exemption from military service.

In 1899, the Doukhobors arrived in Canada and took possession of lands in Saskatchewan, and they soon built 57 villages. These lands were selected by representatives of the Doukhobors before the peasant settlers arrived in Canada. Through negotiations with the Canadian government, the Doukhobors had obtained four large blocks of land totalling 750,000 acres. The four colonies were located just west of Swan River, Manitoba (North Colony), and at Prince Albert (Saskatchewan Colony) and Yorkton, Saskatchewan (South Colony and Good Spirit Lake Annex). Ottawa had allowed them to receive blocks of land rather than individual homesteads, thus facilitating the building of villages and the establishment of a communal society. Farming was not only an economic activity, but it was also central to their religious beliefs, which emphasized the value of a simple, communal life. For example, Doukhobors shared in the returns from farming. In fact, no one person owned the land or

Main Ethnic Group	Western Canada	Percentage	Manitoba	Saskatchewan	Alberta
British	971,830	57.2	57.7	54.5	60.2
German	136,968	8.1	4.7	11.9	6.8
Austro-Hungarian	136,250	8.0	8.2	9.1	6.4
French	89,987	5.3	6.1	4.9	4.9
Russian	63,735	3.7	2.9	4.5	3.8
Norwegian	47,449	2.8	0.6	4.2	3.4
Indian	39,147	2.3	2.5	1.7	2.9
Ukrainian	36,103	2.3	4.1	0.7	1.8
Swedish	37,220	2.2	1.4	2.5	2.7
Polish	27,790	1.6	3.0	1.0	0.9
Jewish	23,381	1.4	3.0	0.6	0.6
Dutch	22,353	1.3	1.3	1.4	1.3
Icelandic	15,800	0.9	2.2	0.5	0.1
Danish	9,556	0.6	0.3	0.5	0.9
Belgian	9,084	0.5	0.8	0.4	0.4
Italian	5,348	0.3	0.3	1.0	0.9
Other	26,219	1.5	0.9	1.5	2.3
Total	1,698,220	100.0	100.0	100.0	100.0

Table 3.8   Population and Ethnicity, Western Canada, 1916

Source: Census of the Prairie Provinces, 1916, Table 7.

the tools. In a land of individual landholdings and the pursuit of profit, the Doukhobors were seen as 'out of step' with the surrounding community.

As the land around their villages and allotments was settled by other newcomers, the Doukhobors came into closer contact with their neighbours, who often coveted the uncultivated areas found on the edges of the land reserves of the Doukhobors. Two factors were at play in their not cultivating all of their land. First, the grant of land reserves was quite generous and went far beyond their immediate needs. Second, their compact communal settlements and intensive land cultivation near their villages resulted in a particular land-use pattern that left lands far from the village underutilized. In sharp contrast, homesteaders were required to cultivate land on their quarter-sections and to erect farmhouses, thus creating the checkerboard pattern of rural settlement. Neighbours were also puzzled by and perplexed with their communal way of life that separated them from the rest of the population. As public resentment increased, the federal government took action. In 1905, Frank Oliver succeeded Clifford Sifton as Minister of the Interior. Oliver decided to enforce the Dominion Lands Act, so when the Doukhobors refused to swear an oath of allegiance to the Queen, Oliver had his excuse.

Failure to take such an oath had two implications. First, it suggested that these people were disloyal to the Queen. Second, it meant that the Doukhobors could not obtain title to their homestead lands. Under this pretext, Oliver used the Dominion Lands Act to cancel their right to land. A hard core of Doukhobors remained committed to the communal way of life, most of whom eventually moved to British Columbia, while others abandoned the village life and took title to homesteads. The villages gradually lost members and lands. The South Colony just north of Yorkton, Saskatchewan, was the last holdout, but by 1918 it ceased to exist on Crown land. It persisted in a much reduced area on purchased land until 1938, as did other communal settlements established in the Kylemore and Kelvington areas.

One explanation for the ultimate failure of Doukhobor experiment was that Canada's model of individual settlement was simply too rigid to accept a communal one. Yet with ownership of land, the Doukhobors could establish agricultural villages. Another explanation focused on their unwillingness to swear allegiance to the Queen. In reality, the Doukhobors were successful farmers and their villages were working. However, they could not fit into the existing culture, which required conformity to the laws—and the informal values and lifeways—of the country. Primarily for that reason, the Community Doukhobors were unable to find a place in Western Canada. They represent a classic example of a people being too different—too 'other'—from the majority to be allowed a comfortable space within the cultural landscape. Ironically, the village model of settlement was perhaps the most effective way of settling the Prairies in the late nineteenth and early twentieth centuries. Professor Carl Tracie (1996: xii) puts it this way:

> At the very time when the individual homesteader was struggling with the very real problems of isolation and loneliness, the Doukhobor settlements, whose compact form allayed these problems, were being dismantled by forces which could not accommodate the communal aspects of the group. Also, although the initial

government concern was the survival of the Doukhobors, their very prosperity, based as it was on communal effort, may have worked against them since it illustrated the success of a system diametrically opposed to the individualistic system dictated by government policy and assumed by mainstream society.

The French/English Faultline

Although the ancestors of Aboriginal peoples were the first to occupy North America, two European powers—the French and the British—colonized Canada. Following the British military victories, the Treaty of Paris confirmed British hegemony over a French-Canadian majority. This historic fact underscores the dominating position of the British and a fundamental weakness in the vision of Canada as two founding nations. Relations between the French and English in North America began nearly 400 years ago. These two cultures have come to represent a major faultline in Canadian society. Since 1841, however, these two communities have had to work together, each dependent on the other. This interaction has done much to shape the cultural and political nature of Canada. Over the years, they have accomplished much together. Nevertheless, significant differences between the two communities exist and, from time to time, these differences flare into serious misunderstandings. Without a doubt, Canadian unity depends on the continuation of this relationship and the need for compromise. A brief examination of that relationship, as outlined in the following pages, will lead to a fuller understanding of the contemporary version of French/English differences, conflicts, and compromises.

The serious nature of this rift has profound geopolitical consequences for Canada. It is therefore crucial that we understand the origins and nature of this faultline. A well-known Canadian political columnist, Jeffrey Simpson, wrote:

> We can also hope that, in the 1980s, Canadians gained a deeper understanding of the faultlines running through their society, and that they will avoid measures that widen them, thereby concentrating on making new arrangements and reforming old ones, so that what the rest of the world rightly believes to be a successful experiment in managing diversity will endure and prosper. (Simpson, 1993: 368)

The beginning of formal French-English relations in Canada stretches back to the British conquest of the French on the Plains of Abraham in 1759, an event that remains a dark page in French-Canadian history. In 1760, the French Canadians watched the remnants of the French army and the French élite board ships to return to France. The French Canadians had no thought of leaving as they were born in the New World, but what would happen to them under British military rule? Would they, like their Acadian brethren, be deported to other British colonies? Britain did not need to take such drastic action as there was no French threat to Britain's North American possessions. In the 1763 Treaty of Paris, France ceded New France to Britain, which placed the French-Canadian majority under the British monarchy. While the English lived in cities in Québec and dominated the Québec economy, French Canadians lived mostly in rural areas where they successfully maintained their culture within a British North America, an achievement that was made possible because of two factors. First, the French Canadians were a large homogeneous population that occupied a contiguous geographic area. Second, Britain forged a close relationship with the former élites of New France (the Roman Catholic clergy and the landed gentry) to ensure the French Canadians' loyalty because Britain wanted to secure its northern colony against its restless American colonies to the south. This relationship between Britain and the French Canadians would be strengthened with the Québec Act of 1774.

The Québec Act, 1774

With the Québec Act of 1774, the individuality and separateness of Québec was recognized, thus ensuring its unique place in British North America. This Act is sometimes described as the Magna Carta for French Canada.[7] Its main provisions ensured the continuation of the aristocratic seigneurial landholding system and guaranteed religious freedom for the colony's Roman Catholic majority, and, by implication, their right to retain their native language.[8] This gave the most powerful people in New France a good reason to support the new rulers. The Roman Catholic Church was placed in a particularly strong position. Not only was the Church allowed to collect tithes and dues but its role as the protector of French culture went unchallenged. Therefore, the clergy played an extremely important role in directing and maintaining a rural French-Canadian society, a role further enhanced by the Church's control of the education system. The *habitants* (farmers) were at the bottom of French-Canadian society's hierarchy. They formed the vast majority of the population and continued to cultivate their land on seigneuries, paying their dues to their lord (seigneur) and faithfully obeying the local

priest and bishop. The British granted another important concession, namely, that civil suits would be tried under French law. Criminal cases, however, fell under English law.

The seigneurial system formed the basis of rural life in New France and, later, in Québec. In 1774, there were about 200 seigneuries in the St Lawrence Lowlands. This type of land settlement left its mark on the landscape (the long, narrow landholdings and the vast estate of the seigneur) and on the mentality of rural French Canadians (close family ties, strong sense of togetherness with neighbouring rural families, and staunch support for the Church). A *habitant*'s landholding, though small, was the key to his family's prosperity, and by bequeathing the farm to his eldest son the *habitant* ensured the continuation of this rural way of life. In 1854, the *habitant* was allowed to purchase his small plot of land from his seigneur, but the last vestiges of this seigneurial system did not disappear until a century later. Even today, the landscape in the St Lawrence Lowlands shows many signs of this type of landholding.

While the heart of this new British territory was the settled lands of the St Lawrence Lowlands, its full geographic extent was immense. Essentially, the Québec Act of 1774 recognized the geographic area of former French territories in North America. Québec's territory in 1774 was extended from the Labrador coast to the St Lawrence Lowlands and beyond to the sparsely settled Great Lakes Lowlands and the Indian lands of the Ohio Basin.

The Loyalists

The American War of Independence changed the political landscape of North America. Within the newly formed United States, a number of Americans, known as the Loyalists, remained loyal to Britain. Like the French-speaking people in North America, most of these Loyalists were born and raised in the New World. For them, North America was their homeland. During the revolution, they had sided with the British. They were hated by the American revolutionaries and lost their homes and property. As they were not welcome in the new republic, many Loyalists resettled in the remaining British colonies in North America, where Britain offered them land. The majority (about 40,000 Loyalists) settled in the Maritimes, particularly in Nova Scotia. About 5,000 relocated in the forested Appalachian Uplands of the Eastern Townships of Québec. A few thousand, including Indians who had supported Britain, took up land in the Great Lakes Lowlands along the northern shores of Lake Ontario. These Loyalists strengthened Britain's hold on its North American possessions, but those who settled in the major cities of Québec and in the Eastern Townships came from a different cultural world than the local francophone residents. Social and political tensions arose from time to time between the two cultural groups.

Within a few decades, the English-speaking settlers in the Great Lakes Lowlands grew in number. Soon they sought to control their own affairs where they could have a more British government with British civil law, British institutions, and an elected assembly. In the Constitutional Act of 1791, Québec was split into Upper and Lower Canada.

The Constitutional Act, 1791

In 1791, the British government passed the Constitutional Act (Canada Act) in an attempt to satisfy the political needs of its French- and English-speaking subjects. These were the main provisions of the Act: (1) the British colony of Québec was divided into the provinces of Upper and Lower Canada, with the Ottawa River as the dividing line, except for two seigneuries located just southwest of the Ottawa River; and (2) each province was governed by a British lieutenant-governor appointed by Britain. From time to time, the lieutenant-governor would consult with his executive council and acknowledge legislation passed by an elected legislative assembly.

In 1791, Lower Canada had a much larger population than Upper Canada. At that time, about 30,000 colonists lived in Upper Canada, most of whom were of Loyalist extraction, with about 10,000 Indians, some of whom had fled northward after the American Revolution. Lower Canada's population was about three times larger. It consisted of about 100,000 French Canadians, 10,000 English Canadians, and perhaps as many as 5,000 Indians.

Following the Constitutional Act, Upper and Lower Canada each had an elected assembly, but the real power remained in the hands of the British-appointed lieutenant-governors. In Lower Canada the lieutenant-governor had the support of the Roman Catholic Church, the seigneurs, and the Château Clique. The **Château Clique**, a group consisting mostly of anglophone merchants, controlled most business enterprises and, as they were favoured by the lieutenant-governor, wielded much political power. In Upper Canada the **Family Compact**—a small group of officials who dominated senior bureaucratic positions, the executive and legislative councils, and the judiciary—held similar positions in commercial and political circles. While these two elite groups promoted their own political and financial well-being, the rest of the population grew more and more dissatisfied with blatant political abuses, which included patronage and unpopular policies that favoured these two groups. Attempts to obtain political reforms leading to a more democratic political system failed. Under these circumstances, social unrest was widespread.

In 1837 rebellions broke out. In Lower Canada Louis-Joseph Papineau led the rebels, while William Lyon Mackenzie headed the rebels in Upper Canada. Both uprisings were ruthlessly suppressed by British troops. The goal of both insurrections was to take control by wresting power from the colonial governments in Toronto and Québec and putting government in the hands of the popularly elected assemblies. In Lower Canada the rebellion was also an expression of Anglo-French animosity. While both uprisings were unsuccessful, the British government nevertheless sent Lord Durham to Canada as Governor-General to investigate the rebels' grievances. He recommended a form of responsible government and the union of the two Canadas. Once the two colonies were unified, the next step, according to Durham, would be the assimilation of the French Canadians into British culture. When Durham left in 1838, a second rebellion broke out in Lower Canada, but it was as unsuccessful as the first.

The Act of Union, 1841

In response to Durham's report, in 1841 the two largest colonies in British North America, Upper and Lower Canada, were united into the Province of Canada. This Act of Union gave substance to the geographic and political realities of British North America. The geographic reality was that a large French-speaking population existed in Lower Canada, while an English-speaking population was concentrated in Upper Canada. The political reality was twofold. Both groups had to work together to accomplish their political goals and neither group could achieve all its goals without some form of compromise. When the two cultures were forced to work together in a single legislative assembly, a new beginning to the French/English faultline surfaced.

The new governor, Sir Charles Bagot, was appointed by and reported to the Colonial Office in London. The governor had the authority to appoint members to a Legislative Council and an Executive Council. The only representative body was a Legislative Assembly. Even though Lower Canada (after union known as Canada East) was somewhat larger in economic strength and population size (670,000) than Upper Canada (after union known as Canada West, with a population of 480,000), each elected 42 members to the Legislative Assembly. The vast majority of inhabitants in Canada East and Canada West lived in the rural countryside. For example, in 1841, the principal towns of Montréal and Toronto had populations of about 40,000 and 15,000 respectively.

Demographic Shifts

By the time of the Constitutional Act in 1791, the balance of French- and English-speaking inhabitants of British North America had begun to tilt more and more in favour of the English. This demographic shift began after the American Revolution when thousands of Loyalists from the former American colonies relocated in British North America. In 1791, the European population of British North America was about 225,000 (mostly French Canadians). Some 162,000 (72 per cent) lived along the St Lawrence River in what is now the province of Québec. Perhaps as many as 50,000 (22 per cent) lived in Atlantic Canada. The remaining 15,000 (6 per cent) were scattered along the north shores of Lake Erie and

Lake Ontario in what is now part of the province of Ontario. By this time, the Aboriginal population of Upper Canada, Lower Canada, and Atlantic Canada had declined substantially to about 25,000.

Within 50 years, not only had this geographic pattern changed but the balance of demographic power had shifted. While their numbers increased due to high fertility rates, French Canadians were no longer the majority in British North America because of the flood of British immigrants. In 1841, British North America had about 1.5 million inhabitants, 45 per cent of them located in Lower Canada, about 33 per cent in Upper Canada, and 12 per cent in the Atlantic colonies. Over the next 30 years, the balance continued to swing in favour of English-speaking regions, thanks to massive immigration from the British Isles. By 1871, Québec's population was only 34 per cent of Canada's population (Table 3.9).

As the country expanded its boundaries and more land was settled, Canada's English-speaking population grew, while Québec's French-speaking population diminished in proportion. Manitoba joined Confederation in 1870 with a population of almost 12,000, which was comprised largely of French- and English-speaking Métis. In 1871, British Columbia, with an estimated population of 28,000 British subjects, became a member of the Dominion of Canada. Beyond these provinces, Indian and Inuit peoples inhabited the land. The total Aboriginal population of all territory that would eventually become Canada was about 100,000 at the time of Confederation. In the subsequent decades, these new lands would be settled by Canadians, Europeans, and Americans. With few exceptions, English became the adopted language of these settlers. For a while, Manitoba was an excep-

tion, but when the English-speaking majority gained control of the government and the public institutions, the Métis found it difficult to maintain their culture and language.

Strained Relations

During these formative years, the Dominion underwent several events that seriously strained relations between its two founding peoples. Two cultures, French and English, were in firm opposition to each other. In the settling of the West, these two cultures clashed over language and religious rights. Four events illustrate the intensity of this power struggle:

- the Red River Rebellion, 1869–70;
- the North-West Rebellion, 1885;
- the Québec Jesuits' Estates Act, 1888;
- the Manitoba Schools Question, 1890.

The Red River Rebellion, 1869–70

In 1868, the British government passed the Rupert's Land Act, which would transfer the Hudson's Bay Company lands to the Crown, and in November of the following year the HBC signed the deed of transfer. At first, nothing changed for the Aboriginal peoples inhabiting the vast grasslands west of the Great Lakes, but Ottawa's plan was to settle these lands. With the arrival of surveyors and then settlers, the world of the hunters and trappers soon disappeared. The first to sense this threat were the French-speaking Métis who lived near the Red River. They became alarmed by the arrival of the land survey teams from Ottawa. The Métis of Assiniboia feared that they might lose their culture, religion, and freedom to hunt buffalo on the open Prairies, so they reacted swiftly. The Métis rebellion, led by Louis Riel, soon became a national issue, reopening differences between English, Protestant Ontario and

Table 3.9	Population by Colony or Province, 1851–1871 (percentages)		
Colony/Province	**1851**	**1861**	**1871**
Ontario	41.1	45.2	46.5
Québec	38.5	36.0	34.2
Nova Scotia	12.0	10.7	11.1
New Brunswick	8.4	8.1	8.2
Manitoba			< 0.1
British Columbia			< 0.8
Total per cent	100.0	100.0	100.0
Total population	2,312,919	3,090,561	3,525,761

Source: Wayne W. McVey and W.E. Kalbach, *Canadian Population* (Toronto: Nelson Canada, 1995), 38. Reprinted with permission of ITP Nelson.

French, Roman Catholic Québec.[9] Québec considered Riel a French-Canadian hero who was defending the Métis, a people of mixed blood who spoke French and followed the Catholic religion. Ontario, on the other hand, considered Riel a traitor and a murderer. For Canada, the larger issue was the place of French Canadians in the West. A compromise was achieved in the Manitoba Act of 1870. Accordingly, the District of Assiniboia became the province of Manitoba. Under this Act, land was set aside for the Métis, and the elected legislative assembly of Manitoba provided a balance between the two ethnic groups with 12 English and 12 French electoral districts. Equally important, Manitoba had two official languages (French and English) and two religious school systems (Catholic and Protestant) financed by public funds.

The North-West Rebellion, 1885

During the 1870s, many Ontarians settled in Manitoba while some Métis sought a new home on the open Prairie. Seeking to remain hunters, one group settled along the South Saskatchewan River where they established a Métis colony around Batoche, about 60 km northeast of what is now Saskatoon. Batoche became the new centre of the French-speaking Métis in Western Canada. As settlers spread into Saskatchewan, the Métis again feared for their future. In 1884, a party of Métis went to Montana where Louis Riel, their old leader, was living and pleaded with him to return to Batoche and lead them again.

Convinced of his destiny, Riel accepted this challenge. Late in 1884, Riel sent a petition to Ottawa with various demands for all the inhabitants of the North-West—Indians, Métis, and whites. After Ottawa ignored his petition, Riel established a provisional government and began to organize the Métis into armed bands. Ottawa responded by sending a militia to suppress the rebellion. The militia travelled from Ontario to Saskatchewan in eight days, thanks to the new railway. Within a relatively short period, the Canadian militia captured Batoche, took Riel prisoner, and defeated the Indians, who were led by Plains Cree Chiefs Poundmaker and Big Bear. For

Québec, the defeat of the Métis and the subsequent hanging of their leader, Louis Riel, not only represented a defeat for a French presence in the West but also widened the gulf between French and English Canadians.

The Québec Jesuits' Estates Act, 1888

In the nineteenth century religious and linguistic intolerance was widespread. For example, Protestant extremists in Ontario were ready to pounce on any perceived injustice to their cause. The Jesuits' efforts to obtain financial compensation for lands that the British took from them in 1763 and later transferred to Lower Canada proved to be such a case.

The Jesuit estates, which were granted under the French regime and used for schools and missions, were appropriated by Britain after the British conquest and given to Lower Canada in 1831. In 1838, Catholic bishops petitioned unsuccessfully for the return of the Jesuit estates. After Confederation, the ownership of the estates passed to the Québec government, with which the Jesuits began negotiating in 1871 for financial compensation. However, the archbishop of Québec argued that the money should be divided among Catholic schools rather than given in its entirety to the Jesuits, who wanted to establish a university in Montréal that would compete with Québec's Université Laval. Québec Premier Honoré Mercier asked Pope Leo XIII to act as arbiter in the dispute among the Roman Catholic hierarchy. In 1888, Québec's Legislative Assembly passed the Jesuits' Estates Act, which determined the division of the financial compensation: $160,000 went to the Jesuits, $140,000 went to the Université Laval, and $100,000 went to selected Catholic dioceses.

Ontario's Orange Order, a Protestant fraternal society, opposed the settlement, arguing that the arbiter, Pope Leo XIII, was an intruder into Canadian affairs and that public funds should not be used to support Catholic schools. In March 1889, the House of Commons debated the motion to disallow the Québec Jesuits' Estates Act and eventually voted against this motion. Similar anglophone, Protestant, anti-Catholic sentiment surfaced in Manitoba regarding the Manitoba Schools Question.

The Manitoba Schools Question, 1890

The British North America Act of 1867 established English and French as legislative and judicial languages in federal and Québec institutions. The remaining three provinces (New Brunswick, Nova Scotia, and Ontario) had only English as the official language. The question of French language and religious rights in acquired western territories first arose in Manitoba.

The French/English issue became the focal point for the entry of the Red River Colony (now Manitoba) into Confederation. Local inhabitants—mostly French-speaking, Roman Catholic Métis, and the less numerous English-speaking Métis—were determined to have some influence over the terms that would include their community as part of Canada. One of their concerns was language rights, which was ultimately resolved when a list of rights drafted by the provisional government became the basis of federal legislation. When the Red River Colony entered Confederation in 1870 as the province of Manitoba, it did so with the assurance that English- and French-language rights, as well as the right to be educated in Protestant or Roman Catholic schools, were protected by provincial legislation.

During the 1870s and 1880s, a large number of Anglo-Protestant settlers, mainly from Ontario, moved to Manitoba, causing the proportion of Anglo-Protestants in the population to increase and the proportion of French and Roman Catholic inhabitants to decrease. This change in the demographics created a stronger Anglo-Protestant culture in Manitoba. In 1890, the provincial government ended public funding of Catholic schools. From Québec's perspective, this legislation led to the most significant loss of French and Catholic rights outside of Québec.

In 1897, the Prime Minister, Sir Wilfrid Laurier, negotiated an agreement with the government of Manitoba. The compromise, which satisfied neither group, allowed for the teaching of Catholic religion in a public school when there were sufficient Catholic students. Similarly, if there were sufficient French-speaking students, classes could be taught in French.

One Country, Two Visions

The greatest challenge to Canadian unity comes from the cultural divide that separates French- and English-speaking Canadians and their respective visions of the country. In the early years of Confederation, events such as the Red River Rebellion, the North-West Rebellion, the Jesuits' Estates Act, and the Manitoba Schools Question widened the French/English faultline. For French Canadians these events demonstrated the 'power' of the English-speaking majority and their unwillingness to accept a vision of Canada as a partnership between the two founding peoples. The root of each vision lies in the history of Canada.

One vision of Canada is based on the principle of two founding peoples. This vision originated in French-Canadian historical experiences and compromises that were necessary for the sharing of political power between the two partners. This vision began with the conquest of New France in 1760, but its true foundation lies in the formation of the Province of Canada in 1841. From 1841 onward, the experience of working together resulted in a Canadian version of cultural dualism.

Henri Bourassa, an outstanding French-Canadian thinker (and Canadian nationalist) in the early twentieth century, was a strong advocate of cultural dualism. He wrote, 'My native land is all of Canada, a federation of separate races and autonomous provinces. The nation I wish to see grow up is the Canadian nation, made up of French Canadians and English Canadians' (quoted in Bumsted, 1998: 253). Bourassa argued that a 'double contract' existed within Confederation. Even today, Bourassa's 'double contract' is an essential element in the two founding peoples concept. He based the notion of a double contract on a liberal interpretation of section 93 of the BNA Act, which guarantees denominational schools. Bourassa expanded the interpretation of the religious rights to include cultural rights for French- and English-speaking Canadians. In more practical terms, Bourassa regarded Confederation as a moral contract that guaranteed French/English duality, the preservation of French-speaking Québec, and the protection of the language and religious rights of French-speaking Canadians in other provinces.

From a geopolitical perspective, Canada is a bicultural country. In one part the majority of Canadians speak English, and in another part the majority speak French. For example, French culture predominates in Québec and has a strong position in New Brunswick. In addition to provincial control over culture, two other geopolitical factors ensure the dynamism of French in those provinces. One factor is the large size of Québec's population—the vitality of Québécois culture is one indication of its success. The second factor is the geographic concentration of French-speaking Canadians in Québec and adjacent parts of Ontario and New Brunswick. For instance, in New Brunswick the French-speaking residents, known as Acadians, constitute over one-third of the population. Federal bilingualism policies instituted in the late 1960s and 1970s also helped rejuvenate francophone minorities. Before these policies, assimilation into the much larger English-speaking culture had seriously weakened the position of francophones in all the provinces, except Québec, and in the two territories. While the attraction of joining the dominant anglophone culture remains, financial support from Ottawa for French educational and cultural facilities in the English-speaking provinces has ensured a place for bilingualism in all provinces and territories.

The Royal Commission on Bilingualism and Biculturalism was an attempt to bridge the gap between English and French Canadians. This Commission, set up in 1963, examined the issue of cultural dualism, that is, an equal partnership between the two cultural groups. But by the 1960s, Canada's demographics revealed a third ethnic force and the concept of duality no longer reflected reality. English-speaking Canada had changed. English-speaking Canada had evolved from a predominantly British population to a more diverse one with several large minority groups who also spoke other languages besides English, namely, German and Ukrainian. Ottawa, in searching for a compromise, announced two policies, bilingualism and multiculturalism.

In the second vision, Canada consists of 10 equal provinces—yet this, too, is complex. On the one hand, this vision represents the simple notion based on provincial powers granted under the British North America Act, which ensured that Canada consists of a union of equal provinces, all of which have the same powers of government. Nonetheless, by assigning provinces, including Québec, powers over education, language, and other cultural matters within their provincial jurisdictions, Québec's French culture was secure from political tampering by the anglophone majority in the rest of Canada. Confederation then provided a form of collective rights for French culture within Québec. Under Canada's federal system, the powers of government are shared between the federal government and 10 provincial governments. But are all provinces really equal? As noted earlier, population size, geographic extent, and financial strength vary considerably, which is reflected in the need for equalization payments (Vignette 3.5).

The vision of 10 equal provinces may reflect an English-Canadian nationalism. For some time, English-speaking Canadians have been searching for their cultural identity. Before World War I, Canadians saw themselves as part of the British Empire. By the end of World War II, this perspective began to change. The Maple Leaf flag, adopted by Parliament in 1964, and 'O Canada', the new national anthem approved by Parliament in 1967 and officially adopted in 1980, were signs of this cultural change. While the Québécois culture was flourishing, thanks in part to generous provincial funding for the arts, English-speaking Canadians continued to lean heavily on American culture. Some looked with envy at the cultural accomplishments of the Québécois and wondered aloud if similar achievements in English-speaking Canada were possible. The answer was yes, providing the provincial governments offered similar financial support for the arts.

Compromise

Given the incompatibility of the two visions—two founding peoples versus 10 equal provinces—and the historical development of the country, Canadian politicians have had the unenviable task of trying to accommodate demands from different groups—especially French Canadians, new immigrants, and Aboriginal peoples—and from different regions without offending other groups or regions. As in the past, politicians have continued to struggle with this Canadian dilemma, but in reality there is no perfect solution, only compromise. With this object in mind the federal government has made many efforts in search of the elusive middle ground.[10] It seems the search for an acceptable compromise between the two opposing visions of Canada will never end, and perhaps that is a good thing because the process is more important then the end result. To understand the current struggle for compromise, it is important to understand the political, economic, and cultural developments that have taken place in Québec over the past five decades. These, and their effects on the English/French faultline, are outlined in the following pages.

Resurgence of Québec Nationalism

After World War II, Québec broke with its past. A rise of Québec nationalism had begun much earlier but gained political momentum during the **Quiet Revolution** of the early 1960s. This development was the result of four major events. The most important was the resurgence of ethnic nationalism, that is, a pride in being a Québécois. The second was Québec's joining the urban/industrial world of North America and the subsequent expansion in the size of its industrial labour force and business class. The third was the removal of the old elite. This reform movement was profoundly anticlerical in its opposition to the entrenched role of the Church in Québec society, particularly the Church's control over education. In many ways, this reform was based on the aspirations of the working and middle classes in the new Québec economy. The fourth was the state's aggressive role in the province's affairs.

With the election of Jean Lesage's Liberal government in 1960, the province moved forcefully in a new direction. It created a more powerful civil service that allowed francophones access to middle and senior positions that were often denied them in the private sector of the Québec economy, which was controlled by English-speaking Quebecers and American companies. It nationalized the province's electric system, thereby creating the industrial giant known as Hydro-Québec, now a powerful symbol of Québec's revitalized economy and society. In turn, Hydro-Québec built a number of huge energy projects that demonstrated the province's industrial strength. By 1968, this Crown corporation had constructed one of the largest dams in the world on the Manicouagan River. Called Manic 5, this dam demonstrated Hydro-Québec's engineering and construction capabilities. To Quebecers, Hydro-Québec was a symbol of Québec's economic liberation from the years of suffocation associated with Maurice Duplessis and his Union Nationale government, which had been closely tied to big businesses owned by English-speaking Canadians and Americans. Clearly, Lesage's political goal of becoming '*maîtres chez nous*' (masters in our own house) had materialized with the success of Hydro-Québec, thus sparking a growth in Québec nationalism. Québec's desire for more autonomy in its own affairs intensified with increased confidence. In short, a new society had arisen in Québec, a society that wanted to chart its destiny. Charles Taylor (1993: 4) summed up this new feeling as 'a French Canada which, after a couple of centuries of enforced incubation [under London and then Ottawa], was ready to take control once more of its history.' The political question Taylor raised is a simple one: Would this 'control' take place within the framework of Canada's political system or outside it?

Separatism

Separatism grew out of the Quiet Revolution. It is a form of ethnic nationalism that is popular with francophones but unpopular with anglophones and allophones (those whose first language is neither French nor English). In 1967, French President Charles de Gaulle visited Québec and ignited the forces of French-Canadian nationalism with his now famous words, '*Vive le Québec. Vive le Québec libre.*' From that moment on, separatism gained support and took on a political form. By the time of the first referendum on independence in 1980, the separatists formed a substantial minority within Québec's population, with perhaps as many as 20 per cent dedicated separatists and another 20 per cent strongly dissatisfied with their place within Canada.

Separatism has had two distinct branches—the Front de libération du Québec (FLQ) and the Parti Québécois (PQ). The FLQ was a small fringe group within the separatist movement. It sought political change through revolutionary means, including bombing, kidnapping, and murder. However, the vast majority of separatists sought change through democratic means. The PQ was committed to a democratic solution by means of a referendum

followed by negotiations with the rest of Canada. Referendums, the process of referring a political question to the electorate for a direct decision by general vote, are notoriously tricky political instruments, but Québec Premier René Lévesque offered Quebecers what he thought was a clear choice—the unpopular status quo or a bold new beginning under sovereignty-association. **Sovereignty-association** meant political separation but a new economic association with the rest of Canada. Lévesque recognized that the integrated nature of Canada and its east-west economic axis made continued economic ties with Canada essential for the survival of Québec.

Prior to the referendum, the PQ vigorously tackled challenging economic problems and critical cultural issues, all of which had three purposes:

- to accelerate the modernization processes that began with the Quiet Revolution;
- to promote the Québécois culture;
- to demonstrate that a PQ government could run the affairs of an independent state.

The provincial government was involved in the marketplace, often through Crown corporations and government assistance for francophone business operations. The provincial government also promoted Québécois culture in a variety of ways. Under the Liberal government of Robert Bourassa, the French language had been declared the sole official language in 1974, but a PQ government, first elected in 1976, went much further with Bill 101 in 1977.[11] This bill made it necessary for most Quebecers, regardless of background or preference, to be educated in French-language schools, and was a key measure in ensuring the supremacy of the French language in the province. Among its many goals, Bill 101 was designed to ensure that the children of new immigrants went to French schools and thus to guarantee that the French-speaking population of Québec would continue to grow.

In 1980, Québec voters rejected sovereignty-association in a referendum. Almost 60 per cent voted to remain in Canada, which suggests that just over half of the francophone voters stood with the *Non* side, along with almost all the English-speaking residents. The rest of Canada responded with a collective sigh of relief, but separatism was far from dead. Several political events renewed separatist sentiment. One was the Constitution Act of 1982, which patriated the Constitution and gave Canadians the Charter of Rights and Freedoms. The Trudeau government accomplished this political feat at the cost of poisoning relations with Québec City by including the Charter of Rights and Freedoms in the Constitution. The Charter curtailed the power of the Québec government, and the Constitution was patriated without the approval of the Québec government. There were also the failed attempts of Brian Mulroney's Conservative government to achieve provincial unanimity for constitutional reform. The first attempt was the Meech Lake Accord, a package of constitutional revisions incorporating Québec's 'minimum' demands for political reform. This Accord was agreed to in principle by Ottawa and the 10 provinces in 1987, but two provinces, Manitoba and Newfoundland, failed to pass it within the required three years. Quebecers felt humiliated and rejected by the rest of Canada. In 1991, the Mulroney government attempted a second round of constitutional negotiations culminating in the Charlottetown Accord, which would give Québec distinct society status, the provinces more power and input on the selection of judges,

Aboriginal peoples the entrenched right to self-government, and the reform of the Supreme Court and the Senate. In 1992, the Charlottetown Accord was roundly rejected in a national referendum and only narrowly approved in four provinces (but not Québec).

These three political misadventures revived the spirits of the separatists, led by Jacques Parizeau and Lucien Bouchard. Parizeau's party, the PQ, returned to power in 1994, promising a referendum on sovereignty. While the referendum question referred to a new partnership with the rest of Canada, Parizeau believed that such an arrangement was impossible and saw only one solution—an independent Québec. In the previous year, the Bloc Québécois (a new federal party representing Québec interests), led by Lucien Bouchard (a former cabinet minister in the Mulroney government), formed the official opposition in the House of Commons. The Québec public expressed their dissatisfaction with Ottawa by rejecting traditional political parties. This left Québec federalists in a vulnerable position where they grew steadily weaker and more disorganized. The federal Liberals could offer little help. In fact, Jean Chrétien, himself a Quebecer, was extremely disliked by many in Québec for a variety of reasons, including his role in the patriation of the Constitution in 1982. He had become a '*tête de turc*' (a scapegoat for federal policies), a symbol of those Quebecers who, as federal ministers, put Québec 'in its place'. (For further discussion of the French/English faultline in Québec, see Chapter 6.)

The results of the 1995 referendum vote in Québec were extremely close. 'No—by a Whisker' screamed the headline of the *Globe and Mail* on the morning after the referendum of 30 October 1995. Québec came within 40,000 votes of approving the separatist

Vignette 3.7 — No—by a Whisker

The Results of the 30 October 1995 Referendum

The Question: 'Do you agree that Québec should become sovereign, after having made a formal offer to Canada for a new Economic and Political Partnership within the scope of the Bill respecting the future of Québec and of the agreement signed on June 12, 1995?'

The Answer (at 10:30 p.m. Eastern Time, 21,907 of 22,427 polls):

	Number	Per cent
No	2,294,162	49.5
Yes	2,254,496	48.7
Rejected	83,340	1.8
Total	4,631,998	100.0

Source: Globe and Mail, 31 Oct. 1995, A1.

dream of becoming an independent state (Vignette 3.7).

Moving Forward

The 1995 referendum was a low point in French/English relations, and its after-effects were many and varied. English Canada, dazed by the outcome, attempted to respond. Ottawa reacted almost immediately after the October referendum by passing a unilateral declaration that recognized Québec as a distinct society. The leader of the separatist forces, Jacques Parizeau, had shocked all Quebecers on the night of the referendum when he blamed the *Oui* side's loss on 'money and the ethnic vote'. In an electrifying moment, the dark side of ethnic nationalism had been revealed. Other centrifugal forces were released, too, including the partitionists, who argued that 'if Canada is divisible, so is Québec.' The Cree in northern Québec threatened secession, and partitionists pressured dozens of municipalities around Montréal and Hull to declare their allegiance to Canada.

By 1996, the federal government had decided to take a hard line with Québec, which included having the Supreme Court of Canada determine the conditions of separation. In the same year, the premiers attempted to address the unity issue. In a much more conciliatory manner, they announced the Calgary Declaration: 'the unique character of Québec society with its French-speaking majority, its culture and its tradition of civil law is fundamental to the well-being of Canada.' In a typically Canadian decision, the premiers added to their Declaration that 'any power conferred to one province in the future must be available to all.' This Declaration was the third attempt at reconciliation with Québec since the patriation of the Constitution in 1982.[12] The next step was for each provincial government to pass the appropriate legislation, giving this declaration legal status. By July 1998, all provinces (except Québec) and territories had passed this resolution in their legislatures.

Changes are taking place in Québec, too. Separatism, while not gone, had lost its spark. Like other Canadians, Quebecers have grown weary of referendums, political bickering, and the resulting unsettling effect on the national and Québec economies. After a decade of weak economic performance, Québec and the rest of Canada are mainly concerned with economic matters, particularly high levels of unemployment and insufficient funds for education and health. Then, too, the threat of the English is a thing of the past. Quebecers, confident in their language and culture, were more comfortable and secure than ever before. By electing a Liberal government on 15 April 2003, Quebecers have charted a new path.

Towards the Future

History and geography explain the basis of the French/English relationship. This dynamic relationship goes to the heart of the nation. Geography compels Canadians to recognize that Québec represents a distinct region of Canada in which a different language and culture dominate. History teaches Canadians that compromise leads to national unity, while conflicts drive a wedge between the two founding peoples of Canada.

Canada's history reflects over 200 years of French/English relations. Through conflicts and compromises, these innumerable interactions, both large and small, have shaped the essential components of the national character, namely, the capacity and willingness to find solutions to complex questions. Both Québec and the rest of Canada are different places and those changes have shaped the French/English faultline and the attempts at compromise in this relationship.

On the surface, reconciliation seems an impossible task, but political realities demand some form of compromise or at least a willingness to search for a solution. Perhaps Paul Villeneuve (1993: 104) was correct when he observed that 'for Canada, survival lies in the travelling toward an identity.' For most Canadians, this implies a recognition of the deep divide between French and English Canada but also an acceptance of the importance of each vision. Canadians have also learned that the political gains achieved by having the dominant English-Canadian society impose its will over the minority French-Canadian society are short-term and will eventually weaken national unity.

Summary

Canada is both an old and a young country. The first people arrived in Canada about 30,000 years ago. Much later, Europeans reached its Atlantic shores and England and France established colonies. In 1760, Britain gained control of New France, thus ending France's dream of an empire in the New World. As a result French Canadians' world changed. Their future was uncertain under British rule. The geographic reality of British North America prevailed, however, and in 1774 the Québec Act granted the French-speaking population its basic rights. Both the Act of Union in 1841 and the British North America Act of 1867 recognized Québec's French heritage.

But the real story of Canada begins in 1867. From the early days of Confederation to 1949, Canada grew from a small country of four provinces to the second largest political state in the world. Its territorial expansion only ceased in 1949 when Newfoundland joined Canada. Canada's vast territorial extent is divided into a series of regions. While envied by most peoples of the world, Canada's vast size, physical structure, and human geography have often led to regional tensions, divides, or faultlines.

Since Confederation, Canadian governments have struggled to overcome the country's difficult geography in order to bind the country together and balance regional differences. This struggle is one element that shapes the Canadian identity. The federal government has a difficult role to play in this struggle, and disagreements with various regions and groups of people are common. Occasionally, tensions turn into bitter and ongoing disputes. These disputes, or faultlines, often take four forms: (1) Aboriginal/non-Aboriginal tensions, (2) centralist/decentralist tensions, (3) newcomers/old-timers, and (4) French/English tensions. Of these forces, the French/English divide has the most serious implications for Canada's future as a nation. If a future Québec referendum points conclusively to sovereignty, a reorganization of Canada's territorial boundaries becomes a real possibility. An important element of Canada's identity is the willingness to compromise, which is the only way the country can make any progress towards resolving that most divisive issue: French/English differences over the nature of Confederation.

Notes

1. Although national, provincial, territorial, and municipal governments exist in Canada, only the federal and provincial governments have powers that no other level of government can usurp.

2. Under the terms of the British North America Act, the Dominion of Canada was composed of four provinces (Ontario, Québec, New Brunswick, and Nova Scotia). Modelled after the British parliamentary and monarchical system of government, the newly formed country had a Parliament made up of three elements: the head of government (a governor-general who represented the monarch), an upper house (the Senate), and a lower house (the House of Commons). This Act was modified several times to accommodate Canada's evolving political needs and its gradual movement to independent nationhood. The patriation of Canada's Constitution in 1982 removed the last vestige of Canada's political dependence on the United Kingdom, although Canada still recognizes the British monarch as its symbolic head.

    The British North America Act was based on the highly centralized government of the United Kingdom in the 1860s. However, this Act assigned specific powers to the provinces in order to satisfy Québec's demand for control over its culture. The Canadian political system that emerged, therefore, allowed for regionalized politics. For example, political parties in the House of Commons sometimes serve regional interests. In the 1920s, the Progressive Party represented the concerns of farmers in Western Canada, while the pro-independence Bloc Québécois, which was formed in 1990, not only serves the interests of Québec but is also active in the separatist movement. Furthermore, while the House of Commons is based on the principle of representation by population, Senate membership is based on the principle of equal regional representation. However, because senators are appointed by the Prime Minister and not elected by the people in the different regions of the country, the Senate fails to provide a regional counterweight to the House of Commons.

3. A group of Irish Americans, known as Feni-

ans, was struggling for Irish independence. They believed that attacking British possessions in North America would advance the cause of a free Ireland. Between 1866 and 1870, the Fenians launched several raids across the border into Canada. The United States did not encourage these raids and eventually forced the Fenians to disband. By the end of the American Civil War, Anglo-American relations again were strained because of Britain's tacit support for the Confederacy in the American Civil War. For that reason, the United States withdrew from the Reciprocity Treaty in 1866. This treaty, a free trade agreement between British North America and the United States, began in 1854; the subsequent years were prosperous ones for British North America. Its loss forced the Province of Canada to seek an alternative economic union with the other British colonies in North America.

4. Representation in the House of Commons is readjusted after each decennial (10-year) census in accordance with the Constitution Act, 1867 (formerly the BNA Act) and the Electoral Boundaries Readjustment Act (1985, as amended). On 13 June 1992, following the release of the population figures from the 1991 census, the Chief Electoral Officer of Canada published in the *Canada Gazette* the results of the calculations required by the Constitution Act, which meant an increase in the number of seats in the House of Commons from 295 to 301. The next readjustment should take place in 2004 and the number of seats may increase to around 305, with most new seats assigned to Ontario, Alberta, and British Columbia. Federal electoral district boundaries were subsequently revised to reflect changes and movements in Canada's population. The most recent representation order was proclaimed on 8 January 1996 and took effect at the dissolution of Parliament on 27 April 1997. The formula for determining the number of seats for each province and territory is available at:<http://www.elections.ca/scripts/fedrep/federal_e/repform_e.htm>.

5. The Manitoba Act of 1870 recognized the legal status of farms and other lands occupied by the Métis as 'fee simple' private property. As well, the Act provided that 1.4 million acres (566,580 ha) be reserved for the children of Métis. The land was allocated to these Métis in 240-acre (97-ha) parcels, plus 160 acres (65 ha) in 'scrip' for each adult head of a family. These lands were distributed after 1875, but much of the 'scrip' land was sold and then occupied by non-Métis. For more on this subject, see Tough (1996: ch. 6).

6. Pontiac, the Odawa chief in the Ohio Valley, led a successful uprising against the British in 1763. By capturing the forts in the Ohio Territory, he exposed Britain's precarious hold on this region, which the British had just obtained from the French. However, Pontiac and his followers could not hold these forts against the British because they could no longer obtain ammunition and muskets from the French. Pontiac concluded that his best move would be to make peace with Britain. The British came to the same conclusion, though for other reasons. Without the help of Pontiac and the other chiefs in this region, Britain would lose control over these lands. Britain therefore had to form an alliance with them. With that objective in mind, George III announced an important concession to these Indians in the Royal Proclamation of 1763, namely, that the King recognized them as valued allies and that the land they used to hunt and trap was 'Indian land' within the British Empire.

7. In 1215, King John of England was forced to sign the Magna Carta. In this charter, he promised to stop interfering with the Church and the law and to consult regularly with the country's leaders before collecting new taxes.

8. While the French language was not recognized by the Québec Act, 1774, the Governor made use of the French language in conducting his business with local officials. For example, the judges appointed by the Governor had to know both languages in order to facilitate the business of the court. In short, while English was the official language of British North America, the British colony of Québec functioned in both the French and English languages.

9. Louis Riel was the Métis political and spiritual leader in the late nineteenth century. This controversial figure is considered both a Father of Confederation and a traitor to the country. Riel, who was born in the Red River Colony in 1844, studied for the priesthood at the Collège de Montréal. The founder of Manitoba and the central figure in both the Red River Rebellion (1869–70) and the North-West Rebellion (1885), he was captured shortly after the Battle of Batoche, where the Métis forces were defeated. After a trial in Regina, the jury found Riel guilty of treason but recommended clemency. Appeals were made to Manitoba's Court of Queen's Bench and to the Judicial Committee of the Privy Council. Both appeals were dismissed. A final appeal went to the federal cabinet, but the government of John A. Macdonald wanted Riel executed. Riel was hanged in Regina on 16 November 1885. His body was interred in the cemetery at the Cathedral of St Boniface in Manitoba.

   Riel's execution has had a lasting effect on Canada. In Québec, French Canadians felt betrayed by the Conservative government and federalism. Riel's execution was proof for Canada's French-speaking population that they could not count on the federal government to look after French-Canadian interests. It was also a blow against a francophone presence in the West. In Ontario the hanging of Louis Riel satisfied the anti-Catholic and anti-French majority. For the Orange Order (the Protestant fraternal society that blamed Riel for the death of one of their members, Thomas Scott, who was executed by a Métis firing squad during the Red River Rebellion), Riel's execution was long overdue. In the West, Riel's hanging resulted in the marginalization of both the Métis and Indian tribes, especially those who participated in the uprising.

10. A recent example is the political fallout from the Québec referendum. Prime Minister Chrétien sought to fulfill his verbal promises made in the closing days before the vote on the 1995 referendum. In a House of Commons resolution, the federal government proposed three concessions to Québec: (1) a veto over constitutional changes; (2) recognition of Québec's distinct society status; and (3) devolution of federal powers to Québec. In the case of the veto, Ottawa was prepared to 'lend' its constitutional veto to Québec, Ontario, Atlantic Canada, and the four western provinces. Not only was the federal government committing itself to seeking permission from the four regions before putting its stamp of approval on any constitutional change, it was also recognizing that Canada consisted of four major regions. The premiers of Alberta and British Columbia reacted negatively to that concept of regionalism. British Columbians in particular saw this arrangement as another example of Ottawa's failure to recognize the west coast as a 'distinct and powerful' part of Canada. The federal government retreated from this issue and quickly amended its resolution to extend the veto to British Columbia. In December 1995, this resolution passed in both the House of Commons and the Senate. It then became the law of the land that Canada consists of five major regions!

11. In 1974, the Québec Liberal government passed Bill 22 (Loi sur la langue officielle), which made French the language of govern-

ment and the workplace. English was no longer an official language in Québec. In 1977, the Parti Québécois government introduced a much stronger language measure in the form of Bill 101 (Charte de la langue française). This legislation eliminated English as one of the official languages of Québec and required the children of all newcomers to Québec to be educated in French. Four years later, Bill 178 required all commercial signs to use only French. The French language has made modest gains outside of Québec. In 1969, New Brunswick passed an Official Languages Act, which gave equal status, rights, and privileges to English and French, and the federal Parliament passed the Official Languages Act, which declared the equal status of English and French in Parliament and in the Canadian public service.

12. Separatists argue that Québec does not have enough powers, that is, Québec is subordinate to Ottawa. For separatists, the solution lies in independence, whether achieved through Parizeau's 'chicken-plucking' strategy or Bouchard's 'winning' referendum strategy. A federalist counter-argument is that, as a member of a federation, Québec has many 'exclusive' powers, such as power over language and education. However, circumstances may force Ottawa to make a decision that adversely affects some provinces while favouring others. In 1982, the patriation of the British North America Act, renamed the Constitution Act, 1867, was such a decision. The Constitution Act, 1982, which was entrenched at the same time, added to the British North America Act in several ways, but without a doubt the most important addition has been the Charter of Rights and Freedoms. These rights and freedoms strengthen the rights of individuals and weaken collective rights. Prime Minister Trudeau, who conceived of society as an agglomeration of individuals (not collectivities), whose rights accrued to them as individuals, saw the Charter as protecting individuals from governments that try to suppress individual rights. Then, too, there is the Supreme Court of Canada's changed role, which has become more proactive with the adjudication of Charter cases.

Key Terms

Aboriginal peoples

All Canadians whose ancestors lived in Canada before the arrival of Europeans; includes status and non-status Indians, Métis, and Inuit.

Château Clique

The political elite of Lower Canada; composed of an alliance of officials and merchants who had considerable political influence with the British-appointed governor; similar to the Family Compact in Upper Canada.

comprehensive land-claim agreements

Agreements based on territory claimed by Aboriginal peoples that was never ceded or surrendered by treaty. Such agreements extinguish the Aboriginal land claim to vast areas in exchange for a relatively small amount of land, capital, and the organizational structure to manage their lands and capital.

culture area

A region within which the population has a common set of attitudes, economic and social practices, and values.

Family Compact

A group of officials who dominated senior bureaucratic positions, the executive and legislative councils, and the judiciary in Upper Canada.

Inuit

People who are descended from the Thule. The Thule migrated into Canada's Arctic from Alaska about 1,000 years ago. The Inuit do not fall under the Indian Act, but are identified as an Aboriginal people under the Constitution Act, 1982.

Loyalists

Colonists who supported the British during the American Revolution. About 40,000 American colonists who were loyal to Britain resettled in Canada, especially in Nova Scotia and Québec.

Métis

People who have a mixed biological and cultural heritage, usually either French-Indian or English/Scottish-Indian. This 'mixing' between Indians and Europeans took place during the fur trade and continues today. Originally the term was more narrowly applied to French-Indian people who settled in the Red River area and who developed a distinct hunting economy and society based on the French language and the Roman Catholic religion.

non-status Indians

Those of Amerindian ancestry who are not registered as Indians under the Indian Act.

Quiet Revolution

A period in Québec during the Liberal government of Jean Lesage (1960 to 1966) that was characterized by social, economic, and educational reforms and by the rebirth of pride and self-confidence among the French-speaking members of Québec society, which led to a resurgence of francophone ethnic nationalism. During this time, the secular nationalist movement gained strength.

reserve

Under the Indian Act, reserves are defined as lands 'held by her Majesty for the use and benefit of the bands for which they were set apart; and subject to this Act and to the terms of any treaty or surrender'.

sovereignty-association

A concept designed by the Parti Québécois under the Lévesque government and employed in the 1980 referendum. This concept was based on the vision of Canada as consisting of two 'equal' peoples. Sovereignty-association called for Quebec sovereignty but with a partnership with Canada based on an economic association.

specific land-claim agreements

Agreements designed to rectify shortcomings in the original treaty agreement with a band or that seek to redress failure on the part of the federal government to meet the terms of the treaty. Many of these have involved the unilateral alienation by the government of reserve land.

status Indians

Aboriginal peoples who are registered as Indians under the Indian Act.

treaty Indians

Aboriginal peoples who are descendants of Indians who signed a numbered treaty and who benefit from the rights described in each treaty. All treaty Indians are status Indians, but not all status Indians are treaty Indians.

Bibliography

Anderson, Robert B., and Robert M. Bone. 1995. 'First Nations Economic Development: A Contingency Perspective', *The Canadian Geographer* 39, 2: 120–30.

Asch, Michael, ed. 1997. *Aboriginal and Treaty Rights in Canada: Essays on Law, Equality, and Respect for Difference*. Vancouver: University of British Columbia Press.

Bonnichsen, Robson, and Karen L. Turnmire, eds.

1999. *Ice Age Peoples of North America: Environments, Origins, and Adaptations of the First Americans*. Corvallis: Oregan State University Press.

Brownlie, Robin Jarvis. 2003. *A Fatherly Eye: Indian Agents, Government Power, and Aboriginal Resistance in Ontario, 1918–1939*. Toronto: Oxford University Press.

Bumsted, J.M. 1998. *A History of the Canadian*

Peoples. Toronto: Oxford University Press.

Canada 1882. *Census of Canada 1880–1*, vol. 1. Ottawa: MacLean, Rogers & Company.

Canada 2003. Elections Canada. At: <http://www.elections.ca/scripts/fedrep/federal_e/repform_e.htm>. Searched 15 Feb. 2003.

Carter, Sarah. 2004. '"We Must Farm To Enable Us To Live": The Plains Cree and Agriculture to 1900', in R. Bruce Morrison and C. Roderick Wilson, eds, *Native Peoples: The Canadian Experience*. Toronto: Oxford University Press.

Coates, Ken. 1992. *Aboriginal Land Claims in Canada: A Regional Perspective*. Toronto: Copp Clark Pitman.

Cook, R. 1969. *Provincial Autonomy, Minority Rights and the Concept Theory, 1867–1921*. Studies of the Royal Commission on Bilingualism and Bilculturalism, no. 4. Ottawa: Queen's Printer.

———. 1993. *The Voyages of Jacques Cartier*. Toronto: University of Toronto Press.

Dickason, Olive Patricia. 2002. *Canada's First Nations: A History of Founding Peoples from Earliest Times*, 3rd edn. Toronto: Oxford University Press.

Fife, Robert, and Joël Bellevance. 2000. 'Cabinet vows to win over the West', *National Post*, 1 Dec., A6.

Fremlin, Gerald, ed. 1974. *The National Atlas of Canada*. Ottawa: Macmillan.

Garreau, Joel. 1981. *The Nine Nations of North America*. Boston: Houghton Mifflin.

Globe and Mail. 1995. 'No—by a Whisker', 31 Oct., A1.

Harris, R. Cole, and John Warkentin. 1991. *Canada Before Confederation: A Study in Historical Geography*. Ottawa: Carleton University Press.

Kerr, Donald, and Deryck W. Holdsworth, eds. 1990. *Historical Atlas of Canada, Volume III: Addressing the Twentieth Century 1891–1961*. Toronto: University of Toronto Press.

Lower, J. Arthur. 1983. *Western Canada: An Outline History*. Vancouver: Douglas & McIntyre.

McVey, Wayne W., and W.E. Kalbach. 1995. *Canadian Population*. Toronto: Nelson Canada.

Marsh, James H., ed. 1988. *The Canadian Encyclopedia*, 2nd edn. Edmonton: Hurtig.

Mitchell, Robert D., and Paul A. Groves. 1987. *North America: The Historical Geography of a Changing Continent*. Totowa, NJ: Rowman & Littlefield.

Moffat, Ben. 2002. 'Geographic Antecedents of Discontent: Power and Western Canadian Regions 1870 to 1935', *Prairie Perspectives* 5: 202–28.

Morton, D. 1983. *A Short History of Canada*. Edmonton: Hurtig.

Nemni, Max. 1994. 'The Case Against Quebec Nationalism', *American Review of Canadian Studies* 24, 2: 171–96.

O'Handley, Kathryn, ed. 1994. *Canadian Parliamentary Guide*. Toronto: Globe and Mail Publishing.

Pielou, E.C. 1991. *After the Ice Age: The Return of Life to Glaciated North America*. Chicago: University of Chicago Press.

Riendeau, Roger. 2000. *A Brief History of Canada*. Markham, Ont.: Fitzhenry & Whiteside.

Simpson, Jeffrey. 1993. *Faultlines: Struggling for a Canadian Vision*. Toronto: HarperCollins.

Smith, P.J. 1982. 'Alberta Since 1945: The Maturing Settlement System', in L.D. McCann, ed., *Heartland and Hinterland: A Regional Geography of Canada*. Scarborough, Ont.: Prentice-Hall.

Statistics Canada. 1998. *The Daily*—1996 Census: Ethnic Origin, Visible Minorities, 17 Feb. 1998 [on-line database], Ottawa. Searched 16 July 1998: <http://www.statcan.ca/Daily/English/>.

———. 2003. *Historical Statistics of Canada*. Catalogue no. 11–516–XIE [on-line database], Ottawa. Searched 15 February 2003: <http://www.statcan.ca/english/freepub/11-516-XIE/sectiona/sectiona.htm>.

Taylor, Charles. 1993. *Reconciling the Solitudes*. Montréal and Kingston: McGill-Queen's University Press.

Tough, Frank. 1996. *As Their Natural Resources*

Fail: *Native Peoples and the Economic History of Northern Manitoba, 1870–1930*. Vancouver: University of British Columbia Press.

Tracie, Carl J. 1996. *Toil and Peaceful Life: Doukhobor Village Settlement in Saskatchewan, 1899–1918*. Regina: Canadian Plains Research Centre, University of Regina.

Villeneuve, Paul. 1993. 'Allocution présidentielle: L'invention de l'avenir au nord de l'amérique', *Le géographe canadien* 37, 2: 98–104.

Warkentin, John, ed. 1968. *Canada: A Geographical Interpretation*. Toronto: Methuen.

Williams, Glyndwr. 1983. 'The Hudson's Bay Company and the Fur Trade, 1670–1870', *The Beaver* (Autumn): 4–81.

Wright, James V. 1995, 1999. *A History of the Native People of Canada*, vols 1 and 2. Ottawa: Canadian Museum of Civilization.

Wynn, G. 1990. *People, Places, Patterns, Processes: Geographic Perspectives on the Canadian Past*. Toronto: Copp Clark Pitman.

Further Reading

The historical geography of Canada recalls past events. Maps play a large role in this rediscovery of Canada's past. In 1970, several geographers and historians explored the idea of preparing a major Canadian historical atlas focused on social and economic themes. Four editors (R. Cole Harris, Volume I; R. Louis Gentilcore, Volume II; and Donald Kerr and Deryck W. Holdsworth, Volume III) brought this enterprise to a successful conclusion in 1993. These three volumes weave together the various historic strands that comprise Canada's historical geography and represent a rich legacy for Canadian scholars and students. For those studying *The Regional Geography of Canada*, these three volumes parallel the discussion found in this chapter and provide a wonderful source of essay topics.

Volume 1: Harris, R. Cole, ed. 1987. *Historical Atlas of Canada, Volume I: From the Beginning to 1800*. Toronto: University of Toronto Press.

Professor Harris and his collaborators have produced a detailed account of the arrival of Canada's First Peoples and then the French and British settlers. Each map and descriptive explanation provides an in-depth record of a particular historic event taking place in one region. Students will discover many interesting maps, including The Last Ice Sheet (Plate 1); Population and Subsistence (Plate 18); The Newfoundland Fishery, 18th Century (Plate 25); Maritime Canada, Late 18th Century (Plate 32); and The Seigneuries (Plate 51).

Volume 2: Gentilcore, R. Louis, ed. 1993. *Historical Atlas of Canada, Volume II: The Land Transformed 1800–1891*. Toronto: University of Toronto Press.

Under Professor Gentilcore's direction, a series of maps with text capture the essential historic events in nineteenth-century Canada. This volume examines agricultural settlement to mid-century as the building of a nation. Canada's economy remained heavily dependent on extraction and exportation of natural resources to Great Britain and the United States, though the development of a manufacturing base in Central Canada was supported, especially after 1879, by the high tariffs of the National Policy. Examples of the wide range of topics found in Volume II include Timber Production and Trade to 1850 (Plate 11); Unrest in the Canadas (Plate 23); and An Emerging Urban System, 1845, 1885 (Plate 45).

Volume 3: Kerr, Donald, and Deryck W. Holdsworth, eds. 1990. *Historical Atlas of*

Canada, Volume III: Addressing the Twentieth Century 1891–1961. Toronto: University of Toronto Press.

The main historic themes examined by Professors Kerr and Holdsworth are the transformation of Canada from an agrarian society to an urban/industrial society and the crises of the Great Depression, World War II, and the immediate post-World War II period. Canada was changing into an industrial country within a North American context. American investment in Canada was growing and American branch plants dominated our manufacturing sector. During this time, Canada's population grew from 5 million to 18 million and this population was composed primarily of Canadians with a British, French, or European background. The number of Indians, Métis, and Inuit was relatively small, under half a million. Topics covered include The Changing Structure of Manufacturing (Plate 7); The Grain Handling System (Plate 19); and Population Changes (Plate 59).

The Historical Atlas of Canada is an excellent reference for scholars and students. Four more recent major events, however, are not covered—the rise of Aboriginal political power, the threat of separation of Québec from the rest of Canada, the influx of non-European immigrants to Canada, and the signing of the Canada–US Free Trade Agreement. These issues are discussed in Chapter 4.

Montréal, Québec
(Jean Bruneau/Valan Photos)

Canada's human geography is changing in many ways. While the population has surpassed 30 million, the growth rate is slowing. Natural increase has fallen below the replacement level, the population is aging, and more people are living in urban places. Population growth is now driven primarily by immigration. Canada is a multicultural society with more than 200 ethnic groups recorded in the 2001 census. As a result of trade liberalization, Canada functions economically more and more within the North American marketplace. With all regions increasing their exports to the US, the north–south flow of trade now rivals the traditional east–west flow. The chapter ends with a detailed discussion of population trends and demographic faultlines.

- Discuss Canada's current demographic, economic, and social changes in terms of the core/periphery demographic transition, and stages of development concepts.
- Comment on the rise of Aboriginal political power, the threat of Québec separation, the role of immigration in Canada's population increase, and the effect of American concerns regarding internal security.
- Argue that the national economic model has been overtaken by a continental version.
- Recognize that demographic, economic, and social changes are not occurring equally across Canada's six geographic regions.
- Analyze the demographic shifts within Canada's four faultlines.

Chapter 4 Canada's Human Face

■ Introduction

Canada is home to just over 30 million people. Just as an economic core/periphery pattern exists in Canada, so does a demographic one. Ontario and Québec account for 62 per cent of Canada's population (Table 4.1). This is not surprising, given their historic position in Canada, their geographic advantages, and their economic strength as measured by GDP.

Beyond population size and distribution, Canada is undergoing profound demographic and social changes (Vignette 4.1). As the 2001 census revealed, Canada's society is not only bigger and older, but it is moving in new social directions. For the first time, Canada's population exceeds 30 million and, equally important, its ethnic composition and cultural diver-

sity now reflect fresh elements in our society. Canada has a substantial number of adherents to Islam, Sikhism, and Hinduism, its larger cities are home to many so-called visible minorities, and Chinese forms the third most commonly spoken language in Canada. Clearly, immigration plays a major role in these demographic and social changes. Other forces of demographic and social change include the rapid increase in Canada's Aboriginal population, the place of Québec within Confederation, and Canada's role within the North American community. These factors have sparked an economic and social revolution that has yet to run its course. The transformation of Canada's transportation system, especially its major

Table 4.1	Population Change by Geographic Regions, 1951–2001				
Geographic Region	1951		2001		Percentage Change
	Population (000s)	Per cent	Population (000s)	Per cent	
Ontario	4,598	32.8	11,410	38.0	5.2
British Columbia	1,165	8.3	3,908	13.1	4.8
Territorial North	25	0.1	93	0.3	0.2
Western Canada	2,548	18.2	5,073	16.9	(1.3)
Atlantic Canada	1,618	11.6	2,286	7.6	(4.0)
Québec	4,056	29.0	7,238	24.1	(4.9)
Canada	14,009	100.0	30,007	100.0	

Source: Adapted from Dominion Bureau of Statistics, Ninth Census of Canada, vol. 1, *Population: General Characteristics* (Ottawa: Queen's Printer, 1953), Table 1; Statistics Canada (2002a).

Vignette 4.1   Demography

Demography is the study of population. It addresses questions concerning size, distribution, and composition of human populations as well as their implications for society. Demographic analysis deals with variables such as fertility, mortality, migration, population composition, population size, and distribution. The demographic transition theory provides a theoretical explanation for changes in population growth over a longer period.

trucking and railway routes, from an east/west orientation to a north/south one marks a visible sign of the shift from a national economy to a continental one (Figure 4.1).

What are the root causes of these changes? In the late twentieth century, Canada experienced a social revolution and an economic transformation. The social revolution came from the liberalization of Canadian society in the 1960s and was marked by the passing of the Constitution Act, 1982. During the 1970s, the spectre of Québec separation haunted Canada, voices of Aboriginal leaders became louder and more demanding, and the massive

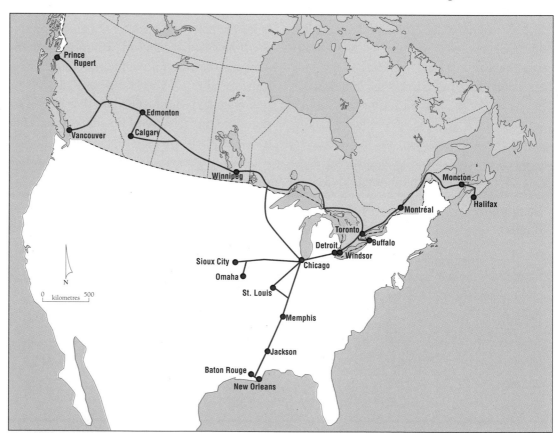

Figure 4.1 The CN rail system. With its purchase in 1998 of Illinois Central, Canadian National Railway extended its access to markets to the populous American Midwest and the American South.

influx of non-European immigrants altered the very fabric of Canadian society. Aboriginal power is expressed in land claims and is made more forceful by its rapidly growing population (Vignette 4.2); Québec's power is expressed politically by the Bloc Québécois and the Parti Québécois; and the massive number of newcomers is necessary to keep Canada's population increasing and to meet the need for semi-skilled and skilled workers.

Vignette 4.2	**From a Colonial Straitjacket to Aboriginal Power**

Until 1969, Canada's Aboriginal peoples were largely invisible to the rest of Canadian society. They were outside the political process and thus denied access to the political decision-making process. In fact, status Indians did not receive the right to vote in federal elections until 1960. Most were geographically separated from urban Canadians, as many status Indians lived on reserves. Métis were in isolated communities, and Inuit still roamed the Arctic. Shunned by Canadian society, these marginalized peoples had been subjected to assimilation policies for more than a hundred years. In 1969, Ottawa made one last attempt to assimilate the Indian people of Canada through its *Statement of the Government of Canada on Indian Policy*, more popularly known as the White Paper. The federal government proposed to eliminate the legal distinction between Indians and other Canadians by repealing the Indian Act and amending the British North America Act to remove those parts of it that called for separate treatment for Indians, and by abolishing the Indian Affairs Branch. Ottawa believed that the separation of Indians from other Canadians was not only divisive but also made the Indians dependent on government and thereby held them back. The remedy was 'equality'. In the context of the 1960s, when oppressed people in other countries, including blacks in the United States and in South Africa, fiercely sought equality and the American Indian Movement railed against colonial suppression, Prime Minister Pierre Trudeau believed that the White Paper was the answer to Canada's Indian problem. Some Indian leaders supported this solution, but many others did not. Reaction was swift. In the same year, Harold Cardinal published *The Unjust Society* and the following year, under his leadership, the Alberta chiefs published a formal rebuttal to the White Paper, titled *Citizens Plus: A Presentation by the Indian Chiefs of Alberta to the Right Honourable P.E. Trudeau*.

Over the next decade, the debate over the place of Indians in Canadian society took several different directions. First, there was legal support for the Indian position, beginning with the *Calder* case in 1973 when the Supreme Court held that the Nisga'a had Aboriginal rights. Second, the election of the Parti Québécois in 1976 called for 'nation-to-nation' discussions between the province and the federal government. Aboriginal leaders seized the opportunity to present their demands in the same constitutional language. Third, recognition of the Aboriginal peoples and their rights in the 1982 Constitution Act dramatically enhanced their status and bargaining power. Fourth, the Constitution Act did not define Aboriginal Rights, leaving that task to negotiations or the courts. The courts have been active in defining Aboriginal rights. In 1997, the Supreme Court's landmark decision in the *Delgamuukw* case overturned the earlier decision denying that Indians in British Columbia had Aboriginal title. Furthermore, the Court ruled that Aboriginal title means that Indians have the right to the resources on those lands.

| Vignette 4.3 | Stages of Development |

Technological change and economic development have led to major changes in the economic structure of industrial states. These changes are sometimes called stages of development. Often, modern industrial nations have gone through three stages of development: pre-industrial, industrial, and post-industrial. In the **pre-industrial stage**, **primary activities**, led by agriculture, dominate the economy and occupy most of the labour force. The **industrial stage** occurs when manufacturing activities surpass those of the primary sector of the economy. In this stage, most of the workers are involved in **secondary activities** (such as high-technology activities processing raw materials and manufacturing goods) and **tertiary activities** (such as providing a variety of services to the public). The last stage is the **post-industrial stage**, in which the economy is dominated by tertiary and **quaternary activities**. Table 4.14 (p. 194) provides a fuller description of the types of economic structures.

Canada, like other industrial countries, has gone through several stages of development (Vignette 4.3). The latest stage has seen an economic transformation of Canada into a North American industrial power that had its beginning in the Auto Pact of 1965, but the transformation came into its own in 1989 when the Canada–US Free Trade Agreement (FTA) came into force. Within a decade, Canada's economy had become heavily integrated into the American economy. Since the FTA, increased trade with the United States provides a measure of the degree of integration. Exports to the United States soared from 75 per cent in 1989 to 85 per cent in 2002. Energy, lumber, and mineral products found a ready market in the US, but manufactured exports, except for automobile exports, had a more difficult time. As exports to the United States increased, Canada's economy became more and more integrated and dependent on the American market. The expansion of the CN rail system exemplified this trend towards a North American economy. In 2002, Paul Tellier, the former CEO of CN, stated that 'We're a Canadian-based North American company', with half of its assets located in the United States and half of its revenues earned in the

United States or from Canadian exports crossing the US–Canada border (Francis, 2002: FP6). Back in the 1990s, Tellier knew that railways were built to follow trade.[1] With the sharp rise in Canadian exports to the United States, the Canadian railway system had to adjust quickly to the new North American economy. Led by the Canadian National Railway, the first North American railroad system was created with the purchase in 1998 of the Illinois Central Railroad by CN. Its rail network now stretches from Halifax on the Atlantic Ocean to New Orleans on the Gulf of Mexico to Vancouver on the Pacific Ocean (Figure 4.1).

Canada's six geographic regions have responded differently to the economic and demographic changes of recent years. Ontario's manufacturing sector has benefited enormously from participating in the North American economy. Aboriginal political power is most effectively expressed in the Territorial North, where the new territory of Nunavut was created; the separatist movement, while blunted by the election of a provincial Liberal government in 2003, remains a force in Québec politics; and visible minorities, by concentrating in Canada's major cities, but especially those in Ontario, Québec, and British Columbia, have created two Canadas— a more cosmopolitan urban Canada and a more traditional small-town and rural Canada.

In this chapter, Canada's population size, density, and distribution are examined first, followed by its natural growth, immigration, age/sex structure, and age dependency. Next, cultural characteristics of Canadians (as measured by ethnicity, multiculturalism, language, and religion) are investigated. The importance of Canada's Constitution, with its emphasis on bilingual, pluralistic, and civic nationalism, becomes readily apparent in this discussion. The chapter also describes the economic sectors that make up Canada's industrial structure and employment issues that affect Canada's human geography, such as labour force trends, unemployment, and earnings of Canadians, including differences between male and female workers. Finally, demographic issues and trends are examined from the perspective of the four faultlines discussed in Chapter 3.

Canada's Population

Demography is a significant factor affecting social and economic change. Such changes are often reflected in and affected by the country's population. For instance, some of the significant population trends being experienced today in Canada include a decline in the rate of natural increase, the aging of the population, and the high fertility rate among Aboriginal peoples. Three measures of Canada's population—size, density, and distribution—are outlined below.

Population Size

Since Confederation, Canada's population has increased from 3.4 million to 30.0 million, a nearly tenfold increase. Three primary factors account for this growth: natural increase, population gained from territorial expansion, and immigration. This remarkable population growth has transformed Canada into a medium-size country. For instance, approximately three-quarters of the nations of the world have smaller populations than Canada. Canadians, however, do not measure their population size against all the countries of the world but against the United States, which is far more populous than Canada. Canadians also take into consideration the geographic size of their country. As the second largest country in the world by geographic area, Canada's population density is one of the lowest in the world. Based on these comparisons, Canada's population seems small.

Population Density

Population density is determined by dividing the total number of people by the total land area. Canada has a population density of three people per square kilometre, which means the country has an extremely low population density (but not the lowest in the world). Australia has a slightly lower population density of two people per square kilometre. All other countries of the world have higher population densities. The United States, for example, has 29 people per square kilometre, while Bangladesh, the most densely populated country, has 900 people per square kilometre.

The explanation for these variations is simple. Land varies greatly in its capacity to support human settlement. Most land in Canada lies beyond the northern limits of agriculture, and land in the Territorial North especially has a low capacity to support human life. Wildlife, historically, constituted most of the food. The region's population density is only 0.03 people

Table 4.2	Population Density, 2001			
Geographic Region	Population (000s)	Population (%)	Land Area (000 km²)	Population Density
Territorial North	93	0.3	3,778	0.03
Atlantic Canada	2,286	7.7	502	4.6
Western Canada	5,073	16.9	1,756	2.9
British Columbia	3,908	13.1	893	4.2
Québec	7,238	24.0	1,358	5.4
Ontario	11,410	38.0	917	12.7
Canada	30,007	100.0	9,203	3.3

Source: Statistics Canada (2002a).

Founded in 1642, Montréal, Québec, is one of Canada's oldest cities and its most cosmopolitan city. Currently Canada's second most populated city, until the 1970s Montréal was the centre of commerce and trade in the country. (Francis Lépine/Valan Photos)

Toronto, Ontario, with a population approaching 5 million, is Canada's most populous city and serves as the economic engine for southern Ontario and the financial capital for Canada. Toronto has also become the nation's most culturally diverse city due to the immigration of new Canadians. (Search4Stock Inc.)

per square kilometre (Table 4.2). On the other hand, Ontario has a much larger population living in a much smaller geographic area. Consequently, Ontario's population density is the highest among the six geographic regions.

Population density figures are more meaningful if they are expressed as the amount of arable land per person. This measure is called

Vancouver, British Columbia, is Canada's third most populated city and serves as Canada's portal to Asian markets. In the foreground, the Cambie Street Bridge spans False Creek and leads to BC Place; to the west of the bridge is the Granville Island Public Market. (Al Harvey/The Slide Farm)

In the heart of Ottawa, Ontario, the Rideau Canal is both an urban recreational zone and a waterway for pleasure craft travelling from Kingston on Lake Ontario. Completed in 1832, the canal was originally designed as an alternative military route between Montréal and Kingston. Parliament Hill and the Château Laurier are two landmarks that now sit on either side of the canal. (Ivy Images)

physiological density. Accordingly, Canada's physiological density is similar to that found in the United States.

Population Distribution

Population distribution is the dispersal of people within a geographic area. Canada's population is extremely unevenly distributed across the country. In fact, few nations have so much of their population concentrated in such a relatively small area of the country, while the rest of the country is almost vacant. So pronounced is this population distribution that an American geography text described Canada's population as if it were 'drawn by a magnet toward the giant neighbor on the south, for they [Canada's inhabitants] are strikingly concentrated along the United States border' (Trewartha et al., 1967: 542). Canadian scholars often view this same distribution as consisting of a national population core surrounded by a sparsely populated hinterland. The population core is sometimes described as Canada's national ecumene (inhabited area).

The greatest population increases have occurred in Canada's population core, particularly in its cities. Some of the increase was due to rural-to-urban migration, while some resulted from people leaving have-not provinces for large urban centres in more prosperous provinces. But most of the increase has been attributed to the influx of immigrants to Toronto, Montréal, and Vancouver.

Population by Geographic Region

Among the six geographic regions of Canada, notable differences exist with respect to population patterns. For instance, the country's uneven population distribution is illustrated in the variation between each region's percentage of Canada's population (Table 4.1). The smallest population is in the Territorial North. In 2001, the Territorial North had fewer people (only 0.3 per cent of Canada's population) than the census agglomeration of Peterborough, Ontario. At the other extreme, Ontario (with 38 per cent of the national population) has the largest population. Québec takes second place, with 24 per cent of Canada's population. Together, Ontario and Québec command a dominating demographic position within Canada. In combination, Central Canada (Ontario and Québec) contains 62 per cent of Canada's population. Even so, British Columbia and Western Canada are slowly gaining a larger share of the national population at the expense of Atlantic Canada, Québec, and the Territorial North.

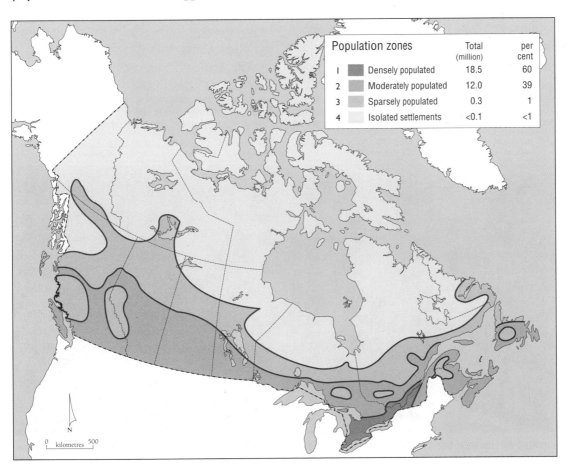

Figure 4.2 Canada's population zones. Canada's population is heavily concentrated in southern Ontario and southern Québec, where a favourable physical geography and an advantageous geographic location have resulted in a dense population. A secondary belt of population spans a southern strip of Canada. Together the densely and moderately populated zones account for 99 per cent of Canada's population. (Further resources: Student Web site, National Atlas section, Map 16. Web site instructions are found on p. xxi)

Table 4.4	Population Zones, 2001			
Zone	Population Size (millions)	Percentage of Total Population	Major City	Population of Major City
1	18.5	60	Toronto	4,682,897
2	12.0	39	Vancouver	1,986,965
3	0.3	1	Fort McMurray	41,466
4	<0.1	<1	Labrador City	7,744

Sources: Based on Statistics Canada (2002a, 2002b).

From 1996 to 2001, the three fastest growing provinces were Alberta at 10.3 per cent, Ontario at 6.1 per cent, and British Columbia at 4.9 per cent.

Population Zones

Population zones provide another picture of the spatial patterns of population in Canada (Figure 4.2). Four population zones reinforce the image of a highly concentrated population or core surrounded by a more dispersed periphery (Table 4.3).

The primary zone, a densely populated area in the Great Lakes–St Lawrence Lowlands, contains the bulk of Canada's population. As the most densely populated area in Canada, the Great Lakes–St. Lawrence Low-

Located on the west coast of Vancouver Island, Port Alberni remains an important resource town with a population of 25,396 (2001). The basis of its economy is the forest industry. In recent years, Port Alberni, like other forest-based resource towns, has fallen on hard times due to declining demand for forest products and to the automation of its production system. As a result, Port Alberni's population dropped by nearly 6 per cent from 1996 to 2001. Similar population declines in other resource towns reflects the troubled forest economy, resulting in a serious and perhaps ongoing population retreat from British Columbia's hinterland. (J.A. Wilkinson/Valan Photos)

lands contains nearly 20 million people (60 per cent of the national population) and almost three-quarters of Canada's metropolitan cities. For example, Toronto, Montréal, Ottawa/Gatineau, Québec City, Hamilton, Oshawa, and London are located in this population core. With a strong economy and rich farmlands, the primary zone continues to attract people and is the fastest growing zone.

The secondary zone surrounds the population core and extends across southern Canada. In general, its boundaries correspond with those enclosing Canada's remaining arable lands. About 12 million Canadians (over one-third of the country's total population) live in this moderately populated zone. Canada's remaining major cities are located within this zone, including Vancouver, Edmonton, Calgary, Winnipeg, and Halifax. Within the secondary zone, some areas, such as southern Alberta and the Lower Mainland of British Columbia, are growing quickly while other areas are subject to population losses, such as Newfoundland and Saskatchewan. As a result, the population of the secondary zone is increasing slowly and unevenly.

The tertiary zone lies within the Subarctic climatic zone. Less than 1 per cent of all Canadians (about 300,000) live in this zone. None of Canada's major cities are in this zone. The

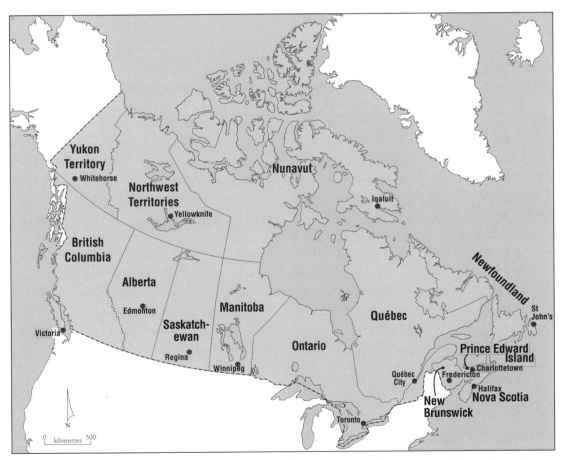

■ **Figure 4.3 Capital cities.** With five exceptions, the capital cities of the 10 provinces and three territories are the largest urban centres in each political unit. The largest city in New Brunswick is Saint John; in Québec, it is Montréal; in Saskatchewan, it is Saskatoon; in Alberta, it is Calgary; and in British Columbia it is Vancouver.

Table 4.4 — Percentage of Urban Population by Region, 1901–2001

Region	1901	1921	1941	1961	1981	1991	1996	2001
Ontario	40.3	58.8	67.5	77.3	81.7	81.8	83.3	84.7
British Columbia	46.4	50.9	64.0	72.6	78.0	80.4	82.1	84.7
Québec	36.1	51.8	61.2	74.3	77.6	77.6	78.0	80.4
Western Canada	19.3	28.7	32.4	57.6	71.4	74.4	74.4	75.7
Atlantic Canada*	24.5	38.8	44.1	50.1	54.9	54.1	52.8	53.9
Canada	34.9	47.4	55.7	70.2	76.2	77.2	77.9	79.7

*Newfoundland was not included in Atlantic Canada's figures until 1961. While comparable statistics are not available for the Territorial North, two observations are possible: (1) prior to the 1950s, few people in this region lived in settlements, and (2) by 2001, only three communities, Whitehorse (21,405), Yellowknife (16,541), and Iqaluit (5,236), had populations exceeding 5,000.
Sources: McVey and Kalbach (1995: 149); Statistics Canada (1997c, 2002a).

populations of the largest cities range between 10,000 and 40,000. Whitehorse and Yellowknife, as the capital cities of Yukon and the Northwest Territories, are administrative centres and **regional service centres**, since they also provide most of the service functions for their areas. **Resource towns** contain a single industry, such as a mine. Fort McMurray, Alberta, is the largest city in the tertiary zone with a population exceeding 40,000 in 2001. From 1996 to 2001, the tertiary zone has suffered a slight population decline.

The quaternary zone falls within the Arctic climatic zone. This zone has less than 100,000 inhabitants and is composed of small, isolated settlements. With a small population and a large geographic area, the quaternary zone has the lowest population density of the four zones. From a resource perspective, it is the least productive area in the country. Most commercial activities centre on exploitation of non-renewable resources, including mineral bodies and petroleum deposits. Unlike in the other zones in Canada, Aboriginal peoples are the majority in this zone. In spite of a high rate of natural increase among the Aboriginal population, the quaternary zone is affected by a net out-migration. Urban centres are small, most with populations under 5,000. Single-industry resource towns are the largest urban centres. The iron-mining town of Labrador City has the largest population with just under 8,000 inhabitants.

Urban Population

Canada is an urban country with nearly 80 per cent of its population living in cities and towns. Canadians prefer to live in an urban setting where amenities are readily available. In 2001, for example, 79.7 per cent of Canadians lived in urban centres of 10,000 people or more compared with 77.9 per cent in 1996. This trend began over 100 years ago (Table 4.4). However, cities are important for other reasons. They are at the cutting edge of technological change and capital cities provide a special administration role (Figure 4.3). In the new knowledge economy, it is not the factories that determine a city's prosperity, but rather its

Table 4.5	Population and Percentage Increase for Census Metropolitan Areas, 1996–2001		
Rank	Census Metropolitan Area	2001 Population	% Change,1996–2001
1	Toronto	4,682,897	9.8
2	Montréal	3,426,350	3.0
3	Vancouver	1,986,965	8.5
4	Ottawa/Gatineau	1,063,664	6.5
5	Calgary	951,395	15.8
6	Edmonton	937,845	8.7
7	Québec City	682,757	1.6
8	Winnipeg	671,274	0.6
9	Hamilton	662,401	6.1
10	London	432,451	8.2
11	Kitchener	414,284	3.8
12	St Catharines/Niagara	377,009	1.2
13	Halifax	359,183	4.7
14	Victoria	311,902	2.5
15	Windsor	307,877	7.3
16	Oshawa	296,298	10.2
17	Saskatoon	225,927	3.1
18	Regina	192,800	(0.4)
19	St John's	172,918	(0.7)
20	Chicoutimi/Jonquière	154,938	(3.4)
21	Sudbury	155,601	(6.0)
22	Sherbrooke	153,811	2.8
23	Kelowna	147,739	8.2
23	Abbotsford	147,370	8.0
24	Kingston	146,838	1.6
25	Trois-Rivières	137,507	(1.7)
26	Saint John	122,678	(2.4)
27	Thunder Bay	121,986	(3.7)

Source: Statistics Canada (2002a).

creativity, education, and culture. The post-industrial society arose in cities and spread to other parts of the country. Urban centres have changed their geographic shape from a compact city form to a highly dispersed one. Along with this urban and suburban growth have come environmental and social problems—traffic congestion, the homeless, water con-

tamination, and garbage disposal.

The process of urbanization has had a profound impact on the geography of population. In 1901, for example, approximately 35 per cent of all Canadians lived in an urban setting (Table 4.4). By 1961, the percentage had doubled to 70 per cent. In the most recent census (2001), urban dwellers constituted nearly 80 per cent of the Canadian population.

The rate of urbanization varies across Canada. The highest level of urban dwellers is in the industrial heartland of Canada. In 2001, Ontario, with 85 per cent of its population living in urban places, led all other geographic regions (Table 4.4). Statistics Canada classifies the largest cities in Canada as census metropolitan areas (CMAs). A **census metropolitan area** is an urban area of at least 100,000 people. Toronto, for instance, has a population of over 4.7 million, which is roughly 14 per cent of Canada's population (Table 4.5). In terms of **population strength**, Toronto is more 'powerful' than three geographic regions (the Territorial North with 0.3 per cent of Canada's population, Atlantic Canada with 7.7 per cent, and British Columbia with 13.1 per cent). In addition to Toronto, Ontario has several other large

Estrie, formerly called the Eastern Townships, lies in the Appalachian physiographic region. Dairy farms are found in the rolling countryside, which is surrounded by wooded uplands. Hay is the principal crop and it is used as feed for diary cows. (Clara Parsons/Valan Photos)

census metropolitan areas, including Ottawa and Hamilton.

By 2001, British Columbia had caught up to Ontario with an urban percentage of 85 per cent (Table 4.4). Most live in Vancouver. In 2001, just over 2 million Canadians called this west coast city home (Table 4.5). Québec is in third place with 80 per cent of its population living in urban areas, particularly Montréal and Québec City. With a population of 3.4 million, Montréal has most of the province's urban population, while Québec City has a population close to 690,000. Western Canada is close behind Québec, with 76 per cent of its population in such urban centres as Edmonton, Calgary, and Winnipeg. Edmonton has a population of 938,000, Calgary has 951,000, and Winnipeg 671,000. Atlantic Canada, with only 53 per cent of its population living in villages, towns, and cities, has the smallest urban population. Its major city, Halifax, has a population of about 359,000.

In this discussion of urban population, the Territorial North has not been included in the comparison because to do so would be misleading. The Territorial North has not followed the same settlement process found in other geographic regions because of cultural and economic differences, a cold environment, and the relocation of Aboriginal peoples into settlements some 50 years ago. While almost everyone lives in a settlement, only two centres (Whitehorse and Yellowknife) have populations over 10,000. Using 10,000 inhabitants as the definition of an urban place, the Territorial North would have 43 per cent of its population classified as urban.

Census Metropolitan Areas

The emergence of census metropolitan areas (CMAs) across Canada is the latest outcome of urbanization. The proportion of Canada's population residing in census metropolitan areas has increased from 30.3 per cent in 1931 to 64 per cent in 2001. In 1931, there were 10 CMAs: Halifax, Hamilton, Montréal, Ottawa, Québec, Saint John, Toronto, Vancouver, Windsor, and Winnipeg. By 2001, 27 census metropolitan areas existed (Table 4.5). Beyond their demographic importance, these large cities serve as centres for business and government. Most of Canada's corporate headquarters are located in Toronto while Calgary is recognized as the energy corporate headquarters.

Since 1951, the populations of the five largest CMAs have increased remarkably. Over this 50-year period, Toronto has gained the greatest number of people and has consistently ranked in the top five cities by growth rate. By 1971, Toronto surpassed Montréal as Canada's largest metropolitan centre, and Toronto, with a higher growth rate, continues to outdistance Montréal in population size. From 1996 to 2001, for instance, Toronto's growth rate was 9.8 per cent while that of Montréal was only 3 per cent. Yet, the greatest rate of increase among the census metropolitan areas—a phenomenal 15.8 per cent—took place in Calgary. In the previous five-year period, Vancouver held first place but then slipped to fifth place in the 1996–2001 period. The five leading census metropolitan areas by growth rates for the last five years were Calgary, Oshawa, Toronto, Edmonton, and Vancouver (Table 4.5).

Population Change

Population change has three components: births, deaths, and migration. **Population increase** is the sum of natural increase and net migration over a given period. The term **population growth** is used when this increase is

expressed as a rate, that is, as a percentage change over time. **Natural increase** is the difference between the **crude birth rate** (CBR) and the **crude death rate** (CDR). CBR is the number of live births per 1,000 people in a given year. CDR is the number of deaths per 1,000 people in a given year. **Net migration** is the difference between in- and out-migration. Canada has a large influx of migrants, but a number of Canadians also leave the country, often for the United States. While reasons for out-migration vary, many Canadians move to the United States because of job opportunities

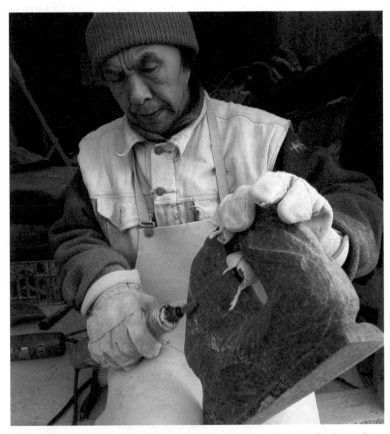

Inuit commercial sculpture descended from the ancient craft of carving. The nomadic peoples of the Arctic, including the Dorset, Thule, and finally the Inuit, carved small objects—such as amulets, fetish figures, and dance masks—that could be carried from place to place. Carving for commercial purposes was firmly established by the 1960s as the federal government had begun to encourage the development and promotion of Inuit sculpture. (CP/Paul Chiasson)

and/or because of a more temperate climate. This trend may slow due to the concern of Americans for internal security following the attacks on New York and Washington in 2001, the American declaration of war on terrorism, and the invasion of Iraq.

Since 1851, Canada has enjoyed continuous population growth (Table 4.5). At first, high rates of natural increase and high levels of immigration propelled this growth. When Canada became an industrial country, its rate of natural increase slowed as parents opted for smaller families. After World War II, however, the baby boom occurred. During the 1950s and 1960s, the rate of natural increase rose sharply and the average family size increased, accounting for almost 70 per cent of Canada's population growth. Combined with high levels of immigration, Canada's rate of growth reached record levels when the average annual rate of increase soared over 2 per cent per year during the 1950s (Vignette 4.4). Since Confederation, this rate was exceeded only once—in 1911, when it reached 3 per cent (Table 4.6).

In the last decade of the twentieth century, the annual rate of population growth fell below 1 per cent for the first time in Canada's history, and most of this growth is now related to immigration rather than natural increase. Since the end of the baby boom in the 1970s, population growth has slowed. In the 1980s, natural increase accounted for just over half of Canada's population growth. Since then, immigration has been the main force behind Canada's population growth rate. Natural increase made up only 27.3 per cent of Canada's population growth in 2001–2. In that census year, Canada's population increased by 352,843, with natural increase accounting for 95,955 and immigration for 255,888 (Statistics Canada, 2003d; see Table 4.7).

Vignette 4.4   The Baby Boom Effect

After Confederation, fertility and mortality declined until the late 1940s, when the birth rate suddenly increased. Demographers refer to this aberration as the baby boom. By the late 1960s, however, fertility rates decreased sharply. This long-term downward trend is associated with industrialized countries and described in the demographic transition theory. The end of the baby boom is sometimes described as the baby bust.

The baby boom lasted for almost 20 years. During that time of high fertility, there were almost 10 million births, which created a bulge in the age structure of Canadian society that continues to have both economic and social implications. Thus, while this demographic phenomenon was short-lived, it has left its mark on Canadian society (Foot, 1996). As consumers of goods and services, baby boomers have had a decided impact on the economy as they move through their life cycle. Companies have geared their products to meet the strong demand for goods and services created by baby boomers. In the early 1950s, the emphasis was on baby products and larger houses. In the 1960s, a similar age-related pressure was exerted on school facilities, creating a demand for more schools and teachers. As the baby boomers approach old age, the demand for health-care services is expected to rise. Companies are already targeting their advertisements at the growing number of retirees from the baby boom generation.

Table 4.6	Population Increase, 1851-2001		
Year	Population (000s)	Percentage Change	Average Annual Rate
1851	2,436.3	—	—
1861	3,229.6	32.6	2.9
1871	3,689.3	14.2	1.3
1881	4,324.8	17.2	1.6
1891	4,833.2	11.8	1.1
1901	5,371.3	11.1	1.1
1911	7,206.6	34.2	3.0
1921	8,787.9	21.9	2.0
1931	10,376.8	18.1	1.7
1941	11,506.7	10.9	1.0
1951	14,009.4	21.8	1.7
1956	16,080.8	14.8	2.8
1961	18,238.2	13.4	2.5
1966	20,014.9	9.7	1.9
1971	21,568.3	7.8	1.5
1976	22,992.6	6.6	1.3
1981	24,343.2	5.9	1.2
1986	25,309.3	4.0	0.8
1991	27,296.9	7.9	1.6
1996	28,846.8	5.7	1.1
2001	30,007.1	4.0	0.8

Sources: McVey and Kalbach (1995: 42); Statistics Canada (1997c, 2002a).

Natural Increase

From 1851 to 1971, most of Canada's population growth was due to natural increase. For most of this time, the majority of Canadians lived in rural settings where large families were the rule (Table 4.7). As most available farmland became occupied, children of farm parents had no choice but to seek their fortunes elsewhere. Many went to the cities in search of work. Birth rates were extremely high, while death rates were low, allowing for a high rate of natural increase. In the 1870s, natural increase exceeded 2 per cent per year. At that rate, Canada's population would double every 30 to 35 years. Like many social phenomena, such rapid growth failed to sustain itself because circumstances changed dramatically. As Canada became an industrial country, fertility rates declined. Except for a short time after World War II, the crude birth rates declined steadily from about 45 births per 1,000 people per year to approximately 13 births per 1,000 people. Over the same

Table 4.7	Canada's Rate of Natural Increase, 1851–2002			
Year	Crude Birth Rate	Crude Death Rate	Natural Increase (per cent)	Natural Increase (000s)
1851	45	20	2.5	61
1861	45	20	2.5	81
1871	42	20	2.2	81
1881	40	19	2.1	90
1891	38	18	2.0	97
1901	35	16	1.9	97
1911	32	14	1.8	129
1921	29.3	11.6	1.8	160
1931	23.2	10.2	1.3	138
1941	22.4	10.1	1.2	145
1951	27.2	9.0	1.6	255
1961	26.1	7.7	1.8	335
1971	16.8	7.3	1.0	205
1981	15.2	7.0	0.8	200
1991	14.3	7.0	0.7	207
1995/1996	12.6	7.1	0.6	163
2001/02	10.5	7.4	0.3	96

Data on births and deaths are now recorded from 1 July to 30 June.
Sources: Adapted from Statistics Canada (1997d, 2003d); McVey and Kalbach (1995: 268, 270).

period, the crude death rate dropped from about 20 deaths per 1,000 people per year to about seven. While there is no single explanation for these changes, public health measures such as water purification, improved nutrition, medical advances, and the development of public health-care systems across the country account for most of the sharp reduction in the mortality rate. Yet, the recent rebound in mortality rates to 7.4 per cent in 2001–2 signals the impact of Canada's aging population. The decline in the fertility rate is more complex. A series of social and economic changes caused parents to opt for smaller families. Among these changes, two stand out: the shift of people from rural areas to towns and cities,

and the sharp increase in the number of women in the labour force.

By the twenty-first century, Canada's crude birth and death rates were extremely low. These vital rates follow those found in other industrial countries. Consequently, Canada has a low rate of natural increase. By 2001–2, the natural rate was 0.3 per cent, and every indication points to a further decline in the coming years (Table 4.7). These trends in birth and death rates, which are consistent with the experience of other industrialized countries, are accounted for in the demographic transition theory.

Table 4.8	Phases in the Demographic Transition Theory	
Phase	**Birth and Death Rates**	**Rate of Natural Increase**
Pre-industrial	High rates often exceed 40 births and deaths per 1,000 people.	Little or no natural increase, though population fluctuations caused by a temporary rise in the death rate do occur.
Early industrial	Falling death rates and a continuation of high birth rates.	Extremely high rates of natural increase.
Late industrial	Falling birth rates and low death rates.	Extremely high rates of natural increase.
Post-industrial	Low rates often below 10 births and deaths per 1,000 people.	Little or no natural increase, though population fluctuations caused by a temporary rise in the birth rate do occur.

The Demographic Transition Theory

The **demographic transition theory** is the most widely accepted theory that describes population change in industrial societies. This theory is based on the assumption that changes in birth and death rates occur as a society moves from a pre-industrial to an industrial economy. These demographic changes occur in four phases, each of which has a distinct set of vital rates that coincide with the phases in the process of industrialization (Table 4.8).

A cursory examination of Canada's birth and death rates over the last 150 years reveals strong similarities to the early industrial, late industrial, and post-industrial phases. In the nineteenth century high birth rates and lower death rates were common. These vital rates are similar to those in the second phase of the demographic transition theory. Then Canada entered the late industrial phase, characterized by low death rates and a declining birth rate. By the 1980s, birth rates had declined significantly, resulting in little natural increase. Such rates characterize the post-industrial phase of

Vignette 4.5   The Concept of Replacement Fertility

The concept of replacement fertility refers to the level of fertility at which women have enough daughters to replace themselves. If women have an average of 2.1 births in their lifetime, then each woman, on average, will have given birth to a daughter and a son. The number 2.1 was determined to represent the minimum level of replacement fertility because, on average, slightly more boys than girls are born. In 1961, the Canadian fertility rate was 3.8 births per woman of child-bearing age (15–49). By 1991, it had dropped below replacement level to 1.8 births per woman of child-bearing age, and it remains well below replacement level today.

the demographic transition theory. Based on a more precise measure of fertility, demographers argue that Canada's rate of natural increase has fallen below its replacement level (Vignette 4.5).

Age and Sex Structure

Age and sex are the most basic characteristics of a population. Every region in Canada has a different age and sex composition and this structure can have considerable impact on its demographic and socio-economic behaviour. However, the greatest differences are found in age structure. Some regions have populations that are younger than the national average while others are older than the national average. Québec has the highest median age while the Territorial North has the lowest (Table 4.9). In the case of Québec, this means that a low proportion of the population is under the age of 15 while the opposite is true for the Territorial North. The explanation for this age difference is related to fertility rates—Québec has one of the lowest fertility rates in Canada while the Territorial North has the highest rates.

Since Confederation, Canada's population

has grown older. Aging of a population is associated with the process of industrialization and urbanization. In general terms, aging is explained by the demographic transition theory. In 1901, for example, nearly 35 per cent of the nation's population was under the age of 15. By 2001, however, this figure had dropped below 20 per cent due to the decline in fertility rates. Only the Territorial North still has a young population. In 2001, approximately 30 per cent of the Territorial North's population was under the age of 15. Again, the explanation lies in the higher fertility rate in this region compared to the national fertility figure. High fertility rates in the North are attributed to Aboriginal peoples, who constitute around 40 per cent of the northern population.

Canada's population is almost equally divided into males and females. In 2001, Canada had slightly more females than males. Since the probability of a woman giving birth to a female or male baby is slightly higher for males, the gender structure of a population should be slightly weighted in favour of males (Vignette 4.5). But other factors, such as migration and shorter life spans for males, can create an imbalance. Therefore, some regions

Table 4.9	Median Age, Percentage over 65, and Sex Ratio, by Region		
Geographic Region	**Median Age**	**Percentage over 65**	**Sex Ratio**
Québec	38.8	13.3	95.4
Atlantic Canada	38.5	13.3	94.9
British Columbia	38.4	13.6	96.5
Ontario	37.2	12.9	95.6
Western Canada	35.8	12.5	98.3
Territorial North	29.7	4.4	104.5
Canada	37.6	13.0	96.1

Sources: Statistics Canada (2002b, 2002c).

have populations with a higher proportion of males, while others have more females. In demographic terms, the **sex ratio** is the number of males divided by the number of females multiplied by 100. Accordingly, the sex ratio indicates the number of males for every 100 females in a population. A sex ratio of 100 represents perfect balance between the sexes, while a sex ratio greater than 100 would indicate more males than females. From 1871 to 1971, Canada's sex ratio exceeded 100. In 1911, it was nearly 113! This imbalance was due to the large number of male immigrants coming to Canada in the first decades of the twentieth century. Since 1981, the sex ratio has remained just below 100. In 2001, Canada's sex ratio was 96.1 (Table 4.9).

Within the geographic regions of Canada, only the Territorial North has a greater proportion of males, with a sex ratio of 104.5. Atlantic Canada has the higher proportion of females with a sex ratio of 94.9 (Table 4.9). The high proportion of males to females in the North reflects its frontier nature, that is, the tendency for more male migrants to move to the North for employment in the resource industry. Capital cities often have a higher proportion of females to males. This phenomenon is believed to be caused by the in-migration of large numbers of young female adults who are attracted to the job opportunities within the public sector.

Age Dependency

Canadian society is aging rapidly. For instance, the number of people 65 years old and over has increased sharply. In 1961, there were 1.4 million senior Canadians, but by 2001 this number had reached 3.9 million, forming 13 per cent of the total population. This trend, however, began long ago in Cana-

da's history. Since Confederation, Canada's demographic structure has changed from a youthful one in 1867 to a mature one in 2001. Social and economic implications of such shifts are far-reaching. One such impact is measured by age dependency, another by old-age dependency.

The **age dependency ratio** is the ratio of persons in the 'dependent' age groups (under 15 and over 64 years) to those in the 'economically productive' age group (between 15 and 64 years). The assumption is that productive members of society are those between the ages of 15 and 64, while unproductive members are either too young (under 15) or too old (over 64) to make an economic contribution. The purpose of the age dependency ratio is to compare the number of dependants with the number of economically productive members of society. While the assumptions underlying this measure are not perfect, the ratio provides an approximation of changes in the age structure of a population over time.

In Canada, the age dependency ratio declined substantially from 1961 to 2001. In 1961, the ratio recorded 70 dependants per 100 people of working age. By 2001, it had declined to about 48 per 100, which is where it remains today. This decline suggests that the costs borne by the working population have diminished. However, this decline has been significantly affected by the large number of baby boomers who have entered the workforce (leaving the under-15 group) over this period and, to a lesser degree, by immigrants to Canada. For instance, the percentage of Canadians under 15 years fell from 34 per cent in 1961 to just under 20 per cent in 2001.

The **old-age dependency ratio** measures only the number of people over 64 as dependants and compares it to the number of those in the productive age group. Given Canada's aging population, age dependency is an important issue. From 1961 to 2001, the old-age dependency ratio increased from 14 elderly Canadians per 100 people of working age to 21 per 100. This form of dependency occurs because, with an aging population, the proportion of people actively participating in the labour force decreases and the cost of supporting the elderly increases. The burden of paying for public services, such as health services, becomes greater for those still in the workforce.

Population Trends and Canadian Society

Canada has evolved from a dualistic society that was characterized by French and British peoples into a pluralistic society characterized by a much wider mix of the world's peoples. In this new Canada, the Constitution Act of 1982 has greatly liberalized society and, in doing so, has made it easier for all Canadians to find a place within the society. This change in Canadian demography began in earnest in the late nineteenth century with efforts to settle the Canadian West when a significant number of immigrants came from non-English-speaking countries in Central and Eastern Europe. More recently, the majority of immigrants have come from non-European countries.

Overall, then, the 13.4 million immigrants coming to Canada from 1901 to 2001 have transformed the country's demography, added new dimensions to its social and religious values, and created a pluralistic society. A pluralistic society, by definition, is composed of many groups of people of different cultural, linguistic, and racial backgrounds. People who speak different languages, hold different religious beliefs, and belong to different ethnic groups have transformed the cultural composition of Canadian society. Without a doubt, this demographic change can be attributed to immigration and, to a much lesser degree, to the growth of the Aboriginal population. The net outcome is a much more heterogeneous Canada. The role of immigration in redefining Canadian society and the nature and significance of ethnicity, religion, and multiculturalism are explored in the following sections.

Immigration

People have various reasons for moving from one place to another. 'Push' factors (such as wars, natural disasters, a declining local economy, limited job prospects, or political oppression) are unfavourable aspects that prompt people to relocate. 'Pull' factors (such as better career opportunities, proximity to family and friends, or a more favourable climate or neighbourhood) may make another location seem more desirable to inhabit. Whatever the reasons for relocating, successive waves of immigrants have not only increased the Canadian population but have also changed its ethnic composition and stimulated its industrial development by bringing labour, capital, and innovative ideas to Canada. In doing so, the newcomers have changed the human face of this country. For example, at the time of Confederation only 1.5 per cent of the Canadian population claimed non-European origin. By 2001, Canadians of non-European origin comprised 20 per cent of the total population.

Canada encourages immigration. Its immigration policy has three streams: family class, refugee class, and economic class. Most enter Canada under the economic class, which has been very successful in attracting skilled newcomers to Canada. Since the 1970s, the sheer volume of immigration has altered the country's demographic character. By 2001, 5.4 million people, or 18.4 per cent of the total population, were born outside the country (Statistic Canada, 2003c: 2). Immigrants have been an important component of society throughout Canada's history. At the beginning of the twentieth century, foreign-born individuals formed only 13 per cent of Canada's population. Between 1901 and 1921, this doubled to nearly 26 per cent due to large numbers of immigrants settling the western provinces. In the last half of the twentieth century, the volume and diversity of immigrants had a particularly powerful impact on Canadian society (Bourne and Rose, 2001: 107–13). Because of the growing number of immigrants who have come to Canada from Asia, Africa, and Latin America, the ethnic and racial composition of Canadian society has changed.

Early Waves of Immigration

As we saw in Chapter 3, the first wave of people came from Asia as long ago as 30,000 years. The descendants of these and other ancient hunters who followed the land bridge into Alaska, Yukon, and eventually the rest of the Americas became the Indian tribes. The second wave came from the coastal lands along the Bering Sea and their descendants form the Inuit of the Canadian Arctic.

The next wave is associated with the colonization of North America that brought European settlers to Canada. The French and British settlements in Atlantic Canada and Québec began in the seventeenth century. After the Conquest of New France by the British, almost all immigrants came from the British Isles and the former British colonies in America. The British authorities wanted to assimilate the French-speaking Canadians and they attempted to do so by settling English-

speaking immigrants in Québec. At the time of Confederation, the British formed the majority of the population in Montréal and a sizable minority in other cities in Québec. After Confederation, the annual influx of immigrants to Canada was usually less than 50,000 people. Few were from France because France no longer had a surplus population and Québec no longer had arable land available for new settlers. In sharp contrast, Western Canada represented the last agricultural frontier in the New World. With the opening up of land in the Prairies, the annual number of immigrants quickly surpassed 50,000, reaching a record number of 400,870 in 1913. By the 1920s, most arable land was occupied or was held by the Hudson's Bay Company and the Canadian Pacific Railway. From then to the 1950s, the annual number of new Canadians fluctuated, but rarely exceeded 100,000. For example, during the Great Depression of the 1930s, the level of immigration fell well below 50,000 people per year. Since World War II, the number has grown and now exceeds 200,000 per year (Figure 4.4). In 2000, Ottawa set the maximum number of foreign nationals to enter Canada as either landed immigrants or refugees at 1 per cent of the total population. In 2001, the maximum figure was 300,000, but the actual number of immigrants only reached 256,000 (Table 4.12, p. 185; for more on immigrants, see Appendix to this book).

The Origins of European Immigrants

The first European settlers were from France and the British Isles. In the first half of the nineteenth century, nearly 1 million immigrants came from Britain. All were English-speaking and most were Protestant. By 1840, most English-speaking immigrants came from

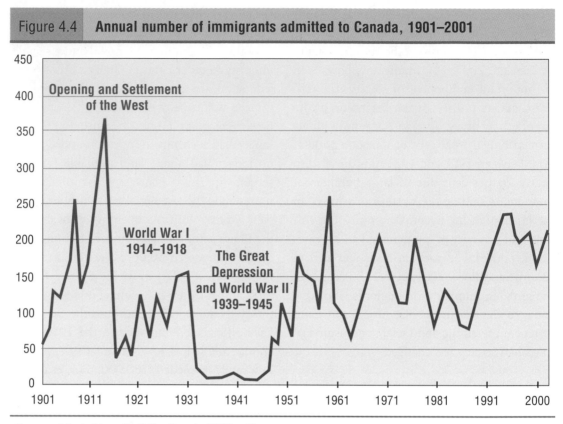

Figure 4.4   **Annual number of immigrants admitted to Canada, 1901–2001**

Sources: Adapted from Statistics Canada (2003c : 2).

famine-stricken Ireland. Many were Irish Catholics. Hundreds of thousands settled in the British colonies of Nova Scotia, New Brunswick, Prince Edward Island, Lower Canada, and Upper Canada. After American settlers from New England occupied most of the Ohio Basin, some Americans moved northward to Upper Canada where land was still available. Again, most American immigrants were English-speaking Protestants.

By the end of the nineteenth century, immigrants were coming from other European countries. The attraction was land in the Canadian West. Between 2 and 3 million homesteaders moved to the Canadian Prairies. Some came from Ontario, but most were from foreign countries, especially Europe and the United States. Over 1 million immigrants

came from Europe, including Britain, Germany, Scandinavia, Central Europe, Russia, and the Ukraine. Even today, the rural landscape in Western Canada reflects this ethnic influx with distinctive place names (Humboldt), church architecture (Greek Orthodox), and even landholdings (Hutterite colonies). By the 1920s, Canada's agricultural lands were occupied, causing the pattern of immigrant settlement to change once again. Most immigrants were now drawn to the cities to seek work in factories or to the hinterland to work in the forests and mines.

Recent Immigration

By the twenty-first century, immigration was oriented to Canada's cities. Equally important, the countries of origin of new immigrants had changed. Canada has become a favourable destination for people from around the world as a result of Canada's new immigration policy. In 1967, a change in immigration regulations heralded a new era. Immigration laws were relaxed to encourage immigration from all countries. In one gesture, the new regulations removed the 'preferred nationalities' clause, which favoured immigrants from the British Isles and other European countries, and replaced it with a universal point system for assessing a candidate's suitability for entry into Canada regardless of geographic origin or ethnic background. Known as the economic class or stream of immigrants, the main purpose was to help meet the need for skilled workers. Economic immigrants comprise nearly 60 per cent of all immigrants, while the family reunification and refugee classes make up just over 25 per cent and 15 per cent respectively.

Since 1971, a majority of immigrants have come from Asia with a lesser number from Africa, the Middle East, the Caribbean, and Latin America. This trend has become more and more dominant. For example, between 1981 and 1990, 70 per cent of new Canadians came from these world regions, and then from 1991 to 2001 this increased to 78 per cent (Table 4.10).

The non-European stream of immigrants has caused a dramatic shift in the ethnic and

| Figure 4.5 | **Proportion of immigrants born in Europe and Asia by period of immigration** |

Source: Adapted from Statistics Canada (2003c : 3).

religious composition of Canada's population (Figure 4.5). Before 1961, the major sources of immigrants were the United Kingdom, Italy, and Germany. For the period 1991–2001, the leading five countries were the People's Republic of China, India, the Philippines, Hong Kong, and Sri Lanka (Table 4.10). This shift in the origins of immigrants—from European countries to Asian, African, and other countries—has been the result of Canada's more liberal immigration policy.

The impact of this new immigration policy is illustrated in the following statistics: before 1961, 6 per cent of immigrants were born outside Europe, whereas from 1991 to 2001, 80 per cent of immigrants were of non-European origin (Table 4.10). As a result of this more open immigration policy, Canada has a more diverse society that includes non-European languages and non-Christian religions.

While most immigrants come to Canada under the economic class, Canada also has a policy of allowing refugees into the country: people who are fleeing the turmoil of war or political oppression are allowed into Canada if they declare themselves as refugees and if their claim is verified by Canadian officials. Refugees constitute about 15 per cent of the total number of immigrants.

The Growing Demographic Importance of Immigration

One measure of the importance of immigration to Canada is the proportion of immigrants in the national population. In 1996, immi-

Table 4.10	**Newcomers by Place of Birth**					
	Immigrated before 1961				**Immigrated 1991–2001[1]**	
	Number	**%**			**Number**	**%**
Total immigrants	**894,465**	**100.0**	**Total immigrants**		**1,830,680**	**100.0**
United Kingdom	217,175	24.3	China, People's Republic of		197,360	10.8
Italy	147,320	16.5	India		156,120	8.5
Germany	96,770	10.8	Philippines		122,010	6.7
Netherlands	79,170	8.9	Hong Kong, Special Administrative Region		118,385	6.5
Poland	44,340	5.0	Sri Lanka		62,590	3.4
United States	34,810	3.9	Pakistan		57,990	3.2
Hungary	27,425	3.1	Taiwan		53,755	2.9
Ukraine	21,240	2.4	United States		51,440	2.8
Greece	20,755	2.3	Iran		47,080	2.6
China, People's Republic of	15,850	1.8	Poland		43,370	2.4

1. Includes data up to 15 May 2001.
Source: Statistics Canada (2003c).

| Table 4.11 | Immigration by Place of Birth |

| | Before 1961 | | 1961–70 | | 1971–80 | | 1981–90 | | 1991–2001[1] | |
	Number	%	Number	%	Number	%	Number	%	Number	%
Total immigrants	894,465	100.0	745,565	100.0	936,275	100.0	1,041,495	100.0	1,830,680	100.0
United States	34,805	3.9	46,880	6.3	62,835	6.7	41,965	4.0	51,440	2.8
Europe	809,330	90.5	515,675	69.2	338,520	36.2	266,185	25.6	357,845	19.5
Asia	28,850	3.2	90,420	12.1	311,960	33.3	491,720	47.2	1,066,230	58.2
Africa	4,635	0.5	23,830	3.2	54,655	5.8	59,715	5.7	139,770	7.6
Caribbean, Central and South America	12,895	1.4	59,895	8.0	154,395	16.5	171,495	16.5	200,010	10.9
Oceania and other countries	3,950	0.4	8,865	1.2	13,910	1.5	10,415	1.0	15,385	0.8

Period of Immigration

1. Includes data up to 15 May 2001.
Source: Statistics Canada (2003c).

grants constituted 17.4 per cent of Canada's population, up from 16.1 per cent in 1991. By 2001, this proportion shifted even more in favour of immigration because the annual intake of immigrants increased. Over the course of the 1990s, 'approximately 2.25 million newcomers came to Canada, generating an increase of nearly 10 per cent to the national population' (Hiebert, 2000: 27). By 2001, immigrants comprised nearly 20 per cent of Canada's population. The reason for this increase is twofold: the rate of natural increase has declined, and Ottawa's quota for landed immigrants has increased.

A second measure of its importance is its growing impact on Canada's population growth. Until the 1990s, Canada's natural increase (the difference between births and deaths) exceeded immigration. As Figure 4.6 illustrates, the change over the last 30 years has been dramatic. In 1976, natural increase (births minus deaths) represented over 80 per cent of the demographic growth in Canada. Today the situation is almost reversed, with immigration representing close to 70 per cent of the population growth. Since these two demographic trends are expected to continue, it follows that the proportion of Canada's population growth attributed to natural increase could drop to 20 per cent and immigration increase to 80 per cent before the end of the first decade of the twenty-first century (Table 4.12).

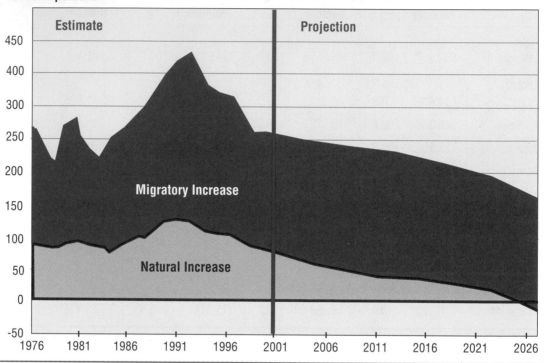

Figure 4.6   Immigration: An increasingly important component of population growth in Canada

Note: The projection is based on medium assumptions: fertility rates at 1.5; immigration at 225,000 new immigrants per year; a gradual increase in life expectancy from 77 to 80 years for men and 83 to 85 for women.

Sources: Adapted from Statistics Canada (2003k).

A third measure of the importance of immigration is its impact on the ethnic character of Canada. According to McVey and Kalbach (1995: 104), 'In the area of sociopolitical and cultural concerns, immigration affects the character of the resident population in direct proportion to its relative size, the degree to which its characteristics differ from those of the host society, and the nature of the interactions that take place between the two populations.' Prior to the 1980s, British immigrants and other European immigrants comprised the main immigration flow to Canada. They assimilated quickly into the existing cul-ture, which had its roots in Europe and many cultural links with Europe—the English and French languages and the British parliamentary system, to name but two links. Since the 1980s, but especially in the years since 1990, the flow of immigrants has shifted from Europe to Asia. The cultural and political impact on Canadian society has been significant, especially in the major cities. Newcomers, by concentrating in major cities, have modified the appearance of these places by bringing certain aspects of their culture, including language, dress, and foods, and by altering the architecture of buildings and

Table 4.12	Demographic Impact of Immigration				
Year	Births	Deaths	Natural Increase	Immigration	Difference
1992–3	403,107	201,808	201,299	265,405	64,106
1993–4	389,286	207,528	181,758	227,860	46,102
1994–5	382,870	212,830	170,040	216,988	46,948
1995–6	372,453	209,746	162,707	224,881	62,174
1996–7	357,313	217,220	140,093	193,452	53,359
1997–8	345,406	217,298	128,108	173,210	45,102
1998–9	338,963	222,538	116,425	205,469	89,044
1999–2000	333,954	229,138	104,816	200,000	95,184
2000–1	327,187	231,232	85,955	255,999	169,933

Source: Adapted from Statistics Canada (2003d).

houses. As well, interest in the traditional Canadian concerns, such as the bicultural nature of Canada and Aboriginal rights, may be of peripheral interest to these new Canadians. Such changes have produced tensions, but for the most part new Canadians not only have found a home in Canada but also have seen some aspects of their cultures being absorbed into the majority society.

In the years to come, immigration is likely to play an even larger role in Canada's population growth. The demographic implications for Canada are:

- The increase in immigration will offset the losses in natural increase, thus keeping Canada's population growth stable.
- Immigrants will form a larger proportion of the national population, thus further increasing the cultural diversity of Canada's pluralistic society.
- Canada's increasing diversity will call into question the validity of the vision of two founding peoples, and could affect the position of Aboriginal peoples within Canadian society and create an ethnic divide between rural and urban Canada.

Ethnicity

Canadian society is composed of many ethnic groups.[3] In fact, more than 200 different ethnic origins were reported in the 2001 census question on ethnic ancestry (Statistics Canada, 2003c). In the same census report, the 10 largest ethnic origins based on responses were Canadian (39.4 per cent), English (20.2 per cent), French (15.8 per cent), Scottish (14.0 per cent), Irish (12.9 per cent), German (9.3 per cent), Italian (4.3 per cent), Chinese (3.7 per cent), Ukrainian (3.6 per cent), and North American Indian (3.4 per cent). An **ethnic group** is made up of members of a population who share a culture that is distinct from that of other groups. Each group has a common identity, shared values, and cultural/linguistic/religious bonds and symbols. **Culture** is the learned collective behaviour of a group of people.

In spite of these varied origins, time erodes the ethnic connection with the country of origin. For Canadians born and raised in this country, the connection with their original (foreign) ethnic origins is tenuous at best (Beaujot, 1991: 297). Place, as cultural geographers insist, plays a critical role in the development of a regional/national identity. This phenomenon is well known. Consider, for example, the attachment of the French born in New France who had no interest in returning to France in 1763, but whose lives were centred on their part of the New World. They were no longer French but had become French Canadian. Not surprisingly, then, the ethnic selection of 'Canadian' by nearly 40 per cent of Canadians is attributed to the geographic notion of place overriding the concept of ethnicity (Table 4.13). To put it differently, the resettlement of people in a new place causes

their ethnicity to fade with time, especially if marriages take place with members from other groups.

The dominance of the British and French ethnic groups in Canada weakened in the years before World War I with the flood of non-British and non-French migrants settling the Prairies, and some of these people eventually moved to the towns and cities of industrial Canada. After World War II, immigrants from all over the world came to Canada in ever-increasing numbers. Canadian society now consisted of a number of ethnic groups whose members shared a sense of identity based on descent, language, religion, tradition, and other common experiences. Overarching these shared ethnic experiences was a sense of belonging to Canada, and since 1971 Ottawa has promoted multiculturalism, the idea that a cultural identity and a national one are not

Table 4.13   Ethnic Origin of Canadians

	2001 Number	%		1996 Number	%
Total population	29,639,030	100.0	Total population	28,528,125	100.0
Canadian	11,682,680	39.4	Canadian	8,806,275	30.9
English	5,978,875	20.2	English	6,832,095	23.9
French	4,668,410	15.8	French	5,597,845	19.6
Scottish	4,157,215	14.0	Scottish	4,260,840	14.9
Irish	3,822,660	12.9	Irish	3,767,610	13.2
German	2,742,765	9.3	German	2,757,140	9.7
Italian	1,270,369	4.3	Italian	1,207,475	4.2
Chinese	1,094,700	3.7	Ukrainian	1,026,475	3.6
Ukrainian	1,071,055	3.6	Chinese	921,585	3.2
North American Indian	1,000,890	3.4	Dutch (Netherlands)	916,215	3.2

Note: Table shows total responses. Because some respondents reported more than one ethnic origin, the sum is greater than the total population or 100%.
Source: Statistics Canada (2003c).

mutually exclusive. The results of this policy are reflected in Table 4.13.

In Québec, a strong sense exists among the people of their French origins, but it is tempered by centuries of life in North America. The term 'Québécois' more accurately reflects their ethnic background and continues to drive that province's society and politics. History and geography are the causes of this sense of ethnic nationalism. Nationalism is nurtured by Québec's place, not in Canada but in North America. As a French enclave in the English-speaking North American continent, Québec naturally fears for its cultural existence. Its close proximity to the United States underscores its precarious position in North America much more than its 'sheltered' position within Canada. The United States exerts an enormous cultural impact on the world through the mass media and now, increasingly, through the Internet. Within North America, American culture poses the major threat to French- and English-Canadian cultures. This fear of assimilation is the basis of ethnic nationalism in Québec, a nationalism that is generally expressed by the Québécois in one of two ways: (1) Québec society is distinct within Canada; (2) it is only through independence that Québec will be fully protected and emancipated.

Dual loyalties exist in all geographic regions. In provinces like Alberta, a sense of togetherness exists, though it is not as strong as in Québec. Disagreements with Ottawa—whether over oil, gun control, or participation with the United States in the Iraq War—often unite Albertans. Québécois take an even stronger position vis-à-vis Ottawa, and, given two referendums, the possibility of separation from the rest of Canada must be taken seriously. Besides performing as a provincial government, the Québec government is also the protector of their culture and language and a major participant in the Québec economy.

Language

Language is a key component of ethnicity. Indeed, language represents the most durable link to the past and the tool for maintaining a culture. Canada's two official languages are English and French. Approximately 85 per cent of Canadians use one of the two official languages. The remainder, who speak other languages, eventually choose English and/or French. New Canadians who speak neither official language hold the key to the future balance between English and French. Where they settle and what official language they learn impacts on the French and English societies. From a political perspective, issues surrounding the two official languages are a delicate matter. In fact, this topic is the basis of a population faultline discussed later in this chapter.

A number of French organizations across Canada aim to protect and promote their language and culture. Such organizations have received federal funding to finance their cultural and educational activities. This public support, plus the presence of French-language radio and television programs, has resulted in a rejuvenation of these provincial organizations.

Many languages, other than French or English, are spoken across the country, especially in larger cities where new immigrants tend to settle. Without the official-language status enjoyed by French and English, however, other languages are often difficult to maintain. The survival of these other languages in Canada is bolstered somewhat by continued immigration and aided by the federal government, which provides funds for heritage language classes.

However, Canadian-born members of ethnic groups—the second and subsequent generations of immigrants—usually lose the language of their parents and grandparents. The same loss of language has affected Aboriginal Canadians. Newhouse (2000: 402–3) sees the emergence of English as the lingua franca among Aboriginal peoples. By 2001, only three of the 50 languages spoken by Canada's Aboriginal peoples had a secure future. They are Cree, with 80,000 speakers (down 7,500 from 1996), Inuktitut, with 29,700 (up nearly 2,000), and Ojibwa, with 23,500 (down 2,400).

Religion

Religion is a key element of culture. The changing nature of Canada's religious composition is indicative of the changing cultural mix of people. The recent shift in Canada's religious makeup is linked with the large number of immigrants arriving from non-Christian countries. Over the past 30 years, the number of Canadians who subscribe to religions such as Islam, Hinduism, Sikhism, and Buddhism has increased substantially.

Culture is not only a durable link to the past but also provides the institutional organization to preserve ethnicity. Religious organizations provide an institutional structure that consolidates people of similar beliefs. One example is the Roman Catholic Church. In the early history of Canada, the Roman Catholic Church helped sustain Catholicism, the French language, and the French-Canadian way of life. While its role has diminished in recent decades, the Church was the dominant cultural force among francophones from the conquest of New France to World War II. Not only did it give spiritual direction to French Canadians within Québec but it also organized its parishes to sponsor group immigration to

more isolated areas of Québec (such as the Clay Belt in northwestern Québec), to other provinces, and to New England. In these group migrations, priests provided the leadership for the move and for organizing the new settlement, including its educational, social, and religious structures.

Religious freedom is the right of all Canadians. This right was inscribed in the British North America Act and is contained in the Constitution Act of 1982. In fact, many immigrants have come to Canada seeking religious freedom. The Doukhobors, a sect that broke away from the Russian Orthodox Church, are one such religious group. They immigrated to Canada in 1899 to escape persecution in Russia, settling initially in Saskatchewan and later moving to an isolated area in the southern interior of British Columbia, where they could live a communal life, practise their religion, and speak Russian. While most Doukhobors have integrated into Canadian life, a minority remain committed to their religion, a communal lifestyle, and pacifism.

Canada is thought of as a Christian country. This image was certainly true in 1867, but today Canada is a much more religiously diverse nation. It is true that most Canadians are Christians. For example, in the 1960s nearly 90 per cent of Canadians declared themselves to be Christian (though some may not have been active church members). By 2001, this figure had dropped below 80 per cent (Statistics Canada, 2000j). At the same time, the number of Muslims, Hindus, Buddhists, and Sikhs had more than doubled from 1991 to 2001, and by 2001 these four religions had followers who made up 5 per cent of Canada's population (ibid.).

Roman Catholics remain the largest group of Christians, comprising in 2001 44 per cent of all Canadians, followed by Protestants at 29 per cent, while other Christians comprise another 4 per cent. According to the 2001 census, most non-Christians adhered to the Islamic faith, followed by Judaism (ibid.). By 2001, Canada's spiritual world had become much more diversified. There has also been a revival of the spiritual practices of Aboriginal Canadians that sometimes play a role in the healing of past injustices and in rehabilitating Aboriginal people who are serving time in prisons.

Religion and Education

Religion has been an important, though sometimes controversial, component of education in Canada. Under the British North America Act, education was assigned to the provinces. This decision was largely due to Québec's desire to have the Roman Catholic Church administer its education program for its French-speaking population and thereby ensure the future of the French language and the Catholic religion. Since each province had a somewhat different demographic mix of Catholic and Protestant residents, public funding of education in the 10 provinces has varied somewhat to meet those demographic situations. One solution has been dual education systems that saw public funding for a public school board that served Protestant families and another school board for Catholic families. Upon entering Confederation, Québec supported both Catholic and Protestant school boards, while Ontario, Alberta, and Saskatchewan funded both public and Catholic school boards. New Brunswick, Nova Scotia, and Prince Edward Island provided funds for only public schools. When British Columbia joined Canada, it also funded only public schools. Since the 1970s, British Columbia has provided tax support to private schools, while the Maritime governments have given public funds to support Catholic schools

when there is a local demand. Even as late as 1949 when Newfoundland became a province of Canada, this provincial government insisted that, under the Terms of Union with Canada, Newfoundland would have three types of denominational school boards (one Roman Catholic and two Protestant). However, by the turn of the twenty-first century, Newfoundland and Labrador had moved to a single 'public' education system, while Québec had replaced the Catholic and Protestant school

The Basilica of Notre Dame was opened in 1829 in what is now 'Old Montréal'. At that time, the Basilica was the largest religious edifice in North America. The interior of the Basilica features finely sculpted polychrome wood decorated in gold leaf and a star-spangled vault. While now surrounded by tall buildings, the physical presence of the Basilica of Notre Dame is reminder of the Roman Catholic Church's powerful role in the history of French-Canadian society. (Kennon Cooke/Valan Photos)

boards with secular boards based on the English and French languages.

The link between religion and education remains a sensitive topic. With the enormous growth of Islam and other non-Christian religions and the secularization of public schools, the link between education and religion has become more complicated. With public schools no longer providing a Christian environment within which education is provided to students, private Protestant and non-Christian-run schools have grown in number and many have sought public funding. Since the dual education solution that emerged in the nineteenth century does not fit the demographic realities of the twenty-first century, the demand to use public funds for religious-run schools has taken on a new direction.

Multiculturalism

Canada has become a multi-ethnic and multi-cultural society. In 2001, as noted above, more than 18 per cent of Canada's population was foreign-born, as compared to 11 per cent in the United States. While multiculturalism is an official policy of the federal government, it is also a demographic reality stemming from Canada's immigration policy.

Multiculturalism as official government policy was announced by Prime Minister Pierre Trudeau in 1971. This policy recognizes all Canadians as full and equal participants in Canadian society. Multiculturalism is a form of cultural pluralism. The federal government announced its support for multiculturalism as part of its response to the Report of the Royal Commission on Bilingualism and Biculturalism, *Book IV* of which was titled *Cultural Contributions of the Other Ethnic Groups*. In fact, the federal government was simply responding to a new reality, namely, that Canada had evolved

into a more complex society composed of many ethnic groups, such as German and Ukrainian Canadians. At the same time, the number of immigrants coming from non-European and non-Christian countries was increasing. Within a decade, multiculturalism took on a new direction—one that focused on newcomers from the global community outside of Europe. While Ottawa has recognized Canada's pluralistic society, this recognition is nonetheless based on the assumption that most immigrants will integrate into Canadian society. Language is a key in this integration process as newcomers realize that their economic and social future depends on fluency in one or both official languages, as well as on a recognition of the broad cultural values found in Canadian society.

The ultimate goal of the 1988 Canadian Multiculturalism Act is to encourage greater human understanding and stronger bonds among Canadians of different cultural backgrounds and ethnic origins. Multiculturalism is a distinctively Canadian approach to social equality in nation-building, encouraging respect for cultural diversity. In sum, multiculturalism is the shared vision of people of diverse cultural backgrounds seeking to live together in harmony.

Multiculturalism, however, has its detractors. Some Canadians fear that multiculturalism might increase ethnic group identification at the expense of Canadian social cohesion. Ottawa, however, has maintained that multiculturalism binds Canada's diverse populations together and the government sees no conflict between Canadians having two identities—a cultural identity and a national one. With most immigrants settling in Toronto, Montréal, and Vancouver, the impact of new cultures and religions is most visible in those cities. As these ethnic groups have put their cultural stamp on the cityscape of Canada's largest cities, they have created more diverse and more cosmopolitan cities. Nevertheless, tensions have arisen from time to time as the new cultures express themselves by altering the traditional architecture of Canada's cities, by imposing new functions within these cities, and by creating new social networks.

Multiculturalism, Québec, and the Separatist Movement

Québec remains a strong advocate of a Canada based on the principle of two founding peoples. In this vision of Canada, multiculturalism presents a third political force that poses a challenge to the two-nations vision of Canada. In 1971, Premier Robert Bourassa of Québec wrote to Prime Minister Trudeau, arguing that multiculturalism contradicted the principle of 'the equality of the two founding peoples'. He further stated that Québec would not be introducing multiculturalism in its jurisdictions but would continue to ensure the well-being of the French language and culture in its territory and beyond (McRoberts, 1997: 129). Claude Ryan, a leading spokesman for Québec in the late twentieth century, maintained that multiculturalism 'omits a central fact: the two official languages of Canada, far from existing in the abstract as the subject of juridical definitions, are the expression of two cultures, two peoples, two societies which give Canada its distinctive shape' (ibid.). Multiculturalism, from the perspective of Québec leaders of the 1970s, widened the gulf between French- and English-speaking Canadians, and posed a threat to the French language and the Québécois culture. Ironically, Québec's low birth rate has caused its government to seek immigrants from French-speaking countries in Africa and the Caribbean. In this way, Québec has added a new dimension to its French-speaking com-munity, but one that has a different historic origin and that adds quite different cultural and racial components to Québec society. The commitment and contribution of these newcomers to Québec culture and the French language are well demonstrated, but their support for the separatist movement is tenuous at best and remains a key political issue within Québec society.

The Economy

Canada ranks as a 'middle' economic power in the world. Canada has the smallest economy of the G-8 countries (United States, Japan, United Kingdom, German, France, Italy, Canada, and Russia). Still, in 2002, Canada was the only G-8 country that had a budgetary surplus even though Ottawa embarked on the biggest public spending expansion in 20 years. Most new funding was directed at health care and other social programs, leaving very little for Canada's armed forces.

Since the mid-1990s, Canada's economy has grown and, consequently, employment levels have increased. From 1997 to 2002, the number of people employed increased from 13.7 million to 15.4 million, while the unemployment rate dropped from 9.6 to 7.7 per cent (Figure 4.7). By 2002–3, Thorpe (2003: FP4) reported that the main economic highlights were:

- Canada recorded the only fiscal surplus among the G-8 countries.
- Canada's foreign debt as percentage of GDP is at its lowest level in more than 50 years and will soon be lower than the percentage of foreign debt in the United States.
- Federal debt has dropped from a high of $550.3 billion in 1996–7 to $500 billion in 2002–3.

• Debt-servicing costs have dropped from about 36 cents of each revenue dollar in 1995–6 to around 20 cents in 2002–3.

Growth in the Private Sector

Driven by exports to the United States, the private sector, led by manufacturers and retailers, recorded the second highest corporate profits in 2002 when profits reached $156.3 billion (Figure 4.8 and Statistics Canada, 2003b: 1). Automobile production provided much of this profit. Automobile companies and automobile parts firms saw profits improve by a third over

2001 figures. Strong US demand and brisk sales in Canada due to low interest rates and incentive for customers accounted for the high profits and exports. The highest annual corporate profit took place in 2000 when profits in energy, electronics, and telecommunications drove profits to $176 billion (Statistics Canada, 2003b: 1).

In 2002, the resource sector saw record energy exports to the United States, but wood and paper producers' operating profits fell by a third, to $3.4 billion. Profits were squeezed by the imposition of a duty, now averaging 27 per cent, on Canadian lumber entering the United States. The duty was imposed because US lumber producers felt that Canadian lumber imports were hurting their business and demanded that their government impose trade restrictions. Ironically, Canadian producers reacted by increasing production to lower per unit costs, thus allowing them to compete even more effectively in the US lumber market. In spite of NAFTA and the World Trade Organization ruling in May 2003 that the US was wrong in imposing duties against Canadian softwood lumber, the lumber trade dispute shows no signs of ending.

The North American Market

Three trade agreements—the Auto Pact of 1965, the FTA of 1989, and the North American Free Trade Agreement of 1994—led to a realignment of Canada's economy. First, Canada's automobile manufacturing moved from supplying a national market to a North American one. The result was twofold: a shift from producing an entire range of vehicles to a more specialized form of production that led to economies of scale and a reorientation of its transportation routes from an east/west automobile trade axis to a north/south one. Sec-

| Figure 4.7 | **Unemployment rate for Canada, 1997 to March 2003** |

Unemployment Rate (seasonally adjusted)

Sources: Adapted from Statistics Canada (2003i).

ond, Canada's trade saw more and more manufactured goods exported to the United States, thus breaking the old pattern of exporting primarily low-value unprocessed or semiprocessed resource products. Third, the volume of exports to the United States grew dramatically. Fourth, in spite of these trade agreements, Washington is prepared to defend US business interests by imposing trade barriers, as has been the case with duties on softwood lumber and grain, to restrict the natural flow of certain Canadian goods into the United States. The purpose of these duties is twofold: (1) to protect US farmers and forest companies in the short run by imposing duties on targeted Canadian exports, and (2) to force Canada to accept a long-term agreement that will limit its grain and lumber exports.

For better or worse, Canada's economic interests (some would say dependency) have tilted more and more towards those of the United States (Norcliffe, 1996: 44). Restructuring of Canada's manufacturing firms is one consequence. During the early 1990s, restructuring placed enormous pressure on companies and their workers. To become more efficient, many businesses reduced the number of their employees and specialized in a smaller product range. Other firms simply could not compete in the Canadian environment with its high labour costs and moved their operations to countries where labour is cheaper. Mexico attracted many such firms because Mexico, as a member of the North American Free Trade Agreement, offers easy access to North American markets.

From a transportation perspective, the Canadian National Railway represents this new duality as 'North America's Railroad'. With a continental economy in place, a North American core/periphery spatial framework has been superimposed on the traditional

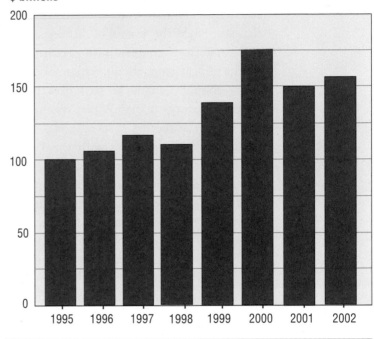

 Figure 4.8   **Annual profits, 1995–2002**

$ billions

Sources: Statistics Canada (2003b: 2).

Canadian version (Figure 4.1). The increasing integration of the North American economy has had both positive and negative impacts. Improved access to the American market has boosted Canadian production and trade and, consequently, Canadian living standards. One telling measure of the degree of that integration is revealed in Canada's international merchandise trade. In 2002, nearly 85 per cent of Canadian exports went to the United States while 71 per cent of its imports came from the US (Statistics Canada, 2003a: 1). Yet, trade disputes have hurt Canada's economy and certain regions have felt the full weight of these trade restrictions. The ongoing lumber dispute has affected the forest industry, especially small producers in British Columbia and Québec. For that reason, tariff and non-tariff barriers still has a substantial effect on Canada–US trade in certain industrial sectors.

Trade with the United States is critical for Canada's economy. This economic dependency raises a host of questions related to Canada–US relations. Since the events of 11 September 2001, America's concern with security has raised a critical geopolitical question for Canadians—should Canada be part of a North American security zone? While Washington puts security at the top of its agenda, Ottawa remains focused on trade with the United States. Therefore, if the United States tightens its border regulations, the implications for Canadian goods moving more slowly across a less open border are enormous. As pointed out in Chapter 1, delays in border crossing would greatly hamper trade, especially with manufacturing firms dependent on just-in-time shipments of parts. Even more important for the long run, international companies, such as Toyota, could divert capital investment in new manufacturing plants to the United States. The federal government, while seeking closer eco-

nomic relations with the United States, does not want to lose its political independence. In 2003, America led the attack on Iraq and Washington sought support from other countries. Without the approval of the United Nations for the war against Iraq, Ottawa decided not to take part in the US/British/Australian invasion force. While the consequences of Ottawa's political decision remain unknown, the long-term interests of both countries call for ever expanding trade between the two countries within the framework of the North American market.

Economic Structure

Like other modern industrial countries, Canada's economic structure has evolved from an agrarian to an industrial economy. By structure, we mean the relative share of primary, secondary, tertiary, and quaternary economic activities in Canada's economy at different times (Table 4.14). In the early nineteenth century Canada's economy was agrarian, meaning that most people were engaged in primary

Table 4.14	Types of Economic Structure
Economic Sectors	**Characteristics**
Primary	Economic activities that are concerned with natural resources of any kind, such as fishing, forestry, mining, and trapping.
Secondary	Economic activities that process, transform, fabricate, or assemble the raw materials derived from primary activities, or that are high-tech knowledge-based activities that reassemble, refinish, or package manufactured goods. Examples would include an automobile factory, a meat-packing plant, a pulp mill, and a software company.
Tertiary	Economic activities that involve the sale and exchange of goods and services, including professional services, retail sales, and education.
Quaternary	Economic activities that deal with the handling and processing of knowledge and information that lead to decision-making by companies and governments. Such activities are often found in the research and development units of companies engaged in primary, secondary, and tertiary activities.

Vignette 4.6   Restructuring Continues

Since the Canada–US Free Trade Agreement in 1989, companies have had to find ways to remain competitive in the North American marketplace. One approach is to reduce the number of workers in Canada, where wages are much higher than in Mexico and many Asian countries. Such reductions are known as restructuring and restructuring has made life particularly difficult for Canadian workers and their families. High taxes, reduction in public services, and the continued closure of small firms and downsizing of the workforces of large companies eroded Canadians' sense of well-being. In the early and mid-1990s, the federal and provincial governments cut deeply into spending on health care, education, social services, and environmental protection in order to balance their budgets. At the same time, trade liberalization fostered strong competition between firms and this competition kept salaries and wages in check.

While the 1990s had some bright spots, particularly in the automobile and high-technology sectors, other industrial sectors remained under pressure to compete in the global marketplace. To do so, firms often reduced the size of the labour force, went out of business, or moved to Mexico. Such restructuring continued into the twenty-first century. Here are some examples. Falconbridge, despite a favourable outlook for profits in 2000, planned to cut 140 of its 1,740 clerical, technical, and other staff at its mining operations in Sudbury (Robinson, 1999: B3), while Canadian Pacific slashed 10 per cent of its workforce (some 1,900 jobs). Earlier, Canadian National announced 3,000 job cuts (Chase and Jang, 1999: A1). These cuts, the companies argue, are necessary to stay efficient in a very competitive marketplace. In the same year, labour-intensive manufacturing firms failed due to fierce foreign competition. The Bata shoe manufacturing plant at Batawa, Ontario, was closed because shoe manufacturing is only profitable in countries with very low labour costs (McArthur, 1999: B1). Rennie Inc., a manufacturer of men's shirts for 44 years in Guelph, Ontario, is unable to compete against low labour costs in less-developed countries. The company employs about 500 workers. The owner, John Rennie, said, 'These are super people—some have been here from day one. It's not a comfortable feeling, having to let these people go' (National Post, 1999a: C3). American branch plants in Canada continue to feel the pressure. Paragon Trade Brands announced the closure of its Canadian plant at Brampton, Ontario, because it is too small and too far from the company's US customers (National Post, 1999b: D2). While the 113 employees manufactured a high-quality product in an efficient operation, distance and economies of scale worked against it. Levi Strauss & Co. closed 11 of its 22 North American plants, including one at Cornwall that employed some 500 workers, because of lower wages in other countries. Bob Kilger, Liberal MP for Cornwall, said, 'It's a real sad day in the community. The plant has been here for 27 years' (Steinhart, 1999: C1). Noma Industries is one of 350 Canadian companies operating in Mexico. Speaking for Noma, its chief financial officer, Douglas Shields, said, 'For us, it's really the access to lower labour rates and to be able to get our products back to our US customer base in one or two days' (McKenna, 1998: A1). The Canadian plants located in Mexico are heavily concentrated in three fields—automotive parts, electronics, and textiles—all of which are labour-intensive activities whose products are sold in the United States and Canada.

activities (see Vignette 4.3). As Canada developed an industrial base in the late nineteenth century, many of its workers were involved in secondary activities. However, this workforce was concentrated in major cities, especially those in the manufacturing belt of Central Canada. The rest of the country was involved in primary activities such as agriculture, forestry, and mining. This created a geographic division in Canada's economic structure, which took on the appearance of an industrial core and a resource hinterland.

After the industrial boom of the post-World War II era, the economic structure of Canada, like that of other modern industrial states, changed again, especially beginning in the 1970s, marked by the US dollar going off the gold standard in 1971 and the OPEC oil crisis of 1973, when the Organization of Petroleum Exporting Countries agreed to reduce production to increase the price of a barrel of oil. The widespread introduction of personal computers in the 1980s and incredible growth of the Internet from the 1980s into the 1990s furthered this economic shift. This time the change was from a manufacturing-dominated economy to a service-oriented one, marking Canada's entrance into the post-industrial world. This phenomenon is also evident in each region of Canada, though manufacturing and service activities remain concentrated in Central Canada and major cities like Vancouver, Calgary, and Halifax. Other important changes, particularly related to the labour force, included an increase in the number of female workers and in the tertiary sector.

Canada's Economy and Labour Force

Canada's economy is growing and changing. During the last decade, four factors have shaped the character of the nation's economy and its labour force. First, the increasing demand for skilled workers, especially in the fields of computer technology, is a key factor. Second, the working-age population is now composed of older workers who, when they retire, will create a demand for experienced workers that may be hard to fill. Third, immigrants will fill more and more positions in the labour force. Fourth, the continuing pressure to keep costs low has forced firms to continue to reduce their relatively high-cost labour forces (Vignette 4.6).

By 2002, the size of the Canadian labour force had reached 16.7 million. As Canada's economy became more complex and interdependent, its labour force adjusted to these changing circumstances by increasing work specialization. Consequently, the type of work and the participation of different members of the population in the workplace have changed dramatically. Aspects of the labour force are examined in this section from these perspectives:

- structure of the labour force;
- domestic market;
- employment and unemployment rate;
- earnings of Canadians;
- women in the labour force.

Structure of the Labour Force

Major shifts in the percentage of workers in the primary, secondary, tertiary, and quaternary sectors are revealed in Table 4.15. These shifts mark three fundamental transitions in the nature of the Canadian economy as discussed at the beginning of this chapter—from agrarian or pre-industrial to early industrial, from early industrial to late industrial, and from late industrial to post-industrial. The

Table 4.15	Major Sectors of Canada's Labour Force			
Development Stage	Year	Primary	Secondary	Tertiary*
Agricultural	1881	51	29	19
Early industrial	1901	44	28	28
Late industrial	1961	14	32	54
Post-industrial	1991	6	21	73
	2001	4	21	75

*Because data for the quaternary sector is not available, percentages for the tertiary sector conventionally include both tertiary and quaternary jobs.

Sources: Adapted from McVey and Kalbach (1995: Table 10.3); Statistics Canada, CANSIM II, Table 282–0008, Ottawa, 2002.

post-industrial economy is marked by the growth of high-technology industries such as telecommunications.

The country's post-industrial economy is characterized by an expanding tertiary/quaternary workforce that accounts for three-quarters of the workers, a slowly growing secondary or manufacturing sector that employs 21 per cent of the labour force, and a shrinking primary or resource sector that now accounts for only 4 per cent of the labour force (Table 4.15). As shown earlier in Table 4.14 and discussed in Vignette 4.3, the primary sector in Canada includes resource activities, the secondary sector focuses on manufacturing activities, and the tertiary/quaternary sector provides a wide variety of services. Additional statistics on the Canadian labour force and the economy are included in the Appendix to this book.

Primary Sector

The primary sector forms a tiny but important part of Canada's economy. In the nineteenth century, the primary sector dominated Canada's economy and accounted for approximate-

ly half of the labour force (Table 4.15). As workers' wages rose, companies looked to ways to increase output per worker. This was achieved by substituting machinery for labour. Two developments then took place—output increased and the labour force decreased. Today, 4 per cent of the labour force is employed in primary industries. Yet, the resource economy continues to play an important role on the world stage by providing close to 30 per cent of the value of Canada's exports (Table 4.16). Resource industries are found in all six geographic regions. However, their importance to these regional economies varies. The primary sector forms a larger part of the economies of peripheral regions, with the Territorial North most dependent on its resource industry.

Secondary Sector

With the Canada–US Free Trade Agreement, the manufacturing industry underwent severe restructuring. Some jobs were lost as firms could not compete with the foreign firms. Firms with labour-intensive operations, such as those producing textiles, shoes, and gar-

Table 4.16	Changes in Percentage Distribution of Canada's Exports, 1995–2002		
Export	1995	2002	Difference
Agriculture and fishing products	7.9	7.5	(0.3)
Energy products	7.7	12.0	4.3
Forest products	13.9	9.0	(4.9)
Industrial goods and materials	19.2	16.9	(2.3)
Machinery and equipment	21.1	23.5	2.4
Automobile products	23.7	23.4	(0.3)
Other consumer goods	3.1	4.2	1.1
Special transactions and adjustments	3.4	3.5	0.1
Total (billion$)	265.3	414.3	

Source: Statistics Canada (2004).

ments, saw their sales drop as retail stores purchased lower-priced imported goods. The problem was simple: labour costs in China and many other foreign countries were only a fraction of Canadian wages. For the companies, the solution was simple: shift operations to countries with low wages. Other jobs disappeared because the rationale of American branch plants in Canada ceased when Canadian tariffs on imported goods dropped. As a result, many American branch plants in Canada ceased to operate because it was more efficient to consolidate their operations within the United States.

In the same decade, two positive developments occurred in the secondary sector. First, automobile assembly production and automobile parts operations expanded, generating many new jobs directly and indirectly. The now defunct Auto Pact was largely responsible for this expansion in Canada's manufacturing industry, though the General Motors of Canada plant in London, Ontario, that produces lightweight combat vehicles is dependent on US military contracts. Second, a new manufacturing activity emerged around the application of new technology to Canadian firms. As Barnes et al. (2000: 11) wrote:

The impact of computer and information technology is visible at every turn in Canada's urban centers, and even the countryside bristles with receiver and transmitter towers. Behind these visible markers of the knowledge economy are a series of industries that produce hardware and software, and firms that incorporate such technology into their products and production processes.

New technology takes the form of high technology (high tech), which is well established in four major areas: biotechnology, information technology, materials technology, and transportation technology. Biotechnology involves harnessing the power of DNA, the double-helix-shaped chemical strand at the centre of each living cell. Genetically modified canola seeds, for example, are widely used by farmers in Western Canada because of higher yields and resistance to disease. Information technology includes information-based, computer-driven, and communications-related activities. Ottawa, the Silicon Valley of the North, specializes in telecommunications but it also has many companies that produce other forms of hardware, such as computer chips, and a variety of software. The software (program) makes use of the hardware (computer) to generate a product, such as text. Materials technology refers to new metal alloys, plastic-coated metals, and laminated glass. Many of these products are produced in Ontario and Québec for various manufacturing companies ranging from automobile to aircraft companies. Transportation technology has created high-speed trains, super tankers, and modern passenger aircraft. Québec-based Bombardier produces both high-speed trains and modern passenger aircraft.

All these high-tech companies have three important qualities: (1) they are value-added industries; (2) they employ highly skilled and highly paid workers; and (3) they depend on and frequently conduct or support pure research (sometimes in conjunction with research scientists at universities or government research centres) to develop a commercial product.

Most high-tech firms are located in the major cities of Canada's industrial heartland. The three leading centres are Toronto, Montréal, and Ottawa. In the case of Ottawa, high-tech firms such as Nortel are located on its outskirts, at Kanata. High-tech firms employ large numbers of computer programmers, and the geographic location of these programmers reveals that nearly half are working in Ontario. Québec has 29 per cent, followed by Western Canada at 13 per cent, British Columbia at 9 per cent, and Atlantic Canada at 3 per cent (Gower, 1998). While many high-tech firms

were adversely affected by the economic downturn that took place from 2000 to the middle of 2003, sales have increased and prospects for 2004 seem bright.

Tertiary Sector

The tertiary sector is expanding rapidly and taking on new functions. In fact, the service industry fuelled job growth in Canada for much of the twentieth century (Table 4.15). Today, three-quarters of Canadian workers are engaged in tertiary functions. Only the United States exceeds this figure and is slightly more 'tertiarized' than Canada.

As society has become more urban and its labour force more specialized, the tertiary sector has grown and diversified. The service industry includes accommodations, communications, education, finance, health, insurance, personal services, public administration, and transportation. As these service functions have grown, some scholars have split this section into two parts, adding the quaternary sector to include decision-making occupations such as research and senior management jobs. Since research and senior management positions are part of most companies, the task of recording such positions as belonging to the 'quaternary sector' is difficult. Other approaches divide the tertiary sector into public and private employment or by low- and high-skilled jobs. Low-skilled employees are found in both the public and private sectors. In the private sector, low-paid workers are common in retail stores, fast-food outlets, and gasoline service stations. Often these are part-time workers with no pension plan or job security. Clearly, they represent the low end of the service industry. Furthermore, such jobs are not restricted to those with limited education. Sessional lecturers at Canadian universities also fall into the low end of the service industry. The high end of the service industry in terms of pay and job security falls to those with skills valued by society. Professional and senior public officials earn high wages and look forward to a secure future. Such jobs provide stability for the individual, his or her family, and the larger community. Professionals include accountants, dentists, doctors, engineers, and lawyers, who occupy the top pay category, followed by professors, teachers, and nurses.

Canada's Labour Force

The Canadian economy is growing at a pace well ahead of other G-8 countries, but how is this growth affecting its labour force, employment levels, and the unemployment rate? Over the five-year period ending in 2002, Canada's labour force increased from 15.4 million to 16.7 million, or by nearly 10 per cent (Table 4.17). This remarkable rate of increase was far ahead of any other G-8 country, including the United States. During this five-year period, the unemployment rate dropped from 8.3 per cent

to 7.7 per cent, though the lowest rate of 6.8 per cent occurred in 2000. Despite the job growth in 2001 and 2002, the unemployment rate increased because of the large number of new workers entering the labour force.

With economic growth across Canada uneven, regional differences exist in employment and unemployment rates. Most economic growth took place in Alberta, Ontario, and British Columbia. Large numbers of migrants from other provinces to those three provinces suggest that many are attracted to job opportunities. The unemployment rates by provinces reflect a regional pattern with the highest rates in Atlantic Canada and the lowest rates in Western Canada (4.19).

Earnings of Canadians

Over the past five years, economic prosperity has translated into higher wages for workers and has resulted in a modest reduction in the income gap between the rich and poor. This improvement in earnings represents a dramatic shift from 1990 to 2000 (Figure 4.9). In fact,

Table 4.17	**Labour Force Characteristics, 1998–2002 (thousands)**				
	1998	**1999**	**2000**	**2001**	**2002**
Population 15 years and over	23,671.1	23,969.0	24,284.9	24,617.8	24,945.1
Labour force	15,417.7	15,721.2	15,999.2	16,246.3	16,689.4
Employed	14,140.4	14,531.2	14,909.7	15,076.8	15,411.8
Full-time	11,466.6	11,849.2	12,208.1	12,345.2	12,528.2
Part-time	2,673.8	2,681.9	2,701.6	2,731.6	2,883.7
Unemployed	1,277.3	1,190.1	1,089.6	1,169.6	1,277.6
Participation rate (%)	65.1	65.6	65.9	66.0	66.9
Unemployment rate (%)	8.3	7.6	6.8	7.2	7.7

Source: Statistics Canada, CANSIM II, tables 282–0002 and 282–0022 and Catalogue no. 71F0004XCB.

Table 4.18	Annual Unemployment Rates by Province, 1986–2002				
Province	**1986–9**	**1996**	**1997**	**2000**	**2002**
Newfoundland	18.4	25.1	18.8	16.7	16.9
Prince Edward Island	15.3	13.8	14.9	12.0	12.1
New Brunswick	12.6	15.5	12.8	10.0	10.4
Nova Scotia	12.1	13.3	12.2	9.1	9.7
Québec	11.2	11.8	11.4	8.4	8.6
British Columbia	10.1	9.6	8.7	7.2	8.5
Ontario	7.9	9.1	8.5	5.7	7.1
Manitoba	8.3	7.9	6.6	4.9	5.2
Alberta	8.6	7.2	6.0	5.9	5.3
Saskatchewan	7.4	7.2	6.0	5.2	5.7
Canada	9.5	10.1	9.2	6.8	7.7

Sources: Scotiabank (1998: 12); Statistics Canada (2003e).

prior to 1997, Canadian workers were less well-off in terms of earnings than they were in the 1980s (Little, 1997: B1, B5). In addition to this dilemma, various levels of government raised their taxes, leaving less money in the pockets of working Canadians. By 2000, the situation for Canadian workers had improved. Not only did the overall average employment income of individuals reached $31,757, an increase of 7.3 per cent over the past 10 years (Statistics Canada, 2003g), but the rate of inflation remained low and federal income taxes were reduced.

Earnings vary by education. The 2001 census confirms that higher education leads to higher earnings. For example, the census reported that more than 60 per cent of the workers in the lowest earnings category did not have more than a high school education in 2000, while more than 60 per cent in the highest category had a university degree (ibid.). As shown in Figure 4.10, university-educated Canadians earn substantially more than less educated Canadians. Those without a high school degree fall to the bottom income category.

Figure 4.9 | Average earnings, 1980, 1990, and 2000

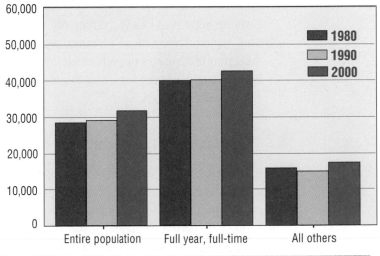

Average earnings in 2000 dollars ($)

Source: Statistics Canada: <http://www12.statcan.ca/english/census01/Products/Analytic/companion/earn/canada.cfm>.

Figure 4.10 Earnings by education level

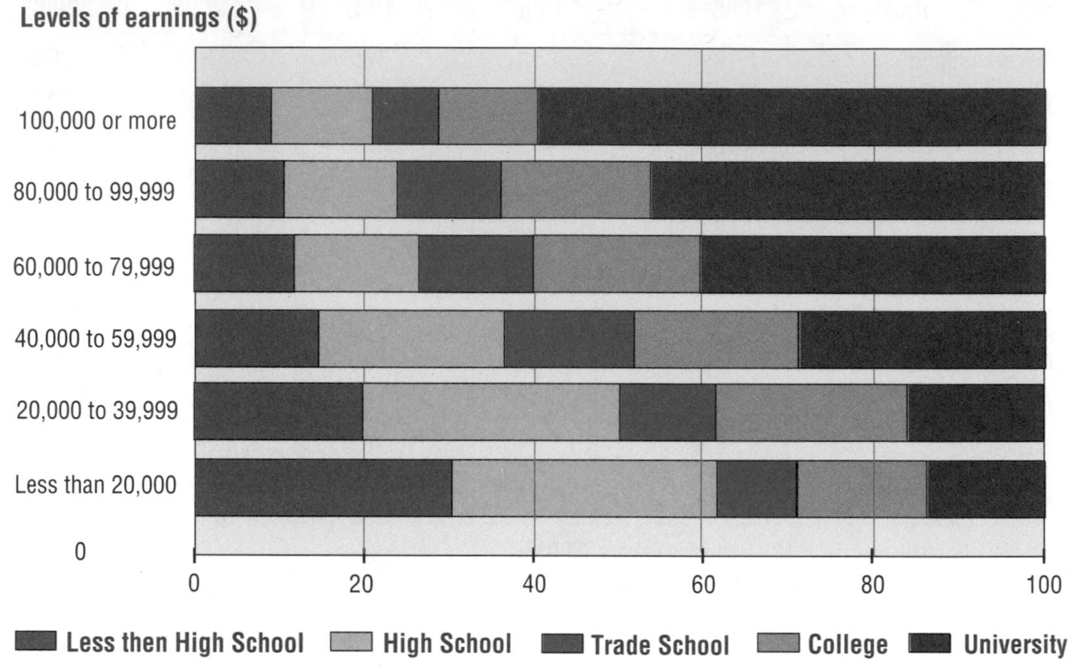

Levels of earnings ($)

Legend: ■ Less then High School   ■ High School   ■ Trade School   ■ College   ■ University

Source: Statistics Canada: <http://www12.statcan.ca/english/census01/Products/Analytic/companion/earn/canada.cfm>.

Earnings vary by geographic region (Table 4.19). Average earnings were greatest in Ontario and lowest in Atlantic Canada, where earnings were well below the national average. This regional pattern of earnings is another reflection of Canada's uneven economic development that takes the form of a core/periphery spatial pattern.

Women in the Labour Force

In the second half of the twentieth century, significant numbers of women began to break away from the traditional homemaker role and entered the workforce. By the 1980s, working wives and mothers became a common and necessary feature of Canadian society. From 1951 to 1996, the percentage of females participating in the wage economy increased steadily. In 1951, for example, the female labour force participation rate was only 24 per cent, but by 1991 it had increased to 60 per cent (McVey and Kalbach, 1995: 251). With so many husbands and wives working, the term 'dual-earner family' was coined. A measure of the growth of dual-earners is revealed by the fact that, in 1961, less than 20 per cent of all families had both the husband and wife employed; by 1991 this figure had more than tripled to 60 per cent (ibid., 253). The dramatic and sudden increase in the participation of women, particularly married women, in the labour force has altered traditional family roles. In a traditional North American nuclear family, the father was regarded as the 'breadwinner' and the mother as the 'homemaker'. One demographic consequence

Table 4.19	Average Earnings for All Earners, Canada, Provinces, and Territories, 1980, 1990, and 2000		
	1980	**1990**	**2000**
Canada	$29,229	$29,596	$31,757
Newfoundland and Labrador	$23,530	$22,017	$24,165
Prince Edward Island	$20,210	$21,546	$22,303
Nova Scotia	$24,422	$25,587	$26,632
New Brunswick	$23,501	$24,173	$24,971
Québec	$29,285	$28,516	$29,385
Ontario	$29,360	$32,181	$35,185
Manitoba	$25,988	$25,859	$27,178
Saskatchewan	$27,460	$24,159	$25,691
Alberta	$31,857	$29,241	$32,603
British Columbia	$31,950	$30,170	$31,544
Yukon	$33,554	$31,402	$31,526
Northwest Territories*	—	—	$36,645
Nunavut*	—	—	$28,215

*The Northwest Territories was divided into two parts in 1999. The western half remained as the NWT; the eastern half became Nunavut.
Source: Statistics Canada: <http://www12.statcan.ca/english/census01/Products/Analytic/companion/earn/canada.cfm>.

of this social change has been a delay in family formation and smaller family units.

Unfortunately, this growing number of women in the workforce has not yet resulted in full pay equity between the sexes (Figure 4.11). Although women's average earnings have risen, a substantial income gap remains between women and men in the same occupations and with the same education. While income for women does increase as their level of education rises, it remains below incomes obtained by men.

Population Trends and Demographic Faultlines

Having established the predominant demo-graphic and socio-economic characteristics of Canadian society as a whole, we will now examine some key demographic shifts as they relate to the four major faultlines discussed in Chapter 3. The various demographic shifts being experienced in Canada will have significant long-term implications for the residents of the various regions. For example, the population of Atlantic Canada is expected to continue to decline. The number of Aboriginal Canadians is growing rapidly, while the French-Canadian population is growing very slowly. The annual number of newcomers to Canada now outnumbers the natural increase of Canada's population and most settle in Canada's largest cities. Sociologists such as Herberg (1989) and Li (2003) see immigration

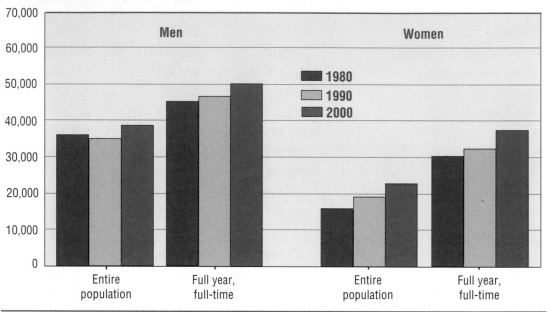

Figure 4.11   **Comparison of male and female earnings**

Average earnings in 2000 dollars ($)

Source: Statistics Canada: <http://www12.statcan.ca/english/census01/Products/Analytic/companion/earn/canada>.

as a tension between maintaining the sense of a common identity and heritage and the need to incorporate some elements of the cultures of the immigrants into mainstream Canadian culture. Geographers such as Hiebert (2000) and Ley (1999) stress the impact of immigration on the city landscape.

Newcomers and Old-Timers

Immigration has not only made Canada a pluralistic society but it has also divided Canadian society into three elements—Aboriginal peoples, whose ancestors came from Asia many thousands of years ago; descendants from previous generations who have come to Canada since the seventeenth century; and recent immigrants who were born overseas. Since Aboriginal peoples are discussed as a

faultline later in this chapter, our focus here is on non-Aboriginal people born in Canada and recent immigrants. The former are broadly referred to as 'old-timers' and the recent immigrants as 'newcomers'.

The primary reason for tension between the two groups is that those born in Canada perceive immigrants as either a benefit to them (and Canada), or as a concern that negatively affects their way of life. As Peter Li (2003: 9) pointed out, 'Over. time, earlier immigrants and their descendants become old-timers of the land, and their charter status places them in a privileged position vis-à-vis the newcomers who must accept the conditions of entry and rules of accommodation laid down by those who came before them.' Since newcomers often come from a different cultural world, they may bring with them language, customs, foods, and religious beliefs that differ sharply from those found in mainstream Canada. While these cultural imports often enrich Canadian society, a few have caused a cultural clash between some members of the two groups. On top of these cultural differences, since many recent immigrants come from non-European countries, racism can add another element to these cultural tensions. Visible minorities make up a major and growing share of the populations of Canada's major cities. Bourne and Rose (2001: 115–16) argue that 'the challenge of overcoming exclusionary practices, including those based on racism in all its forms, and in particular the impact of these practices on labour market opportunities for youth, becomes all the more urgent.'

Immigration brings benefits, including maintaining Canada's population, supplying much needed skilled labour to Canada's workforce, and bringing investment capital. As well, the newcomers add to the cultural richness of Canadian society. Yet, another side to immigration exists in the minds of the old-timers. The two primary concerns are that the newcomers will somehow lessen their economic position in Canadian society, and that, coming from a different cultural, linguistic, and racial background, they will remake Canadian society in their image. This latter fear is reinforced by the concentration of immigrants in Canada's three major cities, where they may create cultural enclaves that prevent their integration into Canadian society. Since 11 September 2001, a new concern has emerged—that terrorists may slip through Canada's immigration screening system. While a small number of criminals, war criminals, and other undesirables have entered Canada in the past, the vast majority are contributing to the well-being of Canada. Indeed, the history of immigration has shown two remarkable developments over time. First, Canadian society absorbs some of the cultural imports and multiculturalism is a political expression of the cultural contribution made by newcomers who arrived in the twentieth century. Second, recent immigrants, especially their children, have shown a remarkable capacity not only to integrate into Canadian society but also to reshape it.

Regional Patterns of Immigration

Immigrants are not evenly dispersed across Canada. The vast majority live in Ontario (over half of all immigrants reside here), British Columbia, Québec, and Alberta. Immigrants within these four provinces have especially been drawn to the major cities. Approximately 94 per cent reside in Canada's census metropolitan areas. The regional pattern of immigration therefore takes on an urban/rural dichotomy. Ottawa has expressed some concern about this dichotomy (Vignette 4.7).

Canada's three major cities attract most immigrants. By 2001, nearly three-quarters of immigrants arriving in the 1990s resided in Toronto, Montréal, and Vancouver (Figure 4.12). Large ethnic concentrations exist within these cities, forming neighbourhood communities with names such Little Italy and Chinatown. Toronto attracts the largest share of new immigrants. Over 2 million immigrants live in the census metropolitan area of Toronto, making up 42 per cent of its population. The trend towards settling in Toronto is increasing. In 1991, 40 per cent of immigrants who arrived in the 1980s came to Toronto. A similar trend is found in Vancouver, but this is not so true for Montréal. Close to 21 per cent of immigrants had settled in other census metropolitan areas in 2001. Nearly 4 per cent of

| Vignette 4.7 | **Must Immigrants Spread Out?** |

Most new Canadians have relocated in Canada's major cities, especially Toronto, Vancouver, and Montréal. This geographic concentration represents the newcomers' attempt to maximize their chances of economic and social success in their adopted country. Ottawa sees two problems with this pattern of settlement. First, such a concentration puts a strain on social services, schools, and housing in these three cities. Second, communities in other parts of the country are facing shortages of skilled workers, including doctors and nurses. In October 2002, Denis Coderre, the federal Immigration Minister, proposed that prospective immigrants sign a social contract promising to live for three to five years in cities, towns, and rural communities outside of Canada's three major cities in return for Canadian citizenship. Such immigrants will be offered jobs and the right to bring their families to Canada. The underlying assumption is that the newcomers will fit into these communities and decide to stay after their contractual period expires (Fife, 2002).

| Figure 4.12 | **Share of immigrants in Montréal, Toronto, and Vancouver, 1981, 1991, and 2001** |

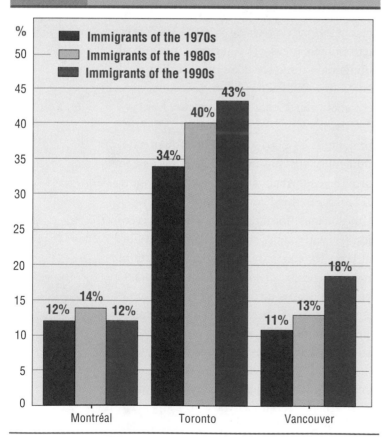

Source: Statistics Canada (2003c: 6).

new immigrants came to Ottawa–Gatineau and Calgary, while Edmonton received 2.5 per cent and Hamilton nearly 2 per cent. A more complete analysis of immigration to Canada's major cities is found in the three regional chapters of Ontario, Québec, and British Columbia.

Aboriginal Population

The population figures for Aboriginal peoples have undergone great changes since European contact. Prior to contact, Aboriginals within the territory that became Canada were quite numerous, especially among the sedentary peoples of the Northwest Coast and in parts of what would become southern Ontario, where rich resources (salmon on the Northwest Coast) and agriculture supported sizable communities. However, less than 100 years ago, the Aboriginal population had declined to about 100,000 because of the effects of European diseases and the loss of vast hunting grounds that had previously been their means of survival. But during the past five decades, Aboriginal peoples have experienced a population explosion. The total Aboriginal popula-

Table 4.20   Major Phases for the Aboriginal Population in Canada

Phase	Characteristics
Pre-Contact (1000–1500)	The Aboriginal population in Canada was at least 200,000 and possibly as large as 500,000. This population may have varied in size due to the carrying capacity of the land, which, in a hunting society, is controlled by the availability of game (food). For instance, natural conditions, especially weather, could affect the size and migration routes of animal populations.
Early Contact (1500–1800)	Aboriginal peoples who came into contact with Europeans were exposed to new diseases. Population losses were greatest among Indian tribes living in Atlantic Canada, Québec, and Ontario because of direct contact with British and French settlers and because of their involvement in the colonial wars. The fur trade became a central feature of their economy. Loss of hunting lands and the over-exploitation of the beaver also added to their demise. By the end of the eighteenth century, the Indian population in these three regions may have dropped below 50,000.
Late Contract (1800–1940)	During this period, direct contact was made with Indian tribes in Western Canada and in British Columbia. Except for the Klondike gold rush and the fur trade, Indians had little contact with outsiders. During this phase, Aboriginal population ceased to decline and, beginning in the twentieth century, the Aboriginal population began to increase.
Modern Contact (1940 to present)	The Aboriginal population began to increase rapidly with natural rates of increase greatly exceeding the national figures. By 2001, total population exceeded 1.3 million. High fertility and low mortality account for this remarkable population rebound. The net result was a population explosion in the latter half of the twentieth century. While this explosion has just begun to slow, the Aboriginal population is expected to reach 1.5 million by 2006 (see Figure 4.13).

tion in Canada now exceeds 1 million. The general population trends from pre-contact to the present illustrate three major demographic phases (Table 4.20).

When Jacques Cartier sailed into Baie de Chaleur in 1541, the Indian and Inuit population in Canada may have been as high as 500,000. The exact figure will never be known. What we have are only estimates. By reconstructing the land's capacity to support wildlife and therefore also hunting societies, anthropologist James Mooney (1928: 7) estimated that about 220,000 Indians and Inuit lived in Canada at the time of contact. More recently, scholars have revised this figure upwards. Dickason (2002: 45) and Denevan (1992: 370) estimate that the number of Aboriginal peoples living in Canada was closer to half a million. Whatever the exact figure, initial contact with Europeans resulted in a rapid depopulation of Aboriginal peoples. Factors include loss of hunting grounds and therefore food shortages, increased warfare, and the spread of new diseases from Europe among the Indian tribes. Communicable diseases, such as smallpox, caused great suffering and many deaths among Indian tribes. Epidemics sometimes quickly reduced the size of tribes by half. The result was that by the time of Confederation, Aboriginal peoples numbered about 100,000.

A demographic rebound of the Aboriginal population began in the 1940s and has just shown signs of slowing down (Figure 4.13). In 2001, there were approximately 1.3 million Canadians of Aboriginal descent (by Aboriginal identity, however, the figure was just under

one million). The outcome of this increase has not only meant greater numbers but also a greater proportion of Aboriginal peoples within Canadian society. In 1951, Aboriginal peoples comprised less than 2 per cent of Canada's population. By 2001, this proportion had reached 4.3 per cent. What, then, led to this recovery in the Aboriginal population? Many factors contributed to the population's increase; especially important was their relocation to settlements where food supplies and medical care were available.

Since World War II the rate of natural increase of Aboriginal peoples has greatly exceeded the national figure. For much of that time, the rate of natural increase was about 3 per cent thanks to high birth rates and low

Figure 4.13   Aboriginal population by ancestry, 1901–2001

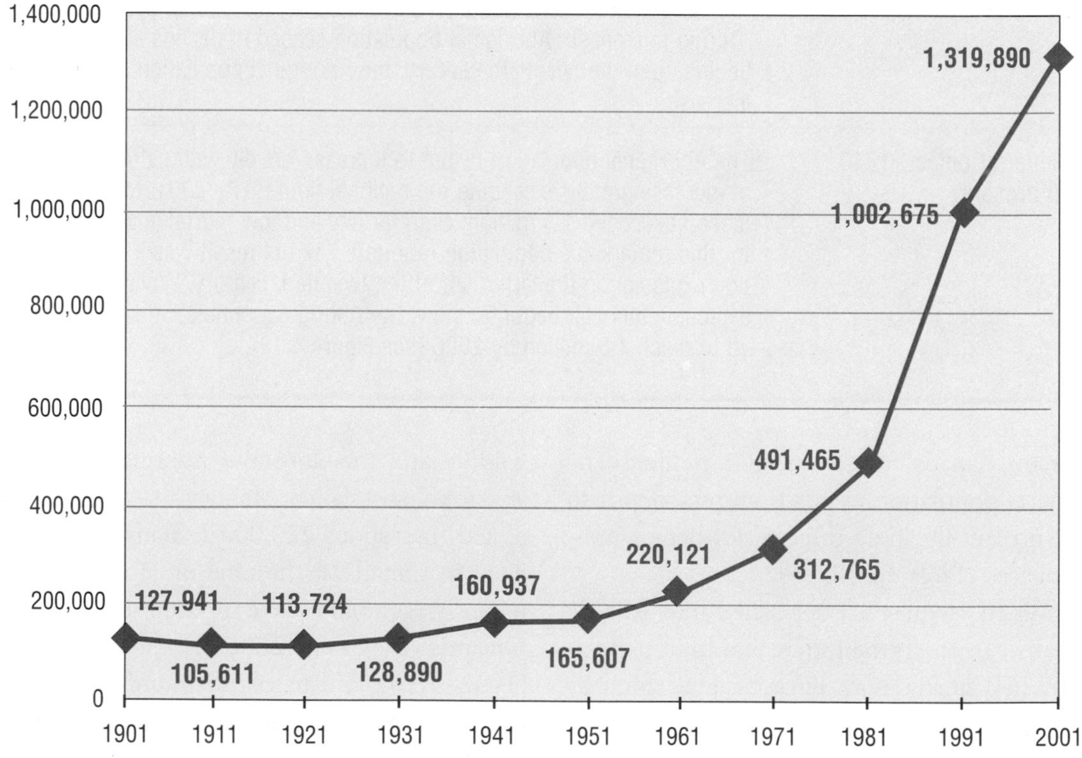

Source: Statistics Canada (2003h).

death rates. The death rate for Aboriginal peoples dropped swiftly with the introduction of modern medicine. While communicable diseases, such as tuberculosis, continue to plague Native peoples, their mortality rates have declined below the national average. However, this low rate is somewhat misleading because the Aboriginal population is so much younger than the national average. Access to medical services and food supplies provides a partial explanation for the drop in the mortality rates.

As the Aboriginal death rate declined, the birth rate remained high. The fertility rate for Aboriginal Canadians remains more than double the national rate. While the national fertility rate is below replacement level, the rate for Aboriginal peoples is much higher. In the past decade, the national rate was about 13 births per 1,000 people while the Aboriginal rate was close to 30 births per 1,000 people. This fertility difference between the national and Aboriginal populations has led to other demographic differences, including a more youthful Aboriginal demographic structure. While only 21 per cent of the Canadian population is under the age of 15, the figure for Aboriginal peoples is 38 per cent—almost double. While their rate of natural increase is high, Native Canadians face very high infant mortality and suicide rates, which indicate a social malaise that has affected the general health of that population. Nevertheless, based on the current demographic trends in Aboriginal society, the Aboriginal population is likely to continue increasing, though at a slower rate. As Aboriginal parents opt for smaller family units, the gap between the birth and death rates will diminish, as will the high rate of natural increase.

The political implications of this high rate of natural increase are profound. Already this rapid population increase has dispelled the myth of the 'vanishing Indian', instead assuring the continued existence of Aboriginal peoples. Aboriginal political power is, in part, linked to population size, so with the percentage of Aboriginal peoples in Canada increasing (perhaps exceeding 5 per cent by 2006), the federal, provincial, and territorial governments will have to heed the political voice of Native leaders more carefully. A rapid population increase can also have negative impacts, such as placing greater demands on scarce resources. For example, the collective resources of Aboriginal peoples are already facing increased housing demands. Another implication of population increase has a regional component. Since the

Simone Arnatsiaq plays with friends from a daycare on the tundra-covered hills looking over Iqaluit, the capital of Nunavut. One of the challenges facing Nunavut is its rapidly growing population. Nunavut has an extremely young population with nearly 40 per cent of its people under the age of fifteen. Comparatively, in the rest of Canada, 20 per cent of the population is under the age of fifteen. (CP/Paul Chiasson)

Vignette 4.8   Aboriginal Ancestry and Identity

The 2001 census produced two population figures for Aboriginal peoples. One figure derived from the census question on ethnic origin. For 2001, the Aboriginal population based on this census question was 1,319,890. The second figure of 976,301 came from the census question based on Aboriginal identity. The identity question consisted of three main categories: North American Indian, Métis, and Inuit. A fourth category included other Aboriginal identities, such as 'registered Indian'.

	2001	1996	Percentage growth 1996–2001
Total: Aboriginal ancestry [1]	1,319,890	1,101,960	19.8
Total: Aboriginal identity	976,305	799,010	22.2
North American Indian [2]	608,850	529,040	15.1
Métis [2]	292,310	204,115	43.2
Inuit [2]	45,070	40,220	12.1
Multiple and other Aboriginal responses [3]	30,080	25,640	17.3

(1) Also known as Aboriginal origin.

(2) Includes persons who reported a North American Indian, Métis, or Inuit identity only.

(3) Includes persons who reported more than one Aboriginal identity group (North American Indian, Métis, or Inuit) and those who reported being a registered Indian and/or band member without reporting an Aboriginal identity.

Source: Statistics Canada (2003g).

distribution of Aboriginal peoples across Canada is uneven, the greatest increases in their numbers will take place in Western Canada, followed by Ontario and British Columbia (Figure 4.14). The five provincial governments in those regions will face considerable pressure from those status Indians seeking a resolution to their land claims and from the non-status Indians and Métis who seek to better their way of life.

French/English Balance

French- and English-speaking Canadians represent the traditional duality of Canadian society. This dual relationship has deep historic roots. With the establishment of the Province of Canada in 1841, the relationship flowered into a partnership with English-speaking Canada West and French-speaking Canada East sharing political power. In 1867, the British North America Act declared that both French and English were official languages of this new Dominion. The assignment of language rights to provinces ensured that the Québec government could protect and promote the French language.

Language is the essential element of French heritage and ensures the preservation of French culture. It is from this perspective that

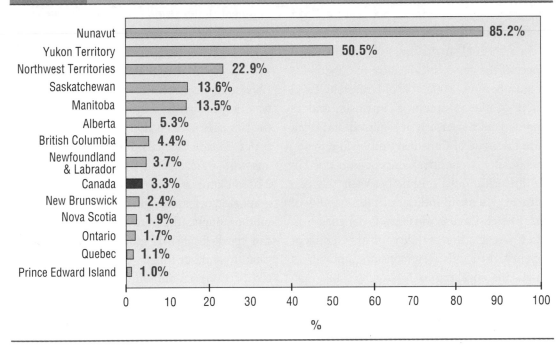

Figure 4.14   **Percentage of Aboriginal population by self-identification, provinces and territories, 2001**

Source: Statistics Canada (2003h).

francophones interpret the social and political significance of the French/English balance. If the census of Canada reports that the proportion of French-speaking Canadians is declining, then francophones become concerned. If a Royal Commission (such as that on Bilingualism and Biculturalism, 1963–71) provides evidence that the French language is in 'deadly danger', then francophones become alarmed and want their governments to take action.

Linguistic rights for both French and English as the official languages give them a special place in Canadian society. One implication is that new Canadians whose language is neither French nor English gravitate to one linguistic group or the other. Within Canada, English is the dominant language. English is also the business language of North America and, to a large degree, the world. Immigrants

therefore have a compelling reason to learn English. Also, most immigrants settle in English-speaking areas and so are naturally drawn into the anglophone linguistic group. Outside of Québec and New Brunswick, the French language is losing ground. Anne Gilbert (Gilbert, 2001: 173) points to the recent decision of the government of Ontario to reject the concept of recognizing both English and French as official languages as a missed opportunity to showcase Canada's serious commitment to its dual languages.

Recent Statistics Canada figures (2001) support this argument. The proportion of Canadians outside of Québec speaking English at home was 85.6 per cent, the proportion speaking a non-official language was 11.7 per cent, and those speaking French was 2.7 per cent. However, a number of Canadians speak both English and French. The number of bilingual Canadians is increasing. From 1951 to 1996, the percentage of Canadians who could speak both official languages increased from 12 per cent to 17 per cent (Statistics Canada, 1997b: 11).

French/English dualism is a fundamental aspect of the geopolitical nature of Canada. As Jacques Bernier (1991: 79) of Université Laval stated: 'Canada's duality is intrinsic, and as long as it is not clearly recognized and dealt with, the issue of Canadian unity will remain.' This duality is a political concept embedded in the historical relationship between the two cultures. The main indicator of the stability of this French/English dualism is language; in other words, the stability of this concept depends on a relatively constant number of Canadians speaking each language. But how should we measure duality? Should mother tongue or household language hold the key? Perhaps the population size of the minority language group should be the determining factor? Or does the number of bilingual Canadians hold sway?

The French/English balance can be measured by a number of factors relating to language use. In the 2001 census, Canadians were asked three questions about language—their mother tongue, home language, and bilingual skills. Mother tongue is the first language learned that is still understood, while home language is the language most often spoken in the home. In that year, 17.5 million Canadians declared that their mother tongue was English, while 20 million indicated that the language spoken at home was English (Statistics Canada, 2002d). Furthermore, Canadians who were able to speak English totalled 22.5 million, or 83 per cent. Since most immigrants today learn English but have a different mother tongue, the number of Canadians who can speak English is much higher than indicated by mother tongue or home language. In comparison, 6.8 million Canadians stated that their mother tongue was French, but only 6.5 million spoke French at home. Many Canadians can speak both official languages and 8.5 million or 32 per cent of Canada's population can speak French.

French-speaking Canadians are distributed in three major geographic areas. The principal one is the province of Québec, where 83.1 per

Table 4.21	Population with French Mother Tongue, 1951–2001					
Year	Canada (000s)	Québec (000s)	Rest of Canada (000s)	Canada (per cent)	Québec (per cent)	Rest of Canada (per cent)
1951	4,069	3,347	722	29.0	82.5	7.3
1961	5,123	4,270	854	28.1	81.2	6.6
1971	5,794	4,867	926	26.9	80.7	6.0
1981	6,178	5,254	924	25.7	82.5	5.2
1991	6,562	5,586	976	24.3	82.0	4.8
1996	6,637	5,747	970	23.5	81.5	4.5
2001	6,782	5,802	980	22.9	81.4	4.4

Sources: Harrison and Marmen (1994: Table 2.1); Statistics Canada (1997b; 2002d)

Table 4.22	Canada's Population by Mother Tongue, 1951–2001		
Year	English	French	Other
1951	59.1	29.0	11.8
1961	58.5	28.1	13.5
1971	60.2	26.9	13.0
1981	61.4	25.7	13.0
1991	60.4	24.3	15.3
1996	59.9	23.5	16.6
2001	59.1	22.9	18.0

Sources: Harrison and Marmen (1994: Table 2.1); Statistics Canada (1997b, 2002d).

cent of the population speaks French in the home (Statistics Canada, 2002d). A second francophone area is in New Brunswick, where Acadians and other French-speakers form 30.3 per cent of the population. A minor cluster exists in the third area, Ontario, where 2.7 per cent speak French at home. Most of these live in northeastern Ontario. The number of French-speakers is increasing in Canada but most of this increase has occurred in Québec (Table 4.21). Québec is the only geographic region where the number of people speaking French at home is greater than those declaring French as their mother tongue. This difference suggests that some immigrants living in Québec are adopting the French language and assimilating into the francophone culture.

The number of French-speaking Canadians is increasing, but at a rate slower than the national rate of population growth. Since 1951, the number of Canadians whose mother tongue is French has increased from 4 million to 6.8 million. Within Québec, the increase is even greater, jumping from 3.3 million in 1951 to 5.8 million in 2001. Even in the predominantly English-speaking areas of Canada, the number of francophones has

increased from 722,000 to 980,000. Over the same time span, however, the proportion of Canadians indicating that French is their mother tongue has dropped from 29 per cent to below 23 per cent (Statistics Canada, 2002d). Finally, the number and percentage of bilingual Canadians have increased sharply. In 1981, 3.7 million Canadians spoke both languages, while nearly 5 million were in this category by 2001.

The French/English duality remains strong at a national level, but the increasing numbers of Canadians whose mother tongue is neither French nor English is a factor of the arrival of immigrants who speak neither official language (Table 4.22). Within Québec, the low fertility level among francophones means that the adoption of French by newcomers is extremely critical.

Between 1996 and 2001, the English-speaking population in Canada grew by about 2.6 per cent compared with only 1.1 per cent for the French-speaking population. As a result, the proportion of Canadians who were French-speaking slipped to 22.9 per cent by 2001. Still, the number of French-speaking Canadians within Québec is rising. Two reasons account for this increase in Québec: the natural increase of the Québécois and the increasing numbers of allophones who become French-speaking. Until the Québec government passed a series of language laws restricting the use of English, most allophones were choosing to learn English. The reason was simple—it gave them greater economic mobility because they recognized that English is spoken in other parts of Canada and the United States.

Among the Québécois, language laws are very popular because they ensure the place of French in Québec, particularly in Montréal. While such legislation runs against the current of bilingualism in the rest of Canada, Québec governments (both Liberal and Parti Québécois) support the principle of a unilingual province in order to ensure the survival of the French language in Canada and North America. While language laws are not popular with all new immigrants, they are required by Québec law to send their children to French schools. As Québec has an extremely low birth rate, immigration and the requirement that children of immigrants learn French keep the French-speaking population growing.

Québec language laws, particularly the law requiring all signs to be in French, have annoyed many English-speaking citizens of the province. In 1988, the Supreme Court of Canada ruled that Québec's French-only sign regulation violated the Charter of Rights. Francophones, ever concerned about the preservation of their culture, reacted strongly, taking to the streets, declaring 'Ne touchez pas à la loi 101.' The Québec Premier, Robert Bourassa, finding himself on the horns of a dilemma, chose to circumvent the Supreme Court's ruling by invoking the 'notwithstanding' clause (section 33) in the Charter. Later, the provincial government amended its sign legislation so that signs could be in both languages as long as the French is larger than the English. Nevertheless, ever since the language legislation favouring the French language was

Table 4.23	Regional Population of Canada by Percentage, 1901–2001							
Region	**1901**	**1921**	**1941**	**1961**	**1981**	**1991**	**1996**	**2001**
Ontario	40.6	33.4	32.9	34.2	35.4	37.0	37.3	38.0
Québec	30.7	26.9	29.0	28.8	26.4	25.3	24.8	24.0
British Columbia	3.3	6.0	7.1	8.9	11.3	12.0	12.9	13.1
Western Canada	7.8	22.2	21.0	17.5	17.4	16.9	16.6	16.9
Atlantic Canada	16.7	11.4	9.8	10.4	9.2	8.5	8.1	7.7
Territorial North*	0.9	0.1	0.2	0.2	0.3	0.3	0.3	0.3
Canada (millions)	5.4	8.8	11.5	18.2	24.3	27.3	28.9	30.0

*In 1901, the Yukon population included many associated with the Klondike gold rush. By 1911, most had left. This accounts for the abrupt change in population for the North after 1901. Also, even though the present provinces of Alberta, Saskatchewan, and much of Manitoba still belonged to the Northwest Territories in 1901, their populations were assigned to the Prairie region.

Sources: McVey and Kalbach (1995: 46); Statistics Canada (1997c, 2002a).

Figure 4.15   **Net migrants by province, 1996–2001**

000s

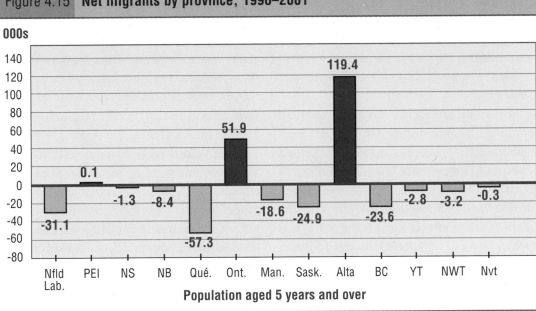

Population aged 5 years and over

Source: Statistics Canada (2003h).

passed, English-speaking Quebecers have felt less comfortable living in that province. In fact, these language laws, combined with two referendums, have caused many anglophone Quebecers to migrate to other provinces. In 1971 there were 789,000 anglophones in Québec, but 30 years later the number had dropped below 600,000 (Statistics Canada, 2002d).

Core/Periphery Populations

Canada's population is unevenly distributed across the country's regions. The population remains concentrated in Central Canada, though population shifts have occurred in Western Canada and British Columbia. Atlantic Canada, on the other hand, has lost ground. Therefore, the hinterland has increased its share of the national population, but that increase has been geographically uneven—the West, including British Colum-

bia, has experienced a relative increase in its share of the national population, while Atlantic Canada has seen a decrease (Table 4.23). Ontario has continued to grow rapidly while Québec's rate of population increase has slowed. The reasons for this distribution in population and population growth can be found in the country's historical migration and settlement patterns, which can be characterized as east to west and rural to urban. Population shifts driven by interprovincial migration are shown in Figure 4.15.

The early history of Canada partly accounts for the powerful population cluster in Central Canada. In 1867, approximately 80 per cent of all Canadians lived in Ontario and Québec, where geographic conditions were most beneficial for settlement. The remainder of the population lived in New Brunswick and Nova Scotia. By 2001, however, these figures had declined to 62 per cent in Ontario and Québec, and 8 per cent in Atlantic Canada (even with the addition of Prince Edward Island and New-foundland). The principal reasons for this relative decline over the past 130 years are: (1) the large population increase in Western Canada and British Columbia; (2) the slow rate of population growth in Québec and Atlantic Canada.

In the late nineteenth and early twentieth centuries, the lure of 'free' land drew homesteaders from eastern Canada, the United States, and Europe to the Canadian Prairies. At the time of Confederation, only marginal agricultural land remained in Atlantic Canada, Québec, and Ontario. Before the West was opened up for settlement, a rural-to-urban migration was already underway in Ontario and Québec, driven by rural overpopulation and employment opportunities in the new factories. Following the completion of the Canadian Pacific Railway, the Canadian Prairies began attracting large numbers of land-hungry migrants. By 1921, Western Canada included 22 per cent of Canada's population, while British Columbia had 6 per cent.

Beginning with the drought of the 1930s, dubbed the Dust Bowl, new migration patterns formed. Families abandoned their homesteads on the grasslands of the three Prairie provinces. Many migrated to urban centres, including those in Ontario and British Columbia, looking for a more prosperous life. Farm labourers also joined this urban migration stream. They had been an invaluable part of the agricultural workforce in the Canadian Prairies, but mechanization would eventually mean that fewer farmers were required. Since only those farmers with large landholdings could maximize the use of expensive farm implements, rural areas rapidly lost much of their population. By 1991, the rural depopulation of Western Canada had slowed. Saskatchewan suffered the greatest population loss, which illustrates the

limited capacity of the province's cities to absorb its surplus rural population. The net result, even accounting for the fact that since the 1970s Alberta has experienced significant economic growth and in-migration, is that Western Canada in 2001 had a smaller share of the national population, down to 16.9 per cent, than it did in 1921 (22.2 per cent) or 1941 (21.0 per cent) (Table 4.23).

After World War II, however, a second wave of immigrants went to the West. Some were attracted by the mild west coast climate, while others sought work, particularly in the resource industries. Population growth varied regionally. Manitoba and Saskatchewan did not fare as well. Their populations grew very slowly, mainly because out-migration was so high that it almost equalled the rate of natural increase. British Columbia and Alberta led in population growth. The attraction of western provinces had changed over time from free land in the Prairies to jobs in Alberta and British Columbia. In British Columbia the forest industry needed many workers, while in Alberta the petroleum industry attracted many. During this time, British Columbia became the favourite destination for Canadians seeking to relocate.

Today, Alberta, with a population increase of 10.3 per cent from 1996 to 2001, is Canada's fastest-growing province, thanks to its booming economy. Ontario (6.1 per cent) and British Columbia (4.9 per cent) also experienced high rates of population increases over that five-year period. These three provinces and Nunavut (8.1 per cent) grew at a greater rate than the national increase of 4 per cent. By 2001, the magnitude of this population shift revealed that the West accounted for 30 per cent of Canada's population, with Western Canada contributing 16.9 per cent and British Columbia 13.1 per cent.

All indications suggest that current regional trends will continue. Certainly, if Atlantic Canada continues to experience high levels of out-migration, its share of the national population will continue to decline. How much of its population will relocate to Ontario or further west will be determined by job opportunities. Ontario and British Columbia are expected to make the greatest population gains, while Western Canada's increase, driven by Alberta, is expected to be more modest. Together, British Columbia and Western Canada could form one-third of the national population by the next census in 2006. Inevitably, the population shift to the West will translate into more political power in Ottawa, but until it threatens the balance of power, the core regions of Ontario and Québec will continue to dominate the political landscape. With this in mind, the discussion of the six geographic regions begins with the two largest regions by population, namely Ontario and Québec, followed by British Columbia, Western Canada, Atlantic Canada, and the Territorial North.

Summary

Since World War II, the human face of Canada has changed. Much change took place after the 1960s when a new economic and social climate arose. Trade liberalization triggered a new economic arrangement within North America. The separatist movement in Québec, the influx of Asian and Caribbean migrants, and the search by Aboriginals for a place in (or apart from) Canadian society have changed the social and political dynamics of Canada. Like other developed countries, Canada's economy has transformed from an agrarian base to industrialization (manufacturing, resource extraction) to a post-industrial stage where most of the jobs are to be found in the growing service sector. But this transformation has taken place within a regional context whereby certain more favoured regions have enjoyed greater economic development and experienced faster-growing populations than less favoured regions. This transformation has also been affected by Canada's geographic position within North America. Having the United States, the most powerful nation in the world, as a neighbour and trading partner has resulted in close economic ties and a strong dependency on the American economy.

Major economic changes in Canada stem from the country's economic integration with the United States and, to a lesser degree, from increased involvement in the global economy. Economic restructuring, sometimes painful, took place following the Free Trade Agreement in 1989. It continues under NAFTA and the World Trade Organization, but at a slower pace. Economic restructuring has also affected the Canadian labour force. Three notable labour changes have taken place: more women are employed, more workers are engaged in tertiary occupations, and more people are now unemployed.

Equally important changes stem from Canadian social and political forces, such as the Constitution with its emphasis on bilingual, pluralistic, and civic nationalism. Also, changes in Canada's demography have resulted in the reduction of Canada's rate of natural increase. Canadian politicians have extended immigration opportunities to people from all over the world and increased the annual number of immigrants coming to Canada. In turn, new Canadians, particularly those from non-European and non-Christian countries, have altered the ethnic and religious makeup of Canada. In response, Ottawa initiated a policy of multiculturalism to create harmonious relations among Canada's various ethnic groups and to promote social equality. Such a policy runs counter to the concept of duality between the two founding peoples, the French and the English, and makes francophones, especially in Québec, feel their cultural status and political power are threatened.

Changes to Canada's economy and society have affected Canada's four critical tensions or faultlines: (1) Aboriginal/non-Aboriginal, (2) core/periphery, (3) French/English, and (4) newcomers/old-timers. Key demographic changes are occurring along these divides, reflecting internal forces and revealing population and political shifts. The first faultline is characterized by a much higher rate of natural increase among the Aboriginal population. The second faultline involves changes in the sizes of Canada's regional populations. Since Confederation, population growth has occurred in all regions but at varying rates. The Ontario, Québec, and Atlantic Canada shares of the national population have declined, while those of British Columbia, Western Canada, and the Territorial North have increased. The third faultline has experienced a faster rate of population increase among English-speaking Canadians than French-speaking Canadians; the explanation lies not in fertility rates but in immigration. The fourth faultline—between new Canadians and those born in Canada—is reflected in the federal policy of multicultural-

ism and is played out daily in Canada's major urban centres, where the vast majority of new Canadians settle.

The following chapters, 5 through 10, focus on the regional nature of Canada, with each chapter devoted to exploring one of Canada's six geographic regions. Friedmann's version of the core/periphery model provides a guide for the ordering of these six regions (Table 1.4). The regional discussion therefore begins with Ontario and Québec, which represent Canada's economic core. These two chapters are followed by chapters on British Columbia, Western Canada, Atlantic Canada, and the Territorial North. While many of the same topics and issues appear in each of the six geographic regions, each is examined from the perspective of the region under discussion. Each regional chapter will illustrate the physical diversity within each geographic region and show how this diversity affects human activities and settlement. Each chapter also has a section on the region's historical development to enrich our sense of the present by providing a link with the past. One recurring theme is the relationship among the six regions. This relationship is cast in the core/periphery model, but it is changing due to continental economic integration (NAFTA) and global trade (the WTO).

Notes

1. The American railway industry has been going through a major restructuring process since the 1970s. Canada joined this process as the flow of goods between Canada and the United States increased in the 1990s. This pressure came from the increased trade stemming from the Free Trade Agreement in 1989 and later with the North American

Free Trade Agreement in 1994. While the vast majority of the trade is between Canada and the US, trade with Mexico is anticipated to increase in the twenty-first century. After CN purchased Illinois Central it then merged with the US railway, Burlington Northern Santa Fe Corporation in 1999. The new CN offers its customers shorter transit times and access to key ports in North America. Canadian exports to the US make up most of CN's traffic. These exports include petroleum and chemicals, forest products, automobiles and automobile parts, grain, coal, minerals, and fertilizer.

2. Canadian National Railway had its origins in the amalgamation of five financially failing railways: from 1917 to 1923, the Grand Trunk, the Grand Trunk Pacific, the Intercolonial, the Canadian Northern, and the National Transcontinental were combined to form the publicly operated Canadian National Railway. In 1993, the Canadian government privatized this Crown corporation.

3. Race, unlike ethnicity, is based on physical characteristics. Racial types are frequently assigned a set of social characteristics, which is known as 'stereotyping'. Sociologists define race as the socially constructed classification of persons into categories on the basis of real or imagined physical characteristics such as skin colour. Others consider race a means of creating major divisions of humankind on the basis of distinct physical characteristics.

4. As Canada reacts to the new economic realities, industrial restructuring is not without economic pain. In the first years of 'free trade', Canadian companies that could not compete in the North American market failed. Others have relocated to Mexico where labour costs are almost 10 times lower. In the first few years after NAFTA, job losses in the manufacturing sector amounted to 21 per cent (Jackson, 1993: 109). These losses were due largely to the closure of branch plants of American parent corporations (Healy, 1993: 287–94) and to Canadian-owned firms relocating to the United States and Mexico. Canada is not alone in this painful restructuring process. David Harvey (1989: 256–78) refers to global economic restructuring as one element in the worldwide transformation of industrial society from modern to postmodern.

Key Terms

age dependency ratio
The ratio of the economically dependent sector of the population to the productive sector, arbitrarily defined as the ratio of the elderly (those 65 years and over) plus the young (those under 15 years) to the population of working age (those 15 to 64 years).

census metropolitan area
A large urban area with a population of at least 100,000, together with adjacent smaller urban centres and even rural areas that have a high degree of economic and social integration with the larger urban area.

crude birth rate
The number of births per 1,000 people in a given year.

crude death rate
The number of deaths per 1,000 people in a given year.

culture
The sum of attitudes, habits, knowledge, and values shared by members of a society and passed on to their children.

demographic transition theory
The historical shift of birth and death rates from high to low levels in a population. The decline in mortality precedes the decline in fertility, resulting in a rapid population growth during the transition period.

demography
The scientific study of human populations, including their size, composition, distribution, density, growth, and related socio-economic characteristics.

ecumene
The portion of the land that is settled.

ethnic group
The association of people who have shared

awareness of a common identity and who identify themselves with a particular culture.

fertility rate

The number of births per 1,000 people in a given year; also called crude birth rate (not to be confused with the 'general fertility rate', which is the number of live births per 1,000 women who are of child-bearing age—15 to 44 years—in a given year).

industrial stage

Each major change in the evolution of the capitalist economic system is called a stage. The industrial stage marks the shift from a predominantly agricultural economy to an industrial one.

mortality rate

The number of deaths per 1,000 people in a given year; also called crude death rate.

natural increase (decrease)

The surplus (or deficit) of births over deaths in a population in a given time period.

net migration

The net effect of immigration and emigration on a country's population in a given period.

old-age dependency ratio

Similar to the age dependency ratio except the old-age dependency ratio focuses only on those over 64.

population density

The total number of people in a geographic area divided by the land area; population per unit of land area.

population distribution

The dispersal of a population within a geographic area.

population growth

The rate at which a population is increasing or decreasing in a given period due to natural increase and net migration; often expressed as a percentage of the original or base population.

population increase

The total population increase resulting from the interaction of births, deaths, and migration in a population in a given period of time.

population strength

Equates population size with economic and political power.

post-industrial stage

Each major change in the evolution of the capitalist economic system is called a stage. The post-industrial stage marks the shift from an industrial economy based on manufacturing to an economy in which service industries, particularly high-technology industries, become the dominant economic activities.

pre-industrial stage

Each major change in the evolution of the capitalist economic system is called a stage. The pre-industrial stage identifies an economic system that predates capitalism.

primary activities

Activities involving the direct extraction of natural resources such as agriculture, fishing, logging, mining, and trapping.

quaternary activities

Activities involving the collection, processing, and manipulation of information.

regional service centres

Urban places where economic functions are provided to residents living within the surrounding area.

resource town

An urban place where a single economic activity focused on resource extraction (e.g., mining, logging, oil drilling) dominates the town's economy; a single-industry town.

secondary activities

Activities that process and transform raw materials into finished goods; the manufacturing sector of an economy.

sex ratio

The ratio of males to females in a given population; usually expressed as the number of males for every 100 females.

tertiary activities

Activities that engage in services such as retailing, wholesaling, education, and financial and professional services.

urban population areas
Communities with economic and social functions that differentiate them from rural places; the common practice of defining urban population is by a specified size that assumes the presence of urban economic and social functions. In 1996, Statistics Canada considered all places with a combination of a population of 1,000 or more and a population density of at least 400 per square mile to be urban areas. People living in urban areas make up the urban population. People living outside of urban areas are considered rural residents and, by definition, constitute the rural population.

Bibliography

Badets, Jane, and Tina W.L. Chui. 1994. *Canada's Changing Immigrant Population*. Catalogue no. 96–311E. Ottawa: Statistics Canada.

Barnes, Trevor J., John N.H. Britton, William J. Coffey, David W. Edginton, and Glen Norcliffe. 2000. 'Canadian Economic Geography at the Millennium', *The Canadian Geographer* 44, 1: 4–24.

Bataïni, Sophie-Hélène, and William J. Coffey. 1998. 'The Location of High Knowledge Content Activities in the Canadian Urban System, 1971–1991', *Cahiers de Géographie du Québec* 42, 115: 7–34.

Beaujot, Roderic. 1991. *Population Change in Canada: The Challenges of Policy Adaptation*. Toronto: McClelland & Stewart.

Bernier, Jacques. 1991. 'Social Cohesion and Conflicts in Quebec', in Guy M. Robinson, ed., *A Social Geography of Canada*. Toronto: Dundurn Press.

Bissoondath, Neil. 1994. *Selling Illusions: The Cult of Multiculturalism*. Toronto: Penguin Books.

Bourne, Larry S., and David F. Ley, eds. 1993. *The Changing Social Geography of Canadian Cities*. Montréal and Kingston: McGill-Queen's University Press.

———— and Damaris Rose. 2001. 'The Changing Face of Canada: The Uneven Geographies of Population and Social Change', *The Canadian Geographer* 45, 1: 105–19.

Britton, John N.H., ed. 1996. *Canada and the Global Economy: The Geography of Structural and Technological Change*. Montréal and Kingston: McGill-Queen's University Press.

Burgess, Bill. 2000. 'Foreign Direct Investment: Facts and Perceptions about Canada', *The Canadian Geographer* 44, 2: 98–113.

Buzzelli, M. 2000. 'Toronto's Postwar Little Italy: Landscape Change and Ethnic Relations', *The Canadian Geographer* 44, 3: 298–301.

Cairns, Alan C. 2000. *Citizens Plus: Aboriginal Peoples and the Canadian State*. Vancouver: University of British Columbia Press.

————, John C. Courtney, Peter Mackinnon, Hans J. Michelmann, and David E. Smith, eds. 1999. *Citizenship, Diversity and Pluralism: Canadian and Comparative Perspectives*. Montréal and Kingston: McGill-Queen's University Press.

Cameron, Duncan, and Mel Watkins, eds. 1993. *Canada under Free Trade*. Toronto: James Lorimer.

Cardinal, Harold. 1969. *The Unjust Society*. Edmonton: Hurtig.

Cartwright, Don. 1996. 'The Expansion of French Language Rights in Ontario, 1968–1993: The Uses of Territoriality in a Policy of Gradualism', *The Canadian Geographer*, 40, 3: 238–57.

Ceh, S.L. Brian. 1997. 'The Recent Evolution of Canadian Inventive Enterprises', *Professional Geographer* 49, 1: 64–76.

Chartrand, Paul L.A.H., ed. 2003. *Who Are Canada's Aboriginal Peoples?* Saskatoon: Purich Publishing.

Chase, Steven, and Brent Jang. 1999. 'CP chops 1,900 rail jobs, warns of more carnage', *Globe and Mail*, 22 July, A1.

Coffey, W.J. 1994. *The Evolution of Canada's Metropolitan Economies*. Montréal: Institute for

Research on Public Policy.

——— and Richard G. Shearmur. 1997. 'The Growth and Location of High Order Services in the Canadian Urban System, 1971–1991', *Professional Geographer* 49, 4: 404–18.

——— and ———. 1998. 'Factors and Correlates of Employment Growth in the Canadian Urban System, 1971–1991', *Growth and Change* 29: 44–66.

Cohen, A. 1990. *A Deal Undone: The Making and Breaking of the Meech Lake Accord*. Vancouver: Douglas & McIntyre.

Denevan, William M. 1992. 'The Pristine Myth: The Landscape of the Americas in 1492', *Annals of the Association of American Geographers* 82, 3: 369–85.

Dickason, Olive Patricia. 2002. *Canada's First Nations: A History of Founding Peoples from Earliest Times*, 3rd edn. Toronto: Oxford University Press.

Dominion Bureau of Statistics. 1953. *Population: General Characteristics*. Vol. 1, Ninth Census of Canada. Ottawa: Queen's Printer.

Fife, Robert. 2002. 'Migrants must spread out: Ottawa', *National Post*, 22 June, A1, A9.

Flanagan, Tom. 2000. *First Nations? Second Thoughts*. Montréal and Kingston: McGill-Queen's University Press.

Fleras, Augie, and Jean Leonard Elliott. 1999. *Unequal Relations: An Introduction to Race, Ethnic, and Aboriginal Dynamics in Canada*, 3rd edn. Scarborough, Ont.: Prentice-Hall.

Fong, Eric, and Milena Gulia. 1997. 'The Effects of Group Characteristics and City Contexts on Neighbourhood Qualities among Racial and Ethnic Groups', *Canadian Studies in Population* 24, 1: 45–66.

——— and ———. 2000. 'Neighbourhood Changes within the Canadian Ethnic Mosaic, 1986–1991', *Population Research and Policy Review* 19: 155–77.

——— and Kumiko Shibuya. 2000. 'The Spatial Separation of the Poor in Canadian Cities', *Demography* 37, 4: 449–59.

Foot, David. 1996. *Boom, Bust and Echo: How to Profit from the Coming Demographic Shift*. Toronto: Macfarlane, Walter & Ross.

Francis, Diane. 2002. 'Tellier keeps his eye on horizon', *National Post*, 24 Aug., FP1, FP4.

Frideres, James S., and René R. Gadacz. 2001. *Aboriginal Peoples in Canada: Contemporary Conflicts*, 6th edn. Toronto: Prentice-Hall.

Gartley, John. 1994. *Focus on Canada: Earnings of Canadians*. Catalogue no. 96–317E. Ottawa: Statistics Canada.

Gilbert, Anne. 1999. *Espaces franco-ontariens, essai*. Ottawa: Le Nordir.

———. 2001. 'Le français au Canada, entre droits and géographie', *The Canadian Geographer* 45, 1: 175–9.

——— and Joan Marshall. 1995. 'Local Changes in Linguistic Balance in the Bilingual Zone: Francophones de l'Ontario et Anglophones du Québec', *The Canadian Geographer* 39, 3: 194–218.

Gower, Dave. 1998. 'The Labour Market for Computer Programmers', *The Daily*, Statistics Canada, 10 June: <http://www.statcan.ca/Daily/English/980610/d 980610.htm>.

Halli, S.S., and L. Driedger, eds. 1999. *Immigrant Canada: Demographic, Economic and Social Challenges*. Toronto: University of Toronto Press.

Harrison, Brian, and Louise Marmen. 1994. *Focus on Canada: Languages in Canada*. Catalogue no. 96–313E. Ottawa: Minister of Industry, Science and Technology.

Harvey, David. 1989. *The Urban Experience*. Baltimore: Johns Hopkins University Press.

Haydamack, Brent. 1998. 'Regional Trajectories of Technological Change in Canadian Manufacturing', *The Canadian Geographer* 42, 1: 2–13.

Healy, Theresa. 1993. 'Selected Plant Closures and Production Relocations: January 1989–June 1992, Ontario', in Cameron and Watkins (1993: Appendix I).

Heberg, Edward N. 1989. *Ethnic Groups in Canada: Adaptations and Transitions*. Toronto: Nelson.

Hébert, Raymond M. 1998. 'Identity, Cultural Production and the Vitality of Francophone Communities Outside Québec', in Leen d'Haenens, ed., *Images of Canadianness: Visions on Canada's Politics, Culture, Economics*. Ottawa: University of Ottawa Press.

Hewitt, Kenneth. 2000. 'Safe Place or Catastrophic Society? Perspectives on Hazards and Disasters in Canada', *The Canadian Geographer* 44, 4: 325–41.

Hiebert, Daniel. 1994. 'Canadian Immigration: Policy, Politics, Geography', *The Canadian Geographer* 38, 3: 254–8.

———. 2000. 'Immigration and the Changing Canadian City', *The Canadian Geographer* 44, 1: 25–43

Holmes, John. 1997. 'In Search of Competitive Efficiency: Labour Process Flexibility in Canadian Newsprint Mills', *The Canadian Geographer* 41, 1: 7–25.

Hynddman, Jennifer, and Margaret Walton-Roberts. 2000. 'Interrogating Borders: A Transnational Approach to Refugee Research in Vancouver', *The Canadian Geographer* 44, 3: 244–58.

Indian Chiefs of Alberta. 1970. *A Presentation by the Indian Chiefs of Alberta to Right Honourable P.E. Trudeau*. Edmonton: Indian Association of Alberta.

Ip, Greg. 1996. 'Jobs Cut Despite Hefty Profits', *Globe and Mail*, 6 Feb., A4.

Jackson, Andrew. 1993. 'Manufacturing', in Cameron and Watkins (1993: ch. 5).

Kalbach, Madeline A., and Warren E. Kalbach. 1995. 'Ethnic Diversity and Persistence as Factors in Socioeconomic Inequality: A Challenge for the Twenty-first Century', *Toward XXI Century: Emerging Socio-Demographic Trends and Policy Issues in Canada*, 147–60. Proceedings of the 1995 Symposium Organized by the Federation of Canadian Demographers.

Kobayashi, Audrey. 1993. 'Multiculturalism: Representing a Canadian Institution', in J. Duncan and David Ley, eds, *Place/Culture/Representations*. New York: Routledge, 205–31.

———. 1994. *Women, Work, and Place*. Montréal and Kingston: McGill-Queen's University Press.

——— and B. Ray. 2000. 'Civil Risk and Landscapes of Marginality in Canada: A Pluralist Approach to Social Justice', *The Canadian Geographer* 44, 4: 401–17.

Ley, David. 1996. *The New Middle Class and the Remaking of the Central City*. Toronto: Oxford University Press.

———. 1999. 'Myths and Meanings of Immigration and the Metropolis', *The Canadian Geographer* 43, 1: 2–18.

_____ and Daniel Hiebert. 2001. 'Immigration Policy as Population Policy', *The Canadian Geographer* 45, 1: 120–5.

Li, Peter S. 1988. *Ethnic Inequality in a Class Society*. Toronto: Wall and Thompson.

———, ed. 1990. *Race and Ethnic Relations in Canada*. Toronto: Oxford University Press.

———. 1996. *The Making of Post-War Canada*. Toronto: Oxford University Press.

———. 1998. *The Chinese in Canada*. Toronto: Oxford University Press.

Little, Bruce. 1997. 'Are We Better Off under the Liberals?', *Globe and Mail*, 26 Apr., B1, B5.

Long, David, and Olive Patricia Dickason, eds. 2000. *Visions of the Heart: Canadian Aboriginal Issues*. Toronto: Harcourt Canada.

McArthur, Keith. 1999. 'Bata pioneers feel the pain of plant closing', *Globe and Mail*, 20 Oct., B1.

McCallum, J. 1995. 'National Borders Matter: Canada–U.S. Regional Trade Patterns', *American Economic Review* 85, 3: 615–23.

Mackenzie, Suzanne, and Glen Norcliffe. 1997. 'Restructuring in the Canadian Newsprint Industry', *The Canadian Geographer* 41, 1: 2–6.

McKenna, Barrie. 1998. 'More Firms Flock to Mexico: number of Canadian factories setting up in duty-free zone has tripled since 1994', *Globe and Mail*, 8 July, A1.

MacLachlan, Ian, and Ryo Sawada. 1997. 'Measures of Income Inequality and Social Polarization in Canadian Metropolitan Areas', *The*

Canadian Geographer 41, 4: 377–97.

McRoberts, Kenneth. 1997. *Misconceiving Canada: The Struggle for National Unity*. Toronto: Oxford University Press.

McVey, Wayne W., and W.E. Kalbach. 1995. *Canadian Population*. Toronto: Nelson Canada.

Marsh, James H., ed. 1988. *The Canadian Encyclopedia*, 2nd edn. Edmonton: Hurtig.

Monture-Angus, Patricia. 1999. 'Considering Colonialism and Oppression: Aboriginal Women, Justice and the "Theory" of Decolonization', *Native Studies Review* 12, 1: 63–94.

———. 2000. 'Lessons in Decolonization: Aboriginal Overrepresentation in Canadian Criminal Justice', in Long and Dickason (2000: 361–86).

Mooney, James. 1928. *The Aboriginal Population of America North of Mexico*. Smithsonian Miscellaneous Collections. Washington: Smithsonian Institution.

Nash, Alan. 1994. 'Some Recent Developments in Canadian Immigration Policy', *The Canadian Geographer* 38, 3: 258–61.

National Post. 1999a. 'Shirt maker closes after 44 years', 1 Feb., C3.

National Post. 1999b. 'Paragon to stop making diapers at Canadian plant', 1 May, D2.

Newhouse, David R. 2000. 'From the Tribal to the Modern: The Development of Modern Aboriginal Societies', in Ron F. Laliberte et al., eds, *Expressions in Canadian Native Studies*. Saskatoon: University of Saskatchewan Extension Press, 395–409.

Nicol, Heather, and Greg Halseth, eds. 2000. *(Re)Development at the Urban Edges*. Publication Series No. 53, Department of Geography. Waterloo, Ont.: University of Waterloo.

Norcliffe, Glen. 1996. 'Foreign Trade in Goods and Services', in Britton (1996: ch. 2).

Norris, Mary Jane. 2000. 'Aboriginal Peoples in Canada: Demographic and Linguistic Perspectives', in Long and Dickason (2000: 167–236).

Olson, R., and Audrey Kobayashi. 1993. 'The Emerging Ethnocultural Mosiac', in Bourne and Ley (1993: 138–52).

Parker, Paul. 1997. 'Canada–Japan Coal Trade: An Alternative Form of the Staple Production Model', *The Canadian Geographer* 41, 3: 248–66.

Peters, Evelyn J. 2000a. 'Aboriginal People and Canadian Geography: A Review of the Recent Literature', *The Canadian Geographer* 44, 1: 44–55.

———. 2000b. 'Aboriginal People in Urban Areas', in Long and Dickason (2000: 237–70).

Pooler, James A. 2000. *Hierarchical Organization in Society: A Canadian Perspective*. Burlington, Ont.: Ashgate.

Preston, Valerie, and Lucia Lo. 2000. 'Canadian Urban Landscape Examples—21: "Asian Theme" Malls in Suburban Toronto: Land Use Conflict in Richmond Hill', *The Canadian Geographer* 44, 2: 182–90.

Randall, Stephen J., and Herman W. Konrad, eds. 1996. *NAFTA in Transition*. Calgary: University of Calgary Press.

Rashid, Abdul. 1994. *Focus on Canada: Family Income in Canada*. Catalogue no. 96–318E. Ottawa: Statistics Canada.

Reid, Scott. 1992. *Canada Remapped: How the Partition of Quebec Will Reshape the Nation*. Vancouver: Arsenal Pulp Press.

———. 1993. *Lament for a Notion: The Life and Death of Canada's Bilingual Dream*. Vancouver: Arsenal Pulp Press.

Reitz, Jeffrey, and Raymond Breton. 1995. *The Illusion of Differences*. Ottawa: C.D. Howe Institute.

Robinson, Allan. 1999. 'Falconbridge to cut 140 jobs despite better profit outlook', *Globe and Mail*, 20 Oct., B3.

Royal Commission on Aboriginal Peoples. 1996. *Report of the Royal Commission on Aboriginal Peoples*, 5 vols. Ottawa: Minister of Supply and Services.

Royal Commission on Bilingualism and Biculturalism. 1970. *Report. Book IV, Cultural Contributions of the Other Ethnic Groups*. Ottawa: Queen's Printer.

Scotiabank. 1998. *Global Economic Outlook*.

Toronto: Bank of Nova Scotia.

Sharpe, Bob. 2000. 'Geographies of Criminal Victimization in Canada', *The Canadian Geographer* 44, 4: 418–28.

Slocombe, D. Scott. 2000. 'Resources, People and Places: Resource and Environmental Geography in Canada 1996–2000', *The Canadian Geographer* 44, 1: 56–66.

Statistics Canada. 1991. *Canada Year Book 1992.* Ottawa: Minister of Industry.

———. 1992a. *Immigration and Citizenship.* Catalogue no. 93–316. Ottawa: Minister of Supply and Services.

———. 1992b. *Age, Sex and Marital Status.* Catalogue no. 93–310. Ottawa: Minister of Industry.

———. 1992c. *Census Divisions and Subdivisions.* Catalogue no. 93–303. Ottawa: Minister of Supply and Services.

———. 1992d. *Census Metropolitan Areas and Census Agglomerations.* Catalogue no. 93–303. Ottawa: Minister of Supply and Services.

———. 1993a. *Ethnic Origin.* Catalogue no. 93–315. Ottawa: Minister of Supply and Services.

———. 1993b. *Mobility and Migration.* Catalogue no. 93–322. Ottawa: Minister of Supply and Services.

———. 1996a. *Canadian Economic Observer: Historical Statistical Supplement 1995/96.* Catalogue no. 11–210. Ottawa: Minister of Supply and Services.

———. 1996b. *Canada Year Book 1997.* Ottawa: Minister of Industry.

———. 1997a. *The Daily*—1996 Census: Immigration and Citizenship, 4 Nov. 1997 [on-line database], Ottawa. Searched 11 Aug. 1998: <http://www.statcan.ca/Daily/English/>.

———. 1997b. *The Daily*—1996 Census: Mother Tongue, Home Language and Knowledge of Languages, 2 Dec. 1997 [on-line database], Ottawa. Searched 14 July 1998: <http://www.statcan.ca/Daily/English/>.

———. 1997c. *A National Overview: Population and Dwelling Counts.* Catalogue no. 93–357–XPB. Ottawa: Minister of Industry.

———. 1997d. *Mortality—Summary List of Causes, 1995.* Catalogue no. 84–209–XPB. Ottawa: Minister of Industry.

———. 1997e. Selected Income Statistics for Individuals, Families and Households, 1991 and 1996 Censuses [on-line database], Ottawa. Searched 9 Mar. 2001: <www.statcan.ca/english/Pgdb/People/Families/famil61c.htm>.

———. 1998a. Labour Force, Employment and Unemployment [on-line database], Ottawa. Searched 12 Aug. 1998: <http://www.statcan.ca/english/Pgdb/Economy/Economic/econ10.htm>.

———. 1998b. Recent Immigrants by Country of Last Residence [on-line database], Ottawa. Searched 17 Aug. 1998: <http://www.statcan.ca/english/Pgdb/People/Population/demo08.htm>.

———. 1998c. Single and Multiple Ethnic Origin Responses, 1996 Census [on-line database], Ottawa. Searched 17 Aug. 1998: <http://www.statcan.ca/english/Pgdb/People/Population/demo28a.htm>.

———. 1998d. *The Daily*—1996 Census: Education, Mobility and Migration, 14 Apr. 1998 [on-line database], Ottawa. Searched 11 Aug. 1998: <http://www.statcan.ca/Daily/English/>.

———. 1998e. *The Daily*—1996 Census: Ethnic Origin, Visible Minorities, 17 Feb. 1998 [on-line database], Ottawa. Searched 16 July 1998: <http://www.statcan.ca/Daily/English/>.

———. 1998f. 1981–1996 Census: Labour Force Activity, 1998 [on-line database], Ottawa. Searched 2 Sept. 1998: <http://www.statcan.ca/english/census96/mar17/labour/table6 t6p59a.htm>.

———. 2000. *Canadian Economic Observer: Historical Statistical Supplement 1999/00.* Catalogue no. 11–210–XPB. Ottawa: Minister of Industry.

———. 2002a. Census of Canada 2001—Census Geography. Highlights and Analysis: Canada's 2001 Population: <http://www12.

statcan.ca/English/census01/>.

———. 2002b. Census of Canada 2001: Age and Sex for Canada, Provinces and Territories 2001: <http://www12.statcan.ca/English/census01/>.

———. 2002c. Census of Canada 2001: Population by Sex for Canada, Provinces and Territories, 2001: <http://www12.statcan.ca/English/census01/>.

———. 2002d. *The Daily*—Census of Population: Language, Mobility and Migration, 10 Dec. [on-line database], Ottawa. Searched 12 Jan. 2003: <http://www.statcan.ca/Daily/English/02120/d021210a.htm>.

———. 2003a. *The Daily*—Canadian International Merchandise Trade, 20 Feb. 2003 [on-line database], Ottawa. Searched 27 Feb. 2003: <http://www.statcan.ca/Daily/English/030220/d030220a.htm>.

———. 2003b. *The Daily*—Financial Statistics for Enterprises, 27 Feb. 2003 [on-line database], Ottawa. Searched 28 Feb. 2003: <http://www.statcan.ca:80/Daily/English/030227/d030227b.htm>.

———. 2003c. Census of Canada 2001—Canada's Ethnocultural Portrait: The Changing Mosaic. Analytical Series 96F0030XIE 2001008, 21 Jan. 2003 [on-line database], Ottawa. Searched 28 Jan. 2003: <http://www12.statcan.ca/english/census01/products/analytic/companion/etoimm/canada.cfm>.

———. 2003d. Components of Population Growth, 28 Feb. 2003 [on-line database], Ottawa. Searched 28 Mar. 2003: <http://www.statcan.ca/English/Pgdb/demo33a.htm>.

———. 2003e. Labour Force, Employed and Unemployed, Numbers and Rates [on-line database], Ottawa. Searched 4 Apr. 2003: <http://www.statcan.ca/english/Pgdb/labor07a.htm>.

———. 2003f. *The Daily*—Overcoming Distance, Overcoming Borders: Comparing North American Regional Trade. 16 Apr. 2003:

<http://www.statcan.ca/Daily/English/030416/d030446h.htm>.

———. 2003g. *The Daily*—Census of Population: Earnings, Levels of Schooling, Field of School Attendance, 12 Mar. 2003: <http://www.statcan.ca/Daily/English/030311/d030311a.htm>.

———. 2003h. Census of Canada 2001—Aboriginal peoples of Canada: A demographic profile Analysis series 96F0030XIE2001007, 21 Jan 2003 [on-line database], Ottawa. Searched 28 Jan. 2003: <http://www12.statcan.ca/english/census01/products/analytic/companion/abor/contents.cfm>.

———. 2003i. *The Daily*—Labour Force Survey, 12 Apr. 2003:

<http://www.statcan.ca/Daily/English/030404/d030404a.htm>.

———. 2003j. Census of Canada 2001—Religions in Canada. Analysis series 96F0030XIE2001015 [on-line database], Ottawa. Searched 16 Aug. 2003:

<http://www12.statcan.ca/english/census01/products/analytic/companion/rel/contents.cfm>

———. 2003k. Learning Resources: Civics and Society—Declining Birth Rate and the Increasing Impact of Immigration. Searched 28 June 2003: <http://www.statcan.ca/english/kits/issues/charts/chart3.htm>.

———. 2003l. Profile of the Canadian Population by Mobility Status: Canada, a Nation on the Move: <http://www12.statcan.ca/english/census01/products/analytic/companion/mob/contents.cfm>.

———. 2004. Exports of goods on a balance-of-payments basis. Canadian Statistics, International Trade. 22 Jan. 2004: <http://www.statcan.ca/english/Pgdb/gblec04.htm>.

Steinhart, David. 1999. 'Levi to shut plants in Cornwall, U.S.: Brantford loses jobs too', *National Post*, 23 Feb., C1.

Thériault, J. Yvon, ed. 1999. *Francophonies minoritaires au Canada: état des lieux*. Moncton: Éditions d'Acadie.

Thorpe, Jacqueline. 2003. 'Red tide stops at 49th

parallel', *National Post*, 27 Feb., FP1, FP4.

Trewartha, G.T., A.H. Robinson, and E.H. Hammond. 1967. *Elements of Geography*, 5th edn. New York: McGraw-Hill.

United Nations. 1994. *Human Development Report 1994*. New York: Oxford University Press.

Young, R.A. 1989. 'Political Scientists, Economists, and the Canada–US Free Trade Agreement', *Canadian Public Policy* 15, 1: 49–56.

Further Reading

Bunting, Trudi, and Pierre Filion, eds. 2000. *Canadian Cities in Transition: The Twenty-First Century*, 2nd edn. Toronto: Oxford University Press.

Canada is a highly urban society. Its cities are complex, dynamic, and open to change. In *Canadian Cities in Transition*, Professors Bunting and Filion have brought together Canada's leading urban geographers to discuss why and how Canadian cities work or don't work. The underlying theme is change and the range of topics is broad.

Cities, as a reflection of society, are constantly changing in appearance, function, and problems. Cities are also a reflection of national and global forces, such as economic restructuring, telecommunications advances, and international migration. The book has six main sections: urban systems, city processes, city structure, land use, governance and planning, and pressing issues. Students might be attracted to the comparison of Canadian and American cities found in Chapter 3. John Mercer and Kim England challenge the idea of a 'North American city', arguing that Canadian cities are unique. Another interesting—but distressing—chapter deals with a social problem found in large cities—urban homelessness. Since urban poverty is not discussed in *The Regional Geography of Canada*, reading Chapter 23 would expose the student to this persistent problem. Tracy Peressini and Lynn McDonald call urban homelessness a national disgrace for a country that claims to care for all its citizens. The authors state that destitute urban dwellers, who are often casualties of our market economy, are a permanent feature of urban Canada. They examine the reasons for this urban phenomenon and explore several solutions, including public housing.

The Ambassador bridge in Windsor,
Ontario (Ivy Images)

Ontario accounts for nearly 41 per cent of Canada's industrial output and 38 per cent of its population. It consists of an industrial core in the south and a downward transition region in the north. Southern Ontario contains over three-quarters of Ontario's population, is the major sector of Canada's industrial core, and is on its way to becoming a major industrial area in North America. Northern Ontario, on the other hand, with a much more modest population, resembles a declining resource hinterland. Southern Ontario's automobile industry is examined in this chapter's *Key Topic*. The economic integration of auto manufacturing in the North American market began under the Auto Pact (1965) and intensified under the Free Trade Agreement (1989) and the North American Free Trade Agreement (1994). Ontario's primary trade axis has been reoriented from east–west to north–south.

- Describe Ontario's physical geography and historical background.
- Present the basic characteristics of Ontario's population and resources.
- Identify Ontario's dichotomy—an industrial core in the south and a resource hinterland in the north.
- Examine the importance of the manufacturing sector, especially the automobile industry, to Ontario's economy.
- Analyze the significance to Ontario of the Auto Pact and the North American Free Trade Agreement.
- Focus on Ontario's changing role within Canada and North America.

Chapter 5 Ontario

■ Introduction

Ontario is often described as the 'Province of Opportunity', but Ontario is a geographic paradox because it has two dissimilar areas, southern Ontario and northern Ontario. Each has very different physical conditions that have formed the basis of two distinct economies. The upbeat description of Ontario as the 'Province of Opportunity' applies to only one part, southern Ontario. Southern Ontario's economy has generated jobs and business opportunities. This area's positive economic environment has drawn people from across Canada and around the world to its cities. The manufacture of automobiles has been a significant part of southern Ontario's economy and constitutes the *Key Topic* in this chapter. While southern Ontario is the industrial and population heartland of the province, northern Ontario is an old resource hinterland, where economic and population growth is stalled. Because so many young people are moving to more prosperous areas in Canada, the population of northern Ontario is losing a valuable segment of its labour force, leaving its cities and towns with disproportionately large numbers of older and very young people. In 2001, northern Ontario had a population of just under 800,000, while southern Ontario had a population of nearly 11 million.

Ontario within Canada

Ontario is the leading economic region in Canada (Figure 5.1). In fact, according to Thomas Courchene, one of Canada's foremost economists, Ontario has evolved into one of North America's leading economic regions and may become the new 'heartland' of North America (Courchene, 1998). Four resources—agriculture, forests, minerals, and hydroelectricity—spurred the province's past economic development. Today, the manufacturing and service industries are driving Ontario's economy.

In addition to its economic strength, Ontario is the political linchpin in Confederation. This political role is attributed partly to Ontario's historic role in Confederation and partly to its population size. With 38 per cent of Canada's population, Ontario sends more representatives to the House of Commons than any other province. Since Ontario's economic future lies in expanding trade with the United States, Ontario is interested in having Ottawa pursue trade policies with the United States that facilitate ease of border crossings and that promote the well-being of its manufacturing industry, especially the automobile sector.

Canadians living in less prosperous and less powerful regions often see Ontario in a different light, perhaps precisely because of Ontario's economic success and political power. Westerners and Maritimers have accused Ontario of using its political power to maintain its economic mastery over the rest of the country. Québecers have pointed to the

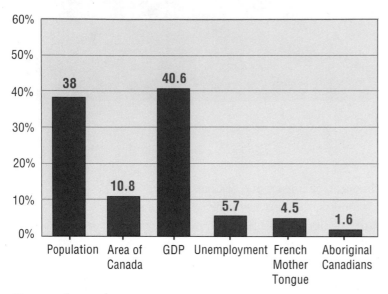

Figure 5.1 Ontario, 2001. Ontario's economic and political strength within Canada is revealed by its share of the nation's GDP and population. Ontario comprises the largest single market in Canada and holds the demographic key to political power in Ottawa, and its economy, by accounting for nearly 41 per cent of the nation's GDP while having only 38 per cent of its population, exhibits a high level of productivity. Franco-Ontarians and Aboriginal peoples form small minorities.

Sources: Tables 1.1, 1.5, and 4.19.

striking difference between the two provinces' unemployment rates as proof that Ottawa favours Ontario. Ontarians reject these criticisms, claiming that Ontario has done its share to shoulder the 'burden of Confederation'. For example, Ontario is the major contributor to equalization payments allocated to have-not provinces.

Ontario's Physical Geography

Ontario is larger than most countries (Figure 5.2), stretching out over 1 million km². Extending over Ontario are three of Canada's physiographic regions (Great Lakes–St Lawrence Lowlands, Canadian Shield, and Hudson Bay Lowland), and three of the country's climatic zones (Arctic, Subarctic, and Great Lakes–St Lawrence) (Figures 2.1 and 2.6). Manitoba lies to its west, Hudson and

James bays to its north, while Québec is on its eastern boundary, marked in part by the Ottawa River. This central location within Canada and its close proximity to the industrial heartland of the United States have facilitated Ontario's economic development.

Southern Ontario is located in the southernmost part of Canada. Windsor, for example, is at latitude 42° N and Toronto is close to 44° N. Southern Ontario lies in the Great Lakes–St Lawrence climatic zone, which has a moderate continental climate. This climate is noted for long, hot, and humid summers followed by short, cold winters. Annual precipitation is about 1,000 mm. The greatest amounts of precipitation occur in the lee of the Great Lakes, where winter snowfall is particularly heavy (see Vignette 5.1). The Great Lakes modify temperatures and funnel winter storms into this region. During the winter, this region experiences a great variety of weather conditions.

Southern Ontario is the most favoured physical area in Canada. As a result, the vast majority of the province's population, nearly 11 million or 93 per cent, live in southern Ontario. The area is underlain by slightly tilted sedimentary rocks, which are covered by a thin deposit of glacial till. Except for the Niagara Escarpment, there is little relief topography. This escarpment provides the most spectacular scenery in this subregion. Formed since deglaciation, the Niagara Escarpment is an erosional remnant of the more resistant sedimentary rocks. A mixed forest vegetation flourishes in this temperate continental climate (Figure 2.7). With a long growing season, ample precipitation, and fertile soils, the southern Ontario lowland has the most productive agricultural lands in Canada. The Canadian Shield, however, marks an abrupt end to agricultural land. At this point, northern Ontario begins.

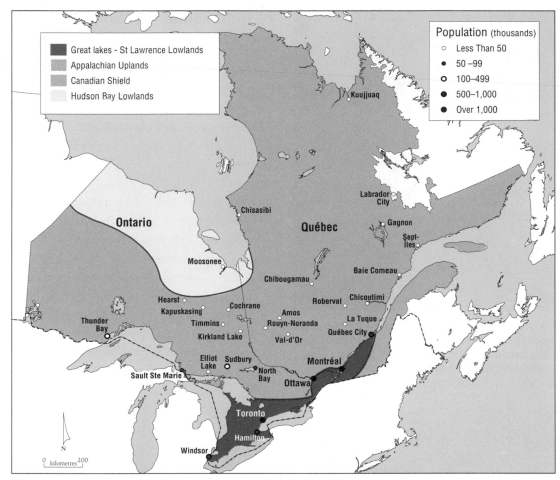

Figure 5.2 Central Canada. Ontario and Québec have the two largest provincial economies and populations in Canada. Toronto and Montréal, the two largest cities in Canada, are located in Canada's manufacturing belt, which stretches from Windsor to Québec City. Canada's most productive agricultural lands are also located in this manufacturing belt. (Further resources: Student Web site, National Atlas section, Map 17. Web site instructions are found on p. xxi)

Northern Ontario has a less favoured physical area and less than 8 per cent of Ontarians live in this part of the province. Northern Ontario lies in two physiographic regions: the Canadian Shield and the Hudson Bay Lowland. Located in much higher latitudes, its winters are longer and colder than those experienced in southern Ontario. For the most part, northern Ontario lies north of 46° N and extends almost as far north as 57° N. Even along its southern edge, which Hamelin calls the Near North, short summers make crop agriculture vulnerable to frost damage. In addition to a difficult climate for agriculture, the rocky Canadian Shield has very little agricultural land. Its rugged, hilly topography is dotted with lakes. Climate, soils, and physiography combine to limit agriculture in northern Ontario. However, this subregion does have vast forests, superb scenery (which supports tourism), and extensive mineral wealth. As a resource hinterland, northern Ontario's economy is almost entirely based on these three resources.

Because Ontario has the largest population of all the provinces, it has the greatest number of seats in the House of Commons—and therefore is seen as holding the balance of power in Canada. After the 37th General Election in 2000, Ontario held one-third of the seats in the House of Commons (103 out of 301 seats). (© Royalty Free/Corbis/Magma)

The Niagara Escarpment, stretching over 700 km from Queenston on the Niagara River to Tobermory on the Bruce Peninsula, is the most significant landform in southern Ontario. The Escarpment reaches over 350 metres above the surrounding land at various points. (Victor Last, Geographical Visual Aids 35664)

Environmental Challenges

Ontario is faced with two major environmental challenges—air pollution and water pollution. Without action by governments, these two problems will only get worse. Both are hidden costs of our industrial world. Toronto, which anchors the highly urbanized Golden Horseshoe, is deeply troubled by smog. Smog becomes a health hazard in the summer months when high temperatures and air inversions combine to keep the smog hanging over the city as a brown haze. The word 'smog' is actually a combination of 'smoke' and 'fog'. Smog is the most visible form of air pollution. It is a brownish-yellow hazy cloud caused when heat and sunlight react with various pollutants emitted by industry, vehicle exhaust, pesticides, and oil-based home products. A study by the Ontario Medical Association estimated that illnesses caused by smog cost Ontario more than $1 billion a year in hospital admissions, emergency room visits, and absenteeism. Even more startling, the Association's study (2001) concluded that about 1,900 people die prematurely in Ontario each year because of smog-related respiratory problems.

Water pollution is an equally great environmental problem. The Walkerton disaster of May 2000, when the local water supply was contaminated by E. coli, causing seven deaths and many illnesses, provided indisputable evidence of the serious nature of water pollution. As one account of this calamity in a small rural town began:

'We have a terrible tragedy here.' With those words, Ontario Premier Mike Harris waded into the Walkerton, Ontario water crisis . . . . He addressed a crowd of reporters and residents in the normally

Vignette 5.1 Ontario's Snowbelts

Ontario's snowbelts are legendary. On the upland slopes facing Lakes Huron and Superior and Georgian Bay, huge snowfalls in the 300 to 400 cm range fall each winter from November to late March. The uplands on the northeastern shores of Lake Superior receive the greatest total snowfall amounts of any area in Ontario (exceeding 400 cm annually). Much of the snowfall in snowbelt areas can be attributed to cold northwesterly to westerly winds blowing off the lakes and ascending the highlands. As the arctic air travels across the relatively warmer Great Lakes, it is warmed and moistened. Snow clouds form over the lakes and, once onshore, intensify as the air is then forced to ascend the hills to the lee of the lakes, triggering heavy snowfalls. Areas on the downslope side of the higher ground to the lee of the lakes receive less than half the annual snow totals of the upslope snowbelt areas. For example, Toronto, Hamilton, and other places to the lee of the Niagara Escarpment are snow-shadow regions with winter amounts of 100 to 140 cm. In the snowbelt regions, snowfall accounts for about 32 per cent of the year's total precipitation; but in the snow-sparse area of southwestern Ontario around Windsor and Chatham, the snow contribution is only about 13 per cent.

quiet town in the heart of Ontario's rural heartland—a part of the province that normally gears up for a flood of fun seekers at this time of year.

Instead, Walkerton began the transition into the town 'where those kids died from E. coli'. It's not what anyone wanted, but it was the end result. Reporters from around North America descended on the area, trying to get to the bottom of what's being described as Canada's worst-ever outbreak of E. coli contamination. Seven people died from drinking contaminated water. Hundreds suffered from the symptoms of the disease, not knowing if they too would die. (CBC, 2000)

The Great Lakes represent both a natural asset and an environmental problem. Within the physical geography of Ontario, the Great Lakes play a significant role as a waterway, a freshwater fishery, and as a popular vacation destination. Ontario shares the Great Lakes with the United States. In many ways, the Great Lakes affect life in the more populated areas of Ontario. Storms track along the Great Lakes into southern Ontario, bringing rain in the summer and snow in the winter. The Great Lakes sustain a commercial fishing industry and recreational fishing. This huge body of fresh water supplies drinking water to more than 9 million Ontarians. Four of the five Great Lakes (Lake Superior, Lake Huron, Lake Erie, and Lake Ontario) straddle the border with the United States (Lake Michigan, wholly in the US, is bordered by the states of Indiana, Illinois, Wisconsin, and Michigan). The Great Lakes suffer from industrial and urban pollution. One reason is that North America's industrial belt and many large cities surround the Great Lakes. In the past, the Great Lakes were taken for granted, but in 1912 the newly formed International Joint Commission, created as a result of the Boundary Waters Treaty of 1909, was asked to investigate the high incidence of typhoid in the Great Lakes region of the US and Canada. The six commissioners—three from Canada and three from the US—produced an exhaustive bacteriological study of contaminants, and subsequent cleanup efforts and, especially, chlorination of urban water supplies alleviated the problem.

Nonetheless, despite the best efforts of the IJC over many years, misuse of the Great Lakes continued. For a long time, industrial plants dumped their toxic wastes into the Great Lakes or buried their chemical wastes, which later began to leak into the lakes. Urban centres also contributed to the pollution problem by discharging untreated sewage into these waters.

By the 1960s, the Great Lakes—and especially Lake Erie—were in serious trouble. Lake Erie was reported to be dying from a vast stew of human waste, industrial garbage, and agricultural chemicals. In 1972, Canada and the United States passed the Great Lakes Water Quality Agreement (GLWQA), revised in 1978 to place greater emphasis on an ecosystem approach to cleaning up these vast boundary waters. Then, in 1987, a protocol to the GLWQA called for Remedial Action Plans (RAPs) to be developed for particularly endangered and polluted areas of the Great Lakes. Of the 43 identified areas requiring RAPs, 17—including Hamilton harbour and Collingwood—were in Ontario (Kreutzwiser and de Loë, 2004). The International Joint Commission has overseen this work, and an investment of $10 billion brought about great improvements (Stewart, 2003: 42). By the end of the 1980s, with the appearance of cleaner water and increasing fish populations, the Great Lakes appeared to be saved. Consequently, funding was reduced. However, the problems have not been fully solved. Today, the major threats to the Great Lakes are:

- The growing levels of phosphorus, which are derived primarily from waste water containing detergents and from runoff water with high levels of chemicals from agricultural fertilizers, are contributing to the creation of a 'dead zone' where only toxic organisms can survive.
- Exotic species such as sea lampreys, Asian carp, and zebra mussels are squeezing out native species and are thereby radically changing the ecosystem of the Great Lakes.

Ontario's Historical Geography

Since its beginning as a British colony, Ontario has been the counterweight to Québec. When the American Revolution began in 1775, Ontario was a densely forested wilderness inhabited by a few fur traders and Indians. After losing the American colonies, loyal British subjects returned to England or relocated in one of Britain's colonies. In 1782–3, Loyalists moved northward to Nova Scotia and New Brunswick, while others resettled in the Eastern Townships of Québec. A smaller number, perhaps as many as 10,000 Loyalists, settled on land along the St Lawrence River upstream from French-speaking *habitants*, colonizing the wilderness area that later became known as Upper Canada. These Loyalists were followed by American, British, and European newcomers, who later flooded into Upper Canada in search of land. In less than a century, this natural landscape was transformed into a British agricultural colony.

Until the War of 1812, many settlers had come from the United States. They, like most other colonists, sought land in the fertile lowlands of the Great Lakes. When hostilities between Britain and the United States escalated, the Americans launched an attack on British North America. The War of 1812 effectively ended the influx of American settlers into Upper Canada. After the war, an increasing number of settlers came from the British Isles, especially Ireland and Scotland. By 1830, Upper Canada's population grew rapidly from natural increase and immigration. Thirty years later, Upper Canada had grown in numbers to nearly 1.4 million. Of these inhabitants, about 33,000 French-speaking Canadians lived along the Detroit River, where their ancestors had put down roots during the French regime in North America.

At first, settlement took place along the coastal edge of Lake Ontario and Lake Erie, where forests were cut down to clear land for farming. Only after 1830 did colonists move northward towards the Canadian Shield, which marked the limits of agricultural land. Within 30 years, virtually all the arable land in the Great Lakes Lowlands was cleared of forest and cultivated. By 1850, settlers occupied the best lands. Within a decade, some turned further northward to try their luck in the few pockets of arable land within the Canadian Shield. By the time of Confederation, there was a severe land shortage in Upper Canada and settlers began to look for land in the western United States.

While Upper Canada was clearly a rural society during the nineteenth century, a network of urban places began to spring up. With the growing shortage of agricultural land, sons and daughters of farmers moved to these urban centres to start a new and different life. In spite of this rural-to-urban migration, the demographic balance of power clearly remained with the rural community—urban areas, such as villages and towns, contained less than 20 per cent of the population of Upper Canada.

When Upper Canada joined Confederation in 1867, it was renamed Ontario (Figure 3.4). At that time, the geographic extent of Ontario

In 1794, York became the capital of Upper Canada. Despite its political status, this frontier village remained on the western edge of British settlement that stretched westward from Lower Canada along the north shore of Lake Ontario. By 1812, York had only 700 residents. This painting (dated 1804) illustrates a group of houses strung along the shore of Toronto Bay. Beyond this narrow strip of cleared land lies the original forest of southern Ontario. (*National Archives of Canada C34334*)

was about 100,000 km²—just a fraction of its present size—but as Canada grew in size, acquiring more territory from Great Britain, both Ontario and Québec benefited by obtaining some of these new lands (Figures 3.5–3.7). In both cases, however, these new lands had little immediate value for economic development and settlement because they were carved out of two physiographic regions (the Canadian Shield and the Hudson Bay Lowland) that were far from markets and had little or no agricultural potential. Since Confederation, the borders of Ontario have been extended three times. These boundary changes have greatly increased the geographic size of the province, but not its agricultural lands. The first expansion occurred in 1874 when Ontario's boundaries were pushed northward to about 51° N and westward towards the Lake of the Woods. The second expansion took place in 1889. Until then, both Manitoba and Ontario claimed the land around the Lake of the Woods. The Ontario government won this dispute and was able to enlarge the province once more. At the same time, Ontario's northwest boundary was adjusted to the Albany River, which flows into James Bay. As a result of this adjustment, Ontario gained access to James Bay. In 1912, the final boundary modification occurred when the District of Keewatin south of 60° N was assigned to Ontario and Manitoba. As a result, Ontario extended its political boundary to the northwest, stretching from northern Manitoba to the coast of Hudson Bay at the latitude of 56°51' N. Through these boundary adjustments, Ontario reached its present geographic extent of 1 million km².

The Birth of an Industrial Core

The geographic essence of Ontario lies in its industrial base. Manufacturing in southern

View of King Street [Toronto], Looking East, 1835 by Thomas Young. By the time of this painting, Toronto had a population of nearly 10,000. In the previous year, York had been renamed Toronto. (*National Archives of Canada 1669*)

Ontario (then Upper Canada) began in the nineteenth century and received a boost in 1854, when Britain and the United States signed the Reciprocity Treaty, which allowed trade between Upper Canada and the United States (primarily the states of Michigan, New York, and Pennsylvania). By 1867, Ontario had the largest population of Canada's four founding provinces and a fledgling industrial base. Small manufacturing outfits existed in small villages and often consisted of less than five employees (i.e., a village blacksmith or miller). Larger manufacturing activities took place in towns and cities, where sawmill, gristmill, and distillery operations were commonplace. At that time, most manufacturing activities were dependent on water power, so most were located near a stream or river.

The Reciprocity Treaty lapsed in 1866, cutting off Upper Canada's access to the American market. Confederation and the 1879 National Policy of Prime Minister Macdonald, however, enabled Ontario to secure the Canadian markets. Under the National Policy high tariffs were imposed on imported manufactured goods, which allowed manufacturing in southern Ontario to flourish. (For more about the Reciprocity Treaty and the National Policy, see Chapters 1 and 3.) The consequences of Confederation and protective tariffs for Ontario were threefold: (1) the creation of a Canadian market for Ontario products; (2) an increase in the size of the more successful manufacturing companies; (3) and the growth of the industrial workforce in Ontario. In the rest of the country, however, prices for manufactured goods (made in Ontario) were generally higher than in the adjacent areas of the United States. The regional price variations reflected two factors: Canadian transportation costs and Ontario's more limited economies of scale in comparison to those of US manufacturers.

Confederation therefore provided the impetus that established a national industrial core in southern Ontario as well as Québec, while the rest of the country was relegated to becoming a domestic market for these manufactured goods. Not until the Free Trade Agreement between Canada and the United States was signed in 1989 did this national core/periphery relationship undergo significant change. (See Chapter 1 for more on the Free Trade Agreement.)

Ontario Today

Ontario remains Canada's economic engine. A measure of Ontario's industrial success is its enormous energy demand. Ontario's powerful economy is fuelled by electricity generated within the province by hydroelectric installations at Niagara Falls and along the Ottawa River, by thermal coal plants using coal from Pennsylvania and West Virginia, and by nuclear generators in plants at Pickering and on the Bruce Peninsula. Still, Ontario's industry requires more energy. To meet that demand, electricity is imported from Québec and petroleum from Alberta.

Ontario is not a homogeneous region. The province contains three physiographic regions: the Great Lakes–St Lawrence Lowlands, the Canadian Shield, and the Hudson Bay Lowland. Each physical environment has a distinct and different economy and settlement patterns. Northern Ontario, for example, has the characteristics of a resource hinterland, while southern Ontario is the epitome of an industrial core. As discussed earlier in this chapter, geography goes a long way towards explaining this regional variation. With this regional dichotomy in mind, the following economic and population statistics for Ontario apply best to southern Ontario:

- Ontario produces 41% of Canada's GDP.
- Its annual output exceeds $300 billion.
- Its average personal income is well above the national average.
- With 38% of Canada's population, it has the largest population of the six geographic regions.

Ontario's economic success has been facilitated by conditions that promoted trade for the province with other provinces and other countries. The St Lawrence Seaway and the Welland Canal have facilitated trade with the United States and other foreign countries (Vignette 5.2).

Vignette 5.2   The Welland Canal

The Welland Canal connects Lake Ontario and Lake Erie, thus allowing ocean ships to enter into the heart of North America. To avoid the Niagara River and its famous falls, the first canal builders faced the daunting task of constructing a canal across the Niagara Peninsula, a distance of some 44 km. In doing so, a series of lift locks were needed to overcome a difference in elevation of nearly 100 metres between Lake Ontario and Lake Erie. The first canal, opened in 1829, was dug by hand. A series of locks made from hand-hewn timbers connected a series of creeks and lakes. As the size of ships increased, the original canal proved inadequate and a new canal was built in 1845. Within 40 years, larger ships required a third renovation, which was opened in 1887. The present canal was completed in 1932. In 1973, a new channel was constructed to bypass the city of Welland. The Welland Canal has been part of the St Lawrence Seaway since 1959 and is operated by the St Lawrence Seaway Management Corporation.

The Welland Canal is a strategic link between Lake Ontario and Lake Erie that provides a water route around Niagara Falls. With ever increasing traffic, the lock system was divided into two at several places, as shown in this aerial photograph. In 2003, approximately 32 million tonnes of goods passed through these locks. The three leading products were grain, iron ore, and coal. (Al Harvey/The Slide Farm)

Trade

Ontario is well positioned to engage in trade, both domestically and internationally. From northern Ontario, forest products and minerals are shipped to both national and international markets. In southern Ontario, manufactured goods ranging from aircraft to steel products are produced for both these markets. Restructuring of Ontario's manufacturing industry took place after 1989 when the Free Trade Agreement was signed. Unlike the Auto Pact, Canadian firms, especially those with high labour costs in their manufacturing process, were faced with a highly competitive market that sometimes resulted in plant closure and/or relocation to the United States or Mexico.

The free trade agreements and Ontario's geographic location within North America have allowed the region's business firms to penetrate the huge US market and that has led to greater integration into the North American economy. For example, Ontario's exports to the United States in the 1980s were about the same value as those to the rest of Canada; but by 1998, Ontario's exports to the US soared to two and a half times the value of exports to the rest of Canada (Courchene, 1998: 276–7). Driven by automobile shipments to the US market, this trade gap continues to widen in the first years of the twenty-first century as the sheer size of the US market has tilted the province's trade in a new direction—a north–south alignment.

The international trade dimension of the Ontario economy is evident in the 2001 export figures. In that year, the leading Canadian exports by value were automobiles and their parts, and the value of motor vehicle exports almost reached $93 billion (Statistics Canada, 2002). Since over 90 per cent of these auto-

motive exports came from assembly and parts plants in Ontario, the importance of this industrial sector to Canada and Ontario cannot be overestimated. Furthermore, a growing automobile industry has a ripple effect on the economy of southern Ontario by creating additional demand for steel, rubber, plastics, aluminum, and glass products. Of course, a slowdown in automobile production would have a negative impact on southern Ontario's economy.

Southern Ontario is, in reality, a northern extension of the North American manufacturing belt. With US automobile plants just across the border in Detroit, it is not surprising that this type of manufacturing spread across the Detroit River into Windsor. Initially this northward diffusion of the automobile industry took the form of branch plants owned by the Big Three (General Motors, Ford, and Chrysler). Until the Auto Pact, Ottawa placed high tariffs on imported automobiles and their parts. In doing so, the federal government encouraged American and other foreign firms to open branch plants that produced similar goods but, because of the size of the Canadian market, were smaller manufacturing plants than those in the United States and various European countries. The net effect was higher prices for Canadian consumers due to higher production costs, the high cost of shipping these goods long distances to regional markets in Canada, and the provincial/federal taxes on automobiles that were higher than those in the United States. Ontario, however, benefited from the expansion of its industrial base.

The Canada–US Auto Pact was designed to integrate Canada's automobile industry into the North American market. At that time, over 90 per cent of Canadian automobile production was controlled by the Canadian branch plants of the Big Three automobile companies.

These companies wanted to rationalize their production and marketing of automobiles within North America and thus be better prepared for competition from foreign producers. While the Auto Pact ended in 2001, this Canada–US agreement not only altered the nature and purpose of automobile manufacturing in Ontario, but it also marked a shift in federal policy towards more liberal trade arrangements and towards economic continentalism in North America (see Chapter 1 for a discussion of continentalism). One important feature of the Auto Pact is that Canadian production could not fall below the 1964 automobile output. Today, Ontario accounts for 16 per cent of automobile production in North America compared to roughly half that figure in 1964.

The New World Economic Order

During the 1980s, two events propelled Canada into a new world economic order. Ottawa had to abandon its closed economy based on high tariffs. Simply put, Canada's prosperity depended on assured access to the US market. The consequences for Ontario were severe, forcing a sometimes painful restructuring of its manufacturing sector. **Restructuring** meant that companies had to become more efficient, and in Canada's high-wage economy, that usually involved reducing the size of their labour force. Many American branch plants consolidated their operations in the United States where these companies could take advantage of a longer production run and therefore reduce production costs through economies of scale.

The first event was the liberalization of world trade. Under the terms of the General Agreement on Tariffs and Trade (GATT), most countries in the world, including Canada,

agreed to reduce barriers to international trade. In 1940, for example, the average tariff on manufactured goods had been about 40 per cent. Since 1947, however, a series of tariff reductions were achieved through agreements made by members of GATT. By the eighth agreement in 1990, the average tariff was 5 per cent (Dicken, 1992: 153). Today the main impediments to trade are not tariffs but non-tariff barriers such as import quotas ('voluntary' export restraints) and health regulations.

The second event to shape the new economic order was the **Free Trade Agreement** (FTA) between Canada and the United States. The FTA committed Canada and the United States to an integrated North American economy. In 1994, the FTA was replaced by a similar agreement, the **North American Free Trade Agreement** (NAFTA), which integrated Mexico in the trade agreement. In the same year a more institutionalized and formal international trade regime, the World Trade Organization (WTO), was created in the final agreement of the Uruguay Round of GATT, which concluded in December 1993. Since tariffs between Canada and the United States are decreasing at a faster rate under NAFTA than under the WTO, NAFTA gives Canadian manufacturing firms an opportunity to adjust to the new North American market before being fully exposed to world trade. By creating larger operations, Canadian firms hope to achieve economies of scale that will decrease their per unit costs of manufacturing and therefore allow them to become more competitive. NAFTA also exposed Canadian manufacturing firms to competition from Mexican-based firms whose labour costs are much lower and whose environmental regulations are less stringent (see Vignette 4.6).

One result of the FTA and NAFTA is that trade between Canada and the United States is increasing, thereby making Canada more dependent on the United States. From this perspective, Canada has hitched its wagon to one horse, which is especially true for southern Ontario (Norcliffe, 1996: 32). The consequences of this trade dependency are three-fold: (1) booms and slumps in the American economy affect Canada more directly than they did before FTA/NAFTA; (2) Canada does not have unlimited access to the US market because Washington can still limit Canadian products from entering its market; and (3) Canada's long-term economic fortunes are even more closely tied to the American economy than was the case before these trade agreements. Canada, but particularly Ontario, has benefited from closer trade ties with the United States. During the late 1990s, for instance, the booming American economy stimulated Ontario's exports of high value, ranging from automobiles to high-tech products. Yet, in 2000, when the American economy entered an economic downturn, Canada's exports to the United States remained at a high level and Ontario's economy continued to grow well into 2003. These trade agreements, while generally considered successful, have raised several controversial issues within the country and led to bitter disputes with the country's main trading partner, the US. For Ontario, plant closures have been a sore point while trade disputes have affected grain and lumber exporting regions of Canada.

Leading up to the FTA, access to the American market was considered crucial for Canada. Therefore, the Canadian government's objectives were the reduction of protective tariffs and the removal of non-tariff barriers—initiatives that would ensure better access to the American market. Non-tariff barriers included quotas, such as on durum wheat, or special countervailing penalties on certain Canadian export industries to protect American produc-

ers of the same goods, such as American grain farmers and American lumber companies. Canada insisted that the FTA have a trade-dispute settlement mechanism to prevent the use of barriers that contravened the trade agreement. As time has proven, even when this mechanism rules in favour of a Canadian firm, intense lobbying by American firms causes the American government to intervene. Since 1989, there have been several bitter disputes between Canadian grain farmers and lumber companies and their American counterparts. Each time, the United States government has pressured Ottawa into accepting a 'voluntary quota' on Canadian exports to the United States. For example, in the 1990s and again in the early 2000s, Washington imposed tariffs on durum wheat and lumber.

Within Canada, the FTA committed the country's manufacturing sector to become more competitive. Firms unable to adjust to the highly competitive marketplace in North America would be forced to close. For most firms, adjusting to greater competition meant lowering the costs of production by creating larger manufacturing plants and by increasing the productivity of workers. As part of this economic restructuring, American branch plants lost their raison d'être (MacLachlan, 1996: 195–214; Norcliffe, 1994: 2–17). A few adjusted their operations by specializing in one aspect of the manufacturing process to serve the North American market. In this way, a few branch plants were able to achieve economies of scale, which reduced their costs of production per unit of output. Most companies simply ceased production, thus allowing the parent plants in the United States to supply the Canadian market.

The outcome, while far from clear, is creating a new geography of manufacturing in southern Ontario. While the main impact was felt in the first years of the Free Trade Agreement, restructuring is an ongoing process. As MacLachlan (1996: 195) stated:

> The full effect of the agreement [NAFTA] on individual regions and labour markets is still conjectural. It appears that southern Ontario stands the greatest risk of employment losses because of its large secondary manufacturing sector and continued dependence on foreign-owned subsidiaries for employment in key manufacturing industries.

The threat of relocation of companies to countries with lower wages continues to have two effects. First, this threat keeps wage increases low and may trigger wage reductions. With plant closures continuing, workers in Ontario's manufacturing belt remain uneasy. In short, the increased competition brought on by NAFTA will continue to challenge and sometimes threaten the province's manufacturers—a sector of the economy that Ontario heavily relies upon. Consequently, a level of uncertainty exists in the labour market.

Ontario's Industrial Structure and Its Geographic Pattern

Geographers regard an economy as having a particular industrial structure based on its economic activities (Table 4.15). These activities traditionally were divided into three categories: primary, secondary, and tertiary. As national economies moved into the post-industrial era, the tertiary or service sector expanded rapidly. In Canada, this sector now comprises 75 per cent of industrial activities by value and employment (Table 4.16). For that reason, geographers have divided the tertiary sector into tertiary and quaternary activities.[1] Tertiary

activities can be seen as limited to service functions, such as the selling of goods and the providing of professional services, while quaternary activities refer to decision-making and innovation functions. Decision-making functions are those that determine the direction of government and industry. Each capital city has such public functions, while Toronto, Montréal, and Calgary house the majority of corporate headquarters.

The allocation of workers into each of these sectors constitutes a measure of the industrial structure of a country or region. In Ontario, the number of workers in the primary sector continues to decline while those in the secondary, tertiary, and quaternary sectors have increased. Unfortunately, data on the industrial sector collected by Statistics Canada do not distinguish workers in the quaternary sector from other workers. Thus, Table 5.1 shows only three industrial sectors—primary, secondary, and tertiary (which includes quaternary).

Employment in Ontario's primary sector is small (1.8 per cent of the total workforce) and shrinking. In 1995, the primary sector accounted for 3 per cent of the total workforce. Of the major resource activities, agriculture is under enormous pressure and is in a state of crisis as a result of stiff competition from foreign producers, the reduction of federal support for farmers, and low prices. As a result, the family farm, the original backbone of Canadian agriculture, is in dire straits. From 1999 to 2002, for instance, the number of agricultural workers dropped from 114,000 to 76,400 (Table 5.1). Similarly, the forest industry, which relies heavily on exports to the American market, is facing trade restrictions imposed by Washington. Lumber companies have had to restructure their operations to seek greater efficiencies, but restructuring has also resulted in downsizing. Companies

Industrial Sector	Workers (000s)	Workers (per cent)
Primary	**111**	**1.8**
Agriculture	76	1.3
Secondary	**1,525**	**25.1**
Manufacturing	1,120	18.5
Tertiary	**4,433**	**73.1**
Trade	921	15.7
Transport	285	4.7
Finance	398	6.6
Professional services	436	7.2
Management	255	4.2
Education	379	6.2
Health care	564	9.3
Culture	288	4.8
Accommodation	364	6.0
Other services	253	4.2
Public administration	291	4.8
Total	6,069	100.0

Table 5.1 **Employment by Industrial Sector in Ontario, 2002**

Source: Adapted from Statistics Canada (2003e).

engaged in fishing and mining have also had to reduce the size of their labour force to remain competitive in the North American and world economy. As a result, the primary sector has suffered the greatest workforce decline.

The secondary sector expanded modestly over the 1995–2002 period, from 23.6 per cent to 25.1 per cent of the total workforce. Ontario contains the bulk of Canada's secondary industry. The secondary sector consists of heavy industry (iron and steel production as well as the processing of raw materials), light industry (automobile, aircraft, and textile manufacturing), and high-technology industry

(fibre optics, software, and Internet applications). The interconnection between these industrial activities is strong. For instance, the automobile industry requires products from heavy industry (steel and nickel) and specialized devices from high-technology firms (digital and computer equipment). The importance of the automobile industry to Ontario is further explored below.

The tertiary sector has become the dominant form of economic activity in all modern industrial states. In 2002, the tertiary sector accounted for 73.1 per cent of Ontario's total workforce. Most of Canada's tertiary industry is located in southern Ontario. The service industry has expanded in many directions to meet the needs of its customers. The complexity of the tertiary sector is revealed in Table 5.1, with tertiary categories ranging from trade to health care. Employment data for 2002 illustrate both the dominance of the tertiary sector in Ontario's economy and the importance of the secondary sector.

Ontario's economic structure takes on a distinctive geographic pattern. Primary activities are found mainly in northern Ontario while secondary and tertiary activities are concentrated in southern Ontario. The resource industries—but especially forestry and mining—are the staple economic activities of northern Ontario. Processing of these resources does take place within this subregion. Examples include transforming timber into lumber and paper products and the smelting of nickel and other ores. In all cases, this form of manufacturing creates a more valuable product that is less expensive to ship to distant markets.

Southern Ontario, because of its fertile soils and its long growing season, is the centre of agriculture in Ontario. Secondary and tertiary activities are concentrated in southern Ontario where there is a large local market and quick access to the larger market in Michigan, Ohio, Pennsylvania, and New York (which comprise the heart of the American manufacturing belt). For example, trucks can reach these American markets in a matter of hours, allowing a just-in-time production system to flourish on both sides of the border. Backups at the major border crossings—especially between Windsor and Detroit—caused by heightened security concerns in the US have become a major problem for the quick and in-time delivery of goods, however.

No longer sheltered by tariffs, Ontario's

manufacturing firms faced stiff competition from countries with much lower labour costs. Adjusting to the global business environment has been difficult and most firms have had to reduce their labour costs. Restructuring has both hurt and benefited firms in this region. Company closures continue to plague towns and cities in southern Ontario. The 1999 closing of the glassware plant in Wallaceburg in southwestern Ontario is typical of the negative impact of NAFTA. Founded as Sydenham Glass some 100 years ago, it was once Wallaceburg's largest employer with more than 2,000 workers in the 1950s (Crone, 1999: C1). The parent company, Libbey Inc. of Toledo, Ohio, transferred production to Mexico and the United States, not because the Ontario plant was not profitable but because greater profits could be made in Mexico. The net result is that some 560 employees lost their jobs. In a similar move, Lennox International Inc. proposed to close its Toronto factory in 2002, moving its production to its US-based factories (Steinhart, 2001: FP5). In 1998, workers at Nike's skate plant at Cambridge, Ontario, also lost their jobs as a result of a plant closure, again, not because the plant was unprofitable but because even higher profits could be made by moving production to Asian countries such as Myanmar and Vietnam, where extremely low wages can be paid and there is little legislation to protect workers, including children. In December 2004, Camco plans to close its appliance plant in Hamilton, putting some 800 long-term employees out of work (CBC, 2003). The reason for the shut-down—lower-cost imports from similar plants in China and Mexico—are troubling to all manufacturing plants located in a high-cost labour market like southern Ontario (Canadian Press, 2003: FP7).

In all cases, the main market remains the United States, but the completion of products is now on either a North American or global pattern of assembly. The automobile industry is not immune to restructuring and plant closures. In 2003, the Navistar heavy-truck assembly plant in Chatham announced plans to close the plant and re-establish operations in Mexico because of lower labour costs (Dabrowski, 2003: FP3).

In spite of these closures, southern Ontario's economy has seen new companies appearing and old companies expanding their operations to take advantage of access to the huge American economy. By the late 1990s, the demand for workers began to outpace job losses and the unemployment rate began to decline. Before then, the annual unemployment rate for Ontario remained above the 1986–9 annual average (Table 4.19). By 2002, the rate had dropped well below the 1986–9 average because the increase of jobs greatly exceeded the job losses. Much of this expansion took place in the automobile industry and in high-technology operations that produce a wide variety of products ranging from computer software to engineering supplies. While the automobile assembly and parts plants depend on a single market (the North American market), high-technology firms can tap into world markets. Hamilton, for example, has seen its workforce in the steel industry drop from one-third of the city's employees to one-fifth over the last 20 years—yet the unemployment rate for 2002 remained low because of new jobs generated by 'new generation industries' such as the pharmaceutical business and environmental technology. The announcement at the end of January 2004 that steel giant Stelco had filed for bankruptcy protection, however, was a 'massive blow for Hamilton' that 'could result in big job losses and lower wages' (Van Alphen and Westhead, 2004: A1). From 1996 to 2002, Ontario's

unemployment rate dropped from 9.1 per cent to 7.1 per cent (Table 4.19). Nonetheless, despite the resurgence of southern Ontario's economy and its low unemployment rate, many 'casualties' of restructuring have not been reabsorbed into the economy. Stelco, for instance, a diversified major manufacturer, has fallen on hard times under pressure from non-unionized mini-mills in the US and overseas that focus on niche markets.

Southern Ontario

Southern Ontario, occupying the southern half of the Great Lakes–St Lawrence Lowlands, is the key agricultural and industrial area of Canada. Residents of southern Ontario enjoy a moderate continental climate while manufacturing firms take advantage of close proximity to the large market just across the border.

Agriculture is an important part of the economy of southern Ontario. Many farmers are engaged in dairy operations while others focus on fruit and vegetable production. As shown in the photograph, many dairy farmers plant hay, which is used for winter feed. (Barrett and MacKay Photography, Inc.)

Agriculture and Climate

Southern Ontario's climate is dominated by its long, warm summer that extends from May to September. Winter, on the other hand, takes hold for three to four months from mid-November to March. In the summer, the warm, moist air masses that originate in the Gulf of Mexico extend over this area, which results in long, hot, and humid summers. In the winter, the frigid Arctic air masses that sweep southward produce a short but sometimes cold winter. Such weather is associated with clear, sunny days. During the early spring and late fall, unsettled, cloudy weather dominates, with minimum temperatures often falling below the freezing point. Proximity to the Great Lakes affects the local weather by funnelling air masses into this region and by increasing local precipitation due to air masses absorbing moisture from the surface of the Great Lakes. During the winter, occasional invasions of Arctic air masses bring cold, stormy weather. High winds funnelled along the Great Lakes storm track often accompany these cold spells, driving the wind chill to a dangerous level. Fortunately, most storms are short-lived and followed by relatively mild weather. Annual precipitation ranges around 100 cm. Warm, moist air masses from the Gulf of Mexico provide most of the moisture for the region, which falls as convectional and frontal precipitation. The greatest amounts occur in the eastern areas of the Great Lakes where winter snowfall is particularly heavy.

Because of its mild climate and fertile soils, southern Ontario accounts for much of Canada's agricultural output by value. Southern Ontario has over half of the highest-quality agricultural land (known as Class 1) in Canada. With a combination of fertile soils and a warm summer, farmers in southern Ontario account for more than $7 billion of agricultural products each year. Located south of 44° N, southern Ontario has abundant moisture and a long, warm-to-hot summer permitting the production of highly specialized crops such as grain corn, soybeans, and sugar beets as well as a wide variety of fruits, grapes, and vegetables that cannot be grown in other parts of Canada (Vignette 5.3). Equally important, the large urban population provides a stable local market for its dairy, livestock, soft-fruit, and vegetable products. In 2001, the census of Canada reported that the area sown to grapes increased by nearly 16 per cent while the area for apples dropped by nearly 21 per cent (Statistics Canada, 2003a). Although the amount of cropland is relatively small, farming operations in southern Ontario are much more intense than in Western Canada, where extensive grain crops dominate farming operations. The difference between the two agricultural areas is revealed in the size of farms and the types of crops. In 2001, the average size of farms in Ontario was 92 ha compared to an average of 519 ha for farms in Saskatchewan (Statistics Canada, 2003b). Grain farming and cattle ranching dominate in Western Canada, while there is a much more diversified agricultural land use in southern Ontario, where corn, barley, and winter wheat are the important crops along with highly specialized ones such as tobacco and vegetables. Corn and grain often serve as fodder for the hog, dairy, and beef livestock farms that operate in Ontario.

Agricultural land use varies within southern Ontario due to subtle but important physical differences. For example, soils are less fertile and the growing season is shorter east of Toronto than southwest of Toronto. There are three highly specialized agriculture zones west of Toronto: the Essex–Kent vegetable area, the

Vignette 5.3   The Niagara Fruit Belt

The Niagara Fruit Belt lies on the narrow Ontario Plain that extends from Hamilton to Niagara-on-the-Lake. The Ontario Plain has fertile, sandy soils suitable for most types of agriculture. Lake Ontario marks its northern limit while the Niagara Escarpment forms its southern edge. This small agricultural zone contains the best grape and soft-fruit growing lands in Canada. While the Niagara Fruit Belt occupies a northerly location for grape vines and soft-fruit trees, local factors have more than offset the threat of frost at 43° N latitude. These factors are:

- Air drainage from the Niagara Escarpment to Lake Ontario reduces the danger of both spring and fall frosts.
- The waters of Lake Ontario are warm in autumn and proximity to the Niagara Fruit Belt helps to moderate advancing cold air masses.
- In the early spring, the cool waters of Lake Ontario keep temperatures low, thereby delaying the opening of the fruit blossoms until late spring when the risk of frost is much lower.

The Niagara Fruit Belt extends about 65 km between Hamilton and Niagara-on-the-Lake. The Niagara Fruit Belt is one of the major fruit and grape producing areas in Canada. Most vineyards are located on the slopes of, or below, the Niagara Escarpment. For many years, hardy vines that produced low-quality grapes resulted in a poor quality (but inexpensive) wine. With the Free Trade Agreement, Canada had to remove its tariffs, making it difficult to compete with foreign wines. Since that time, farmers in the Niagara Fruit Belt have been successful in growing the finest varieties of grapes and have been able to make some of the finest wines in the world. (Barrett and MacKay Photography, Inc.)

Norfolk Tobacco Belt, and the Niagara Fruit Belt. All lie in southwestern Ontario where the growing season is relatively long. These zones, located south of 43° N, are the southernmost lands in Canada.

Agricultural changes, however, are affecting southern Ontario farmers. For instance, prior to 1995, feed grain for Ontario's livestock industry was imported from Western Canada. However, with the elimination of the Crow

Benefit (a transportation subsidy), western feed grain is now too expensive for Ontario farmers, thus causing a major realignment in the livestock industry. Another change is forecast when Canada and Ontario discontinue marketing boards. Though their purposes vary, Canadian and provincial marketing boards generally control their respective markets by ensuring stable prices and incomes for producers and standard market access. Ontario farmers, especially dairy farmers, are accustomed to selling their products through marketing boards. These boards may be forced to cease their operations because, by controlling agricultural trade, marketing boards may be violating free trade rules under NAFTA and the WTO. Without such protection, Ontario dairy farmers may have difficulty competing with American farm products.

Manufacturing

Southern Ontario, with almost half of all manufacturing jobs in Canada, is also the leading financial and service subregion in Canada. In addition, Ontario is the leading region for the high-technology industry (Vignette 5.4). Most Canadian-based corporations have located their head offices in Toronto. Toronto's Bay Street is the centre of the Canadian financial world. Other major urban centres also house important sectors of southern Ontario's manufacturing industry. Hamilton is Canada's major steel city. Two steel companies, Dofasco and

Vignette 5.4   Location in a Hot-Wired World

Across Canada, high-technology companies are found in many places but particularly in Canada's major urban centres. In fact, Britton (1996b: 266) states that 'Canadian urbanization is the key to understanding the location of technology-intensive activities.' Toronto, Montréal, and Ottawa house the bulk of Canada's high-technology industry. 'The reality is there is no death of distance in the information age', said Marc Busch, a business professor at Queen's University, who has puzzled over the question of why some places have become high-tech hotbeds while others have not (Atkin, 2000: C1). The answer is 'clusters'. A cluster is a place where institutions, companies, and individuals have a commitment and enthusiasm for innovative technological research along with the capital necessary to develop and market the product. High-tech clusters are often anchored around a university or a public agency like the National Research Council. Other cluster factors are the presence of creative individuals, including engineering and computer professors and their students, company researchers, and computer specialists. Such clusters are found in many places across Canada, but most are located in the core industrial areas of Canada where economies of association, including the need for personal exchanges, are so important. Most high-technology companies are found in Toronto, Montréal, and Ottawa rather than in Saskatoon, Halifax, or Victoria. The latter do have high-tech operations but they are on the edge of the high-tech world and depend on a niche operation. For Saskatoon, it is agricultural biotechnology. Paradoxically, while the products of high-technology industry are reducing the friction of distance by interconnecting Canadians through e-mail and the Internet, the industry itself is still reliant on proximity and personal contacts, making its location highly concentrated in a few large centres.

Stelco, are based in Hamilton. Auto-assembly plants are in eight urban centres: Windsor, St Thomas, Ingersoll, Cambridge, Brampton, Alliston, Oakville, and Oshawa. All these communities have two features: excellent access to major highways, and proximity to major markets in Michigan, Ohio, Pennsylvania, New York, and Québec.

Historically, four factors led to the successful development of manufacturing in southern Ontario. Using the automobile industry as an example, it becomes apparent how each factor contributed to the growth of manufacturing in Ontario. The first factor is Ontario's geographic advantage—close proximity to America's manufacturing belt, which led American industries to locate branch plants in Canada early in the twentieth century. This spillover effect first took place in the Windsor–Detroit area in 1904, when the Ford Motor Company established an automobile assembly plant in Windsor. The second factor was trade restriction on foreign manufactured goods (imposed by the National Policy in 1879). Automobile parts from Ford's plant in Detroit had to be ferried across the Detroit River and assembled in an old carriage factory in Windsor. Access to the markets in the British Empire provided a third advantage. American branch plants in Canada could take advantage of lower tariffs for Canadian-made products in the British Empire. As a result, Canadian-built Ford cars were sold not only in Canada but also in various parts of the British Empire. General Motors followed this same strategy by opening a branch plant in Oshawa. The size of its domestic market provided southern Ontario with its fourth location advantage for the automobile industry. Most cars were sold in southern Ontario, thus minimizing transportation costs.

With the liberalization of trade, manufacturing in Canada changed. The first step was the Auto Pact. By the early 1960s, the Big Three automobile companies proposed a common market between Canada and the US for the production and marketing of their products. In 1965, Ottawa and Washington passed the appropriate legislation for the Auto Pact. The second step took place in 1989 when the Free Trade Agreement brought other components of the economy under a single Canada–US market, thus removing sometimes high tariffs between the two countries. Since American firms had already benefited from economies of scale, their costs of production were considerably lower than similar Canadian firms. As well, the American firms were well established in the marketplace and dislodging them was a daunting task. Canadian-based manufacturing firms were forced to play catch-up and many did not survive. American firms often closed their smaller and therefore less efficient branch plants and served their Canadian customers from their American factories. Some Canadian firms merged in order to attain a size deemed necessary to compete in the North American and global markets. The purpose was twofold: to compete in the world marketplace and to avoid a takeover by a large foreign company. Mergers, however, often resulted in the closure of offices to avoid duplication and, thus, in the reduction of the number of employees.

Like many towns in southern Ontario, Cambridge had to adjust to this new economic environment. Created in 1973 by the amalgamation of Galt, Preston, and Hespeler, Cambridge is an industrial centre just west of Toronto. Since 1989, the economy of Cambridge has experienced strong growth. While some firms have closed their operations, others have established themselves in Cambridge. With over 400 manufacturing firms, Cambridge has a solid industrial core. These firms range from textile companies to technology-based service industries. In 1985, Cambridge became an important automobile-assembly centre for Toyota Canada. Since then, Toyota has expanded its automobile production. Employment at its assembly plant jumped from 2,000 workers to 3,500 workers, making Toyota the city's largest employer. The presence of an automobile-assembly plant has had several benefits for Cambridge: a stable employment base, high-paying jobs, and spin-off effects due to purchases by the company and its employees.

For Cambridge, the Free Trade Agreement had its downside. Fierce competition resulted in plant closures. Savage Shoes, Inglis, Franklin Manufacturing, InterRoyal Corp., and Dobbie Industries Ltd have shut their doors, laying off hundreds of workers. Savage Shoes simply could not compete with lower-cost shoes from developing countries, where labour costs are a fraction of those in Canada. After the FTA, the American owners of the Inglis household appliance plant, the Inter-Royal furniture factory, and the Dobbie wool-spinning mill decided to close their branch plants and produce their entire output from their much larger plants in the United States. One American branch plant, Rockwell Automation, has managed to survive, but with a smaller labour force. Before the FTA, this firm produced electromechanical devices used in automated factories in Canada. To survive in the North American market, the plant had to find a single product for the global market. Now it is the only Rockwell plant in the world that produces medium-voltage products and electrical generators, 70 per cent of which are exported to the United States and other foreign countries. Niche production for the North American market allowed this plant to achieve economies of scale, and now its prod-

ucts can compete for business in all parts of the world (Campbell, 1996: D1–2).

Key Topic: The Automobile Industry

The automobile industry remains the key manufacturing activity in North America. Peter Dicken (1992: 268) put it best:

> The significance of the industry lay not only in its sheer scale but also in its immense spin-off effects through its linkage with numerous other industries. The motor vehicle industry came to be regarded as a vital ingredient in national economic development strategies.

The rapid expansion of the auto industry in southern Ontario dates back to the 1960s when the Auto Pact was signed. Canada now accounts for 16 per cent of North America's vehicle output and exports 85 per cent these vehicles to the United States. In 2002, Canadian producers contributed 2.5 million cars and trucks to the North American market. The Canadian automobile industry employs half a million Canadian workers and represents 2 per cent of Canada's GDP.

The Auto Pact

The Auto Pact is an example of a successful single-industry production-sharing agreement between Canada and the United States. Before the Auto Pact, the auto industry in Canada had small, high-cost plants, each a smaller version of the much larger plants in the United States. Even though Canadian wages for auto workers were lower than those for American workers, Canadian automobile prices were considerably higher than for the same models in the United

States. The Canada–US Auto Pact was signed in 1965 and ended in 2001 (Vignette 5.5). From Canada's perspective, the Auto Pact served three objectives:

- It secured guarantees that Canadian automobile plants would not close.
- It brought to Canadian plants the advantage of economies of scale by allowing them to specialize in a few types of automobiles that would supply the North American market.
- It reduced the price of cars for Canadian customers.

The agreement called for Canada to eliminate the 15 per cent tariff on imported automobiles and US parts for Canadian manufacturers, and for the US to eliminate its corresponding tariff. The agreement included special protection for Canadian plants. Canada was guaranteed a minimum level of automobile production based on production levels for each type of automobile manufactured in Canada in 1964, that is, the ratio of vehicle production to sales in Canada for each class of vehicle had to be at least 75 per cent or the percentage attained in the 1964 model year, whichever was the highest. Furthermore, **value added** in vehicles produced in Canada had to be no less than the dollar amount achieved in 1964.

The Growth of the Auto Industry

The automobile industry drives the Ontario economy. Nearly one in seven Canadian manufacturing jobs depends directly on this industry. Wages in the assembly plants are relatively high for semi-skilled workers, but somewhat lower for workers in parts firms. Nevertheless, these incomes enable workers to make substantial purchases, thereby stimulating southern Ontario's retail sector. In addition, the automobile industry accounts for

---

Vignette 5.5   The End of the Auto Pact

Under the terms of the Canada–US Automotive Products Agreement of 1965, qualified motor vehicle manufacturers (Ford, General Motors, and DaimlerChrysler) are able to import both vehicles and automotive parts duty-free into Canada. Other motor vehicle manufacturers must pay a 6.1 per cent import duty. For the Japanese automotive companies, this import duty cost approximately $120 million in 1999. In the mid-1990s, the Japanese government requested that Canada either grant the same trade privilege to Honda and Toyota or eliminate the import duty. This request fell on deaf ears. Japan and the European Union then applied to the World Trade Organization. In June 2000, the WTO declared that the Auto Pact discriminates against imported vehicles from Japan and the European Union because these automotive firms must pay the 6.1 per cent import duty. In the fall of 2000, the WTO imposed a timetable for Canada to dismantle the Auto Pact. By July 2001, the Auto Pact agreement that helped build Canada's automobile-assembly and parts industry ceased to exist. Given this change, the Japanese/European share of the North American automobile market is expected to increase. Just how this will affect Canadian automobile production remains unclear because Japanese automobile manufacturers have assembly plants in Canada.

nearly one-quarter of Canada's merchandise exports. (Most of Canada's remaining exports are primary products, such as grain and forest products.)

From 1995 to 2002, exports of automotive products (passenger autos and chassis, trucks and other motor vehicles, and motor vehicle parts) increased in value from $62.9 billion to $97 billion (Statistics Canada, 2003c). The reasons for this rise in exports were twofold. First, Canadian assembly plants had a higher productivity than comparable US plants. In 2001, Canadian workers had a 7 per cent advantage over American workers in the time required to produce an automobile (Brant, 2002: FP5). Second, the low Canadian dollar allowed Canadian exports of cars and trucks to cost less than similar vehicles produced in the United States. The importance of this industry to Ontario is clear, but so is its cyclical nature, indicating that sudden downturns are possible. By 2003, two problems surfaced. First, North America has too much automobile production

capacity, forcing companies to close less productive plants. The North American automobile industry had the capacity to assemble 23 million units a year, but sales in 2002 fell below 18 million units and sales predictions of 17 million units for 2003 pointed to layoffs and plant closures (Keenan, 2003: B4). Adding to the capacity problem, new, more efficient and flexible assembly plants are planned and, by 2005, these additional plants would add another 3 million units of capacity. Second, with the exchange rate on the Canadian dollar rising from 63 cents US in late 2002 to 73 cents in May 2003, the price of Canadian exports to the United States rose by nearly 16 per cent. If the Canadian dollar retains this position, automobile exports to the United States will be affected, as will investments in new assembly plants and auto part firms.

The automobile industry consists of two separate operations: the assembly of automobiles and trucks and the production of their parts. In addition to these two operations are

The automobile assembly plants in Ontario are highly automated and the use of robotic welding units, along with outsourcing and the just-in-time principle, have contributed to a highly rationalized industry. (Ivy Images)

manufacturing firms that supply semi-processed materials. In southern Ontario, fabricating firms produce steel, rubber, plastics, aluminum, and glass parts for automobile-assembly and parts plants in Canada and the United States. Then there are the service firms, ranging from the advertisers and designers to the sales and service staff. In short, the auto industry is a final-product type of manufacturing, and as such, its added value reaches a maximum.

By being highly efficient and strategically located, automobile-parts firms can operate on a **just-in-time principle**—auto components are produced in small batches and quickly delivered as needed to their customers. In turn, this allows the auto-assembly plants to achieve considerable savings by reducing their inventories, warehousing space, and labour costs. By 1999, Canada's automobile industry was in a very healthy state, reaching its highest annual production level at 3.0 million units, which accounted for 17 per cent of North American production. Three years later, North American sales had softened and Canadian output dropped to 2.5 million units, or 16 per cent of North American production.

A similar pattern of growth is emerging in the auto-parts industry. By the late 1980s, this sector was expanding mainly because automobile manufacturers decided to subcontract much of the parts business, a practice called **outsourcing**. Purchasing parts from specialized firms allowed automobile manufacturers to concentrate on assembling automobiles and reducing their costs. Cost savings result from the fiercely competitive bidding process among parts firms and the lower wages that parts firms pay their workers. Their labour force is non-unionized and, as a result, workers receive a much lower pay than unionized workers in the automobile-assembly plants.

This wage differential—and the savings it provides the auto manufacturers—is the main reason why General Motors, Ford, and Daimler-Chrysler continue to divert work to auto-parts firms. In Ontario, the Canadian-based Magna International Inc. has grown into the third largest auto-parts company in North America. Encouraged by Canadian workers' improved productivity and a weak Canadian dollar, automobile firms subcontracted more parts to firms in Canada. By 1996, the average North American vehicle contained about $1,500 of Canadian-made parts, up 27 per cent since the late 1980s (Jestin, 1996: 10). By 2002, this figure had increased slightly. Employment in auto parts production reached a peak in 1999, but with the slowing of demand in 2002 it slipped back to mid-1990s levels. Contraction of automobile production has had a negative impact on auto-parts suppliers and, given the fierce competition between producers, these suppliers are under heavy pressure to reduce prices.

Automobile-assembly plants are concentrated in southern Ontario where transportation links to the major markets of Canada and the United States are readily available and driving distances are short (Figure 5.3). All Canadian-based automobile companies sell most of their cars and trucks in the United States. In 2002, approximately 85 per cent of all vehicles produced in Ontario were sold in the United States. Three companies, General Motors, Ford, and DaimlerChrysler, accounted for 65 per cent of the North American vehicle production in 2002. Ten years earlier, the Big Three dominated the automobile market by controlling 90 per cent of the car and truck sales. Competition from Japanese and Korean automobile companies with assembly plants located in North America, including southern Ontario, is fierce. These Asian-based firms have captured 35 per cent of Canada's market and an equivalent part of the American market.

While Japanese automobile factories are located in the United States, several advantageous features in Canada have attracted substantial Japanese investment in Ontario. These include the availability of a highly motivated workforce, the lower value of the Canadian dollar compared to the American dollar, and the existence of a public medical system. In American-based automobile-assembly plants, the cost of providing medical insurance is often built into the wage agreement.

With the closing of the General Motors plant at Sainte-Thérèse near Montréal in 2002, all auto-assembly plants are located in southern Ontario close to both the Canadian and American markets (Table 5.2). In the long-run, this closure will affect auto-parts firms based in the Montréal area because these suppliers tend to locate their factories near automobile-assembly plants to reduce transportation costs. For example, in 1996 Magna International decided to build a truck frame plant in St Thomas, Ontario, from which truck frames could be delivered to any of three General

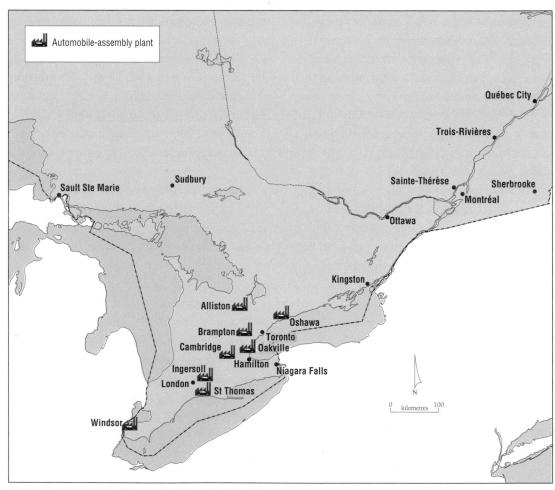

Figure 5.3 Automobile-assembly centres in Ontario. Canada's automobile-assembly plants are located in southern Ontario. The Sainte-Thérèse plant north of Montréal closed in 2002. (Further resources: Student Web site, National Atlas section, Maps 18, 19, and 20. Web site instructions are found on p. xxi.)

Table 5.2	Automobile-Assembly Plants in Southern Ontario, 2002		
Company	**Location**	**Employment**	**Product**
DaimlerChrysler Canada	Windsor Windsor	2,000 5,600	Dodge Ram vans Chrysler minivans
Ford Canada	St Thomas	2,700	Crown Victoria Grand Marquis
Cami/Suzuki Canada	Ingersoll	1,750	Geo Metro Geo Tracker Suzuki Swift Suzuki Sidekick Suzuki Vitara
Toyota Motor	Cambridge	3,500	Toyota Corolla
DaimlerChrysler Canada	Brampton	3,200	Chrysler Intrepid Chrysler Concorde Eagle Vision
Honda Canada	Alliston	4,200	Honda Civic Honda Acura
Ford Canada	Oakville	3,600	Ford Windstar
Ford Canada	Oakville	1,400	F-series half-ton pickup trucks
GM Canada	Oshawa	3,250	C/K half-ton pickup trucks
GM Canada	Oshawa	3,250	Chevrolet Lumina Monte Carlo
GM Canada	Oshawa	3,800	Chevrolet Lumina Buick Regal

Source: Updated version of Bruce Little and Greg Keenan, 'Ontario's Economic Future Is the Sum of Its Auto Parts', *Globe and Mail*, 2 Mar. 1996, A1, A10. Reprinted with permission from The Globe and Mail.

Motors truck-assembly plants within one day. One of these assembly plants is located at Oshawa, Ontario, while the other two are located in Pontiac, Michigan, and Fort Wayne, Indiana.

The Future

The liberalization of trade and the globalization of the automobile industry have created a highly competitive situation in the prized North American market. The first signs of a slowdown in the North American automobile market appeared in the spring of 2000, causing Canadian automobile exports to drop. Over the next three years, the ripple effects quickly spread through the assembly plants and auto-parts firms. The impact on communities with such plants will vary because production by the Big Three is decreasing while output by Honda and Toyota is increasing. In the long run, however, the key is to attract new investment. Windsor and Oakville may see new automobile-assembly plants in their communities, but this possibility has become less certain now that American demand for new automobiles has slowed and the Canadian dollar has increased relative to the US dollar.

By 2003, the automobile industry was going through a rough patch. Contraction of production around the world and in North America is likely to continue. Both assembly and parts production are likely to become more specialized, thus concentrating manufacturing in fewer and fewer places. Within North America, the Big Three are under pressure as they have lost market share over the last 10 years. While the Big Three are trying to maintain their existing market share, they hope to become more competitive by closing older assembly plants and building more efficient and flexible factories. In May 2003, Daimler-Chrysler abandoned its plans to build a state-of-the-art assembly plant at Windsor. At the same time, the company announced that it

would close its Dodge Ram van assembly plant on Pillette Road in Windsor on 12 June 2003, thus eliminating 1,000 jobs (Brent, 2003: FP5). Ford hopes to build a new plant at Oakville but its pickup truck assembly line will close. On the other hand, Honda and Toyota have plans to increase production.

Northern Ontario

Located in two physiographic regions, the Canadian Shield and the Hudson Bay Lowland, northern Ontario comprises 87 per cent of the geographic area of Ontario yet has less than 7 per cent of Ontario's population. Fewer than 750,000 people reside in northern Ontario, most in towns and cities, and almost all of this subregion lies in the Subarctic climatic zone, which for the most part is beyond the zone of commercial agriculture. For that reason, the vast majority live in urban places located along the two major transportation routes that link Montréal and Toronto with

Mining forms a major part of northern Ontario's economy. The Williams gold mine near Marathon is the largest gold-producing mine in Canada. In 2002, Canada accounted for 5.8 per cent of world gold production, which amounted to 6 million ounces. (Barrett and Mackey Photography, Inc.)

Winnipeg. Both of these trans-Canada routes cross the rugged Canadian Shield. The southern route includes the Canadian Pacific Railway and the Trans-Canada Highway. The Trans-Canada Highway connects North Bay with Sudbury, Sault Ste Marie, Nipigon, Thunder Bay, Dryden, and Kenora. The northern route includes the Canadian National Railway and a major highway. This northern highway connects North Bay, Timmins, Kirkland Lake, Cochrane, Kapuskasing, and Nipigon to Thunder Bay.

Like old resource hinterlands elsewhere in Canada, northern Ontario is troubled by a sluggish economy, a declining population base, and high unemployment rates. Consequently, its population has three demographic characteristics that are strikingly different from southern Ontario: (1) an aging population, (2) a net out-migration, especially of younger members of its population, and (3) few immigrants.

Mining, forestry, and tourism are the major economic activities, although many people are employed in the public sector. Like other resource hinterlands, the development of northern Ontario's economy was linked to external markets. Resource exploitation in northern Ontario began after the construction of railways in the late nineteenth century. Soon thereafter, resource development took place along the lines of the Canadian Pacific Railway (CPR) and the Canadian National Railways (CNR). The same pattern of resource development took place along the Temiskaming and Northern Ontario Railway (also known as the Ontario Northland Railway).[2] The railway extends northward from North Bay to New Liskeard, Kirkland Lake, and Cochrane, which is on the CNR main line. Northern Ontario exports minerals and forest products, including gold, nickel, newsprint, and lumber. These resource products contribute less than 10 per cent of the value of Ontario's exports to foreign countries.

Unlike in the northern hinterland of Québec, the natural conditions in northern Ontario are not conducive to major hydroelectric developments. While a number of rivers flow across northern Ontario to James Bay and Hudson Bay, the gentle slope of the land, especially in the Hudson Bay Lowland, does not make the construction of hydroelectric dams feasible. A few established sites for hydroelectric production, such as along the Ottawa River and at Abitibi Canyon about 100 km north of Cochrane, are important, but they cannot meet the energy demands in Ontario. The much higher elevations in the Canadian Shield area of Québec provide the necessary natural drop to drive the turbines that produce hydroelectric power. Since Québec has a surplus of electrical power, some of it is transmitted to southern Ontario.

Northern Ontario has a strikingly different settlement pattern from that of southern Ontario (Vignette 5.6). Long distances separate the four major cities. For example, the distance from North Bay to Sudbury is over 100 km; from Sudbury to Sault Ste Marie is about 300 km; and from Sault Ste Marie to Thunder Bay is about 500 km. The explanation for this oasis-like settlement pattern is due to northern Ontario's physical geography. The rocky terrain of the Canadian Shield discourages continuous settlement; a series of isolated settlements are located at mining sites, pulp plants, and key transportation junctions. Similar to other old hinterlands, the urban centres of northern Ontario are declining and the populations of single-industry towns like Cobalt, Kirkland Lake, and Porcupine rose and fell as the cycle of resource exploitation ran its course.

| Vignette 5.6 | Population Distribution in Northwestern Ontario |

The large, sprawling Thunder Bay census division of northwestern Ontario stretches from Lake Superior to Hudson Bay and James Bay. This division has less than 160,000 people. As is typical of northern Ontario, most people live in a few large urban centres while the remaining territory is virtually empty. Northwestern Ontario's population geography has three distinct zones:

- Zone 1: Most people (71 per cent) reside in the City of Thunder Bay. In 2001, this major population cluster had 125,100 people.
- Zone 2: Most of the remainder (27 per cent) live south of the CN rail line where a number of small towns, reserves, and villages are located.
- Zone 3: North of the CN rail line lies the 'empty' population zone, which is beyond the reach of the national highway system and has only 2 per cent of the region's population.

Beyond these towns and cities lies a sparsely populated area where fewer than 10,000 people live. Hamelin described this area as the Middle North. Cree and Ojibwa Indians live on isolated reserves or in small Native settlements in this vast region of scrub forest and muskeg. Only a few gold mines, trapping cabins, and fishing lodges dot the landscape. For the Indian residents, fishing and hunting are an important source of food, while trapping and guiding provide cash income. Since this cash income usually does not meet their basic needs, transfer payments supplement their income. The recent discovery of the Victor diamond deposit near the Attawaspiskat First Nation reserve might provide an economic boost for the band.

Mining Industry

Northern Ontario's mining industry is centred in the Canadian Shield, which provided ideal geological conditions for the formation of hard-rock minerals such as gold, nickel, and uranium. The annual value of mineral production is around $5 billion. Most production comes from Red Lake (gold), Hemlo (gold), Wawa (gold), Manitouwage (gold), Marathon (gold), Thunder Bay (gold, copper, zinc), and Sudbury (nickel, copper).

The first important mineral discovery took place in 1883 when the copper-nickel ores of the Sudbury area were discovered during the building of the Canadian Pacific Railway. In 1903, a rich silver deposit near Cobalt was detected during the construction of the Temiskaming and Northern Ontario Railway. Rich gold deposits were uncovered near Timmins in 1909 and at Kirkland Lake in 1911. While new mineral finds kept the mining economy of Ontario going, the overall pattern was one of boom and bust. During the 1960s, a large lead-zinc deposit was discovered in Timmins. A decade later, rich gold deposits were found at Marathon (the Hemlo gold mine) and at Pickle Lake. Promising deposits, including diamonds at three locations (Victor, Marathon, and Wawa) are currently being evaluated for commercial production. Overall, however, the labour force is declining. The main reason is increased workforce productivity as a result of greater mechanization of mining processes.

Minerals are a non-renewable resource that is depleted over time. Therefore, mining communities can have a short lifespan. A mine closure can occur without warning, causing the sudden demise of a single-industry centre. Over the years, a number of ore bodies have been exhausted or production has ceased because new, lower-cost mines have been discovered. These two circumstances have forced some mines to close, resulting in a great outflow of miners and their families. For instance, Elliot Lake's uranium mine was closed, not because its ore was exhausted but because open-pit uranium mines in northern Saskatchewan could produce uranium oxide at a much lower price. Ontario Hydro, the principal buyer of Elliot Lake uranium, decided to purchase its supplies from the uranium mines in northern Saskatchewan. Other mines that have closed include an iron mine at Steep Rock and the silver mines at Cobalt. While former mining towns may remain, they undergo a drastic downsizing. With a stock of low-priced houses and an attractive wilderness setting, Elliot Lake made a modest recovery as a retirement centre. However, its population has declined steadily and is now only about 12,000 (see Table 5.6).

Forest Industry

The forest industry of northern Ontario produces approximately $15 billion of products each year, with 60 per cent exported to markets in the United States. From time to time, the free flow of lumber to the United States is affected by US duties. In 1996, the two countries signed the Softwood Lumber Agreement, which limited lumber exports to the United States to 14.7 billion board feet per year. This five-year agreement expired and the United States imposed a 27 per cent duty in May 2002. Washington claimed that the provinces, including Ontario, were subsidizing their lumber producers by offering them timber leases at very low prices (prices are based on stumpage fees). By January 2004 the two sides had not resolved this issue. As a result, Ontario softwood lumber producers, many of which are American-owned firms, slowed or ceased production.

Across Ontario, the northern coniferous forest is classified into two regions: the boreal barrens and the boreal forest regions. The boreal barrens region is found in the Hudson Bay Lowland. This forest region, which represents a transition between the boreal forest and the tundra, has no commercial value. Scattered patches of spruce and tamarack are surrounded by poorly drained land known as muskeg. The commercial forest industry is located in the boreal forest region that extends from Québec to Manitoba. The geographic limits of the boreal forest region closely parallel those of the Canadian Shield. The main species are black and white spruce, tamarack (larch), balsam fir, Jack pine, white birch, and poplar. Some 50 communities depend on the forest industry based on the boreal forest region. By volume of wood cut, Ontario ranks just behind British Columbia and Québec. Ontario's mills produce pulp and paper, lumber, fence posts, and plywood. The pulp and paper industry in northern Ontario accounts for about 25 per cent of the national production.

Ontario is a leading exporter of newsprint and pulpwood to the United States. In fact, most pulp and paper firms operating in northern Ontario are American-owned companies. Road access is an extremely important factor in the forest industry. Trucks must bring the pulpwood to the mill and then the final product must be shipped to market. Pulp and paper mills are located in communities that have access to transportation networks—Thunder Bay, Sault Ste Marie, Sturgeon Falls, Kenora, Fort Frances, Dryden, Marathon, Iroquois Falls, Kapuskasing, and Terrace Bay.

Figure 5.4 US lumber lobby wins again. The Forest industry across Canada has suffered from US duties on softwood lumber. Ontario's lumber producers, located in northern Ontario, have had to lay off workers. In April 2004, the NAFTA dispute panel ruled in Canada's favour—again! Yet, the panel can only recommend that the US comply. The US governemt is reluctant to take action because of the US lumber lobby. (© Adrian Raeside)

Technology in the forest industry has advanced over the years and greatly increased the productivity of workers in the mills and in the woods. When loggers first arrived in northern Ontario, they had only axes and saws. A logger with such equipment might cut five to 10 trees a day. By the middle of the twentieth century, the technology had advanced. Chainsaws then enabled a logger to cut 50 to 70 trees a day. Mechanical harvesters, which can harvest hundreds of trees per day, have now replaced the chainsaw.

Transportation of timber to mills has also changed. In the past, logging often took place in the winter months. Horses pulled the logs to frozen rivers where they were stored until the river melted in the spring. Tractors later took over the task of hauling the logs to the river. In the spring, the logs were floated to sawmills, which were often located at the mouths of the rivers. Today, most timber is used in the pulp and paper industry. These huge plants require a steady supply of pulpwood. This demand and the technological changes that have taken place have altered the nature of logging in four ways: (1) logging has become a year-round affair; (2) most logs are hauled to the mills by trucks; (3) trees are harvested before they reach maturity; and (4) mechanical tree harvesters are employed and usually clear-cut a wooded area.

Today, the forestry industry faces several challenges. The first challenge is to maintain a balance between logging and the regeneration of the forest. Since over 90 per cent of Ontario's forest lands are owned by the provincial government, private companies must obtain forest leases. The Ontario government, like most other provincial governments, has insisted that these private companies take on the responsibility for restoring trees to logged areas through forest management agreements. Efforts to hand-plant seedlings are helping to speed up the process of reforestation, but how successful these efforts have been can only be known 20 years from now. The practice of granting forest companies long-term timber leases is based on the assumption that it is in the companies' self-interest to manage the forest well. It remains to be seen whether this assumption and the corresponding leasing policy will ensure the protection and regeneration of Ontario's forests.

A second challenge is the changing nature of the boreal forest from a predominantly coniferous forest to a broadleaf one. Since the 1950s, the volume of timber harvesting has more than tripled. This level of logging has put so much pressure on the boreal forest that original species are disappearing. As a result, the boreal forest is shifting from coniferous to broadleaf species. Spruce, pine, and fir have been replaced by poplar and birch, which are both pioneer species, well adapted to regenerate in clear-cut portions of the boreal forest. The long-term consequences of this species shift are unknown, but such a shift clearly illustrates how forest companies have replaced forest fires as the main agent of change.

A third challenge facing the forest industry

The boreal forest stretches across most of northern Ontario, providing the basis for a forest industry. The boreal forest, in combination with its lakes and rugged landscape, produces spectacular scenery that attracts many tourists. (© Royalty Free/Corbis/Magma)

is the age of pulp and paper plants. Many mills in Ontario were built before World War II and continue to use old technology, which results in much higher discharges of toxic wastes into the environment. The consequences of such practices can be life-threatening. The worst cases of massive toxic discharges into streams and rivers occurred in the 1960s. At that time, a chemical plant that prepared the bleaching solution for the pulp plant at Dryden was releasing effluents containing mercury into the English-Wabigoon river system.[3] Since then, forestry mills have updated their operations to ensure a cleaner and safer natural environment.

Ontario's Urban Geography

Ontario is the most highly urbanized province in Canada. Almost 9.5 million people—nearly 85 per cent of the province's total population—live in towns and cities. As well, many of Canada's largest cities are located in

Ontario. For example, 10 of the 25 largest cities in Canada are in Ontario. With the exceptions of Sudbury and Thunder Bay, these census metropolitan areas (CMAs) are located in southern Ontario.

The pattern of urban growth from 1981 to 2002 varied considerably for these CMAs (Table 5.3). The fastest-growing cities in Ontario were Oshawa, Toronto, Ottawa, and Kitchener. Over this time period, these four centres had growth rates well above the provincial average of 47.7 per cent for CMAs. London and Hamilton fell just below the provincial average. The remaining four centres (Sudbury, Thunder Bay, Windsor, and St Catharines/Niagara[4]) had much lower growth rates. Sudbury and Thunder Bay are both located in the resource hinterland of the Cana-dian Shield. In fact, from 1996 to 2001, both cities suffered a population decline due to the stagnant economy found in northern Ontario.

The four largest cities in Ontario (Toronto, Ottawa, Hamilton, and London) are located in southern Ontario (Figure 5.5). These four cities form the core of the three urban clusters in southern Ontario, namely, the Golden Horseshoe, southwestern Ontario, and the Ottawa Valley. Each of southern Ontario's three major urban clusters has a population of at least 1 million. As mentioned earlier, northern Ontario has fewer and smaller cities, each sep-arated by long distances. It has two major cities (Sudbury and Thunder Bay) and no sin-gle urban cluster. Sudbury, the largest centre, has only 155,600 residents, while Thunder Bay's population is 122,000.

Table 5.3	Census Metropolitan Areas in Ontario, 1996–2001		
Census Metropolitan Area	Population 1996 (000s)	Population 2001 (000s)	Change (per cent)
Thunder Bay	126.6	122.0	(3.0)
Sudbury	165.6	155.6	(6.0)
Oshawa	268.8	296.3	10.2
Windsor	286.8	307.9	7.3
Kitchener	382.9	414.3	8.2
St Catharines/Niagara	372.4	377.9	1.2
London	416.6	432.5	3.8
Hamilton	624.4	662.4	6.1
Ottawa*	751.7	806.1	7.2
Toronto	4,263.8	4,682.9	9.8
Total	7,659.6	8,257.0	7.8

*The population of the Ottawa–Hull CMA was 743,821 in 1981 and 1,063,664 in 2001. Since an amalgamation in 2001 of Hull, Gatineau, and Aylmer, Québec, the new municipality has been renamed Gatineau. Ottawa had a population of 751,646 in 1996 and 806,096 in 2001.
Source: Statistics Canada (2003d).

The Golden Horseshoe

The Golden Horseshoe obtained its name because of its horseshoe-like shape around the western end of Lake Ontario and its outstanding economic performance over the years. As a result, this tiny area of the Great Lakes Lowlands forms the most densely populated area of Canada. The Golden Horseshoe extends from the US border at Niagara Falls northward to Hamilton, Toronto, and then on to Oshawa. Nearly 7 million Canadians live, work, and play in Canada's largest population cluster, and many visitors come either as tourists or on business trips. Accounting for nearly one-quarter of Canada's population, the Golden Horseshoe contains more than a dozen towns and cities, including Toronto, Hamilton, Oshawa, St Catharines, Niagara Falls, Burlington, Oakville, Pickering, Ajax, and Whitby. Toronto, the largest city in Canada, is its urban anchor, while Hamilton, with its steel plants, is the focus of heavy industry (Vignette 5.7), and Oshawa is Canada's leading automobile-manufacturing city.

Toronto

As the largest city in Canada, Toronto dominates the urban landscape of southern Ontario and plays a key role in Canada's urban hierar-

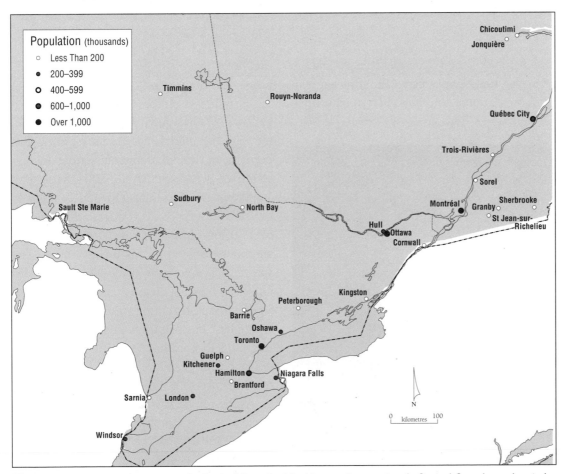

Figure 5.5 Major urban centres in Central Canada. Most large urban centres in Central Canada are located in the Great Lakes–St Lawrence Lowlands, especially in southern Ontario.

Vignette 5.7   Hamilton: Steel City

Hamilton, the third largest city in Ontario, is situated at the west end of Lake Ontario on Burlington Bay. Hamilton is Canada's largest steel producer and ranks high in industrial production.

Originally, however, Hamilton was a textile town. Streams flowing down the Niagara Escarpment provided power for the mills, and the city's location at the head of navigation on Lake Ontario gave it access to other towns and cities.

Steel production flourished with the railway boom on the Canadian Prairies in the early 1900s. Hamilton's steel industry lan-

guished during the Great Depression but during both World War I and World War II, military demand for steel created boom conditions.

After World War II, Hamilton's steel industry shifted its production to the rapidly expanding consumer market, especially for appliances and automobiles. Two of Canada's largest steel firms (Stelco and Dofasco) are located in Hamilton. Stelco, unfortunately, has fallen on hard times. In May 2004, Stelco filed for bankruptcy protection while the company attempts to restructure its operations.

chy. Toronto is the financial capital of Canada. The city houses the main offices of national and international banks and investment firms, as well as the Toronto Stock Exchange. Over

Toronto's financial district includes the head offices of Canada's corporations and major banks, and is the home of the Toronto Stock Exchange. The concentration of corporate and financial power in Toronto reflects the core element of the core/periphery model. (©Royalty Free/Corbis/Magma)

the years, Toronto's population growth outpaced that of most other cities, and by 2002 the population of the Greater Toronto Area had reached 5 million. Over the last decade, the main factor driving Toronto's growth has been the arrival of immigrants from foreign countries. Between 1996 and 2001, nearly half a million new Canadians settled in Toronto. At the same time, people and businesses spilled over Toronto's boundaries, causing a geographic expansion of this urban growth. The spread of Toronto's population and businesses into neighbouring areas is partly due to lower land values and rents. Land values and rents are highest in downtown Toronto. Cities surrounding Toronto have much lower land values and can therefore attract businesses by offering lower office rents.

As the province's major cultural and entertainment centre, Toronto is a hub for the entertainment industry, which ranges from live theatre to professional sports, and plays an essential role in shaping Canadian culture. Tourists flock to Toronto to enjoy these world-class cultural and entertainment activities. Just how dependent is the travel and tourist industry on tourists was revealed in the outbreak of SARS in April 2003 (Vignette 5.8).

Like other major cities in North America, Toronto is trying to cope with a rapidly expanding population, much of which has moved into adjoining urban areas. One effort to deal with the administration of this urban area was to create a super-city known as the Greater Toronto Area (GTA). In 1998, the City of Toronto and the surrounding regional municipalities of Halton, Peel, York, and Durham officially formed a single city government with the hope that solutions to problems arising from rapid population growth and geographic expansion of the urban population could be found.[5] While many opposed this approach to mega-urban government, the objective is to reduce administrative overlap and thereby create a more efficient urban government. Whether or not this single administration approach is succeeding in serving such a large population remains to be seen. However, many problems remain, including traffic congestion. With so many people living out-side of downtown Toronto but working in the city centre, massive numbers of them commute daily, creating traffic jams during the morning and afternoon rush hours. Traffic congestion leads to time lost travelling to and from work. A radical solution would be charging a toll on automobile traffic into and out of Toronto. London, England, imposed such a tax and it resulted in a reduction in traffic flow and a decrease in travel time.[6]

As the most popular destination for immigrants to Canada, Toronto has experienced remarkable ethnic diversification over the past 30 years, to the point where visible minorities are hardly in the minority. Visible minorities comprised 37 per cent of Toronto's population in 2001 compared to 26 per cent in 1991. By 2001, Toronto had over 2 million immigrants, who accounted for 44 per cent of the city's population. In recent years, most have come from China and India. The origin of immigrants by place of birth reveals that, in 2001, nearly 18 per cent of the 792,030 immigrants came from China and Hong Kong, with another 10 per cent from India and 7 per cent from the Philippines (Table 5.4).

Vignette 5.8   Economic Impact of SARS on Toronto

In the spring of 2003, Toronto gained notoriety as one of the world's major sites of the Severe Acute Respiratory Syndrome (SARS) epidemic. On 28 February, a Canadian who had been visiting Hong Kong unknowingly contracted SARS and brought the virus back to Toronto. This individual was placed in a Toronto hospital, but the deadly virus was not identified for two weeks, allowing SARS to spread to others, especially those caring for this patient, who eventually died. While the epidemic was largely confined to hospitals, the world took note that 23 deaths had occurred over a six-week period. On 23 April, the World Health Organization (WHO) issued a travel advisory warning people to stay away from Toronto. By 30 April, WHO had removed the travel warning. But with tourists avoiding Toronto, its travel and tourist industry lost an estimated $1 billion in the month of April. Fortunately, the SARS scare has passed quickly, and tourists have returned to Toronto.

Source: Brean (2003: A6).

Table 5.4	Top 10 Countries by Birth of Immigrants to Toronto, 1991–2001	
Country of Birth	**Immigrated 1991–2001**	**%**
China, People's Republic of	85,345	10.8
India	81,845	10.3
Philippines	54,885	6.9
Hong Kong, Special Administrative Region	54,805	6.9
Sri Lanka	50,425	6.4
Pakistan	39,265	5.0
Jamaica	25,355	3.2
Iran	23,840	3.0
Poland	21,555	2.7
Guyana	20,800	2.6
Total of all immigrants to Toronto, 1991–2001	**792,030**	**100.0**

Source: Statistics Canada (2003f).

Toronto is known as a city of neighbourhoods, partly because immigrant groups have clustered in certain neighbourhoods. Little Italy and Little Portugal are older, well-established ethnic neighbourhoods. More recent Asian, African, and Caribbean neighbourhoods are emerging. Recent immigration into Toronto has had an impact on its cityscape, including 'foreign' architecture and commercial activities designed to meet the demands of these new Canadians. Asian theme malls represent a dramatic change to Toronto's space. Asian theme malls consist of a large number of small retail outlets, many of which are either Chinese restaurants or specialty stores. Unlike other suburban malls, no anchor store exists and, rather than a single developer who leases space, each retail unit is owned by the operator of that retail space. The failed development of an Asian theme mall in Richmond Hill marks the resistance by local residents to a changing urban space with a new ethnic iden-

tity (Preston and Lo, 2000: 182–90). Toronto Island also represents a distinct neighbourhood born out of confrontation with city planners who had decided to transform Toronto Island into public open space. While originally a sandbar attached to the mainland, this district of Toronto now consists of a series of islands, most of which are dedicated to parkland. However, residential communities, both of which resemble bohemian enclaves, exist on two islands (Ward's Island and Algonquin Island).

Ottawa Valley

With a population of just over 1 million, Ottawa–Gatineau is not only a major population cluster in Canada, but it is also an important area where both official languages are used. As the nation's capital, federal government operations are found on both sides of the Ottawa River. In 2001, Ottawa had a popula-

Since the 1960s, immigrants have been coming to Canada from non-European countries in large numbers. In 2001, 16 per cent of immigrants came from China, 11 per cent from India, and 6 per cent from Pakistan. These photos depict two thriving communities in Toronto: Chinatown and Little India, the Southeast Asian community in Toronto's east end. (©Megan Mueller)

tion of just over 800,000 while Hull (now Gatineau) had nearly 260,000 citizens. Together, Ottawa–Gatineau represents the fourth largest metropolitan area in Canada.

Much of its growth has come from in-migration from other parts of Canada and from foreign countries. Many are attracted by the employment opportunities offered by the fed-

Table 5.5	Top 10 Ethnic Origins in Ottawa–Hull*, 2001		
Ethnic Origin	**Total—Single and Multiple Responses****	**Single Responses**	**Multiple Responses****
Canadian	463,280	231,095	232,180
French	272,085	58,455	213,625
English	200,900	39,645	161,255
Irish	183,130	25,010	158,120
Scottish	152,215	18,705	133,510
German	63,295	10,370	52,925
Italian	37,440	16,950	20,485
Chinese	31,595	26,895	4,705
North American Indian	25,780	3,830	21,950
Polish	23,695	7,560	16,135
Total population	**1,050,755**	**596,345**	**454,415**

*Now known as Ottawa–Gatineau.
**Respondents who reported multiple ethnic origins are counted more than once as they are included in the multiple responses for each origin they reported. For example, a respondent who reported 'English and Scottish' would be included in the multiple responses for English and for Scottish.

Source: Statistics Canada (2003f).

eral government and the business community. In 2001, 18 per cent of its population was foreign-born compared to 15 per cent in 1991. Visible minorities accounted for 14 per cent of population, with the three leading groups being blacks, Chinese, and Arabs/West Asians (Table 5.5).

On the Ontario side, Greater Ottawa is the third largest urban cluster in Ontario. The federal government, located in the national capital, is the major employer in the region. The federal government requires a wide variety of goods and services in its daily operations. This demand provides an opportunity for many small and medium-size firms in the Ottawa Valley. By locating its departments and agencies in both Ottawa and Gatineau, the federal government has ensured that Ottawa's eco-

nomic orbit extends to a number of small towns on both the Ontario and Québec sides of the Ottawa River. Most people live on the Ontario side, especially in the city of Ottawa. Other urban centres on the Ontario side include Nepean, Gloucester, Kanata, and the municipalities of Rockcliffe and Vanier. Around Ottawa, in its periphery, are a number of smaller centres, including Pembroke, Cornwall, Brockville, and Kingston. However, with the exception of Pembroke, these cities are more closely tied to economic and social activities in Toronto and Montréal.

As the capital of Canada, Ottawa is the focus of national and international affairs and has subsequently developed into a national administrative centre with a large federal civil service. In its early days, the forest industry played a strong economic role, but it was eventually overshadowed by the activities of the federal government and then the high-tech industry. Still, the geographic location made this an ideal place for a pulp and paper mill: pulpwood was easily harvested from the interior forest lands and then transported along the Ottawa and Gatineau rivers to the Eddy Pulp and Paper Mill in Hull, Québec. An added advantage of this site is the availability of low-cost electrical power from the hydroelectric installation on the Gatineau River near the Chaudière Falls. Today, the E.B. Eddy plant produces fine and coated paper.

Ottawa has become an industrial leader in high technology. Described in the press as the 'Silicon Valley of the North', the key to Ottawa's success has been the emergence of high-technology firms specializing in telecommunications, computers, software and computer services, and electronics (Britton, 1996b: 266). Northern Telecommunications (Nortel), the leading high-tech company, specializes in fibre optics. In the 1990s, the demand for high-tech products seemed limitless and high-tech companies expanded rapidly. For example, in 1997, Nortel announced plans for major expansion in the Ottawa area. Over a four-year period, Nortel planned to invest $250 million and hire nearly 5,000 new employees (McCarthy, 1997: B4). Such a commitment demonstrates the attractiveness of Ottawa as a location for the high-tech industry, but a sharp downturn in demand for its products beginning in 2000 not only put Nortel's expansion plans on hold but put the company close to bankruptcy. By early 2003, Nortel had recovered, but with a much smaller operation and a greatly reduced labour force. Other leading figures in Ottawa's high-tech world were less fortunate and some have been purchased by larger companies outside of Canada. Losing

corporate headquarters is not good news. As James Bagnall (2003: FP5) reported:

> Bit by bit, Ottawa is losing control of its tech industry. With software maker Corel now in the hands of a California-based venture capital firm and fibre-optics star JDS Uniphase set to shift its headquarters next month to San Jose, Ottawa is swiftly reverting to its former role as a technology R&D centre, serving the whims of multinationals based elsewhere.

Southwestern Ontario

Southwestern Ontario represents the third major urban cluster in southern Ontario. About 1 million people live in this area. Though this region does not have a major metropolitan city, it is close to Toronto, Buffalo, and Detroit. The main urban centres within southwestern Ontario are Cambridge, Kitchener, London, and Waterloo. Windsor and Sarnia are on the edge of this urban cluster.

London is the unofficial capital of southwestern Ontario. It provides administrative, commercial, and cultural services for the larger region. London is also the headquarters of several insurance companies, including London Life Insurance Company. Among its economic activities, London is noted for its manufacturing, including the production of armoured personnel carriers and diesel locomotives by General Motors. London, therefore, has a sound and growing industrial foundation based on insurance, manufacturing, and high-tech industries. In 2003, Magna announced plans to build a new auto-parts plant that will supply car seats for the Equinox, a new sports utility vehicle produced at Suzuki/GM's Cami assembly plant in Ingersoll. These manufacturing firms pay relatively high wages to their employees. Consumer spending by these employees supports a strong retail sector.

Automobile-assembly plants are located in Cambridge, Ingersoll, St Thomas, Alliston, and Windsor (Table 5.2). A new Daimler-Chrysler assembly plant is planned for Windsor. Auto-parts plants play an important role in the economy of southwestern Ontario while a number of high-tech firms, particularly in the Cambridge, Kitchener, and Waterloo areas, add to the region's economic diversity. These three cities form what is known as Canada's 'Technology Triangle' where innovative technology is often developed in research institutes or universities. Such technology is then used to create commercial products. Leading technology companies include Research In Motion Ltd, Open Text Corp., Dalsa Inc., Automated Tooling Systems, and Com Dev Ltd. The following description by Bathelt and Hecht

(1990: 225) captured the essence of this technological zone in the 1990s and remains accurate today.

All [high-tech firms] are surprisingly strongly tied, in terms of input and output linkages, to economic activities within a 100-km radius, which includes the Toronto area. Skilled labour and local residence/education are the strongest location factors. The relatively favourable industrial climate in the Cambridge Technological Triangle region should enable key technology industries to flourish in the area.

Cities of Northern Ontario

In sharp contrast to cities in southern Ontario, those urban centres in northern Ontario find their economies stalled and their populations declining. Like most downward transitional regions, the resource base is losing its economic strength for two reasons: (1) as the best resources have been exploited, resource production is now more costly, and (2) as companies introduce more and more technology, fewer workers are required.

Timmins, located in the mineral-rich Canadian Shield, is situated in the centre of a gold belt, while Sudbury is in a world-famous nickel belt, where the mining and smelting of nickel and copper ores are core functions. Both cities began as single-industry towns, but over time they have become important regional centres with many service industries and more diversified economies, factors that have supported community stability. Even with these efforts, however, the two communities have been unable to maintain their populations. For example, from 1996 to 2001, Sudbury's popu-

Ontario leads all regions of Canada in high technological innovations, including Canada's contribution to space exploration. At a plant in Brampton, two workers are inspecting the wrist section of the Canadarm. (CP/Toronto Star/Rene Johnson)

Centre	Population 1996	Population 2001	Change (per cent)
Sudbury	165,336	155,219	(6.1)
Thunder Bay	126,643	121,986	(3.7)
Sault Ste Marie	83,619	78,908	(5.6)
North Bay	64,785	63,681	(1.7)
Timmins	47,499	43,686	(8.0)
Kenora	16,365	15,838	(3.2)
Elliot Lake	13,588	11,956	(12.0)
Kapuskasing	10,036	9,238	(8.0)
Kirkland Lake	9,908	8,616	(13.0)
Total	531,776	509,128	(4.3)

Table 5.6   Urban Centres in Northern Ontario, 1996–2001

Source: Statistics Canada (2002).

lation declined by 6.1 per cent while the population of Timmins dropped by 8 per cent (Table 5.6).

Sault Ste Marie is a steel town located on the Great Lakes between Lake Superior and Lake Huron. The town's major firm is the Algoma steel mill, which has been struggling to survive in a highly competitive marketplace. Algoma is located at some distance from its major industrial customers in southern Ontario and the United States. With high transportation costs, the Algoma mill is at a geographic disadvantage. However, Algoma has made use of local railway and seaway connections to secure raw materials, such as iron ore, and to market its steel products. In 2001, Algoma Steel marked its one hundredth anniversary and, at the same time, filed for protection under the Companies' Creditors Arrangement Act. As part of its new business plan, costs were reduced primarily by having its workers accept a 9 per cent reduction in pay. While Algoma made a small profit in 2002, the weak American economy and the increasing value of the Canadian dollar forced the company to take drastic action by reducing its labour force of 3,400 workers by 18 per cent (Bertin and Keenan, 2003: B1). While a smaller Algoma operation will hurt the local economy, closure would have a devastating effect on the city.

Thunder Bay, situated at the western end of Lake Superior, is a major transshipment point and a key element in the east–west transportation system. Bulky products—particularly grain, iron, and coal—are shipped to Thunder Bay for loading onto lake vessels. In recent years, however, this function has diminished. In the past, for example, iron from mines at Atikokan passed through Thunder Bay on its way to steel mills located along the shores of the Great Lakes. This mining operation is now defunct. Grain from the Canadian Prairies is still the major product handled at Thunder

Bay. The Canadian Wheat Board sends Prairie grain by rail to Thunder Bay and then by grain carrier to eastern ports for export to foreign countries. Until 1996, Ottawa heavily subsidized grain shipments under the Crow Benefit. With the elimination of that subsidy, the advantage of sending grain by rail to Thunder Bay has diminished, leaving its grain transportation role in doubt. (For more on this subject, see Chapter 8.)

Northern Ontario's population has grown very slowly and since 1996 has begun to decline. For example, from 1981 to 1996 population growth in the major cities of northern Ontario amounted to only 4.7 per cent, while from 1996 to 2001, as shown in Table 5.6, the population of these cities decreased by 4.3 per cent. In comparison, cities in southern Ontario have increased at a much higher rate over the same periods—nearly 30 per cent from 1981 to 1996 and, from 1996 to 2001, by 7.8 per cent (Statistics Canada, 1982, 1992, 2002).

Cities and towns in northern Ontario are struggling to diversify their economies. One option being pursued is tourism. Many private firms, such as motels, summer camps, and fishing and hunting lodges, have sprung up (especially in smaller communities) to meet the demand from the growing number of tourists coming to experience the wilderness of northern Ontario. The urgency to diversify is driven by the continuing reduction in the size of the primary labour force. For example, the workforce at the nickel mines in Sudbury has fallen by half over the last 30 years. Sudbury has reacted by aggressively seeking to expand its service industry. The city has met with some success, partly because of its strategic location at the junction of three highways—the Trans-Canada Highway, Highway 69 (which leads to Toronto), and Highway 144 (which connects Sudbury with Timmins)—

and partly because of its efforts to have federal and provincial agencies (the federal National Tax Data Centre and provincial Department of Mines) and institutions (Laurentian University and the Science North Museum) locate in Sudbury.

Ontario's Future

In the twenty-first century, Ontario retains its position with the largest population and economy of the six geographic regions. Within the Canadian economy, Ontario is especially dominant in the financial and manufacturing sector, but its resource industries face troubling times as the quality and quantity of resources diminish. Such a decline is particularly noticeable in the mining industry as the higher-grade ore is mined first. However, even in the exploitation of renewable resources, companies take the most valuable and most accessible resources first. Consequently, with time, resource hinterlands begin the inevitable economic decline. Within Ontario, therefore, the industrial core of southern Ontario's economy remains secure, but great uncertainty lies ahead for the resource hinterland of northern Ontario. For southern Ontario, the key economic development over the next few years depends heavily on maintaining or increasing its share of the automobile production within North America. Northern Ontario's future is less secure and faces two immediate threats—the possible closure of Algoma Steel and reduced lumber exports to the United States if the lumber dispute with the United States is not resolved.

Since the 1989 Free Trade Agreement, some of Ontario's manufacturing firms have gained a major foothold in the US market. While interprovincial trade remains important, most export increases are related to expanding sales in the United States. Much of this change is epitomized by the automobile industry. Before 1965, Canadians purchased their cars from manufacturers in southern Ontario. Today, they obtain their cars from assembly plants in different regions of North America.

Like other regions of Canada, Ontario faces the challenge of remaining competitive with other countries. Dangers lurk in the global economy. Fierce competition from other countries continues to force Ontario-based firms to make hard decisions, such as to cease operations, to relocate to low-wage countries, or to reduce their labour costs. Algoma Steel, for instance, has downsized its operation and obtained wage concessions from its unionized workers. By substituting capital for labour, these manufacturing firms become more efficient and require fewer employees.

Summary

Ontario remains the most powerful province in Canada. Its subregions of southern Ontario and northern Ontario have very different physical and human geographies. Southern Ontario is not only the industrial and population heartland of the province and nation but also an increasingly important sector of the North American economy. Southern Ontario's population is increasing rapidly. Canadians and immigrants are attracted to southern Ontario's robust economy. In sharp contrast, northern Ontario is an old resource hinterland whose economy is stalled. Since 1996, its population has declined as young people move to more prosperous areas of Canada, especially to the major cities in southern Ontario.

Ontario remains the dominant of the six geographic regions in Canada, both economically and by population. Ontario is experiencing growth in its financial and manufacturing activities, but its resource industries face troubling times. With the elimination of tariffs, Ontario will face continued challenges from manufacturing firms and resource companies in other parts of North America and the world. Sustainable development of its resources will also become a crucial challenge.

With the emergence of a powerful north–south trade alignment, Ontario sees its economic future as more broadly defined within North America, thereby diminishing its traditional role as the economic heartland of Canada. Ontario's new economic role may have both economic and political consequences for the rest of Canada.

Notes

1. As a general rule, wages in the primary and secondary sectors are often higher than the national average, while wages in certain tertiary jobs are below the national average. In fact, many jobs in the service sector are associated with low hourly wages and part-time employment. These are often disparagingly referred to as 'hamburger-flipping' jobs or McJobs.
2. At the beginning of the twentieth century, the Ontario government sought to develop its northeastern territories, especially its agricultural lands in the Clay Belt, its forest stands, and its mineral deposits in the Canadian Shield. Between 1903 and 1909, the government-financed Temiskaming and Northern Ontario Railway was built from the Canadian Pacific Railway line at North Bay northward to the small town of Cochrane on the National Transcontinental Railway (now the CNR). Over the next decade, branch lines were extended to Cobalt, Timmins, and Iro-

quois Falls. Overall, the Ontario government was pleased with the railway's impact on resource development. Plans were made to build the railway farther north—into the Hudson Bay Lowland. By 1932, the Temiskaming and Northern Ontario Railway stretched from Cochrane to Moosonee at the southern tip of James Bay, but further developments did not occur in the resource-scarce Hudson Bay Lowland.

3. In 1970, mercury was discovered in the fish near the Grassy Narrows Indian Reserve, which is about 500 km downstream from the pulp mill. Levels of methyl mercury in the aquatic food chain were 10 to 50 times higher than those in the surrounding waterways (Shkilnyk, 1985: 189). These levels were similar to those found in the fish of Minamata Bay, Japan. Over 100 residents of this Japanese village died from mercury poisoning in the 1960s, and over 1,000 people suffered irreversible neurological damage. Because they depend on fish and game, the Ojibwa at the Grassy Narrows Reserve ate fish on a daily basis and many complained of mercury-related illnesses. While, unlike the Minamata incident, no one died, the economic and social impact on the Ojibwa was nevertheless an industrial tragedy of immense proportions.

4. Statistics Canada has combined St Catharines and Niagara Falls as a single census metropolitan area although these cities still exist as separate political jurisdictions.

5. Toronto has tried various forms of regional government. About 40 years ago, the Ontario government established a form of regional government by combining the City of Toronto with the surrounding centres of Etobicoke, North York, Scarborough, York, and East York into the municipality of Metropolitan Toronto (Metro Toronto). These six jurisdictions became one large urban area for regional planning purposes, but each retained its city government. As Metro Toronto continued to grow, it spread into adjacent jurisdictions. In 1988, the Ontario government created the Greater Toronto Area (GTA). Often called the Metro Toronto Region, it consists of Metro Toronto and the regional municipalities of Halton, Peel, York, and Durham. In 1997, the provincial government passed a bill to change the municipal government structure to ensure more efficient, cost-effective services. The legislation came into effect in 1998, amalgamating six former municipalities (Etobicoke, York, Toronto, East York, North York, and Scarborough) and the regional municipalities into the new City of Toronto.

6. Toronto may solve its traffic problem by imposing a tax on automobiles entering the city's downtown. The first world-class city to experiment with a road toll was London, England. In 2002, London began charging drivers of automobiles £5 a day (about $12) to enter and leave the centre of London between 7 a.m. and 6:30 p.m. on weekdays, and this has resulted in a sharp decline in the volume of traffic, a great improvement in urban mobility, and a modest reduction in air pollution.

Key Terms

Free Trade Agreement
Trade agreement between Canada and the United States enacted in 1989.

just-in-time principle
System of manufacturing in which parts are delivered from suppliers at the time required by the manufacturer.

North American Free Trade Agreement
Trade agreement between Canada, the United States, and Mexico enacted in 1994.

outsourcing
Arrangement by a manufacturing firm to

obtain its parts from other firms.

restructuring

Economic adjustments made necessary by fierce competition; companies driven to reduce costs often resort to reducing the num-

ber of workers at their plants.

value added

A difference between a firm's sales revenue and the cost of its materials.

Bibliography

Amrhein, Carl, and Harold Reynolds. 1997. 'Using the Getis Statistic to Explore Aggregation Effects in Metropolitan Toronto Census Data', *The Canadian Geographer* 41, 2: 137–48.

Atkin, David. 2000. 'In a Hot-Wired World, Everything's Personal', *National Post*, 15 Nov., C1, C6–7.

Bagnall, James. 2003. 'Ottawa slipping off the tech map', *National Post*, 22 Aug., FP5.

Barnes, Trevor J., John N.H. Britton, William J. Coffey, David W. Edgington, Meric S. Gertler, and Glen Norcliffe. 2000. 'Canadian Economic Geography at the Millennium', *The Canadian Geographer* 44, 1: 4–24.

Bathelt, Harald, and Alfred Hecht. 1990. 'Key Technology Industries in the Waterloo Region: Canada's Technology Triangle (CTT)', *The Canadian Geographer* 34, 3: 225–34.

Bertin, Oliver. 1998. 'Agreement freezes Boeing wages', *Globe and Mail*, 11 July, B2.

———— and Greg Keenan. 2003. 'Algoma to axe 600, cut costs in latest bid to stay profitable', *Globe and Mail*, 15 May, B1.

———— and Jeff Sallot. 1998. 'Boeing plant set to close', *Globe and Mail*, 4 June, B1.

Blackbourn, Anthony, and Robert G. Putnam. 1984. *The Industrial Geography of Canada*. London: Croom Helm.

Bluestone, B., and B. Harrison. 1982. *The De-industrialization of America*. New York: Basic Books.

Brean, Joseph. 2003. 'Outbreak predicted to cost city $1-billion', *National Post*, 3 May, A6.

Brent, Paul. 2002. 'Car output still higher in Canada', *National Post*, 15 Aug., FP5.

————. 2003. 'Daimler calls off Windsor plant plans', *National Post*, 23 May, FP5.

Britton, John N.H., ed. 1996a. *Canada and the Global Economy: The Geography of Structural and Technological Change*. Montréal and Kingston: McGill-Queen's University Press.

————. 1996b. 'High Tech Canada', in Britton (1996a: ch. 14).

Campbell, Murray. 1996. 'Galt 25 Years Later', *Globe and Mail*, 13 Apr., D1–2.

Canada. 1994. *1993 Canadian Minerals Yearbook: Review and Outlook*. Ottawa: Ministry of Natural Resources.

Canadian Press. 2003. 'Imports threaten Camco Inc. plant', *National Post*, 5 Feb., FP7.

CBC News. 2000. 'Inside Walkerton: A water tragedy', CBC Backgrounder, 26 May [on-line database], Toronto. Searched 15 Nov. 2003: <http://www.cbc.ca/news/indepth/walkerton/>.

————. 2003. 'Camco closing Hamilton plant; 800 jobs cut', 4 Dec. [on-line database], Toronto. Searched 22 Jan. 2004: <http://www.cbc.ca/stories/2003/10/17/camco_031017>.

Ceh, S.L. Brian. 1996. 'Temporal and Spatial Variability of Canadian Inventive Companies and Their Inventions: An Issue of Ownership', *The Canadian Geographer* 40, 4: 319–37.

Crone, Greg. 1999. '560 jobs lost as Libbey closes Ontario plant', *National Post*, 5 Jan., C1.

Dabrowski, Wojtek. 2003. 'Ottawa says no to bailout of Navistar Plant', *National Post*, 17 May, FP3.

Dear, M.J., J.J. Drake, and L.G. Reeds, eds. 1987. *Steel City: Hamilton and Region*. Toronto: University of Toronto Press.

Dicken, Peter. 1992. *Global Shift: The Internationalization of Economic Activity*, 2nd edn. New York: Guilford Press.

Donald, Betsy. 2002. 'The Permeable City: Toronto's spatial shift at the turn of the millennium', *The Professional Geographer* 54, 2: 190–203.

Economist, The. 1996. *Country Profiles: Canada*. London: The Economist Intelligence Unit.

Erwin, Steve. 2001. 'Labour gears up for post-auto pact era', *National Post*, 2 Jan., D3.

Francis, R. Douglas, Richard Jones, and Donald B. Smith. 1996. *Destinies: Canadian History Since Confederation*, 3rd edn. Toronto: Harcourt Brace.

Gad, Gunter. 1991. 'Toronto's Financial District', *The Canadian Geographer* 35, 2: 203–7.

Gertler, Meric S. 1998. 'Negotiated Path or "Business as Usual"? Ontario's Transition to a Continental Production Regime', paper presented at the annual meeting of the Association of American Geographers, Boston, 17 Mar.

Gilbert, Anne, and Joan Marshall. 1995. 'Local Changes in Linguistic Balance in the Bilingual Zone: Francophones de l'Ontario et Anglophones du Québec', *The Canadian Geographer* 39, 3: 194–218.

Harris, Richard. 1999. *Unplanned Suburbs: Toronto's American Tragedy, 1900 to 1950*. Baltimore: Johns Hopkins University Press.

Herod, Andrew. 2000. 'Implications of Just-in-Time Production for Union Strategy: Lessons from the 1998 General Motors-United Auto Workers Dispute', *Annals of the Association of American Geographers* 90, 3: 521–47.

Hill, Bert. 1999. 'Nortel to spend $400 M to stay on fibre-optic cusp', *National Post*, 3 Nov., C3.

Holmes, John. 1983. 'Industrial Reorganization, Capital Restructuring and Locational Change: An Analysis of the Canadian Automobile Industry in the 1960s', *Economic Geography* 59: 251–71.

———. 1996. 'Restructuring in a Continental Production System', in Britton (1996a: ch. 13).

Ibbitson, John. 2001. *Loyal No More: Ontario's Struggle for a Separate Destiny*. Toronto: HarperCollins.

Jack, Ian. 2000. 'EU, Japan gang up on Canada', *National Post*, 12 Aug., D1, D6.

Jestin, Warren. 1996. *Global Economic Outlook*. Toronto: Scotiabank.

Keenan, Greg. 2003. 'Layoffs mount among auto makers', *Globe and Mail*, 26 Mar., B4.

Konadu-Agvemang, Kwadwo. 1999. 'Characteristics and Migration Experience of Africans in Canada with specific reference to Ghanaians in Greater Toronto', *The Canadian Geographer* 43, 3: 400–14.

Kreutzwiser, Reid, and Rob de Loë. 2004. 'Water Security: From Exports to Contamination of Local Water Supplies', in Bruce Mitchell, ed., *Resource and Environmental Management in Canada*, 3rd edn. Toronto: Oxford University Press, ch. 6.

Krueger, Ralph. 2000. 'Trials and Tribulations of the Canadian Fruit-Growing Industry', *The Canadian Geographer* 44, 4: 342–54.

Little, Bruce, and Greg Keenan. 1996. 'Ontario's Economic Future Is the Sum of Its Auto Parts', *Globe and Mail*, 2 Mar., A1, A5.

Lorch, Brian J. 1999. 'New Format Development and its Impact on the Commercial Structure of Thunder Bay', in K. Jones, ed., *Case Studies of the Impact of International Retailing in Canada*. Toronto: Ryerson Polytechnic University, ch. 5.

McCarthy, Shawn. 1997. 'Ottawa Trades Bureaucrats for Bytes', *Globe and Mail*, 30 Aug., B1, B4.

MacDonald, N.B. 1980. 'The Future of the Canadian Automotive Industry in the Context of the North American Industry'. Working Paper no. 2. Science Council of Canada. Ottawa: Minister of Supply and Services.

McKnight, Tom L. 1992. *Regional Geography of the United States and Canada*. Englewood Cliffs, NJ: Prentice-Hall.

MacLachlan, Ian. 1996. 'Organizational Restructuring of U.S.-Based Manufacturing Subsidiaries and Plant Closure', in Britton (1996a: ch. 11).

Marsh, James H., ed. 1988. *The Canadian Encyclopedia*, 2nd edn. Edmonton: Hurtig.

Matthew, Malcolm R. 1993. 'The Suburbanization of Toronto Offices', *The Canadian Geogra-*

pher 37, 4: 293–306.

Norcliffe, Glen. 1994. 'Regional Labour Market Adjustments in a Period of Structural Transformation: An Assessment of the Canadian Case', *The Canadian Geographer* 38, 1: 2–17.

————. 1996. 'Mapping Deindustrialization: Brian Kipping's Landscape of Toronto', *The Canadian Geographer* 40, 3: 266–73.

Ontario Medical Association. 2001. The Illness Costs of Air Pollution in Ontario: A Summary of Finding [on-line database], Toronto. Searched 15 Nov. 2003: <http://www.oma.org/phealth/icap.htm>.

Paterson, J.H. 1994. *North America: A Geography of the United States and Canada*. New York: Oxford University Press.

Polèse, Mario, and Richard Stren, eds. 2002. *The Social Sustainability of Cities and the Management of Change*. Toronto: University of Toronto Press.

Preston, Valerie, and Lucia Lo. 2000. 'Canadian Urban Landscape Examples—21: "Asian Theme" Malls in Suburban Toronto: Land Use Conflicts in Richmond Hill', *The Canadian Geographer* 44, 2: 182–90.

Putnam, D.F., ed. 1952. *Canadian Regions: A Geography of Canada*. Toronto: Dent & Sons.

Relph, Edward. 1991. 'Suburban Downtowns of the Greater Toronto Area', *The Canadian Geographer* 35, 4: 421–5.

————. 2002. *The Toronto Guide: The City, the Fringe, the Region: An Illustrated Interpretation of Toronto's Landscapes*. Revised and updated for the Canadian Association of Geographers' annual meeting, Toronto, May. Toronto: Centre for Urban and Community Studies at the University of Toronto.

Rumney, Thomas. 2001. *The Geography of Canada Bibliography Series: Vol. 4, Ontario*. Plattsburgh, NY: Plattsburgh State University, Center for the Study of Canada.

Shaw, Anthony B. 1999. 'The Emerging Cool Climate Wine Regions of Eastern Canada', *Journal of Wine Research* 10, 2: 79–94.

Shkilnyk, A.M. 1985. *A Poison Stronger Than Love: The Destruction of an Ojibwa Community*. New Haven: Yale University Press.

Shrubsole, Dan, and Milford Green. 1997. 'The Actual and Perceived Effects of Floodplain Land-Use Regulations on Residential Property Values in London, Ontario', *The Canadian Geographer* 41, 2: 166–77.

Smith, Jamie, Beth Lavender, Heather Auld, David Broadhurst, and Tim Bullock. 1998. 'Adapting to Climate Variability and Change in Ontario', in *The Canada Country Study: Climate Impacts and Adaptation*, vol. 4. Ottawa: Environment Canada.

Stanford, Quentin H., ed. 1998. *Canadian Oxford World Atlas*, 4th edn. Toronto: Oxford University Press.

Statistics Canada. 1982. *Census Metropolitan Areas and Census Agglomerations*. Catalogue no. 95–903. Ottawa: Minister of Supply and Services.

————. 1992. *Census Metropolitan Areas and City Agglomerations*. Catalogue no. 93–303. Ottawa: Minister of Supply and Services.

————. 1996. *Labour Force Annual Averages 1995*. Catalogue no. 71–220–XPB. Ottawa: Statistics Canada.

————. 1997a. *Historical Overview of Canadian Agriculture*. Catalogue no. 93–358–XPB. Ottawa: Industry Canada.

————. 1997b. *A National Overview: Population and Dwelling Counts*. Catalogue no. 93–357–XPB. Ottawa: Industry Canada.

————. 1997c. *Canada Year Book*. Catalogue no. 11–402–XPE/1997. Ottawa: Minister of Industry.

————. 1997e. 1996 Census: Nation Tables—Population by Mother Tongue, Showing Age Groups, for Canada, Provinces and Territories, 1996 Census—20% Sample Data, 2 Dec. 1997 [on-line database], Ottawa. Searched 15 July 1998: <http://www.statcan.ca/english/_census96/>.

————. 1997f. *The Daily*—1996 Census: Mother Tongue, Home Language and Knowledge of Languages, 2 Dec. [on-line database], Ottawa.

Searched 14 July 1998: <http://www. statcan. ca/Daily/English/>.

———. 1998a. *Canadian Economic Observer*. Catalogue no. 11–010–XPB. Ottawa: Statistics Canada.

———. 1998b. *The Daily*—1996 Census: Ethnic Origin, Visible Minorities, 17 Feb. [on-line database], Ottawa. Searched 16 July 1998: <http://www.statcan.ca/Daily/English/>.

———. 1998c. *The Daily*—1996 Census: Aboriginal Data, 13 Jan. [on-line database], Ottawa. Searched 14 July 1998: <http://www.statcan.ca/Daily/English/>.

———. 2002. 2001 Census: Population and Dwelling Counts, for Census Metropolitan Areas and Census Agglomerations, 2001 and 1996 Censuses. 16 July [on-line database], Ottawa: <http://www.statcan.ca/english/IPS/Data/93F0050XCB2001013.htm>.

———. 2003a. Canadian Statistics, Census of Agriculture. Top three fruit and field-grown vegetable crops by area, province [on-line database], Ottawa. Searched 12 May 2003: <http://www.statcan.ca/english/Pgdb/econ101a.htm>.

———. 2003b. Canadian Statistics, Census of Agriculture. Total area of farms, land tenure and land in crops, provinces [on-line database], Ottawa. Searched 12 May 2003: <http://www.statcan.ca/english/Pgdb/econ124i.htm>.

———. 2003c. Canadian Statistics, International Trade. Exports of goods on a balance-of-payments basis [on-line database], Ottawa. Searched 3 May 2003: <http://www.statcan.ca/english/Pgdb/gblec04.htm>.

———. 2003d. Canadian Statistics: Population of Census Metropolitan Areas [on-line database], Ottawa. Searched 12 May 2003: <http://www.statcan.ca/english/Pgdb/demo05.htm>.

———. 2003e. Census of Canada 2001—Labour, employment and unemployment, by industry, by province [on-line database], Ottawa. Searched 3 May 2003: <http://www.statca.ca/english/Pgdb/labor21b.htm>.

———. 2003f. Census of Canada 2001—Canada's Ethnocultural Portrait: The Changing Mosaic. Analytical series 96F0030XIE2001008, 21 Jan. [on-line database], Ottawa. Searched 28 January 2003: <http://www12.statcan.ca/english/census01/products/analytic/companion/etoimm/subprovs.cfm>.

———. 2003g. Canadian Statistics: Distribution of Employed People, by Industry, by Province [on-line database], Ottawa. Search 20 May 2003: <http://www.statcan.ca/english/Pgdb/labor21b.htm>.

Stewart, Walter. 2003. 'A Late Great Lake?', *Canadian Geographic* 123, 5: 36–46.

Steinhart, David. 1999. 'Ste-Thérèse truck plant reopens under Paccar label', *National Post*, 3 Aug., C7.

Van Alphen, Tony, and Rick Westhead. 2004. 'Stelco job cuts loom in fight to stay alive', *Toronto Star*, 30 Jan., A1, A18.

Wang, Shuguang. 1999. 'Chinese Commercial Activities in Toronto CMA: New Developments, Patterns and Impacts', *The Canadian Geographer* 43, 1: 19–35.

Weintraub, Sidney, and Christopher Sands. 1998. *The North American Auto Industry under NAFTA*. Washington: Center for Strategic and International Studies Press.

Further Reading

Courchene, Thomas J., with Colin R. Telmer. 1998. *From Heartland to North American Region State: The Social, Fiscal and Federal Evolution of Ontario*. Monograph Series on Public Policy, Centre for Public Management. Toronto: Faculty of Management, University of Toronto.

The premise of this book by two economists is simple: Ontario has become a regional economic force within the North American economy. Ontario's interests now lie with the continental economy, not the national one. Put differently, Ontario benefited from Canada's national economy with its high tariffs but now the province's self-interest lies in the North American economy. How did this happen?

For over 100 years, Ontario stood for a strong central government and the national economy. Both elements worked in Ontario's favour, placing the province at the apex of Canada's core/periphery economy. In fact, these elements were the 'power pillars' that reinforced Ontario's dominant place within the Canadian economy. The province could count on Ottawa to further its interests. For example, the federal government negotiated the Auto Pact with the United States and this agreement has benefited Ontario enormously. Similarly, the National Energy Program worked in Ontario's favour at the expense of three periphery provinces—Alberta, Saskatchewan, and British Columbia. From a spatial perspective, Ontario was the industrial core (along with Québec), while the rest of the country formed the resource hinterland. Federal high-tariff policies encouraged manufacturing in Ontario and discouraged such developments in the rest of the country.

In 1989, the Free Trade Agreement pushed Ontario into the North American economy. Courchene and Telmer argue that Ontario's interests have changed. Ontario lost its economic stranglehold on the rest of the country and now its economic self-interest lies in finding a place in the larger North American economy. To achieve such a goal, Ontario would benefit from a weak central government. Trade figures support Courchene and Telmer's view that Ontario is now more closely allied with the United States. For example, in 1981, Ontario's exports to other provinces were approximately equal to exports to the rest of the world (which effectively means the United States). In 1994, exports to the rest of the world were 2.1 times greater than exports to other provinces. From a spatial perspective, Canada is now witnessing a series of regional economic alliances with other countries. All provinces are importing more manufactured goods from foreign countries, thereby breaking their dependency on Ontario's manufactured goods. At the same time, all provinces, but particularly Ontario, see their trade with the United States growing by leaps and bounds. Economic data clearly show that Ontario's trade has increased greatly with the United States, but much of this increase is due to greater sale of automobiles to American customers. The political reason for such sales is not due to the Free Trade Agreement but to the Auto Pact. Nevertheless, Courchene and Telmer demonstrate that a north–south trade axis links Ontario with the United States. Extending their economic logic across Canada, geography (distance) dictates that a series of regional trade arrangements with the United States and other foreign countries must emerge.

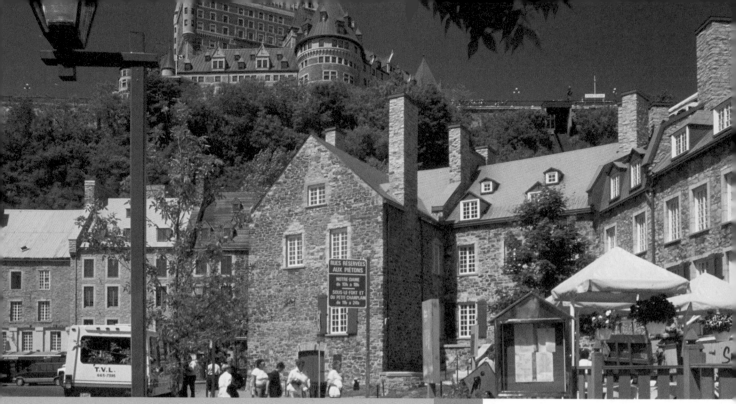

Overview and Objectives

Within Canada, Québec ranks second in economic output and population size, accounting for 21 per cent of the country's GDP and 24 per cent of its population in 2002. It has both a northern resource hinterland and a southern industrial core. Manufacturing and high-technology industries make Québec an important part of Canada's industrial heartland, but it is also seeking a place in the North American economy. As home to most Canadians of French origin, Québec represents a distinct cultural and linguistic region of Canada. Hydro-Québec played a prominent role in the province's economic and social transformation, and the James Bay Project led to the first modern land-claim agreement.

- Describe Québec's physical geography and historical roots.
- Present the basic elements of Québec's population and economy in the context of region's physiography.
- Examine Québec's cultural and economic development within the context of the core/periphery model.
- Discuss Québec's position within Canada, its francophone culture, and its separatist movement.
- Examine the effects of the James Bay Project on the Cree and Inuit.
- Focus on Hydro-Québec's strategy of developing low-cost electrical power in northern Québec to stimulate industrial activities in southern Québec.
- Comment on Québec's changing role within Canada and North America.

Chapter 6 Québec

■ Introduction

By virtue of its geography and history, Québec occupies a special place in Canada. To begin with, the St Lawrence River opened North America to French settlement, exploration, and the fur trade. This great river, which figured so prominently in Québec's geography, continues to play an essential role in Québec's life. The port of Montréal owes much of its past economic prosperity to its role as a major transshipment point for goods entering North America (Vignette 6.1).

History has made this corner of North America the homeland of the French-speaking people in Canada and the United States. The French language and culture have generated a strong sense of belonging among its francophone citizens, forming the basis of ethnic pride, loyalty, and nationalism, which from time to time fuels the desire for an independent political state. Such feelings are particularly strong among the 'old stock' who are descendants of some 10,000 settlers who migrated from France in the seventeenth and eighteenth centuries. While their first loyalty is often to Québec, many Québécois also have a strong attachment to Canada. This sense of dual loyalty is unique in Canada and it accounts for Québec's interest in the concept of a political partnership within Confederation (for a fuller discussion of this concept, see the section, One Country, Two Visions, in Chapter 3). Second, Québec's culture is largely derived from the historical experience of francophones living in North America for over 400 years and of their being Canadians for nearly 140 years. Events of the distant past, whether the British attempts at assimilation of the French or their ruthless suppression of revolt in 1837, or more recent ones, such as the War Measures Act of 1970 and the repatriation of the Canadian Constitution in 1982 over the opposition of the Québec government, fuelled the desire for a separate state. Third, Québec is recognized throughout Canada and the United States as the geographic heart of French language, customs, and heritage in North America. Many French-speaking Canadians in the rest of Canada and Franco-Americans come to visit Québec to renew their sense of ethnicity. In sum, Québécois history and geography give Québec its raison d'être and its vision of Canada as a nation of two founding peoples. Canada's cultural duality has led to tensions between French- and English-speaking Canadians that have often strained the bonds of Confederation. (See Chapters 3 and 4 for a broader discussion of the French/English faultline and the two competing visions of Canada.)

The vast majority of people living in Québec speak French. In 2001, Québec's population was 7.2 million, marking a small increase from 7.1 million in 1996. Over 83 per cent of Quebecers declared French as their mother tongue (Figure 6.1). These Quebecers are known as the Québécois. The remaining

Vignette 6.1   The St Lawrence River

The St Lawrence River provides a natural waterway into the interior of North America. Cities along its shores, particularly Québec City and Montréal, have benefited greatly from the waterway's role as a major trading route. Montréal's favourable location on the St Lawrence gives it an economic advantage and fuels the city's growth.

While waters from the Great Lakes empty into the St Lawrence River, modern ships could not reach the ports along the Great Lakes. For ocean-going ships, two major obstacles had to be overcome: the Lachine Rapids near Montréal, and Niagara Falls, which prevented ships from passing from Lake Ontario to Lake Erie. The building of the Lachine Canal (1825) and the Welland Canal (1829) allowed ships to reach the Great Lakes. Over time, these canals and locks were improved to accommodate the increasing size of ships and barges.

After World War II, there were vast improvements in ocean transportation. This was the era of the 'supertanker'. Even the average size of freighters had increased substantially. These larger ships required much greater depth of water than their predecessors in order to float. Consequently, more and more ocean-going ships were unable to enter the St Lawrence River past Montréal. The solution to this transportation bottleneck was the St Lawrence Seaway, which, when it was inaugurated in 1959, allowed ocean vessels to travel from the Atlantic Ocean to the Great Lakes. While Montréal lost its natural advantage as a transshipment point, Montréal remains an important port, especially for container traffic.

The Port of Montréal became a year-round port in 1964. Since then, Montréal has obtained more and more of the North Atlantic container traffic. In 2003, Montréal handled 38 per cent of container traffic on the Atlantic coast compared to 35 per cent for New York City. (Search4Stock, Inc.)

Quebecers include Aboriginal peoples, anglophones (those whose mother tongue is English), and allophones (those immigrants whose mother tongue is neither English nor French). Anglophones are concentrated in Montréal, Estrie (the Eastern Townships), and Outremont (the Ottawa Valley). Allophones are concentrated mainly in Montréal. In northern Québec, Cree and Inuit form the majority. From time to time, social tensions surface between French- and English-speaking Quebecers, and between the Cree and the Québec government.

Québec is a large geographic region with four of Canada's physiographic regions found in its territory (Figure 2.1). Each physiographic region has a different settlement and land-use pattern. The St Lawrence Lowlands region is the agricultural, industrial, and population core of Québec. The remaining three regions are economic hinterlands. The Canadian Shield extends from the St Lawrence Lowlands to Hudson Strait. It, along with the Hudson Bay Lowland found at the southern tip of James Bay, represents a resource frontier. The Appalachian Uplands, which borders on New Brunswick and three American states—Vermont, New Hampshire, and Maine—lies south of the St Lawrence Lowlands and extends eastward to the Gaspé Peninsula. It is divided into two areas: an agricultural/industrial area in Estrie and a downward transitional resource hinterland in Gaspésie.

Québec within Canada

Québec's economic strength within Canada is considerable. Ranking second among the 10 provinces, Québec accounted for 21.1 per cent of all the goods and services produced in this country and 24 per cent of Canada's popula-

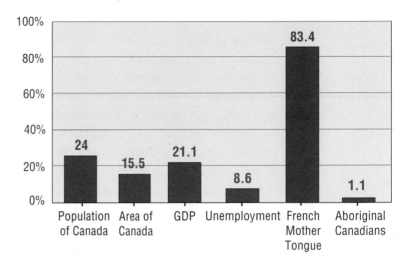

Figure 6.1 Québec, 2001. By population size and gross domestic product, Québec is the second-ranking geographic region in Canada. Its predominantly French-speaking population forms over 84 per cent of the total population while Aboriginal peoples comprise about 1 per cent of Québec's population. *Sources:* Tables 1.1, 1.5, and 4.19

tion in 2001 (Figure 6.1). Besides its resource and manufacturing base, Québec's economy is propelled by a cluster of high-tech firms and an exceptionally strong tourist industry. Both Montréal and Québec City are popular tourist destinations among American and European visitors.

Québec's position within Canada was much stronger at the time of Confederation. Since then, its place within Canada has slipped for six main reasons. First, Québec's geopolitical position within Canada has changed from one of four provinces in 1867 to one of 10 provinces and three territories in 2001. Second, Québec's demographic position has shrunk from 32 per cent of Canada's population in 1871 to 24 per cent in 2001. Third, Québec's economy missed out on the massive expansion of the automobile industry in Canada following the Auto Pact in 1965 largely because of its greater distance from the American automotive industry and market than Ontario. Fourth, Québec's economy was hurt, at least initially, by the rise of separatism in the 1970s and the resulting two referendums in 1980 and 1995. The main damage was due to an unattractive investment climate for many Canadian and foreign companies, and the relocation of a few company headquarters from Montréal to Toronto. Fifth, Québec's economy is now operating within the American and global economic system. Consequently, resource and manufacturing industries are facing strong competition and even trade restrictions from other countries. Three examples are provided. In 2001, Washington introduced restrictions on softwood lumber from Québec entering the United States. These restrictions are designed to ensure that American lumber firms retain their traditional market share, but their impact on the forest industry in Québec has been damaging. Manufacturing companies have also felt the sting of a sluggish world market. Bombardier, one of the world's leading aircraft manufacturers, has faced a slumping market for its business jets. Since 2002, the company has reduced its size and, in doing so, had to lay off hundreds of employees. In 2003, the closure of Noranda's magnesium plant at Danville was due to its inability to compete in the American market with low-cost imports from China. Sixth, Québec's economy has expanded at a slower rate than Ontario's economy, which accounts for out-migration and relatively high unemployment rates. For example, in 2001, Québec had an unemployment rate of 8.7 per cent while Ontario's rate was 6.3 percent.

However, not all the economic news is negative. The liberalization of trade has had many positive impacts on Québec's economy. First, the worldwide trade system has forced existing manufacturing firms to become more efficient and more conscious of quality control. In turn, these firms, including the appar-el industry, have been able to maintain market share in Canada and gain a foothold in the American market. Second, the chief advantage for aerospace, metal refining, and pharmaceutical firms to locate in Québec—especially in the Montréal manufacturing core—is that these companies can easily serve both North American and global customers. The challenge for Québec is to continue on the path to create a more robust economy within its industrial core and under an atmosphere of harmony between its francophone majority and its minority groups. To some degree, the pace of economic progress is linked to resolving the lingering issues affecting French/English communities and to moving forward on a more co-operative type of hydroelectric development in the James Bay region for the benefit of both the province and the Aboriginal peoples for whom northern Québec is their traditional homeland.

Québec's Physical Geography

Québec, the largest province in Canada, has a wide range of natural conditions. Its climate varies from the mild continental climate in the St Lawrence Valley to the cold arctic climate found in Nunavik (Inuit lands lying north of the 55th parallel). For the most part, the weather across Québec falls within a predictable range, but, like other regions of Canada, Québec has had its share of extreme weather events. While Québec commonly experiences heavy rainfall and freezing rain, two recent storms stand out. Extremely heavy rains in July 1996 in the Saguenay region caused extensive flooding and resulted in loss of life, extensive damage, and the evacuation of 15,000 people (Vignette 6.2). Two years later, freezing rain struck southern Québec and eastern Ontario, causing great damage and

Vignette 6.2   The Great Flood

In July 1996, a savage rainstorm took place some 250 km north of Québec City in the Saguenay region. Lasting over a four-day period from 17–21 July 1996, a record 277 mm of rain fell. Flooding was widespread. The tributary streams and rivers of the Saguenay River overflowed their banks and eroded the surrounding land and highway system. Over a hundred houses, farmsteads, and buildings were destroyed, forcing the evacuation of some 15,000 people. Seven were killed in the flood, including two children killed in their sleep when a mudslide buried their house. Dams and roads were washed out. Electricity and telephone services were disrupted. L'Anse Saint Jean, La Baie, Jonquière, Chicoutimi, and Hébertville were hardest hit. The loss of property from this storm was over $600 million.

Vignette 6.3   The Great Ice Storm

On 5 January 1998, the worst ice storm in Canada's history struck eastern Ontario, southern Québec, and the Maritime provinces. It also affected the neighbouring New England states. While freezing rain occurs regularly in Canadian winters, this ice storm was unusually long. On the first day of the storm, about 20 mm of freezing rain fell on the Montréal area. Similar amounts fell in Kingston, Ottawa, Sherbrooke, and Trois-Rivières. These five cities and their surrounding rural communities felt the full fury of nature. But, unlike the one-day ice storm of 1961 that affected the Montréal area, the storm of 1998 continued throughout the next five days. The weight of the ice snapped hydroelectric poles and caused high-voltage transmission towers to collapse. The electrical system in eastern Ontario and southern Québec failed. More than 4 million people were without electrical power for at least 36 hours. About 500,000 people in the Estrie (the Eastern Townships) were without power for several weeks. The physical damage was enormous, perhaps reaching $500 million. Economic losses to farmers and business also totalled millions of dollars. Hydro-Québec faced considerable reconstruction of its damaged transmission system. The storm impacted negatively both human activity and the natural landscape (many trees were lost or cut down as a result).

disrupting the electrical power. The ice storm began on 5 January 1998 and continued for several more days. Over 4 million people were without electricity, many for several weeks (Vignette 6.3).

Four of Canada's physiographic regions extend over the province's territory—the Hudson Bay Lowland, the Canadian Shield, the Appalachian Uplands, and the St Lawrence Lowlands—and each has a different resource base and settlement pattern (Figure 5.2).

The heartland of Québec lies in the St Lawrence Lowlands. Formed from the Champlain Sea some 10,000 years ago, this physiographic region provides the best agricultural land in Québec. Settlers from France began farming along the edge of the St Lawrence River some 400 years ago. New France was established within this physiographic region and, by means of the St Lawrence, spread into the interior of North America. It remains the cultural core of Québec. The St. Lawrence Lowlands offers several natural advantages that give the region its central role in modern Québec. Arable land is one such advantage. By far the best agricultural land is in a small area between Montréal and Québec City. As well, most industrial plants are located in this same area.

The Appalachian Uplands physiographic region is a northern extension of the Appalachian Mountains. Most arable land is in Estrie, where the Loyalists settled in the late eighteenth century and were largely replaced by French-speaking farmers by the end of the nineteenth century. In the Gaspé Peninsula, there are a number of small communities along the coast. Because the local economy is so weak, many people are unemployed. Others combine a resource activity, such as farming, fishing, or logging, with part-time employment in the villages and towns. Because of the area's spectacular scenery, tourism has become an important source of income during the summer in the Gaspé Peninsula. Mining and forestry are other economic activities in this physiographic region. With the exception of the Lake Champlain gap in the Appalachian Uplands, access to the populous parts of New England is blocked by these rugged uplands. The Lake Champlain gap has therefore become a very important north–south transportation link between Montréal and points south in the US, especially New York City.

As the largest physiographic region in Québec, the Canadian Shield occupies over three-quarters of the province's territory (Figure 5.2). The Canadian Shield is noted for its forest products and hydroelectric produc-

tion—it has most of the hydroelectric sites in Canada because of a combination of heavy precipitation, large rivers, and high elevations. Near Montréal and Québec City, the Canadian Shield is a recreational playground where the rolling, rugged, forested upland with its numerous lakes has become a popular site for urbanites and tourists. Further north, single-industry towns based on mining and forestry dot the landscape. Beyond the commercial forest zone lie the lands of the Cree and Inuit, where the Cree must coexist with the massive La Grande hydroelectric project. The creation of hydroelectric reservoirs has resulted in a serious environmental problem for the Cree—the microbial breakdown of large amounts of submerged vegetation creates high levels of methylmercury contaminants. These contaminants enter the aquatic food chain and eventually find their way into fish. The Cree, who traditionally have consumed large quantities of fish, have been confronted with the risk of mercury poisoning. In arctic Québec, a number of Inuit settlements are found along the coasts of Hudson Bay and Hudson Strait. The economies in the Cree and Inuit communities are unable to provide sufficient employment opportunities. Consequently, the Cree and Inuit have high unemployment rates. The Québec government supports a hunting and trapping program, thus encouraging many Cree families to stay on the land. While this program has important cultural and social values, these families tend to catch and eat large quantities of fish. No one is certain about the impact of fish consumption by the Cree, but for cultural reasons they continue to follow their traditional lifestyle (Scott, 2001: 175–202).

The Hudson Bay Lowland extends from Ontario along the southeast edge of James Bay near Rupert River. By far, this lowland is the smallest physiographic region in Québec. Few people live here.

Environmental Challenges

Québec, like other provinces, is confronted with a series of environmental problems, most of which stem from the discharge of agricultural and industrial wastes into the atmosphere and water bodies or from the construction of huge hydroelectric dams. The most populated area of Québec is found near the St Lawrence River where both human and other forms of biological life are threatened by its polluted waters. Once a pristine river, the St Lawrence is now badly contaminated. The reason is simple. Over the years, great quantities of industrial toxic wastes have been deposited directly into the river. The result today is a river with high levels of toxic chemicals—polychlorinated biphenyls (PCBs) and dichlorodiphenyltrichloroethane (DDT), as well as heavy metals such as lead, mercury, and cadmium. More specifically, the waste contaminants from the aluminum and magnesium smelters—known as polycyclic aromatic hydrocarbons (PAHs)—have had a serious impact on the ecological system of the St Lawrence. Take, for example, the impact on the beluga whales. At one time, the population of the St Lawrence beluga whales totalled almost 10,000. Today, the population is under 1,000.

Even more disturbing is the link to the health of the local population. According to a scientific paper by Martineau and others (2002), the human population living along the estuary of the St Lawrence River has rates of cancer higher than those found in people in the rest of Québec and Canada, and some of these cancers have been epidemiologically related to PAHs.

Québec's Historical Geography

Relative to other parts of North America, the history of Québec is long, rich, and complicated by the period of British rule and then the search for its place within Confederation (see Tables 6.1–6.3). Beneath the surface of this search lies the deeply fractured French/English faultline. Québec's historical geography can be divided into three periods: New France, British occupation, and Confederation.

New France, 1608–1760

For France, its introduction to North America began in 1534 when Jacques Cartier sailed into the Bay of Chaleur and set foot on the shores of the Gaspé Peninsula. Yet, the first permanent French settlement in Québec took place in 1608, when Champlain founded a fur-trading post at the site of Québec City, thus establishing the French colony in North America. Although France eventually lost its North American colony, it left a cultural legacy in the form of the French language and the Catholic religion, and a French stamp on the landscape with its unique settlement pattern and arrangement of farms into long lots. Village after village was built around a Catholic church and the surrounding farmland was arranged in long lots based on the French seigneurial system.

During the seventeenth and eighteenth centuries, France had control over vast areas of North America. Its core, however, was the St Lawrence Valley, from which New France developed a vast fur-trading empire. The wealth from the fur trade was enormous and was the reason for France's interest in the New World. Almost every French settler wanted to participate in the fur trade, which left only a few to clear the forest and till the land. For example, several canoes full of beaver pelts could make a man extremely wealthy compared with the meagre returns obtained from the back-breaking toil of clearing land and breaking the soil. Frenchmen who were coureurs de bois (trappers) often lived with the Indians. A few were extremely successful and returned to France to enjoy their good fortune. Others remained in the fur trade or settled in New France.

Geography played a part in New France's success both as a fur-trading empire and as an agricultural colony. The St Lawrence River provided a route to the interior, which gave the French explorers and fur traders an advantage over their English rivals, who had to contend with crossing the Appalachian Mountains. With further exploration of the interior of North America, the French were able to expand their territory and secure more and better fur-trading routes. For example, by portaging from the Great Lakes to the Ohio River and then to the Mississippi, the famous French explorer, René-Robert Cavelier, Sieur de La Salle, reached the mouth of the Mississippi in 1682. Early in the eighteenth century, French fur traders, led by Pierre Gaultier de Varennes, Sieur de la Vérendrye, established a series of fur-trading posts in Manitoba. These trading posts, especially Fort Bourbon on the Saskatchewan River at Cedar Lake (just east of the present-day community of The Pas), made it convenient for Indians to trade their furs with French traders rather than travel farther to the British fur-trading posts along the Hudson Bay coast.

New France also established a successful agricultural society. Once the land was cleared, the fertile soils in the St Lawrence Lowlands provided a solid basis for a feudal agricultural settlement known as the seigneurial system. Farming took hold in New France, particular-

ly following the efforts of Jean Talon, the greatest administrator of New France. By the early eighteenth century, the French had turned to farming, leaving the lure of the wilderness to a relatively small number of more daring souls. By then, New France had existed for over 100 years, and many of its people had been born and raised in the New World. They were no longer 'French' but Canadiens.

By the middle of the eighteenth century, almost all the lands in the St Lawrence Lowlands were under cultivation. Farmlands stretched in a continuous belt from Québec City to Montréal. Like the feudal agricultural system in France, peasant farmers (**habitants**) worked the land, paying their lords (**seigneurs**) both in kind and in labour. Through the efforts of the habitants and their seigneurs, New France became a successful agricultural colony, but one whose system of landownership and rural life was quite different from that of the British colonies along the Atlantic seaboard (a difference that would eventually cause Britain to split Québec into Upper and Lower Canada in 1791 [see Chapter 3]).

The Seigneurial System

When the first intendant of New France, Jean Talon, arrived in New France in 1665, he encountered a population of only 3,000 inhabitants, most of whom were men engaged in the fur trade. Talon had been instructed by Louis XIV to create a feudal agricultural society resembling that of rural France in the seventeenth century. Talon undertook three measures to achieve this goal. First, he recruited peasants from France. Second, he sent for young women—orphaned girls and daughters of poor families in France—to provide wives for the men of the colony. Third, he imposed the French feudal system of landownership, known as the seigneurial system. In the seigneurial system, huge tracts of land were granted to those favoured by the king, namely, the nobility, religious institutions of the Roman Catholic Church, military officers, and high-ranking government officials. The seigneur was obliged to swear allegiance to the king and to have his tenants cultivate the lands on his estate. The tenants owed certain obligations to their seigneur: paying yearly dues (*cens et rentes*) to their seigneur, working the seigneur's land, especially in regard to road maintenance (*corvée*), and paying rent for using the seigneur's grinding mill and bake ovens (*droit de banalité*).

In the days of New France, the seigneurial system offered several advantages: (1) it encouraged the establishment and functioning of an agricultural society; (2) it provided access to the seigneur's mill where the *habitants* could grind their grain; and (3) the mill served

as a centre of defence for the *habitants* when under attack.

By 1760, there were approximately 200 seigneuries. Seigneuries, which were usually 1 by 3 leagues (5 by 15 km) in size, were generally divided into river lots (*rangs*). These long, rectangular lots were well adapted to the St Lawrence Valley for several reasons, the most important of which was that each *habitant* had access to a river, either a tributary of the St Lawrence River or the river itself. At that time, most people and goods were transported along the river system in New France. For that reason, river access was vital for each *habitant* family.

After the British Conquest of New France, the seigneurial system was retained by the British more for political reasons—to gain the support of the principal source of power (the seigneurs and the Roman Catholic Church)—

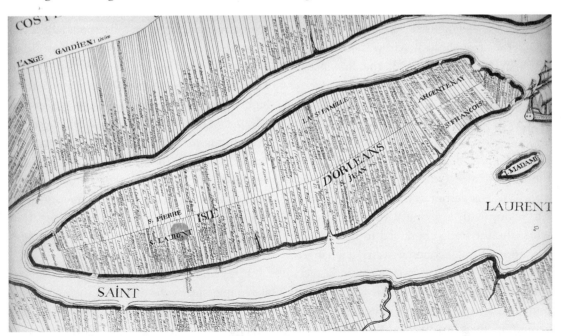

Historic map of Île d'Orléans, which is situated in the St Lawrence River near Québec City. Île d'Orléans was one of the first areas settled by the French and its *habitants* produced food for themselves and their seigneur. Île d'Orléans clearly illustrates the *rang* land tenure system of New France which took the form of long, narrow lots. The purpose of the *rang* was to provide each *habitant* access to the river transportation system. (*National Archives of Canada NM0048248*)

Table 6.1	Timeline: Historical Milestones in New France
Year	**Event**
1534	Jacques Cartier sailed into the Bay of Chaleur and claimed the land for France. The following year, Cartier discovered the mouth of the St Lawrence River, which provided access to the interior of North America.
1608	Samuel de Champlain, described as the 'Father of New France', founded a fur trading post near the site of Québec City. Champlain was instrumental in the development of the fur economy, which provided the initial economic basis for New France.
1642	Paul de Comedey de Maisonneuve established Ville-Marie on Île de Montréal, which is strategically situated at the confluence of the Ottawa and St Lawrence rivers. Later, Ville-Marie was renamed Montréal.
1759	The struggle between France and England over North America saw the British defeat the French army on the Plains of Abraham. The final battle between the French and English forces ended with the capture of Montréal by the British. In 1763, the formal surrender of New France to England took place with the Treaty of Paris.

than for economic reasons. By the early nineteenth century, however, agriculture in Lower Canada had become a commercial venture, making the seigneurial system an anachronism. The seigneurial system was abolished in 1854 by the Legislature of the Province of Canada.

British Colony, 1760–1867

Following the defeat of the French in 1760, the British ruled Québec for over 100 years. The British governor was installed at Québec City along with a regiment of British troops, while the fur trade continued to flourish and the agricultural economy went unchanged. Most French Canadians were peasant farmers. After the Conquest, their life on the land remained much the same. Their social and economic lives revolved around the parish church and a landholding system centred on the seigneuries. Life in the towns and cities, however, changed radically due to a massive influx of British immigrants, the powerful

political position of English Canadians, and their control of the commercial and industrial sectors of urban places. By 1851, French Canadians accounted for only about half of the population of Montréal. From 1851 to 1951, enormous demographic changes took place. The saving grace for the *Canadiens* was their high fertility rate, which was known as the 'revenge of the cradles' (*revanche des berceaux*). Even though this rate began to slow, for 100 years Québec's birth rate remained higher than the birth rate in the rest of Canada. Thus, Québec's slower decline in fertility helped to compensate for the province's out-migration and for the growing English-speaking population.

Land hunger forced many French Canadians to migrate. By the middle of the nineteenth century, many had left the St Lawrence Lowlands due to a land shortage. Birth rates were so high in this rural society that there was not enough land left for the children of farm families. French Canadians migrated in three directions: to the Appalachian Uplands, where they either purchased farms from English-speaking farmers or found jobs in textile mills; to the Canadian Shield, where they tried to exist on extremely marginal agricultural land; or to New England's industrial towns, where most were employed in textile factories. By the early twentieth century large numbers of French Canadians, perhaps as many as one million, had left Québec for the United States, while only a small number settled in the Canadian West that was calling out for homesteaders. Through all of these economic and political changes, the vast majority of French Canadians maintained their language and Catholic religion. They turned to the Roman Catholic Church for both spiritual and political leadership. The Church, in turn, encouraged the people to stay on the land, far from the secularizing influences in towns and cities, where the English Protestants lived.

By the 1830s, political unrest was growing in both Upper and Lower Canada. Each colonial government was headed by a governor who was appointed in England and had absolute power. The governor administered the colony along with leading members of the community. This cozy arrangement not only concentrated power in the hands of a few but also led to blatant abuse by powerful elites. In Upper Canada, the political elite was known as the Family Compact, while in Lower Canada it was the Château Clique. In 1837, rebellions took place in each colony (Vignette 6.4). Both rebellions, which called for political reforms and the curbing of the political power of the elites, were crushed by the British Army.

Britain, however, was determined to remedy the political situation in its two colonies. In an attempt to identify the main sources of discontent in Upper and Lower Canada, the British government sent Lord Durham, a politician, diplomat, and colonial administrator, to

North America. Durham recognized that the political solution lay in an elected government, where power was dispersed among elected representatives rather than concentrated among an appointed elite. Durham also observed the French/English faultline, which he described as 'two nations warring in the bosom of a single state'. In Durham's report he recommended responsible (elected) government and the union of English-speaking people in Upper Canada with the French-speaking settlers of Lower Canada.[1] Lord Durham believed that

Roman Catholic churches are found in most communities across Québec, signifying the dominant role that the Roman Catholic Church once played in the social life and political affairs of the province. By the time of the Quiet Revolution, however, the Church's influence in Québec affairs had greatly diminished. (Search4Stock Inc.)

assimilation of the French was desirable and possible, claiming that the French Canadians were 'a people with no literature and no history' (Mills, 1988: 637). He recommended that English be the sole language of the new Province of Canada, and that a massive immigration of English-speaking settlers be launched in order to create an English majority in Lower Canada. In response to Durham's recommendations, the Act of Union was passed by the British Parliament in 1841, uniting the two colonies into the Province of Canada and thus creating a single elected assembly.

Lower Canada was now known as Canada East, and in spite of the political changes and the flood of immigrants from the British Isles, the French Canadians were not assimilated. Under the new form of British administration, the task of maintaining their French culture and language was not easy, but it was achieved because of several factors, the most important being their strong desire to remain Catholic and French-speaking. A second factor was the institutional support provided by the Roman Catholic Church. By providing spiritual guidance and schooling in French, the clergy played an essential role in cultural preservation. Geography and demography were other factors essential to the survival of French Canada in the nineteenth century. The overwhelming number and concentration of French-speaking people in Canada East provided a critical mass necessary for cultural survival. A high birth rate and high rate of natural increase for the *Canadiens* ensured an expanding population of French Catholics. Other factors were the rural nature of the French-speaking population, which isolated them from English-speaking residents of the major cities, and the emergence of a French-Canadian intellectual elite whose writings preserved the history and literature of French Canada. One of the most popular novels in Canada East was *Jean Rivard*. Written by Antoine Gérin-Lajoie, *Jean Rivard* is the story of a young French Canadian who is advised by

Vignette 6.4   The Rebellions of 1837

In Lower Canada, Louis-Joseph Papineau, a lawyer, seigneur, and politician, was the leader of the French-speaking majority in the Assembly of Lower Canada. In 1834, Papineau issued a list of grievances known as 'The Ninety-Two Resolutions'. At this time, the economy was depressed and tensions between the French-Canadian majority and the British minority were growing. Papineau sought to shift political power from the British authorities to the elected Assembly of Lower Canada. He planned to use his majority in the Assembly to pass legislation, including tax bills. The British government rejected 'The Ninety-Two Resolutions'—it was just a matter of time before an armed uprising broke out. When it did, the British reacted with force. Even with the strong support of rural areas, Papineau and his Patriotes were soundly defeated. Nearly 300 rebels were killed in six battles. Papineau fled to the United States. The British then hanged 12 captured rebels and exiled another 58 to Australia. A rebellion also took place in Upper Canada at the same time and it, too, was crushed. The British government sought to remedy the unrest in both of its colonies. It began this process with a fact-finding mission headed by Lord Durham. The result and Britain's solution was the Act of Union in 1841.

his *curé* on the advantages of becoming a farmer rather than a lawyer. The novel promotes the virtue of living in a harsh and remote land, which is superior (in God's eyes) to living in an urban centre with its many worldly temptations.

Confederation, 1867–present

Confederation, achieved in 1867, sought to unite two cultures—English and French—within a British parliamentary system. For Québec, Confederation provided a political framework offering three benefits: an economic union with Ontario, Nova Scotia, and New Brunswick; a political environment where Roman Catholicism and, to a lesser degree, the French language were guaranteed protection by Ottawa; and provincial control over education and language. George-Étienne Cartier, one of the Fathers of Confederation and a French-Canadian leader, viewed these provin-

cial powers as a way for Québec to shape its own destiny within Confederation. Cartier may have identified a fourth benefit—since Québec and Ontario often had mutual economic interests, they could, by working together, influence federal policies and thereby shape the future of Canada.

Confederation also led to the expansion of the geographic size of Québec (Figures 3.4 and 3.6). Since Confederation, Québec's geographic size has increased greatly. It is now 1.5 million km². As Canada acquired more territory from the British government, Ottawa assigned to Québec parts of Rupert's Land lying north of the St Lawrence drainage basin. In 1898, the Québec government received the first block of Rupert's Land. The second was obtained in 1912.[2] Some of this land, however, was claimed by another British colony, Newfoundland. In its argument, Newfoundland demanded all of Québec's territory that drained into the Atlantic Ocean. Though the

Table 6.2	Timeline: Historical Milestones in the British Colony
Year	**Event**
1763	The Treaty of Paris awards New France to Great Britain.
1774	The British Parliament passed the Québec Act, which recognized that Québec, as a British colony, had special rights, including use of the French language, the Catholic religion, and French civil law.
1791	The British Parliament approved the Constitutional Act that created two colonies in British North America called Upper and Lower Canada
1837–8	French-Canadian patriots revolt against the British regime and are defeated by British troops. In the following year, Lord Durham submitted his report to the British Parliament on resolving the unrest in the colonies of Upper and Lower Canada. Among his many recommendations, Durham advocated the creation of an English majority in Lower Canada by massive immigration from the British Isles and by assimilating French Canadians in areas where they were the minority.
1848	Based on the Durham report, the British parliament passed The Act of Union that reunited Upper and Lower Canada into a single colony and made English the official language of the newly formed province of Canada.

Imperial Privy Council of Britain awarded this land, known as Labrador, to Newfoundland in 1927 (Figure 3.7), to this day the Québec government does not recognize the decision.

Prior to World War II, Québec continued to project an image of a rural, inward-looking, Church-dominated society. In 1960, the Quiet Revolution unleashed the force of change that drove Québec into a modern industrial state. As with other social revolutions, the origin of changes began long before 1960. Yet, the election of the Liberal government of Jean Lesage marked a dramatic transformation in government, French-Canadian society, and the place

of French within Québec. His government initiated major political innovations that accelerated the process of social and economic change within Québec society. By doing so, the provincial government replaced the Catholic Church as the leader and protector of French culture and language in Québec. The Quiet Revolution instilled a sense of pride and accomplishment among Quebecers. (See Chapter 3 for more on the Quiet Revolution.) The main reforms of the Lesage government were hinged on state intervention in the Québec economy through Crown corporations, and on the expansion of a French-speak-

Table 6.3	Timeline: Historical Milestones in Confederation
Year	**Event**
1867	The Dominion of Canada was formed, the new state consisting of Québec, Ontario, New Brunswick, and Nova Scotia.
1898	Ottawa extended Québec's northern boundary to the Eastman River, thus expanding Québec's territory well beyond its core area of the St Lawrence Lowlands into the Cree lands of James Bay in the Canadian Shield.
1912	Ottawa added the Territory of Ungava to Québec, thus extending Québec to Inuit lands of Nunavik. With the addition of these two northern lands in 1898 and 1912, Québec's territory more than doubled.
1927	In settling a dispute between Canada and Newfoundland, Britain rejected Canada's claim that the boundary should be placed just inland from the shore. Instead, Britain declared that the boundary between its two colonies followed the Hudson Bay and Atlantic Ocean watersheds. Québec does not recognize this boundary.
1960	The Quiet Revolution marked the beginning of major economic and social changes, a new period of confrontation with Ottawa over provincial rights, and the re-emergence of the dream of an independent Québec.
1976	The Parti Québécois formed the government of Québec with the goal of separating from Canada but maintaining close economic relations.
1980	The first referendum seeking a new relationship with Canada was defeated with 60 per cent of Quebecers voting against the idea of sovereignty-association with Canada.
1995	A second referendum, calling for the separation of Québec from Canada, was narrowly rejected with just over 50 per cent voting 'no'.

ing provincial civil service. The government's principal achievements were:

- the nationalization of private electrical companies under Hydro-Québec;
- the modernization and secularization of the education system, making it accessible to all;
- the investment of Québec Pension Plan funds in Québec firms, thereby stimulating the francophone business sector;
- and the establishment of Maisons du Québec (quasi-embassies) in Paris, London, and New York, thus signalling to Ottawa that the Québec government wanted to represent Québec interests to the rest of the world.

With these accomplishments behind them, Quebecers felt confident about their future. Lesage's slogan, '*Maîtres chez nous*' (Masters in our own house), became a reality. For federalists in Québec, these achievements proved that a strong Québec could function within Canada, but for separatists they were not enough. The rise of separatism in Québec signalled that some Québécois felt only an independent Québec could adequately represent French-Canadian interests. For them the slogan became 'Le Québec aux Québécois' (Québec for the Québécois). After two referendums and the election of a Liberal government in 2003, Quebecers have, for the time being, turned away from pursuing political separation and are focusing more on economic and social concerns. Then, too, Québec nationalism, which is at the core of the separatist movement, has shifted somewhat from the goals and values of the 'old stock' francophones and has become more inclusive, i.e., embracing all French-speaking Quebecers, including immigrants (allophones) and even bilingual anglophones.

Québec Today

Québec is a modern industrial society effectively operating within the North American and global economies. Immigration has made its society more diversified and has created a substantial allophone community. Québec has a varied resource base, including a large number of untapped hydroelectric sites, a solid manufacturing industry sparked by growth in the new economy (pharmaceuticals, aviation, and the multimedia), and a growing tourist industry. However, Québec faces two challenges. First, Québec's share of Canada's economy and population is slipping. If the current trends continue, both Western Canada and British Columbia will surpass Québec before the midpoint of the twenty-first century. The political implications are considerable. Even though Québec is guaranteed 75 seats in the House of Commons, Québec's political clout will diminish as the total number of seats increases to accommodate population increases in other geographic regions.[3] Second, the creation of a North American market has challenged and will continue to challenge the province, forcing industries to adapt to a continental market.

As a modern industrial society, Québec exhibits the customary economic and social characteristics that other such societies do. For instance, most Quebecers, like their counterparts in the rest of Canada, live in urban settings. Québec also has an extremely low rate of natural increase. Attracting immigrants to the province ensures a positive rate of population growth, but some worry that immigrants could change the linguistic balance between French- and English-speaking Quebecers. In fact, before the language laws were introduced, almost all non-French-speaking immigrants chose to speak English and send their children to English schools. The language laws now require immigrants to send their children to French schools, thereby ensuring the growth of the French-speaking population.[4]

Today, language continues to shape Québec's economy, politics, and its relationship with Ottawa and the other provinces. French is the issue around which most other concerns are framed—in other words, most political, social, and economic issues in Québec are seen from the perspective of language and culture. The French/English faultline is therefore a crucial component of Québec and must be well understood if it is to be adequately addressed.

The French/English Faultline in Québec

The two main components of the French/English faultline in Québec are: relations between the French-speaking Québécois and English-speaking Quebecers, and the state of the Canadian federation and Québec's place in it. Each of these components interacts with and affects the other. As Québécois society continues to evolve within a Canadian political context, the question of Québec's place within the country remains a contentious political issue that deeply divides Québec society along linguistic lines.

Québécois tend to favour a new arrangement with the rest of Canada, while the English-speaking minority in Québec wish to retain the status quo. The division, however, is more complex than it appears. Politically, there are two camps—those who support federalism and those who back separatism. Generally, federalists include both some Québécois and the English-speaking minority. They would like to remain in Canada, but many would like more political powers for their province. Sep-

aratists, which include many Québécois and a small number of the English-speaking minority, seek a new political arrangement with Canada. Like the federalists, they do not hold a single political view. Within separatism, moderates seek a partnership with Canada or, at least, some form of economic union, while hard-liners insist on independence. This ambivalence stems from the dual attachment to both Canada and Québec discussed previously.

Though linguistic tensions date back to the early days of Upper and Lower Canada, linguistic political divisions within Québec have really solidified in the past few decades. A number of events contributed to this division. After the Quiet Revolution, attitudes between French- and English-speaking Quebecers began to harden. The Quiet Revolution had addressed the economic-based linguistic division of a French-speaking majority being led by an English-speaking business minority. A power shift occurred in the province—more and more French-speaking Québécois were running businesses and attaining positions of power. The English-speaking minority no longer had the economic advantage and also had to adapt to new nationalist policies like the language laws.

French/English relations were then made more complex by events outside of Québec, which included a growing sense of nationalism among English-speaking Canadians and, like the Québécois, a new sense of pride. One symbol of that Canadian nationalism was the **patriation** of the Canadian Constitution from Britain, which gave Canada direct control over government decision-making, including the freedom to amend the Constitution.

In 1982, the patriation of the Canadian Constitution was the culmination of a frenzied period of constitutional debate, set in motion by the defeat of the Québec referendum on sovereignty-association in 1980. Patriation of the Constitution achieved a number of things. By giving Ottawa control over constitutional amendments, the Canadian Parliament eliminated the last vestiges of Canada's colonial status. The Canadian Constitution also provided a Canadian formula for future amendments to the Constitution, extended English and French minority-language rights in education, and gave Canada a constitutional Charter of Rights and Freedoms. However, the action of the federal government, and the other provinces, in patriating the BNA Act (renamed the Constitution Act, 1867) and creating the Constitution Act, 1982 (which includes the Charter) had one serious flaw. It was signed over the bitter objections of the Québec government. Since the Trudeau Liberal government had campaigned for the rejection of independence in the 1980 Québec referendum by promising renewed federalism to the Québécois, many people in Québec felt betrayed. For separatists, this was the final straw. In their view, Ottawa and the other provinces should not have acted without the Québec government's consent.

The combination of events surrounding patriation recharged the separatist movement in Québec. It also led Québec federalists to question the relationship between Ottawa and Québec. A long-time federalist, Léon Dion, concluded that 'English Canada will only yield—and even this is not assured—if there is a knife at its throat' (Simpson, 1993: 312). Dion was stating an old idea in a new context, namely, that Ottawa would not negotiate a new relationship for Québec within Canada unless Québec was prepared to break up the Canadian federation. In Dion's metaphor, the threat of separation was the knife and the method was a referendum. The goal was a new

partnership with Canada. Other Quebecers, like Jacques Parizeau, had a different strategy, one that would lead to an independent Québec.

Because patriation had caused such controversy in Québec, the succeeding Conservative government of Brian Mulroney made efforts to find a compromise that would have the Québec government sign the Canadian Constitution. Unfortunately, the two ensuing rounds of negotiations and their respective agreements—the Meech Lake Accord (1987) and the Charlottetown Accord (1992)—ended in failure. However, these attempts and the negotiations, parliamentary committees, and other measures pursued during this period demonstrated three realities of Confederation:

- the difficulty in reaching a political compromise when conflicting interests exist between French- and English-speaking Canadians ;
- Ottawa's commitment in seeking a compromise;
- the support of the majority of Canadians, both English- and French-speaking, for efforts to try to find such an elusive compromise.

Failure to find a compromise further exacerbated the political 'insult' caused by patriation, leading to a resurgence of separatist feelings in Québec.[5] By 1995, a Parti Québécois government, led by Jacques Parizeau, was again appealing to its electorate through a referendum to give it a mandate to pursue sovereignty for Québec.

The 1995 Referendum

The 1995 referendum was a political watershed for Canada. The question was: 'Do you agree that Quebec should become sovereign, after having made a formal offer to Canada for a new Economic and Political Partnership within the scope of the Bill respecting the future of Québec and of the agreement signed on June 12, 1995?' While one can debate the meaning of the referendum question, all would agree that Quebecers came very close to separating from Canada—the 'Non' side won by a very slim majority. If there had been a 'Oui' majority, Premier Parizeau's strategy would have been to immediately declare Québec an independent nation.

The deep language divide within Québec society is shown clearly in the results of the referendum. In 1995, the political map of Québec showed strong support for the 'Oui' side in rural Québec where the francophone population dominates. In fact, support for independence was strongest among Québec

■ **Figure 6.2 Referendum fatigue.** Following the 1995 referendum, Quebecers had grown weary of the political wrangling over Québec's place in Confederation. Economic issues had become more important to the average citizen. One sign of this shift is seen in the drop in the number of seats held by the Bloc Québécois (44 in the 1997 federal election and 38 in the 2000 election). Another is the election of a Liberal government in Québec in 2003. However, while the issue of Québec's separation from Canada remains on the political back burner, this does not mean that the sovereignty movement is dead. (*The Globe and Mail*, 31 Oct. 1995, A24. Reprinted with permission from the Globe and Mail)

francophones, with estimates placing it as high as 60 per cent, and weakest among anglophones and allophones. Only four areas of Québec—Montréal, the Ottawa Valley, Estrie, and northern Québec—supported the federalist side, but these were the areas with the majority of the voters. In the first three areas, there is a high proportion of anglophones and allophones, and in northern Québec, Cree and Inuit peoples make up the vast majority. At that time, the Cree had been particularly outspoken on the issue of sovereignty, arguing that they have the right to determine their political future (just like Québec), whether it means remaining part of Canada or joining an independent Québec.

The issue of separation is a complex political matter not defined in the Constitution of Canada. However, on 20 August 1998, the Supreme Court of Canada ruled that under both Canadian and international law, Québec (or any other province) cannot declare its independence unilaterally, thus bringing to an end the threat of separation by decree. Nevertheless, the Supreme Court accepted the concept that separation was possible through negotiation after a 'clear majority' voted for separation.

The French/English faultline is a prominent feature of Québec politics and society, a feature that will continue to dominate linguistic relations within the province, as well as federal-provincial relations in the country. In short, constitutional events since 1980, combined with already existing linguistic tensions in Québec, led to a national crisis that culminated in the 1995 referendum. Since then, the separatist movement has lost its momentum. Quebecers have grown weary of the political wrangling over Québec's place in Confederation (Figure 6.2). For the average Quebecer, economic, social, and health issues have become more important. Two recent elections support this contention: the 2000 federal election saw the popular vote of the Bloc Québécois slip to 40 per cent, and three years later a provincial Liberal government replaced the Parti Québécois, thus putting the issue of Québec's separation from Canada on the political back burner. According to a poll by the Centre for Research and Information on Canada (2003), the percentage of Quebecers who want independence has dropped from 39 per cent in 1999 to 21 per cent in 2002.

Québec's Economy

Like Ontario, Québec has a strong and diversified economy that is heavily dependent on foreign trade. Access to the American market has had a positive impact on Québec's economy. Whether primary products, such as iron ore, semi-processed materials, such as magnesium, or manufactured products, such as aircraft, most production is exported to the US or world markets. Prior to the Free Trade Agreement in 1989, approximately half of Québec's output went to other parts of the country and half was exported. By 2002, nearly two-thirds of its output is exported, especially to the United States. The intervention of both the provincial and federal governments to support the revitalization of Québec's manufacturing sector and to help it compete in the North American and world markets has paid dividends. This revitalization strategy promoted key industrial sectors (including the aerospace, pharmaceutical, and metal-processing industries), and encouraged industrial development in southern Québec by offering firms low-cost electrical power. Yet, trade is a two-edged sword and governments can no longer shelter their economies from foreign producers. Imports from developing countries with

low wages have undercut Canadian producers and sometimes pushed Canadian firms into bankruptcy. As Canada lowers its tariffs on imports from developing countries, Québec's apparel and textile firms face stiff competition. By 2002, the global economy saw low-priced magnesium imports from China hurting the magnesium refining industry in Québec.

The spatial aspects of Québec's economy are similar to Ontario, and, like Ontario, Québec can be divided into two economic areas: an agricultural and manufacturing core exists in southern Québec while a resource-based periphery is found in northern Québec. Québec's core is associated with the fertile Great Lakes–St Lawrence Lowlands and the hinterland with the Canadian Shield (Figure 2.1). Consequently, Québec has an array of resources and access to the US and global markets similar to that found in Ontario. Yet, there are differences. The climatic conditions in Ontario's part of the Great Lakes–St Lawrence Lowlands provide for a longer growing season, thus giving Ontario an advantage. On the other hand, Québec's part of the Canadian Shield contains much better physical conditions for the production of hydroelectricity. Then, too, the close proximity of the Canadian Shield to Montréal and Québec City provides excellent recreation sites—lakes for summer activities and ski hills for winter ones, thus making the area one of North America's most popular recreational tourist areas.

Québec and Ontario both contain parts of the Canadian manufacturing belt, which extends from Windsor to Québec City, and the St Lawrence Seaway connects these two industrial cores (Vignette 6.5). Proximity to Canada's greatest trading partner, the United States, is extremely important for manufacturing. Windsor and other centres in southern Ontario have had the advantage of proximity to the American automobile-manufacturing centre of Detroit. Montréal, on the other hand, is farther from major American manufacturing cities. Nonetheless, many firms in Canada's manufacturing belt supply products to other manufacturing companies in this belt, indicating a high degree of economic integration. An example of such integration is the auto-parts industry. Over 20 firms in the Montréal area produce automobile parts and then ship them to automobile-assembly plants in Ontario and the United States. General Motors of Canada did have an assembly plant in Sainte-Thérèse that made the Pontiac Firebird and Chevrolet Camaro sports cars, but this assembly plant was closed in 2002.

Over the years, Québec-based firms have specialized in certain industrial sectors, including apparel and textiles, high-tech industries, metal refining, printing, and transport equipment. The apparel and textile industries formed the traditional heart of manufacturing in Québec, but, since the Free Trade Agreement, these industries have faced stiff competition from abroad because of the reduction in tariffs on apparel and textiles, which has allowed imports from low-wage countries to grab a major share of the North American market. The final blow will occur in 2005, when the World Trade Organization agreement calls for quota-free imports from China, the world's biggest and least expensive producer of quality garments.

High-tech industries, which require a highly skilled labour force, have fared much better. In fact, many economic gains in Québec have come from high-tech firms, which account for 38 per cent of Canada's high-tech exports. Because Montréal has a critical mass of high-tech companies, several universities, and strong provincial support, it is the second most important centre for the new economy in Canada. Leading components are found in aerospace, biotechnology, fibre optics, and computers (both hardware and software). Québec provides nearly half of Canada's information technologies, aerospace, and pharmaceutical exports as well as 40 per cent of its biotechnology exports. Its aerospace firms, led by Bombardier and CAE Inc., have been particularly successful in recent years. For example, CAE is a global leader in advanced simulation and control equipment for aircraft and ships. The company employs about 4,000 highly trained workers in the Montréal area and another 2,000 in other

Vignette 6.5   The St Lawrence Seaway

The St Lawrence Seaway is an international waterway that, through a series of canals, dams, and locks, allows ocean-going vessels to reach the industrial heartland of North America. The shipping season extends over 250 days from mid-April to mid-December. The seaway is the preferred transportation system for bulky commodities. Iron ore from northern Québec and Labrador and wheat from the Canadian Prairies are the leading commodities shipped on the seaway. Navigation to Montréal takes place along the St Lawrence River, while ports further inland, such as Toronto, Chicago, and Thunder Bay, fall within the St Lawrence Seaway. Between Montréal and Lake Ontario, seven locks allow large vessels to navigate these waters, while the Welland Canal allows access between Lake Ontario and Lake Erie and the Sault Ste Marie locks provide a water link between Lake Huron and Lake Superior.

parts of the world. Approximately one-third are engaged in research and engineering, designing and testing new products such as a robot that strips paint from aircraft bodies. Its main product, however, is a flight simulator manufactured at its plant in Saint-Laurent near Montréal. Unlike most manufacturing firms, most of CAE's sales do not go to the United States but to customers in Europe, the Middle East, and Asia.

Industrial Structure

As a core region of Canada, Québec's economy has many similarities with that of Ontario

(Table 6.4). First, both regions saw their workforces increase in size. From 1995 to 2002, Québec's workforce increased by 12 per cent while Ontario's workforce increased by 16 per cent. Second, the division of Québec's and Ontario's industrial labour forces into the three principal sectors (primary, secondary, and tertiary) indicates that less than 3 per cent of workers in each province are engaged in the primary sector, around one-quarter of workers are in the secondary or processing industries, and approximately three-quarters are in tertiary industries (Table 6.5). Third, the secondary sectors of Québec and Ontario increased their share of the total labour force

Table 6.4	Employment by Industrial Sector in Québec, 2002			
Industrial Sector	Québec Workers	Québec Workers (per cent)	Ontario Workers (per cent)	Differences (Percentage points)
Primary	**102.8**	**2.8**	**1.8**	**1.0**
Agriculture	62.3	1.7	1.3	(0.4)
Secondary	**843.2**	**23.6**	**25.1**	**(1.6)**
Manufacturing	640.7	18.2	18.5	(0.3)
Tertiary	**2,646.5**	**73.4**	**73.1**	**0.3**
Trade	593.6	16.4	15.9	0.5
Transport	149.8	4.2	4.7	(0.5)
Finance	191.9	5.3	6.6	(1.3)
Professional service	214.2	6.0	7.2	(1.2)
Management	121.4	3.3	4.2	(0.9)
Education	237.2	6.5	6.2	0.3
Health Care	404.1	11.1	9.3	1.7
Culture	159.7	4.5	4.8	(0.3)
Accommodation	206.6	5.8	6.0	(0.2)
Other services	160.8	4.5	4.2	0.3
Public administration	207.4	5.8	4.8	1.0
Total	3,592.7	100.0	100.0	

Source: Statistics Canada (2003).

Table 6.5	Comparison of Ontario and Québec Industrial Structures, 1995 to 2002					
Sector	Ontario % 1995	Ontario % 2002	Percentage Difference	Québec % 1995	Québec % 2002	Percentage Difference
Primary	3.0	1.8	(1.2)	3.5	2.8	(0.7)
Secondary	23.6	25.1	1.5	23.0	23.5	0.5
Tertiary	73.4	73.1	(0.3)	73.5	73.7	0.2
Totals	100.0	100.0		100.0	100.0	
Workers (000)	5,231	6,068		3,204	3,593	

Sources: Adapted from Statistics Canada (1996b, 2003).

from 1995 to 2002 at the expense of their primary sectors.

But what aspect of the Québec and Ontario industrial structures distinguishes them from those of the other geographic regions? The answer lies in their secondary sectors. As core regions, both Québec and Ontario have strong secondary industrial sectors dominated by manufacturing. For instance, in 2002, Québec's secondary sector employed nearly one-quarter of its labour force compared to just over 25 per cent in Ontario (Table 6.5). In comparison, the four other geographic regions, the so-called hinterland regions, had much smaller secondary sectors that employ 17 per cent or less of the labour force. In 2002, Western Canada had 17 per cent of its labour force in the secondary sector, Atlantic Canada had 16.9 per cent, British Columbia had 16.7 per cent, and the Territorial North had approximately 2 per cent. At the same time, the tertiary sectors of Québec and Ontario fall below 75 per cent while the tertiary sectors of the other geographic regions, with the exception of Western Canada, exceed 75 per cent. Within the tertiary sector, by far the biggest rate of increase from 1999 to 2002 occurred in the trade subsector. In Ontario, trade increased

from 14.9 per cent of the total workforce to 15.7 per cent while Québec's trade workforce grew from 15.4 per cent to 16.5 per cent.

Southern Québec

Southern Québec is the economic, social, and political core of Québec, while northern Québec is a sparsely populated resource hinterland. The relationship between the two regions is a provincial version of the core/periphery model and overlaying that model is the francophone presence in southern Québec and an Aboriginal one in northern Québec.

The more physically favoured lands are found in southern Québec, especially in the valley of the St Lawrence River. Southern Québec contains two physiographic regions: the Appalachian Uplands and the St Lawrence Lowlands. Bordered on the north by the Canadian Shield and on the south by the United States, southern Québec is only a small part of the territory of Québec, but it contains over 90 per cent of Québec's population and agricultural lands, and is the industrial heartland of the province. The human and physical characteristics of each physiographic region in southern Québec are presented below.

Appalachian Uplands

The Appalachian Uplands region lies west of the Atlantic Ocean and north of the United States. Its northern edge faces the St Lawrence Lowlands. Because of this geography, the Appalachian Uplands region was settled at different times and in different ways. First settlements took place along the Gaspé coast. In the sixteenth century the waters off the Gaspé Peninsula attracted fishers from Spain, Portugal, England, and France. Even today, fishing in the Atlantic Ocean provides a way of life, though many supplement their income with farming and wage employment in the small coastal settlements. The main settlements are scattered along the Gaspé coast and the south shore of the lower St Lawrence River, leaving the rugged interior with few settlements. The largest urban centres are along the south shore. Rimouski, with a population of nearly 50,000 in 2001, is by far the largest city in the region. Rivière-du-Loup and Matane are medium-size towns with populations roughly half the size of Rimouski. Towns are much smaller along the Gaspé coast, as the area has a very limited resource base. Percé, now an important tourist town, is the largest of the small centres along the Gaspé coast with a population of about 4,000. This French-speaking area has place names like New Richmond, New Carlisle, and Chandler, the legacy of early English settlers, some of whom had relocated along the Gaspé coast after the American Revolution. Over the last half-century, the lack of jobs caused many to migrate out of this area, which seriously reduced the size of English-speaking communities. In 2001, English-speaking residents constituted less than 5 per cent of the region's population.

Estrie (the Eastern Townships) is a pocket of communities in the rolling land of the Appalachian Uplands located east of Montréal. Estrie was settled after the American Revolution by British Loyalists. The British organized the surveyed land into rectangular townships rather than the French long lots found in the St Lawrence Valley. Much of the land was ill-suited for cultivation. Within several generations, the more marginal lands were abandoned when English-speaking owners left to look for jobs in Montréal and Boston or to try homesteading on the American frontier. Because of a land shortage in the St Lawrence Valley, French Canadians began to move into the Eastern Townships. French-Canadian migrants, often sons of farmers living in the St Lawrence Lowlands, either bought or took over abandoned farms from the original English-speaking settlers.

The physical geography of Estrie is much more conducive to economic development and agricultural settlement than are the Gaspé coast and south shore of the St Lawrence. From the 1870s to the 1970s, mining at Thetford Mines and Asbestos was a key sector of the regional economy. Since the 1970s, asbestos mining has fallen on hard times because the use of this mineral has proven hazardous to human life. Still, mining of this deposit continues. Agriculture and forestry have also proven to be durable in this area. Dairy farming takes place in the broad valleys, and logging in the forested uplands. Overall, Estrie is the most prosperous area of the Appalachian Uplands. Sherbrooke exemplifies the relative well-being of the region. Sherbrooke has grown over the years and is the largest urban centre in the Appalachian Uplands. With a population of 154,000 in 2001, Sherbrooke has become an important regional centre (Table 6.6). From 1996 to 2001, its population increased by 2.8 per cent, making it the third-fastest-growing city in

Québec after Gatineau (4.4 per cent) and Montréal (3 per cent). This demographic increase over the years is reflected in its economic growth. Proximity to Montréal has often worked in Sherbrooke's favour, allowing it to engage in the textile industry in the nineteenth century and now in the high-technology industries.

Sherbrooke also demonstrates the displacement of English-speaking Quebecers by French-speaking Quebecers. It was founded in 1802 by British Loyalists from Vermont, but French Canadians came en masse in the 1840s to work in the textile factories. By 1871, the population was half French and half English. Today, nearly 95 per cent of the residents of Sherbrooke are francophones.

St Lawrence Lowlands

Most of Québec's agricultural and industrial production is in the St Lawrence Lowlands. In fact, this area functions as the province's core. It contains Québec's largest market and is close to transportation networks and the St Lawrence River, which facilitate trade with foreign countries.

The warm to hot summer climate coupled with abundant rainfall makes the St Lawrence Lowlands the most favoured region in Québec for agricultural activities. Livestock farms specializing in cattle, hogs, or sheep are common, while some farmers concentrate on dairy, poultry, and egg production. Livestock farmers grow forage crops for winter feed. During the summer, the cattle graze on pastures. Specialized crops, particularly vegetables and fruit, are also popular and are sold mostly in the major urban centres of the province.

Though farmers in this region engage in a variety of agricultural activities, dairy and vegetable farming predominate. With 37 per cent of Canada's one million milk cows, Québec leads in the production of dairy products.

In 2004, approximately 40 per cent of Canada's milk cows were located in Québec. While most of Québec's industrial milk and cream is consumed within the province, some is exported to other provinces, especially Ontario. (Ivy Images)

Under the Canadian industrial milk marketing board, dairy farmers in Québec supply nearly 40 per cent of the Canadian market. Moreover, the processing of agricultural products provides an added benefit for the Québec economy. For instance, large butter and cheese firms rely on milk products from the dairy farms. Often the processed food products are designed for the provincial market. For example, because of the market for unpasteurized cheese products in Québec, a few cheese firms began producing *fromage au lait cru*, cheese made with unpasteurized milk. These popular cheese products compete favourably with European imports, including popular brands from France.

Dairy farmers have fared relatively well under Québec's fluid milk marketing board and a national marketing system that allocates producers a share of the Canadian market for milk. In 2002, cash sales from milk and cream amounted to $1.5 billion in Québec. Under NAFTA and WTO rules, however, Canada is under pressure to dismantle its marketing boards. Without a regulated dairy industry, dairy farmers in Québec would face stiff competition from American dairy imports. In 2002, the World Trade Organization ruled that Canadian dairy exports are illegally subsidized because the national marketing system's price for Canadian milk production is above market price. While 95 per cent of Canada's total milk production is consumed domestically, the value of Canadian dairy exports amounts to about $250 million. According to the WTO ruling, Canada will have to reduce its exports by around half of the current figure of $250 million.

High-technology industries are playing a central role in the new Québec economy. Bombardier, with its RJ Regional jets, is one of the leaders in this field. Yet, the highly competitive and volatile marketplace makes Bombardier and other aircraft companies vulnerable. One sign of this vulnerability is the heavy dependency on financial aid from national governments. Bombardier has received substantial financial aid from the federal government of Canada and provincial government of Québec. (CP/Ryan Remoirz)

Manufacturing is concentrated in the Montréal area but extends eastward to Estrie and northeastward to Québec City. As part of the Canadian manufacturing axis, this industrial zone serves as the engine driving both the provincial and the national economies. The manufacturing sector is very sensitive to global trade. On the one hand, information technologies, aerospace production, pharmaceutical operations, and biotechnology firms benefit from the liberalization of world trade and NAFTA. Since 1989, these firms have expanded, although in the early years of the twenty-first century some businesses, such as the aerospace industry, are contracting due to a reduction in worldwide demand for their products. These high-technology industries employ highly skilled and well-paid workers. Most of Québec's manufacturing sector is in Montréal, where high-tech companies are now the leading edge in manufacturing. These companies were not only responsible for Montréal's impressive economic recovery in the late 1990s but are also transforming the city into one of the leading high-tech centres in North America.[6] This shift from traditional to high-tech manufacturing has two consequences for Montréal's labour force. First, the demand for highly skilled workers is increasing, creating labour shortages and a search for skilled immigrants. Second, the demand for low-skilled workers is decreasing, thereby contributing to the relatively high unemployment rate in the region and province.

The traditional sector, composed of textile, knitting, leather, and clothing firms, was once the mainstay of manufacturing but it has been declining for some time and its future is uncertain. As a labour-intensive industry where education and a good command of English or French is not necessary, many low-income, immigrant families have members, especially female members, working in this industry. Now struggling to compete with foreign imports, the traditional manufacturing firms have sought to lower their costs of production by reducing the size of their labour force, introducing more efficient machines, and seeking niche markets in North America. While Montréal remains the focus of Canada's clothing and textile industry, its grip on market share is slipping for two reasons. First, established firms are either shifting their production from Canada to countries with low wages, or these firms are subcontracting various elements of production in these low-wage countries. Second, Ottawa is continuing to reduce tariffs on textiles, clothing, and related products from developing countries as a means to stimulate economic development in these nations.

With the automobile and aerospace industries facing a slump in demand for their products, both industries have had to reduce their production capacity. In 2002, Québec lost its only automobile-assembly plant when GM closed its plant at Sainte-Thérèse, laying off more than 1,000 workers. In an agreement with the provincial government, GM has agreed to increase its purchases of auto parts from Québec firms to lessen the economic blow caused by the assembly-plant closure and to invest in developing lightweight auto parts using aluminum and manganese materials. Last year, GM purchased $1.6 billion of auto parts from Québec firms and that figure could jump to $2.6 billion in 2003 (MacAfee, 2003: FP5).

Québec's aerospace industry has become a major player on the world scene. Assisted by generous funding from Ottawa and the Québec government, Bombardier has become Canada's leading aerospace firm. The firm accounts for around one-quarter of business jet sales in the world and employs nearly 20,000 workers. Bombardier produces passenger jets (the Challenger and the Regional Jet) and smaller business jets (Learjets). The Regional Jet, a stretched version of the Challenger, is the leading aircraft for short-haul markets in North America and Europe. The success of this firm is due largely to its ability to sell its product in the global market. Bombardier and other aerospace firms face a difficult future in the short run. While jet aircraft sales exploded in the late 1990s and peaked in 2001, sales have declined sharply. Bombardier, for instance, saw its sales plummet from 162 passenger jets in 2001 to 77 in 2002 (Silcoff, 2003: FP5), primarily the result of unexpected events beginning in September 2001 that caused the demand for business jets to suddenly plunge.

On 11 September 2001 the air travel business was dealt a near-mortal blow with the crashing of passenger jets into the World Trade Center in New York and the Pentagon outside Washington by terrorists. With the United States declaring war on terrorists, the volume of business and tourist travel dropped dramatically and public confidence in air travel has been slow to return. More recently, the outbreak and spread of SARS dealt another blow to airline companies, including Montreal-based Air Canada. Like other major airline companies, Air Canada is facing a difficult future with fewer air travellers and stiff competition from smaller airlines like West Jet. In May 2003 and again in May 2004, Air Canada just escaped from bankruptcy—and this escape may not be permanent. Air Canada is cutting costs by reducing the size of its staff and by renegotiating wage rates with its union members.

Under these circumstances, manufacturing in Montréal and the rest of the St Lawrence Lowlands is entering a challenging period of economic adjustment. In the greater Montréal region, both traditional and high-technology firms are under enormous pressure to survive in this most competitive world. In both cases, firms are compelled to reduce the size of their labour force and to require the remaining workers to become more productive as well as to accept lower wages. During this painful period of adjustment, economic growth in the eastern segment of Canada's industrial core is likely to slow down and unemployment rates are likely to rise.

Key Topic: Hydro-Québec

Since the Quiet Revolution, successive Québec governments have played an active role in shaping the province's industrial economy. This approach, initiated by the Lesage government, has pursued two goals—one economic, the other political. Its economic objective has been to stimulate economic growth through state intervention in the marketplace. Its political goal is to increase Québec's ownership and control of its economy within the francophone business community. The most direct way to achieve these two goals was to create public (Crown) corporations. The best example of the success of this industrial development strategy is Hydro-Québec. Now the prize Crown corporation in Québec, Hydro-Québec has played a fundamental role in achieving these two goals and, in so doing, has helped transform Québec's economy and build a strong French-speaking business community. Today Hydro-Québec is Canada's largest electric utility. The success of Hydro-Québec is based on the great natural advantage for producing hydroelectric power in Québec's Canadian Shield (Vignette 6.6).

Created in 1944, Hydro-Québec was a minor force in the Québec economy until the

Lesage government came to power in 1960. At that time, the Québec government announced its intention to purchase the private electricity companies in Québec and place them under the umbrella of Hydro-Québec. The Crown corporation soon extended its activities to cover the whole province by purchasing the shares of nearly all remaining privately owned electrical utilities and taking over their debts. With a virtual monopoly to generate and distribute electricity in the province, Hydro-Québec undertook the task of expanding Québec's hydroelectric capacity. In the 1960s, Hydro-Québec built dams on the Manicouagan River, creating the huge Manic-Outardes hydroelectric complex. The design and construction of this massive project were carried out by Québec firms and workers, thereby creating highly specialized engineering firms

capable of designing and building huge hydroelectric projects.

Hydro-Québec harnessed the vast water resources of the Canadian Shield and transported this electrical energy from the dams on the Manicouagan River to Québec City and Montréal. In order to transmit the complex's annual production of about 30 billion kWh over a distance of nearly 700 km, Hydro-Québec became the first utility in the world to transmit electricity at very high voltages (735 kv), which reduces the amount of electrical energy lost in the transmission process. This breakthrough technology enabled Hydro-Québec to transmit electrical power from Churchill Falls[7] and James Bay to markets in southern Québec and New England.

Following the success of the Manicouagan River complex, Hydro-Québec embarked on

Vignette 6.6   Waterpower Resources

Hydroelectric developments depend on three factors: climate, topography, and access to market. The development of hydroelectric power in Canada varies considerably from region to region. In the 1960s, most power sites were built close to Canadian markets. Since then, high-voltage transmission lines have facilitated the transmission of electricity over long distances, permitting more distant sites to be developed. Several large-scale hydroelectric projects have been developed in more remote areas in the northern parts of Québec and Manitoba. Canada's principal hydroelectric generating stations and their installed capacity in megawatts are:

- LG-2 on La Grande Rivière, Québec (5,328 MW)
- Churchill Falls on the Churchill River,

Labrador (5,225 MW)
- LG-4 on La Grande Rivière, Québec (1,650 MW)
- Gordon M. Shrum on the Peace River, BC (2,416 MW)
- Kettle Rapids on the Nelson River, Manitoba (1,255 MW).

The main advantages of hydroelectric developments are: the long life of the facilities, low operating costs, and zero air pollution. However, there are drawbacks: the initial high capital investment, the long construction period, and the socio-environmental costs. When river valleys used for hunting and fishing by Aboriginal peoples are flooded, the loss of wildlife habitat and fishing grounds has serious impacts on their economy and culture. Such losses represent a very high but hidden social cost.

an even larger project—the James Bay Project (Figure 6.3). As it turned out, the James Bay Project, which includes La Grande, Great Whale, and Nottaway river basins, was the crowning success of Hydro-Québec, but also its *bête noire*. This was the largest hydroelectric project in the world and would transmit power over long distances never attempted before. Québec could now sell massive amounts of power to the United States and, in so doing, recoup much of the construction costs. The natural environment, however, was dramatically altered by river diversions and by creating huge reservoirs. The project negatively affected the fish stocks and the prime wildlife habitat along the river valleys and lakes. The hunting and trapping lifestyle of the local Cree was disrupted by this construction project and the resulting alterations to the natural landscape. Although Hydro-Québec made efforts to minimize the environmental damage and compensated the Cree, Inuit, and Naskapi through the James Bay and Northern Québec Agreement, the Cree strongly opposed further

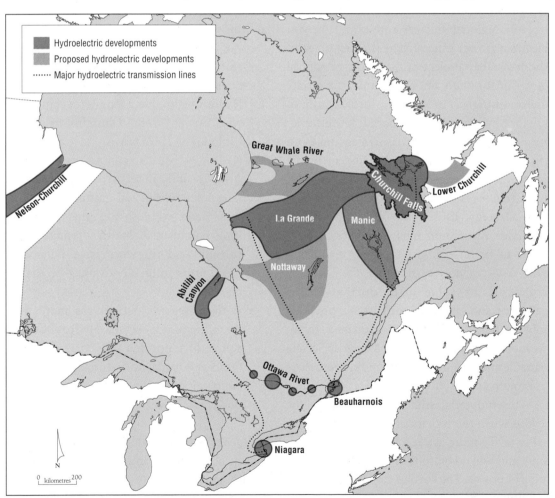

Figure 6.3 Hydroelectric power in Central Canada. Québec's dominant role in the production of hydroelectric power is due to two natural factors: (1) the heavy annual precipitation in northern Québec, and (2) the high elevation of the Canadian Shield in Québec. (Further resources: Student Web site, National Atlas section, Map 17. Web site instructions are found on p. xxi)

hydroelectric developments in northern Québec. In 2001, however, the Cree and the government of Québec signed an agreement to permit further hydroelectric development in the James Bay region (see the section on 'Hydro Power' later in this chapter for further discussion on these topics.)

Industrial Strategy and Regional Linkages

With hydroelectric power developments in Québec's Canadian Shield, an industrial strategy and regional linkages were born. The vast electrical power generated by the first phase of the James Bay Project provided the provincial government with an opportunity to attract energy-hungry industries into southern Québec by offering them special, low electricity rates. Such an industrial strategy takes on the spatial form of the core/periphery model where the hinterland supplies the energy for industrial users in the core. An early version of this strategy took place in 1957 when Reynolds Aluminum built a smelter at Baie-Comeau because Hydro-Québec provided it with low-cost power. In return, the company added to the value of production in the province and, more importantly, it provided jobs in an economically depressed area. Today, Reynolds Aluminum, known locally as Société canadienne des métaux Reynolds, employs approximately 2,500 workers.

This industrial strategy took hold in the 1960s when the provincial government decided to take a more active role in Québec's economy and when Hydro-Québec had a large surplus of power. The surplus came from Québec obtaining the rights to the electrical power produced at Churchill Falls in Labrador and from Hydro-Québec's recently completed Manic-Outardes hydroelectric complex on the

Manicouagan River, and it was used to lure industrial companies into southern Québec. With the completion of the first—La Grande—phase of the James Bay Project in 1985, the provincial government took a more aggressive stance by offering risk-sharing contracts and electric-power rebates to industrial firms requiring large amounts of power in their processing operations, such as metallurgical companies. As a result, industrial firms either expanded their operations or new firms located along the St Lawrence River.

Hydro-Québec is able to provide industrial firms with low-cost energy for four reasons:

- Northern Québec can produce vast quantities of low-cost electrical power.
- Hydro-Québec has a long-term contract to buy power from Churchill Falls in Labrador at 1969 prices.
- Hydro-Québec has control over its price structure and can set extremely low power rates for its industrial customers. In fact, these rates were so low that American firms complained that they represented a subsidy and therefore gave Canadian-based firms an unfair advantage in the US market.
- With prices for natural gas more than doubling since 2000, Hydro-Québec's competitive position in New England has improved because natural gas thermal electrical plants can no longer produce power at a lower per unit cost.

Québec's industrial strategy attracted metallurgical firms. One international firm, Norsk Hydro, took up the government's offer.[8] Norsk Hydro is involved in a large magnesium-smelting operation near Trois-Rivières and such smelting operations require huge amounts of electrical power. In 1988, Norsk Hydro signed

Vignette 6.7   Norsk Hydro Canada Inc.

In 1988 Norsk Hydro signed a 25-year contract with Hydro-Québec for electrical power at special rates. Norsk Hydro buys hydro power at rates that fluctuate with the world price of magnesium. For the first three years, the company's magnesium smelter at Bécancour was eligible for electrical power rebates of up to 50 per cent of consumption. The purpose was to help defray the start-up costs of the smelter during its first three years of operation.

The Norwegian state-owned company spent $600 million to build its magnesium smelter at Bécancour. Production began in 1990. By 1991, Norsk Hydro had captured 24 per cent of the American market. Its magnesium metal products are used by automobile parts firms. An American firm, Magnesium Corp. of America, complained to the US Commerce Department that Norsk Hydro used unfair trade practices by selling its products on the US market below the cost of production. This allegation was based on the fact that Norsk Hydro obtained its electrical power at rates below the cost of producing that power. In 1992, the US Commerce Department placed a 32.7 per cent anti-dumping duty on Norsk Hydro's magnesium exports to the United States. This duty was later withdrawn after Hydro-Québec agreed to remove its electric rebates, but a countervailing duty of 4.48 per cent remains. In spite of the duty, Norsk Hydro made plans in 1997 to double its total capacity at its Bécancour smelter from 43,000 tonnes of magnesium products to 86,000 tonnes by 2000. At that time, employment at the smelter was anticipated to increase from 360 workers to 430. Since 2001, competition from China has forced Norsk Hydro to curtail its expansion plans.

Source: Yakabuski (1997: B6). Reprinted with permission from The Globe and Mail.

a 25-year contract with Hydro-Québec to purchase electrical power at very low rates (Vignette 6.7). By the mid-1990s, the demand for magnesium die castings for automobile engines was growing at an annual rate of 15 per cent and the price of magnesium rose to over $1.30 a pound, causing Norsk Hydro to expand its smelter on the south shore of the St Lawrence River. By 2001, automobile production began to slow and magnesium prices, which had reached $1.50 a pound, began to fall. At the same time, magnesium imports from China began to undercut North American producers. As China grabbed a larger and larger share of the North American market, the magnesium refining industry in Québec was threatened. In 2003, Norsk Hydro's plant at Bécancour and the Noranda state-of-the-art magnesium plant at Danville lost their profit margin because the North American price for magnesium dropped below $1 a pound. For the Danville plant, the lower price was a disaster. In the late 1990s, Noranda assumed that world prices for magnesium would remain around $1.50 a pound. With the Québec government investing 20 per cent into the construction of Noranda's $1.3 billon magnesium plant at Danville, neither company nor government officials foresaw that the price of magnesium would fall so precipitously. In the spring of 2003, Noranda announced that it would temporarily close its plant and not reopen it until the price for magnesium reaches $1.30 a pound. Noranda may have a long wait because China can supply the North American magnesium market at prices well below $1 a pound.

Northern Québec

Northern Québec lies beyond the ecumene of the St Lawrence Valley. As a resource hinterland lying in the Canadian Shield, the main economic activities are forestry and mining. Consequently, the region is sparsely populated and has few other opportunities for development. While tourism flourishes near the major population centres, most of northern Québec is too remote for tourists. Attempts at agricultural settlement have had marginal success in the Lac Saint-Jean area, but further north the short summer and thin soils associated with the Canadian Shield prohibit commercial agriculture (Vignette 6.8). Northern Québec's economy is best suited for forestry, mining, and smelting based on low-cost hydroelectric power. The smelting of bauxite (alumina ore) is a major processing industry in the Saguenay and along the North Shore. Alcan Aluminium, the world's second-largest aluminum producer, continues to expand its production capacity in Québec. Low-cost electrical power and easy access to shipping lanes in the Atlantic Ocean provide key location factors. Three recent developments demonstrate Alcan's commitment to the region where low-cost electrical energy keeps production costs low and where ocean shipping to import the raw material, bauxite, and to export the finished product minimizes transportation costs:

- In 1998, Alcan began building its $2.2 billion aluminum smelter at Alma in the Lac Saint-Jean area. At the time, this investment represented the largest private undertaking in the province's history.

Bauxite is shipped from overseas countries (Australia, Brazil, India, Ghana, and Guinea; imports from Jamaica ceased in 2001) via the Atlantic Ocean and the St Lawrence and Saguenay rivers to Alma.
- The following year, Alcan purchased new equipment worth $200 million for its Alma smelter, making this plant the largest one in the Saguenay and boosting the company's production capacity to 1.9 million tonnes a year.
- In 2002, Alcan and four partners announced plans to expand their aluminum plant (Alouette smelter) at the port of Sept-Îles, making it the largest smelter in the world.

Agriculture

The settlement of the Clay Belt demonstrates the difficulty of farming in the northern area of Québec. The Clay Belt occupies an enormous area of northwestern Québec and northeastern Ontario. The Canadian Shield acquired this relatively thick layer of sand, silt, and clay sediments near the end of the last ice age about 10,000 years ago. The rivers flowing to Hudson and James bays were blocked by the remnants of the Laurentide ice sheet. Gradually, a huge glacial lake called Lake Barlow-Ojibway was formed and lasted for several thousand years over an area of about 200,000 km². Lake sediments, especially minute particles of clay, sand, and silt, were deposited. When the glacial lake drained, much of the land was covered by peat, making it unsuitable for agriculture. In fact, less than 5 per cent of the Clay Belt contains arable land, only a portion of which has been cultivated. As a result, farmland in the Clay Belt is in scattered pockets, separated by large areas of forested country.

Vignette 6.8   Agricultural Settlement of the Clay Belt in Québec

By the late nineteenth century, the growing rural population in the St Lawrence Valley began to move into the Clay Belt of northern Québec in search of arable land. French-Canadian migrants, often sons of farmers in the St Lawrence Lowlands, tried to clear small pockets of land in the Canadian Shield. By the 1880s, settlers had reached the Little Clay Belt around Lake Temiskaming. Agricultural settlement began following the construction of the Grand Trunk Railway (now the Canadian National Railways) from Québec City to the Little Clay Belt and then across the Canadian Shield to Winnipeg.

The Québec government and the Roman Catholic Church encouraged settlers to take up 'free' lands in both the Little Clay Belt around Lake Temiskaming and the Great Clay Belt around Lake Abitibi. Part of their concern was that thousands of French Canadians were migrating to New England and would be assimilated into American society. The Québec government and the Roman Catholic Church interpreted this migration as a 'loss' to French Canada. For that reason, they made special efforts to attract French Canadians to the last remaining agricultural lands in Québec—the so-called Clay Belt—where the French language and heritage could be maintained.

The Clay Belt is located within the Canadian Shield of northern Ontario and Québec. French-Canadian farmers began to settle these lands in the late nineteenth century. Life was hard and many farms were abandoned by the end of the twentieth century. (Victor Last, Geographical Visual Aids 17646)

Efforts to settle land in the Clay Belt began in the late nineteenth century (Vignette 6.8), but received a boost during the Great Depression of the 1930s. At this time, both the Québec and Ontario governments encouraged a 'back to the land' movement in the Clay Belt. In Québec, the Catholic Church took a lead role by organizing settlements in Catholic parishes. Farms were small in size and had little arable land. As they were far from urban markets, these settlers were destined for a hard life with few prospects for developing commercial farming operations.

In Québec, the resettlement program had two goals: to combat unemployment in the cities of southern Québec and to ensure that the settlers followed a French-Catholic lifestyle. The provincial government and the Catholic Church gave little attention to the long-term prospect of commercial farming in the Clay Belt; rather, the emphasis was on defusing the unemployment problem in Québec and 'saving souls'. At its peak, agricultural colonization resulted in fewer than 15,000 farms. Most of these farms were small with little commercial production.

Since World War II, agriculture in the Clay Belt has undergone three changes: (1) the emergence of a local market for dairy products in the nearby urban centres; (2) the establishment of a dairy marketing board, thereby providing price stability for farmers; and (3) the consolidation of most farms into fewer but larger farms. The trend to farm consolidation is revealed by the tenfold reduction in the number of farms in the Clay Belt from 15,000 in 1936 to 1,500 in 2001. Farm consolidation was driven by economic pressures. For instance, a commercial dairy farm needs sufficient land for summer grazing and for the production of fodder crops for winter feed. As well, a commercial family farm must have a minimum of 100 milk cows and modern milking machinery in order to achieve viability according to economies of scale. Though the changes undergone in the Clay Belt have improved the prospects for commercial farmers, with individual farmers benefiting from larger dairy farms and the regular payments from the provincial fluid milk marketing board, the potential for commercial agriculture in northern Québec remains extremely limited.

Forest Industry

Québec has 22 per cent of Canada's productive forest lands. In terms of productive forest, Québec ranks first among Canada's geographic regions, but it ranks second behind British Columbia in terms of total volume of wood cut. In 2000, for instance, Québec accounted for 20 per cent of Canada's harvested logs while British Columbia reached almost 40 per cent (Canada, 2003). However, Québec leads British Columbia in terms of pulp and paper production and the output of newsprint. Québec's advantage is its close proximity to major US cities, especially New York, where the demand for paper and newsprint is extremely high.

In total, Québec accounts for nearly 760,000 km² of forest lands. These lands exist in each of the four physiographic regions of Québec, but the vast majority of commercial forest is in the Canadian Shield south of 53° N. The boreal forest is concentrated in the southern half of the Canadian Shield, while mixed and hardwood forests are found in the St Lawrence Lowlands and the Appalachian Uplands. The Hudson Bay Lowland has few commercial stands. While timber is cut in the northern hinterland, most logs are processed at mills located at the mouths of tributaries flowing into the St Lawrence River. For instance, logs are floated down the St Maurice River to the pulp and paper mill at Trois-Rivière. The main exceptions are the Lac Saint-Jean and Saguenay regions where there are many sawmills and eight large pulp and paper mills. Québec has a total of 64 pulp and paper mills.

The forest industry contributes significantly to the province's economy in both value of production and employment. Forest products make up 21 per cent of Québec's total exports, providing another measure of the significance of the forest industry to the Québec economy. In 2000, the value of forest production was $18 billion and some 250 municipalities depended on forestry (Québec, 2003a). The forest industry generates many jobs, with logging accounting for 9,000 jobs, wood processing for 40,000, and pulp and paper for 34,000 (Québec, 2003b). Over 80 per cent of Québec's forest products are exported, mainly to the United States. With the United States imposing a 27 per cent tariff on Canadian softwood lumber in 2001, the Québec lumber industry has suffered and the value of its exports declined in 2002.

Over the last 15 years, more and more processing of wood into higher-value products has taken place. In fact, wood processing and paper manufacturing represent the most important manufacturing sector in the province by value of production. While the value of forest products has increased, the introduction of advanced woodcutting machinery and computerized mills has reduced demand for forest workers. This inverse relationship between technology and labour is particularly troublesome for workers in northern Québec, where few employment alternatives exist.

Mining Industry

Mining has always been important in northern Québec. While the value of mineral production is only about 8 per cent of Canada's output, northern Québec communities often rely on mining for their existence. The cyclical nature of the mining industry, due to its dependency on world markets, poses a problem for resource communities. Low demand means layoffs and even the closure of mines and ore-processing mills. For example, the iron mine at Gagnon was closed in 1985, spelling the end to this single-industry town. This boom-bust cycle driven by fluctuations in world prices is particularly hurtful to the narrowly based economies of resource hinterlands. For instance, the world demand for mineral products reached a peak in the 1960s, declined sharply in the 1980s, recovered somewhat by the late 1990s, and then began to slide downward in 2002. Such price fluctuations have had a profound impact on Québec's iron-mining industry. Québec iron mines account for much of the iron ore produced in North America. Québec's low-grade iron ore is converted into high-grade iron pellets and these pellets are the main material for North American and European blast furnaces and electrical mini-mills. The iron pellets also allow the mining companies to reduce their transportation costs from the mine site to the nearest ocean shipping point (Port Cartier and Sept-Îles). In 1982, the slowdown in the world economy caused the Iron Ore Company of Canada to close its pelletizing plant at Sept-Îles, thus concentrating its pellet-making operations in Labrador City. In 2000, plans were made to reopen the Sept-Îles plant but so far the company continues to use the Labrador City plant.

Commercial mineral deposits in the hard

rock of northern Québec often consist of gold, iron, and copper. There are two main mining areas within this area of the Canadian Shield. In the west, gold and copper are mined at Noranda and Val-d'Or, the centres for much of this production. In the northeast, iron is mined. Deposits of iron were first reported in 1895 by A.P. Low of the Geological Survey of Canada, the first geologist to investigate the region's mineral potential. However, these iron ore deposits had no commercial value because more accessible mines could supply the needs of the iron and steel companies.

All that changed in the 1940s. American steel producers could no longer count on domestic supplies of iron ore, so they sought more reliable sources, including those in northern Québec and Labrador. Through a process of market integration initiated by US steel companies, Québec's resource hinterland became dependent on a particular group of steel companies in the industrial heartland of the United States for its economic well-being. Two mining companies, Québec Cartier Mining Company and the Iron Ore Company of Canada, developed iron mines in isolated areas of northern Québec and Labrador. By 1947, plans were laid for an open-pit mine in northern Québec near the border with Labrador. The Iron Ore Company built a town (Schefferville) for miners and their families; transmitted power from Churchill Falls to operate the mine and the town; and built a railway (the Québec North Shore and Labrador Railway) to deliver the iron ore to the port at Sept-Îles, from where the ore was eventually transported to supply US steel mills in Ohio and Pennsylvania.

The demand for iron ore rose in the 1960s, resulting in the establishment of three more mining towns—Wabush and Labrador City in Labrador and Fermont, Québec. At the same

time, Québec Cartier Mining Company built a similar iron-mining operation by constructing the Cartier Railway from Port Cartier on the St Lawrence River to the resource town of Gagnon. But, by the 1980s, world steel production had surpassed the demand, causing a severe slump in the demand for iron ore. To add to this economic problem, US steel plants were now less efficient than the new steel mills in Brazil, Canada, Korea, and Japan, and lower-cost iron mines had opened in Australia and Brazil. As lower-priced steel from these countries undercut the price of US steel, American steel companies had to reduce their output, close plants, and sell their shares in the two mining companies. The repercussions for the mine workers in Québec and Labrador were severe. Production from these northern mines fell by half in the early 1980s and hundreds of workers were laid off. In 1983 and 1985, the mines at Schefferville and Gagnon were closed. During the 1990s, both companies restructured their operations to reduce costs. As demand from American iron and steel plants increased in the 1990s, these two companies were ready to supply this North American market. By 2002, the Iron Ore Company of Canada operated a pellet plant at Labrador City and three mines at Fermont, Wabush, and Labrador City, while the Quebec Cartier Mining Company had one mine at Mont-Wright and a pellet plant at Port Cartier.

Hydro Power: The James Bay Project

The James Bay Project is a massive project that calls for the production of hydroelectricity from all the rivers that flow into James Bay from Québec territory. This hydroelectric project was announced in 1971 by Premier Robert Bourassa. The James Bay Project is divided into three separate river basins (La Grande, Great Whale, and the Nottaway-Broadback-Rupert basins). The project involves about 20 rivers and affects an area one-fifth the size of Québec. Construction of the first phase of the James Bay Project, La Grande Project, began in 1972 and was completed in 1985 at a cost of $15 billion (Bone, 2003: 128). The La Grande Project involves diverting waters from three other rivers (East-main, Opinaca, and Caniapiscau) into La Grande Rivière. Electrical energy generated from the three power stations in La Grande Basin is more than 10,000 MW each year.

The first phase of the James Bay Project, however, raised considerable controversy. It evoked an unprecedented response from Aboriginal peoples and environmental organizations. For them, the James Bay Project

The James Bay Project generates huge amounts of electrical energy. At the same time, the project has altered the natural environment by diverting rivers, changing their seasonal flow, and creating enormous reservoirs in Québec's Canadian Shield. (Ivy Images)

unleashed social and environmental problems that remain unresolved. For example, the project resulted in an unexpected high content of mercury in the reservoirs, which has had serious implications for the Cree, who consume fish on a regular basis. Hydro-Québec maintains that the environmental impacts have been mitigated to an acceptable level through modifications to the design of the project. Remaining environmental impacts, such as high mercury content in the waters of the reservoirs, will, they argue, diminish over time.

In 1985, the second phase of the James Bay Project was announced. The Great Whale River project, located just north of the La Grande Basin, was to consist of three power houses, four reservoirs, and the diversion of two rivers. In addition, another generating station (LG-1) was to be located at the mouth of the La Grande Rivière. Opposition from the Cree and the Sierra Club, who jointly mounted a large public relations campaign, had an effect on public opinion in New England and New York, but when a new natural gas pipeline from Alberta to New England came on stream to provide a low-cost alternative the government of Québec announced in 1994 that the project would not proceed until the demand (price) for electricity in New England improved. From the very beginning, the Cree opposed the James Bay Project because of its effect on their hunting grounds. The Cree, joined by the Inuit of Arctic Québec, forced a land-claims settlement known as the James Bay and Northern Québec Agreement.

The James Bay and Northern Québec Agreement

In 1971, 6,000 Cree in northern Québec lived as eight bands scattered across 375,000 km² of rivers and forest. They were under the admin-

istration of the federal Department of Indian Affairs and Northern Development. The James Bay Project threatened to flood their lands. This threat united the eight bands. When construction began in 1972, the Cree asked the Inuit to join them in taking legal action to halt the construction until the Cree and Inuit land claims were addressed. This action forced the Québec government and the Aboriginal claimants to the bargaining table. The result was the **James Bay and Northern Québec Agreement** (JBNQA). Under this agreement, both the federal and Québec governments became responsible for providing the 'treaty' benefits. As the first modern land-claim agreement in Canada, this 1975 agreement provided land, cash, and the power to administer cultural matters (education, health, and social services) to Aboriginal peoples. In exchange, the Cree and Inuit surrendered their Aboriginal claims to northern Québec and agreed to allow construction of the La Grande project to proceed.

In combination, these events—the negotiations, the agreement, and the construction project—have forever altered the lives of the Cree and Inuit. Both groups, now living in settlements, are more involved in the modern industrial society than ever before. Many are employed in businesses run by Cree and Inuit organizations, while others work in construction activities in the growing Cree and Inuit settlements and for Hydro-Québec. Still others are involved in the administration of their cultural affairs through the Cree Regional Authority and the Kativik Regional Government. In comparison with other Aboriginal peoples in Québec, the economic situation of the Cree and Inuit is much improved and certainly much better than that of those Québec Indians who do not have the benefit of such an agreement (Simard et al., 1996). Still, the Cree felt that both Ottawa and Québec had failed to honour their responsibilities under the JBNQA. When Québec announced plans for the second phase of the James Bay Project in 1985, relations between the Cree and the Québec government were so confrontational that the Cree actively opposed the Great Whale River project and took the Québec government to court over a number of issues related to the earlier agreement. Without a doubt, tensions between the Cree and the Québec government reached a low point in the latter part of the twentieth century, marking a particularly strained Québec version of the Aboriginal/non-Aboriginal faultline.

Turnaround

In 2001, the Cree reached an agreement with the Québec government for the economic development of the resources of northern Québec. This agreement opens the door to a major hydroelectric project on the Rupert and Eastmain rivers. The acceptance of such an agreement is an astonishing reversal for the Québec Cree, who have mounted numerous international protests against further hydroelectric developments in their traditional lands. Some argue that the Cree now recognize that their participation in northern economic development is their only option. But feelings run high because efforts to protect the land for a hunting and trapping lifestyle have faltered. Paul Dixon, the Cree trapper representative, put it this way: 'They [Québec] promised the traditional way of life would continue undisturbed. Today, the whole territory has been slated for development' (Roslin, 2001: FP7).

Some, especially younger Cree, have chosen an urban lifestyle. For them, living on the land is no longer a viable option. Some say that the Cree leaders had to make a deal. Faced with a rapidly growing population, high unemployment rates, a critical shortage of public housing, and a desperate need for sewer and water systems, the Cree leaders had to seek an agreement with the Québec provincial government. Québec wanted to develop the northern resources and the Cree needed revenue to operate their communities and to find work for their people. Under the terms of the agreement, the Cree receive $3.6 billion over 50 years (roughly $70 million a year), but these funds release Québec from its obligations for economic and community development associated with the James Bay and Northern Québec Agreement. A newly created Cree Economic Development Agency will administer these funds with a mandate to foster growth of Cree businesses. The agreement also requires that the Cree will drop their lawsuits against the Québec government for failure to meet its obligations under the James Bay and Northern Québec Agreement. Whether or not the Cree will benefit from this model of economic development remains to be seen. What is clear, however, is that similar agreements are taking place across the country and all are designed to allow Aboriginal people to participate in economic development taking place in their traditional lands.

Tourism

In Canada and in the rest of the world, tourism is becoming an extremely important economic sector. Québec is no exception. In fact, Québec combines its natural beauty, historic past, and francophone culture to draw more and more tourists each year. For example, from 1993 to 2000, the number of tourists visiting the province increased every year, and tourist dollars spent in Québec over that eight-year period increased by over 50 per cent

Table 6.6	Major Cities and Census Metropolitan Areas in Québec, 1996–2001		
Census Metropolitan Area	Population 1996 (000s)	Population 2001 (000s)	Change (per cent)
Montréal	3,326.4	3,426.4	3.0
Québec City	671.9	682.8	1.6
Hull*	247.1	257.6	4.2
Chicoutimi–Jonquière	160.5	154.9	(3.4)
Sherbrooke	149.6	153.8	2.8
Trois-Rivières	140.0	137.5	(1.7)
Saint-Jean-sur-Richelieu	76.5	79.6	4.1
Drummondville	65.1	68.5	5.1
Granby	58.9	60.3	2.4
Shawinigan	59.9	57.3	(4.3)
Saint-Hyacinthe	50.0	49.5	(1.0)
Rimouski	48.1	47.7	(0.9)
Sorel-Tracy	43.0	41.0	(1.7)
Victoriaville	40.4	41.2	2.0
Total	4,006.3	4,905.8	22.5

*The population of the Ottawa–Hull CMA was 743,821 in 1996 and 1,081,000 in 2001.
Source: Statistics Canada (2002).

(Québec, 2003c). Most tourists in Québec come from other regions of Canada and from the United States, France, the United Kingdom, Germany, and Japan. Tourism in Québec accounts for over 21,000 businesses that employ more than 250,000 workers (ibid.). Since Québec is ideally located to cater to tourists from New England and Western Europe, its future as a world-class tourist destination seems secure.

The centres of Montréal and Québec City are major attractions for tourists seeking an urban vacation with a francophone atmosphere, while the Laurentides attract visitors looking for a summer or winter playground in the forests and lakes of the Canadian Shield. In addition to the natural beauty, the Laurentides

have several advantages as an all-season recreational area. The Laurentides are but a short distance from Montréal and relatively close to New England and to major American cities such as Boston and New York. The climate in this area is ideal for the tourist business—the summers are hot, and heavy snowfall provides ideal conditions for winter sports. For example, Mount Tremblant Resort is a world-class tourist destination for both winter and summer activities. Tourism is the major business in the Laurentides, and the towns and villages cater to the tourists who flock there to enjoy water sports in the summer, skiing and snowboarding in the winter, and the numerous restaurants and shops all year long.

While Québec has the setting for summer

and winter recreation activities, the tourist industry is vulnerable to external events. Since the terrorist attacks on New York and Washington and the resulting war on terrorism announced by Washington, the tourist industry has suffered greatly. Added to the dismal tourist business in 2001, the war on Iraq in 2003 sent another shiver through the tourist industry. In the same year, the rising Canadian dollar, Toronto's struggle with SARS, and the discovery of one cow with mad cow disease in Alberta combined to discourage tourists from coming to Québec.

Québec's Urban Geography

Over 80 per cent of Québec's population lives in urban centres. Two of the largest metropolitan cities in Canada, Montréal and Québec City, are located in the province. Other major population centres are Gatineau, Chicoutimi–Jonquière, Sherbrooke, and Trois-Rivières. In total, these six urban clusters have a population of 4.9 million (Table 6.6).

As in Ontario and most other regions of Canada, migration has played a major role in Québec's urbanization. Push and pull factors (described in Chapter 4) have attracted rural Quebecers to cities. The key push factors in rural Québec were limited job opportunities, a shrinking labour force in the primary sector, and an increasing number of young people entering the workforce.

From 1996 to 2001, the population of the major cities, led by Gatineau (formerly Hull), Sherbrooke, and Montréal, grew by 2.3 per cent (Table 6.6). Gatineau's population growth was largely due to the employment and business opportunities generated by the federal government. Shawinigan and Chicoutimi–Jonquière saw their populations drop by 4.3 and 3.4 per cent respectively. Historically, Shawini-gan owes its economic growth to hydroelectric production on the St Maurice River. Since the 1960s, the paper, aluminum, and chemical plants that were attracted by electrical power have reduced or ceased their activity. Chicoutimi–Jonquière is located in the resource hinterland of the Lac Saint-Jean region. The two main economic activities are the aluminum plants and the forest mills. As in other resource hinterlands, employment prospects in these two industries are not bright.

Montréal

Montréal, the metropolis of the province, is the industrial, commercial, and cultural focus of Québec. For a century and a half, Montréal was the largest and most prosperous city in the Dominion. Within Canada today, this historic city has slipped behind Toronto in size and importance. Within Québec, however, Montréal remains the dominant city. In 2000, Montréal had a population of almost 3.5 million, making it the largest population concentration in Québec.

Like other great cities, Montréal is surrounded by smaller cities and towns. Such centres are often called satellite towns because they fall within the trade orbit of the larger city. Trois-Rivières, a major urban centre located on the St Lawrence over 100 km northeast of Montréal, is a satellite city. Smaller centres within 50 km of Montréal, such as Sainte-Hyacinthe, St-Jérôme, and Salaberry-de-Valleyfield, are even more closely tied to that metropolis. Laval and Longueuil serve as bedroom communities for Montréal. In most cases, urbanites have fled to the suburbs, attracted by lower-cost housing and the amenities of suburban life. These suburbanites often work in Montréal and live within commuting distance. Each weekday morning and late afternoon, commuters generate huge volumes of slow traffic and crowded buses in and around Montréal.

Given its strategic location and economic size, Montréal serves as the transportation hub of Québec, making it the regional core of the province and the rest of Québec the periphery. At the national scale, Montréal is part of the national core because it is part of Canada's manufacturing belt.

Following the Free Trade Agreement, Montréal's manufacturing sector had to respond to strong foreign competition. Labour-intensive manufacturing firms experienced great difficulty in competing with foreign firms that had substantially lower labour costs. Montréal's manufacturing firms made two major changes: labour-intensive plants substituted machinery for workers to increase productivity, and high-technology firms expanded. By the late 1990s, Montréal's economy had become much more specialized in aerospace, computers, fibre optics, multilingual software, telecommunications, and other areas of industrial research and development. The provincial government has taken a leading role by providing subsidies for high-tech firms that relocate to Montréal and other Québec cities.

Historical Context

Geography and history lie behind Montréal's success over the years. Montréal's strategic location on the St Lawrence River was a key factor in its initial growth as a transportation route and as a source of electrical energy, though today most power comes from tributaries flowing into the St Lawrence. In the 1820s, the Lachine Canal near Montréal provided the energy to drive the grist mills. Early in this century, power was transmitted to Montréal first from the hydroelectric installation on the St Maurice River at Shawinigan and then from the Beauharnois power site on the St Lawrence River.

Montréal benefited from the fur trade. Wealth accumulated by partners in the North West Company provided much of the capital necessary for new industrial ventures in manufacturing, resource development, and transportation. At the time of Confederation, Montréal was the largest and most prosperous city in Canada. With the introduction of high protective tariffs in 1879, Montréal's industrial base increased, thus propelling the city's economic and demographic growth. As Montréal's transportation and manufacturing industries expanded, its role as a financial and cultural centre solidified. By the end of the nineteenth century, Montréal was Canada's transportation hub and financial centre. For example, the construction of railway lines, particularly the Grand Trunk Railway, confirmed Montréal's role as a transportation centre and facilitated the development of railway repair depots and the manufacture of engines and railcars. This dominant role continued for 100 years after Confederation. In 1967, Montréal was the premier city in Canada and the undisputed headquarters of Canada's textile and financial enterprises. In the years that followed, Montréal's demographic and economic growth continued, but at a slower pace than that of Toronto.

Montréal and Toronto

Beginning in the 1970s, Toronto quickly replaced Montréal as the premier city in Canada. Montréal was no longer the largest city and the financial capital of the country. The principal reason for this shift in metropolitan power was the strong economic growth in Toronto and in southern Ontario that was powered by the Auto Pact and the resulting expansion of the automobile industry. A secondary factor was the economic and demographic fallout

Table 6.7	Population Change: Montréal and Toronto, 1951–2001 (000s)		
Year	Montréal	Toronto	Difference
1951	1,539	1,262	277
1961	2,216	1,919	297
1971	2,743	2,628	115
1981	2,828	2,999	(171)
1991	3,127	3,893	(766)
2001	3,426	4,683	(1,257)

Source: Statistics Canada (2002).

from the political unrest in Québec and the very real possibility of Québec separating from Canada. The unsettled political environment leading up to the 1980 Québec referendum worried the anglophone community. Many anglophones and a few corporations moved to Toronto. While the francophone business community and the provincial government kept the Montréal economy growing, by 1981 Toronto had a population of 3.0 million compared to Montréal's 2.8 million (Table 6.7). Three other economic factors that explain Montréal's slow growth and Toronto's much faster growth were:

• Montréal's economy was much more dependent on labour-intensive manufacturing (which was in decline) such as that in the textile industry.
• Montréal's industrialization had begun much earlier than Toronto's and, for that reason, its manufacturing firms tended to be older and less efficient than those in Toronto.
• Toronto has a national hinterland while Montréal has a provincial one and, for that reason, Toronto has a much larger 'external' demand for its wholesale services industry.

For those reasons, Toronto's growth rate exceeded that of Montréal. Even in the 10-year period from 1951 to 1961, when both cities grew at phenomenal rates—Montréal at 4.3 per cent per year and Toronto at 5.1 per cent—Toronto's rate was higher. From 1971 to 1981, the rate of population increase slowed in both cities, with Montréal's close to zero. During that time, Montréal's annual growth rate was only 0.3 per cent, while Toronto's was 1.9 per cent. This gap continued in the early 1990s with the annual rates for Toronto and Montréal at 1.9 per cent and 1.3 per cent, respectively. By 2000, Toronto's population had reached 4.8 million while Montréal's was 3.5 million.

Québec City

Québec City has the next largest urban concentration in the province, totalling nearly 700,000 people. Close to the Laurentides, Québec City has a magnificent physical setting on high banks just above the St Lawrence River. It is the only walled city in North America and features buildings over 300 years old. In 1985, Québec City was selected as a World Heritage Site by UNESCO.

The economic base of Québec City revolves around three functions. First, it is a

Held each year in Québec City, the Québec Carnival is the largest winter carnival in the world, ranking just behind the famous carnivals in Rio and New Orleans. The origins of the Québec Carnival go back to the beginning of New France when a tradition of winter revelry took hold to offer relief from the long, cold winters. The modern version of a winter carnival began in 1955. The Québec Carnival involves many activities ranging from night parades, snow sculptures, and the popular canoe race on the frozen St Lawrence River, as shown above. (CP/Clement Allard)

government and university town. As the seat of government for the province, Québec City employs a large number of civil servants. Second, it has become a world-class tourist centre. The Old World charm of Québec City draws tourists from around the world, while special events such as its Winter Carnival are very popular. Third, Québec City is only minutes away from excellent seasonal recreation areas—from skiing in the winter to water sports in the summer. The economic base of Québec City remains heavily dependent on its political and cultural roles, but it also has a number of other economic functions: it is a port and rail centre, as well as a centre for resource processing, metal fabricating, and manufacturing. High-tech industries now play an important role in Québec's economy,

including the highly specialized photonics and optics sector.

Beyond the Urban Core

Québec's urban geography takes on a distinct regional character. Southern Québec has experienced modest growth in four of its five CMAs. Only Trois-Rivières suffered a population loss from 1996 to 2001. Beyond the urban core, most cities and towns suffered population losses caused by out-migration (Table 6.8). This demographic pattern is associated with weak regional economies. From 1996 to 2001, population declines took place in the urban centres of northern Québec, the North Shore, and Gaspé Peninsula. Two forces were at play.

First, the economies in these outlying areas

Table 6.8	Population of Cities of Northern Québec, the North Shore, and the Gaspé Peninsula		
City by Area	1996	2001	% Change
Northern Québec			
Chicoutimi–Jonquière	160,454	154,938	(3.4)
Rouyn–Noranda	39,096	36,308	(7.1)
Val-d'Or	33,756	32,423	(3.9)
Alma	30,377	30,126	(0.8)
North Shore			
Baie-Comeau	31,795	28,940	(9.0)
Sept-Îles	28,005	26,952	(3.8)
Gaspé Peninsula			
Gaspé	16,527	14,932	(9.6)
Percé	3,993	3,614	(9.5)

Source: Statistics Canada (2002).

of Québec are contracting. Within the new North American marketplace, resource companies that export most of their products are under great pressure to reduce their costs, which translates into reducing the size of the labour force or, in the worst-case scenario, closing their operations. Urban centres are affected by the negative spinoff from industrial restructuring, which results in a smaller population and, therefore, fewer customers. This ripple effect may reduce the size of the market beyond the point of profitability and thus force some businesses to close.

Second, the demographic outcome of a declining regional economy where few job opportunities exist normally leads to out-migration, especially by the younger and more educated members of the local population. Such migration to other places causes a contraction of the local population and a loss of potential community leaders, and signals a general economic malaise that results in a relentless downward spiral of both the regional economies and urban populations.

Québec's Future

Québec remains a central pillar of Canada's industrial axis. Along with Ontario, these two industrial cores produce the bulk of Canada's manufactured goods and contain the majority of Canada's population. As one of the four original British colonies to form Canada in 1867, Québec's position within Confederation is weakening as it loses its share of Canada's population and economic output. The reason is simple: while Québec's economy and population have expanded, other regions of Canada have grown more rapidly. For this reason alone, Québec's unique culture and language take on even more importance. As the heartland of francophones in Canada and North

America, Québec has a special role to play. Cultural events like the St Jean Baptiste festival evoke a sense of ethnic nationalism and a love of the land and its people, which is popularly expressed as 'j'ai le goût du Québec'. Within this cultural context, the French language serves as a linchpin and, for that reason, Québec, in 1977, made French the only official language in the province.

As Québec is one of the principal industrial regions in Canada, the economic challenge it faces now and in the years ahead is to secure a position in the North American marketplace. Over the past several decades Québec's economy has modernized, but its pace of growth lags behind that of Ontario. Within Québec, the resource-based economies of northern Québec, the North Shore, and the Gaspé Peninsula are, like in other hinterland areas of Canada, contracting and hinterland populations are shrinking. Put differently, the outlying areas of Québec have 'une économie en perte de vitesse' (an economy losing steam). Much of this contrast between the industrial core of Québec, centred on Montréal and its hinterlands, is due to the shift towards a more automated resource industry where productivity must rise to keep per unit costs low. The net outcome is fewer workers in the resource sector, and since resource extraction is the mainstay of the hinterland economy, a population decline follows. As in northern Ontario, northern Québec's forestry industry has a fixed resource base and therefore has little room to expand—except to shift to more value-added wood production. Unlike forestry, mining is a non-renewable resource and is subject to a boom-and-bust cycle of development. Although hydroelectric projects boost provincial energy output, they employ few people after their construction.

In short, economic prospects for the Cana-dian Shield and the Appalachian Uplands are dim. The resource-based economies in these regions are likely to generate sluggish economic growth, little population increase, and a small but steady out-migration of young people to places where employment prospects are brighter. The economic performance of the St Lawrence Lowlands remains the bright light of the Québec economy. Its manufacturing sector—especially the high-tech firms—accounts for most of the increases in employment and value of production. Québec is adjusting to the new economic circumstances of a North American marketplace. High-tech industries, such as the aerospace and pharmaceutical industries, have expanded and found a niche in the global market, though by 2002, these export-oriented industries were under pressure as the global and American economies slowed. By the end of 2003, both economies were well on the way to recovery, thus increasing the demand for high-tech products.

Beyond Québec's economy looms a larger question—Québec's place within Canada. After two referendums, the issue of political separation has passed for now, but the issue of independence is never far beneath the surface of Québec politics. For that reason alone, most political leaders outside of Québec have recognized Québec's special place within the country. In the process of seeking solutions, Canadian society and its identity have evolved into what Saul calls a 'soft nation'.

Québec possesses great vitality for fresh and innovative political, economic, and cultural undertakings. Since the Quiet Revolution, the Québec government has become a strong force in shaping its economy and promoting its culture. Through Crown corporations and agencies, such as Hydro-Québec, the provincial government has become a major player in the economic affairs of Québec. With the election of a Liberal government in 2003, the government's role in the economy may diminish somewhat, and its financial support for private business (some call it 'corporate welfare') was greatly reduced in the first Liberal budget. Language laws have solidified the place of French in Québec. Control over immigration has ensured the steady influx of French-speaking immigrants into Québec. The French language and Québécois culture are flourishing. The Québécois regard these developments with pride. For that reason, the rally cry of the separatists, which was based on a threat to the French language and Québécois culture, has lost much of its appeal. With a self-confidence formed from the enormous changes the province has undergone over the past 40 years, Québec is now exploring a new kind of self-government with the Inuit of Nunavik. If successful, this concept of a semi-autonomous state within a province would mark a political breakthrough that may be applied to other northern areas of Canadian provinces (Vignette 6.9).

Bridging the gap between the French and English visions of Canada is not a simple task. The first step was to secure Québec's culture and language. Since the 1960s, each Québec government has passed legislation to ensure that Québec not only survives but thrives as a French collectivity within North America. A view of this nationalism was expressed by Beauregard (1980: 7): 'Le Québec veut, quant à

Vignette 6.9   Mapping the Road to Nunavik

Nunavik extends over a vast arctic land in northern Québec. This land of permafrost is the homeland of some 9,000 Québec Inuit. Like other Aboriginal peoples, the Québec Inuit are seeking a form of regional autonomy that would respond to their needs, desires, and aspirations. To achieve that goal, three formidable challenges must be resolved. First, how can a province contain an autonomous government? This begs the question of the division of powers between the province of Québec and the yet-to-be formed government of Nunavik. Second, how can Nunavik (a non-ethnic government) treat all its residents equally and still promote the Inuit culture? Lastly, how can 9,000 people living in 14 communities scattered over 660,000 km² not only govern themselves but generate sufficient revenue to pay for their government?

The road to Nunavik is difficult but not impossible. Somehow the structure, operations, powers, and design of this new form of government within a Canadian province can be achieved. A treaty among the governments of Canada, Québec, and the Québec Inuit can set the terms for an autonomous government within Québec. At that moment, Québec will be the first province to share political power with a regional government and, by this example, will have created a unique model for other provinces with large Aboriginal populations to follow. Since 2001, negotiations have continued between Ottawa, Québec City, and Makivik Corporation, which represents the Inuit. In 2003, a framework agreement was struck with the purpose of creating a new form of government in Nunavik.

Sources: Report of the Nunavik Commission (2001); personal communication with Donat Savoie, chief federal negotiator for Nunavik, 10 Aug. 2003.

lui, faire reconnaître son particularisme et son destin culturel' (Québec wants, for itself, recognition of its uniqueness and its cultural destiny). The second step is to revitalize Québec's economy and, in doing so, to increase its economic growth rate.

Summary

Québec began as New France. It prospered after the British Conquest but its people struggled to maintain a French presence in British North America, and later in Canada. This ongoing struggle has become the principal source of tension within Québec and Canadian society. Québec, with its French language and society, remains a francophone bastion in Canada and North America. Within Québec, however, anglophone, allophone, and Aboriginal minorities exist. Tensions between these minorities and the francophone majority often spread to the national French/English faultline.

Geography and history have marked Québec with a special sense of place. The St Lawrence Lowlands was destined to play a key role in Québec's past, and today, as the industrial heartland of Québec, this remains the province's dominant economic region and is a major player in the economic affairs of both the province and the country. More recently, the emergence of the Canadian Shield as the primary source of energy for the cities and factories of southern Québec strengthened the economic relationship between the two geographic parts of the province and, with an Aboriginal majority in northern Québec, exposed a different economy and culture. Like Ontario, Québec is part of the industrial heartland of Canada and an important industrial area of North America. As such, Québec faces an economic challenge: how to define its economic place within the North American and global markets.

As the francophone minority within Canada, Québec often expresses dissatisfaction with its political place within Confederation. While most Québécois have a strong attachment to Canada, their first loyalty is often to Québec. Separatism, ever a threat to Confederation, remains an unresolved issue. For that reason, the French/English faultline continues to be the most contentious political issue in Canada. At the same time, the Aboriginal/non-Aboriginal faultline in Québec is largely between the Québec Cree and the Québec government and the key sticking point is the future of the vast James Bay area, the Cree homeland.

Notes

1. By responsible government, Lord Durham meant a 'political system in which the Executive is directly and immediately responsible to the Legislature, in which the ministers are members of the Legislature, chosen from the party which includes the majority of the elected representatives of the people' (Lucas, 1912: I, 138).

2. In 1912, Québec gained northern territories inhabited by the Inuit and Cree. Ottawa ceded these lands to Québec with the understanding that the Québec government would be responsible for settling land claims with the Aboriginal peoples in these territories. The Cree in northern Québec, in response to the separatist claim to territorial independence, have declared that they have the right to secede from Québec. They argue that if Québec has the right to secede from Canada, then the Cree have the right to secede from Québec. From a geopolitical perspective, the partitioning of Canada or

Québec makes sense only to those supporting ethnic nationalism.

3. For the 2000 federal election the House of Commons had 301 seats. Quebecers kept their 75 ridings, although, based on its share of population, Québec should have had only 71 seats. It kept 75 seats because of a law that guarantees each province no fewer seats than it had in 1976. For the June 2004 election, reapportionment to account for increased population added seven more seats to the Commons, primarily for Ontario, BC, and Alberta, for a total of 308 seats.

4. Camille Laurin, the father of Bill 101, declared that French was the province's only official language. Bill 101 required the children of immigrants to go to French schools and made the presence of French compulsory in the workplace and on commercial signs.

5. Many Québécois saw the failures of the Meech Lake Accord and the Charlottetown Accord as the unwillingness of English Canada to respect the 'equal partnership of Québec in the confederation'. In the mid-1980s the Québec Liberal government of the late Robert Bourassa put forth five conditions that had to be met if Québec was to accept the 1982 Constitution Act: (1) constitutional recognition of Québec as a distinct society; (2) a provincial role in immigration; (3) a provincial role in Supreme Court appointments; (4) limitations on federal powers to spend in areas of provincial jurisdiction; and (5) a veto for Québec in future constitutional amendments.

6. BioChem Pharma Inc. is one of Canada's largest pharmaceutical research firms. Its plant is located in Laval, a community just a few kilometres north of Montréal. It was founded as a small research company, but with the discovery of a drug to treat the HIV/AIDS virus it soon became the crowning jewel in Québec's biotech industry. As a world leader in developing drugs for AIDS, cancer, and hepatitis, this Québec high-tech firm was recently sold for $5.9 billion in a share swap with Shire Pharmaceuticals Group PLC based in the United Kingdom. BioChem will continue to function as a research pharmaceutical company at Laval but its products will be marketed through Shire. For Shire, this merger represents a major step towards creating the largest specialty pharmaceutical company in the world. The merger will broaden and diversify Shire's products and revenue base and strengthen the development of new drugs (Leger, 2000: C1, C6).

7. The case of Churchill Falls is an interesting one. The divide between the Atlantic Ocean and Hudson Bay marks the Labrador–Québec boundary. For historical reasons, the Québec government does not formally recognize this boundary, but it does treat the area as part of Newfoundland. Newfoundland owns the large Churchill Falls hydroelectric project, but because of the high cost of transmitting the electrical power from Churchill Falls across the Strait of Belle Isle to Newfoundland, virtually all this power is purchased by Hydro-Québec. Hydro-Québec then transmits it across Québec to markets in the St Lawrence Lowlands and the United States.

8. The 13 companies with risk-sharing contracts are: Norsk Hydro Canada Inc., Aluminerie Alouette Inc., Quebec-Cartier Mining Co., Cafco Industries Ltd, Timminco Ltd, QIT-Fer et Titane Inc., PPB Canada Inc., Reynolds Metals Co., Argonal, Hydrogenal, SKW Canada Inc., ABI Inc., and Aluminerie Lauralco Inc.

Key Terms

habitants

French peasants who settled the land in New France under a form of feudal agriculture known as the seigneurial system. After the British Conquest, the seigneurial system continued until the mid-nineteenth century, marking a significant difference between Upper and Lower Canada.

James Bay and Northern Québec Agreement

The 1971 announcement of the James Bay Project triggered a series of events that quickly led to a negotiated settlement and, in 1975, an agreement. In this modern treaty, Aboriginal title was surrendered by the Inuit and Cree of northern Québec in exchange for specific rights, including self-government and benefits (cash and financial support for the hunting economy). Signatories to the agreement included Québec, the Société d'énergie de la Baie James, the Société de développement de la Baie James, the Commission hydro-électrique de Québec, the Grand Council of the Crees of Québec, the James Bay Cree, the Northern Québec Inuit Association, the Inuit of Québec, the Inuit of Port Burwell, and the government of Canada.

patriation

The act of bringing legislation, especially a Constitution, under the authority of the autonomous country to which it applies.

seigneurs

Members of the French aristocracy who were awarded land in New France by the French king. A seigneur was an estate owner who had peasants (*habitants*) to work his land.

Bibliography

Beauregard, Ludger, ed. 1980. 'Numéro spécial: La problématique géopolitique du Québec', *Cahiers de géographie du Québec* 24, 61: 1–185.

Bone, Robert M. 2003. *The Geography of the Canadian North*, 2nd edn. Toronto: Oxford University Press.

Bothwell, Robert. 1995. *Canada and Quebec: One Country, Two Histories*. Vancouver: University of British Columbia Press.

Bradbury, John H. 1982. 'State Corporations and Resource-Based Development in Quebec, Canada, 1960–1980', *Economic Geography* 58, 1: 45–61.

Bussière, Luc. 1995. 'Une économie en perte de vitesse', *Provincial Outlook: Economic Forecast* 10, 1: 21–9.

Caldwell, Gary, and Eric Waddell, eds. 1982. *The English of Québec: From Majority to Minority Status*. Institute québécois de recherche sur la culture. Sainte-Foye: Les Presses de l'Université Laval.

Canada. 1993. *Canada Year Book 1994*. Ottawa: Minister of Industry, Science and Technology.

Canada. 2003. Statistics Canada: Canadian Statistics—Net merchantable volume of round-wood harvested. Searched 26 June 2003: <http://www.statcan.ca/english/Pgdb/prim41.htm>.

Canada Centre for Remote Sensing. 1998. Saguenay Flood 1996. Searched 3 Oct. 1998: <http://otter.ccrs.nrcan.gc.ca/ccrs/tekrd/rd/apps/hydro/saguenay/saquente.html>.

Centre for Research and Information on Canada. 2003. Portraits of Canada 2002. Searched 6 Dec. 2002: <http://www.cric.ca>.

Chevalier, Jacques. 1993. 'Toronto–Ottawa–Montréal: Concentrations majeures Canadiennes de l'innovation par la recherche-développement', *Le géographe Canadien* 37, 3: 242–57.

Coffey, W.J. 1996. 'Make or Buy: Internalization and Externalization of Producer Service Inputs in the Montreal Metropolitan Area', *Canadian Journal of Regional Science* 19, 1: 25–48.

——— and Réjean Drolet. 1994. 'La décentralisation des services supérieurs dans la région métropolitaine de Montréal, 1981–1989', *Le géographe Canadien* 38, 3: 215–29.

——— and M. Polèse. 1989. 'The Role of Cultural Barriers in the Location of Producer Services: Some Reflections on the Toronto-Montreal Rivalry and the Limits to Urban Polarization', *Canadian Journal of Regional Science* 14: 433–46.

——— and ———. 1993. 'Le déclin de l'empire montréalais: regard sur l'économie d'une métropole en mutation', *Recherches sociographiques* 34: 417–37.

Conference Board of Canada. 2000. *Provincial Outlook*. Winter.

Cook, Ramsay. 1976. *Canada and the French-Canadian Question*. Toronto: Macmillan.

Courville, Serge. 2000. *Le Québec: geneses et mutations du territoire*. Sainte-Foye: Les Presses de l'Université Laval.

De Benedetti, George J., and Maurice Beaudin. 1996. 'Linguistic Minority Communities' Contribution to Economic Well-being: Two Case Studies', *Canadian Journal of Regional Science* 19, 2: 175–92.

Doloreux, David. 1998. 'Politique technopolitaine et territorie: le cas de Laval', *Canadian Journal of Regional Science* 21, 3: 441–60.

———. 1999. 'La pépinière d'entreprises dans le contexte d'un parc scientifique: l'exemple du Centre Québécois d'Innovation en Biotechnologie à Laval, Québec (Canada)', *The Canadian Geographer* 43, 4: 423–32.

Gibbens, Robert. 2000. 'Sept-Iles pellet plant likely set for startup', *National Post*, 4 July, C4.

Hamelin, Louis-Edmond. 1998. 'L'entièreté du Québec: le cas du Nord', *Cahiers de Géographie du Québec* 42, 115: 95–110.

Hecky, Robert E. 1987. 'Methylmercury Contamination in Northern Canada', *Northern Perspectives* 15, 3: 8–9.

Higgins, Benjamin. 1986. *The Rise and Fall of Montreal?* Moncton: Instititut canadien pour le développement regional.

Hodgins, Bruce W., Richard P. Bowles, James L. Hanley, and George A. Rawlyk. 1974. *Canadiens, Canadians and Québécois*. Scarborough, Ont.: Prentice-Hall.

Hornig, James F., ed. 1999. *Social and Environmental Impacts of the James Bay Hydroelectric Project*. Montréal and Kingston: McGill-Queen's University Press.

Joyal, André, and Laurent Deshaies. 2000. 'Réseaux d'information des PME en milieu non métropolitain', *Cahiers de Géographie du Québec* 44, 122: 189–210.

Kaplan, David H. 1994. 'Two Nations in Search of a State: Canada's Ambivalent Spatial Identities', *Annals of the Association of American Geographers* 84, 4: 585–606.

Lasserre, Jean-Claude. 1999. 'Pour comprendre la stagnation et les mutations des traffics sur le Saint-Laurent: une évaluation comparée des portes continentals nord-américaines', *Cahiers de Géographie du Québec* 43, 118: 7–42.

Ledent, J., L. Bourne, and F. Dansereau. 1999. 'Comparative Development in Montreal and Toronto/Une Analyse Comparée du Développement à Montréal et Toronto', Special Issue, *Canadian Journal of Regional Science* 22, 1–2.

Leger, Kathryn. 2000. 'Biochem sold for $5.9-billion', *National Post*, 12 Dec., C1, C6.

Lo, Lucia, and Carlos Teixeria. 1998. 'If Québec Goes . . . The "Exodus" Impact?', *The Professional Geographer* 50, 4: 481–98.

Louder, Dean, and Eric Waddell. 1992. *French America: Mobility, Identity and Minority Experience Across the Continent*. Baton Rouge: Louisiana State University Press.

Lucas, C.P. 1912. *Lord Durham's Report of the Affairs of British North America*, vol. 1. Oxford: Clarendon Press.

MacAfee, Michelle. 2003. 'GM's $400M Quebec pact to spur 3,000 jobs', *National Post*, 1 Mar., FP5.

McCutcheon, Sean. 1991. *Electric Rivers: The Story of the James Bay Project*. Montréal: Black Rose Books.

Manzagol, Claude, and Christopher R. Bryant. 1998. *Montréal 2001: visages et défis d'une métropole*. Montréal: Presses de l'Université de Montréal.

Marsh, James H., ed. 1988. *The Canadian Encyclopedia*, 2nd edn. Edmonton: Hurtig.

Martineau, Daniel, Karin Lemberger, André Dallaire, Philippe Labelle, Thomas P. Lipscomb, Pascal Michel, and Igor Mikaelian. 2002. 'Cancer in Wildlife, a Case Study: Beluga from the St. Lawrence Estuary, Québec, Canada', *Environmental Health Perspectives* 110, 3: 285–92.

Mills, David. 1988. 'Durham Report', in Marsh (1988: 637–8).

National Post. 1999. 'Alcan to invest another $200M in Quebec smelter', 4 Nov., C2.

Nemni, Max. 1994. 'The Case Against Quebec Nationalism', *American Review of Canadian Studies* 24, 2: 171–96.

Noël, Michel. 1997. *The Native Peoples of Québec*. Ottawa: University of Ottawa Press.

Peluso, Tony, Laurie Baker, and Paul J. Thomassin. 1998. 'The Siting of Ethanol Plants in Quebec', *Canadian Journal of Regional Science* 21, 1: 73–88.

Phillips, David. 1993. *The Day Niagara Falls Ran Dry!* Toronto: Canadian Geographic and Key Porter Books.

Polése, Mario. 2000. 'Is Quebec Special in the Emerging North American Economy?', *Canadian Journal of Regional Studies* 23, 2: 187–212.

——— and Martin Roy. 1999. 'La dynamique spatiale des activités économiques au Québec: analyse pour la période 1971–1991 fondée sur un découpage centre-périphérie', *Cahiers de Géographie du Québec* 43, 118: 43–71.

——— and Richard Stren, eds. 2000. *The Social Sustainability of Cities: Diversity and the Management of Change*. Toronto: University of Toronto Press.

Québec. 1998a. Pour donner au monde le goût du Québec: Résumé de la politique de développement touristique [on-line], Gouvernement du Québec, Ministère du Tourisme. Searched 3 Feb. 1999: <http://www.tourisme.gouv.qc.ca/francais/mto/publications/poldevtour.html>.

———. 1998b. Le tourisme au Québec: enjeux et orientations [on-line], Gouvernement du Québec, Ministère du Tourisme. Searched 3 Feb. 1999: <http://www.tourisme.gouv.qc.ca/francais/mto/enjeux.html>.

———. 2003a. Les Forêts, Le Québec forestier/Portrait statistique: Le secteur forestier dans l'économie [on-line], Gouvernement du Québec, Ministère des Ressources naturelles. Searched 23 June 2003: <http://www.mrn.gouv.qc.ca/english/forest/investors/index.jsp>.

———. 2003b. The Québec Forest Industry [on-line], Québec Forest Industries Association. Search 27 June 2003: <http://www.aifq.qc.ca/english/stats/papers01.html>.

———. 2003c. Québec Statistiques Touristiques: Le tourisme au Québec en bref—2001.

Report of the Nunavik Commission. 2001. *Let Us Share*. Québec City: Conseil du Trésor, Gouvernement du Québec.

Richardson, Boyce. 1975. *Strangers Devour the Land: The Cree Hunters of the James Bay Area versus Premier Bourassa and the James Bay Development Corporation*. Toronto: Macmillan.

Rioux, C., J.-C. Michaud, B. Urli, and L. Grossilin. 1998. 'Développement local et décisions collectives: le cas du Québec-côtier', *Canadian Journal of Regional Science* 21, 3: 365–74.

Rose, Damaris, and Marc Villemaire. 1997. 'Reshuffling Paperworkers: Technological Change and Experiences of Reorganization at a Quebec Newsprint Mill', *The Canadian Geographer* 41, 1: 41–60.

Roslin, Alex. 2001. 'Cree deal a model or betrayal?', *National Post*, 10 Nov., FP7.

Rumney, Thomas. 1996. *The Geography of Canada Bibliography Series: Vol. 1, Quebec*. Plattsburgh, NY: Plattsburgh State University, Center for the Study of Canada.

Saint-Laurent, Diane. 2000. 'Approches biogéographe de la nature en ville: parcs, espaces verts et friches', *Cahiers de Géographie du Québec* 44, 122: 147–66.

Salée, Daniel. 2003. 'Quebec's Changing Political Culture and the Future of Federal-Provincial Relations in Canada', in Harvey Lazar and Hamish Telford, eds, *Canada: State of the Federation 2000/01: Canadian Political Culture(s) in Transition*. Montréal and Kingston: McGill-Queen's University Press.

Salisbury, Richard Frank. 1986. *A Homeland for the Cree: Regional Development in James Bay, 1971–1981*. Montréal and Kingston: McGill-Queen's University Press.

Scott, Richard T. 2001. 'Becoming a Mercury Dealer: Moral Implications and the Construction of Objective Knowledge for the James Bay Cree's Aboriginal Autonomy and Development in Northern Quebec and Labrador', in Colin H. Scott, ed., *Aboriginal Autonomy and Development in Northern Quebec and Labrador*. Vancouver: University of British Columbia Press.

Silcoff, Sean. 2003. 'Landry willing to backstop Bombardier', *National Post*, 22 Jan., FP3.

Simard, Jean-Jacques, et al. 1996. *Tendances Nordiques: Les changements sociaux 1970–1990 chez les Cris et les Inuit du Québec: Une enquête statistique exploratoire*. Québec City: GETIC, Université Laval.

Simard, Majella. 1998. 'Les théories de développement régional et la contribution des ressources dans le démarrage des petites localités en voie de dépeuplement: le cas du Bas Saint-Laurent', *Canadian Journal of Regional Science* 21, 1: 127–50.

Simard, Martin. 2000. 'Développement local et identité communautaire: l'exemple du quartier Saint-Roch à Québec', *Cahiers de Géographie du Québec* 44, 1–2: 167–88.

Simeon, Richard. 1988. 'Meech Lake and Shifting Conceptions of Canadian Federalism', *Canadian Public Policy* 14: S7–24.

Simpson, Jeffrey. 1993. *Faultlines: Struggling for a Canadian Vision*. Toronto: HarperCollins.

Stanford, Quentin H., ed. 1998. *Canadian Oxford World Atlas*, 4th edn. Toronto: Oxford University Press.

Statistics Canada. 1982. *Census Metropolitan Areas and Census Agglomerations with Components*. Catalogue no. 95–903. Ottawa: Minister of Industry.

———. 1992. *Census Divisions and Subdivisions*. Catalogue no. 93–304. Ottawa: Minister of Supply and Services.

———. 1996a. *Labour Force Annual Averages 1995*. Catalogue no. 71–220–XPB. Ottawa: Statistics Canada.

———. 1996b. Canadian Statistics: Distribution of Employed People, by Industry, by Province [on-line database], Ottawa. Searched 20 June 1996: <http://www.statcan.ca/english/Pgdb/labor21b.htm>.

———. 1997a. *A National Overview: Population and Dwelling Counts*. Catalogue no. 93–357–XPB. Ottawa: Industry Canada.

———. 1997b. 1996 Census: Nation Tables—Population by Mother Tongue, Showing Age Groups, for Canada, Provinces and Territories, 1996 Census—20% Sample Data, 2 Dec. 1997 [on-line database], Ottawa. Searched 15 July 1998: <http://www.statcan.ca/english/census96/>.

———. 1997c. *The Daily*—1996 Census: Mother Tongue, Home Language and Knowledge of Languages, 2 Dec. [on-line database], Ottawa. Searched 14 July 1998: <http://www.statcan.ca/Daily/English/>.

———. 1998a. *Canadian Economic Observer*. Catalogue no. 11–010–XPB. Ottawa: Statistics Canada.

———. 1998b. *The Daily*—1996 Census: Aboriginal Data, 13 Jan. [on-line database], Ottawa. Searched 14 July 1998: <http://www.statcan.ca/Daily/English/>.

———. 1998c. *The Daily*—1996 Census: Ethnic Origin, Visible Minorities, 17 Feb. [on-line database], Ottawa. Searched 16 July 1998: <http://www.statcan.ca/Daily/English/>.

———. 2002. 2001 Census: Population and

Dwelling Counts, for Census Metropolitan Areas and Census Agglomerations, 2001 and 1996 Censuses, 16 July [on-line database], Ottawa: <http://www.statcan.ca/english/IPS/Data/93F0050XCB2001013.htm>.

———. 2003. Canadian Statistics: Distribution of Employed People, by Industry, by Province [on-line database], Ottawa. Searched 20 May 2003: <http://www.statcan.ca/english/Pgdb/labor21b.htm>.

Thibodeau, Jean-Claude, and Yvon Martineau. 1996. 'Essaimage technologique en région périphérique: étude de cas', *Revue Canadienne des sciences régionales* 19, 1: 49–64.

Thompson, Robert. 2000. 'IBM Canada growing in Quebec', *National Post*, 30 Nov., C6.

Trent, John E., Robert Young, and Guy Lachapelle, eds. 1996. *Québec-Canada: What is the Path Ahead/Nouveaux sentiers vers l'avenir.* Ottawa: University of Ottawa Press.

Turgeon, Pierre. 1992. *La Radissonie: le pays de la Baie James.* Montréal: Libre Expressions.

Vandermissen, Marie-Hélène, Paul Villeneuve, and Marius Thériault. 2001. 'L'évolution de la mobilité des femmes à Québec entre 1977 et 1996', *Cahiers de géographie du Québec* 45, 125: 211–43.

Villeneuve, Paul. 1992. 'Un Québec en révolution tranquille', in Roger Brunet, ed., *Géographie universelle.* Paris: Hachette and Reclus, 357–73.

———. 1997. 'Le Québec et l'intégration continentale: Un processus à plusieurs vitesses et à directions multiples', *Cahiers de géographie du Québec* 41, 114: 337–47.

——— and G. Côté. 1994. 'Conflits de localization et étalement urbain: y a-t-il un lien?', *Cahiers de Géographie du Québec* 38, 105: 397–412.

Yakabuski, Konrad. 1997. 'Norsk to Expand Quebec Plant', *Globe and Mail*, 12 June, B6.

———. 1998. 'Alcan plans $2.2-billion smelter', *Globe and Mail*, 20 Feb., B1.

Further Reading

Bourassa, Robert. 1985. *Power from the North.* Scarborough, Ont.: Prentice-Hall.

Power from the North tells the story of the James Bay Project from the perspective of Premier Robert Bourassa. In the late 1960s, Bourassa grasped the concept of selling low-cost hydroelectricity from northern Québec to energy-deficient New England. The James Bay Hydroelectric Project would develop the water resources of Québec's northern hinterland, provide low-cost power to industries willing to locate along the St Lawrence River, and generate sufficient revenues from long-term contracts with utility companies in New England to pay for the project. In addition, Hydro-Québec, while challenged to build such a massive project, would demonstrate once and for all the ability and capacity of Quebecers to undertake and complete such huge construc-

tion efforts. For Québec, this was the project of the century. So far, only the first of three phases has been completed. The first phase harnessed the waters of La Grande, Eastmain, and Caniapiscau basins, leaving uncompleted the second phase (the Great Whale Basin) and the third phase (the Nottaway, Broadback, and Rupert basins).

Four themes resonate in *Power from the North*:

- a strong pro-development sentiment;
- a continental energy policy benefiting both Québec and New England;
- a provincial development strategy based on supplying low-cost electrical energy to industrial firms willing to relocate to southern Québec;
- a triggering of an angry Aboriginal response to this project resulting in the

James Bay and Northern Québec Agreement.

Students should recognize that the James Bay Project remains very controversial (McCutcheon, 1991; Richardson, 1975; Salisbury, 1986; Turgeon, 1992). Native and environmental groups banded together to fight this project. In this sense, the James Bay Project took on a larger dimension, that is, the choice between development and the well-being of the environment and Aboriginal peoples. For example, microbial mercury methylation resulting from decaying vegetation in newly constructed reservoirs enters the food chain, making the frequent consumption of fish hazardous to a person's health (Hecky, 1987: 9). In 1999, James Hornig edited a collection of papers on James Bay some 20 years after the project was announced by Premier Bourassa. With the benefit of time, these articles present a more balanced assessment of the social and environmental impacts of the James Bay Hydroelectric Project.

The Stanley Park Seawall
(Al Harvey/The Slide Farm)

British Columbia is an emerging economic power in Canada. Until the slump of the late 1990s it outstripped the other regions in economic growth, and this long period of rapid expansion altered its geopolitical position within Canada. The disagreements between Ottawa and BC reflect the centralist/decentralist faultline. Much of BC's economic growth is based on its natural resources, and more recently, its high-technology and service industries. This rapid economic growth has sparked an influx of new residents.

BC has two distinct subregions: its heavily populated southwest corner is an economic core, while the rest of the province is a resource hinterland. While the economy shows signs of maturing, it remains dependent on resource industries, especially the forest industry. The magnitude of Aboriginal land claims marks the ownership of BC as a critical issue that must be resolved.

- Describe British Columbia's physical geography and history.
- Present the basic elements of BC's economy and population.
- Examine these elements within BC's physical setting and the core/periphery model.
- Present arguments for two faultlines in BC: Aboriginal/non-Aboriginal and centralist/decentralist.
- Focus on the forest industry and its changing role in BC's economy.
- Ask whether BC is an upward transitional region or an industrial core.

Chapter 7

British Columbia

■ Introduction

British Columbia is a growing economic and political force within Canada. Much of its economic and demographic growth has taken place in the Greater Vancouver region and this important west coast city continues to drive the BC economy. Gradually, however, the rest of British Columbia participated in this economic growth, especially as a resource hinterland. On the other hand, most high-technology and service industries are located in the Greater Vancouver region. Over the past decade, both high-technology and service industries have expanded, thereby diversifying the province's economy and contributing to Vancouver's continued growth. The tourism industry, by capitalizing on BC's varied and attractive physical geography, has sparked new investment and generated more service jobs. Foreign trade, especially with Japan and other Pacific Rim countries, has also stimulated BC's economy. From the perspective of British Columbians, BC is part of the economic heartland of Canada and North America, yet BC's economy retains many characteristics of Friedmann's upward transitional region. Over the last five years, its economy lost its momentum. This economic slowdown, which began in 1998 due largely to BC's dependency on exports to Asian markets, shows little sign of ending. In 2002, the forest industry was hurt by a 27 per cent duty on softwood lumber exports to the United States. In the following year, forest fires ravaged vast timber stands in the interior of the province. The long-term consequences of these two developments—US duties and forest fires—on the forest industry are difficult to gauge, but both already have had a negative impact on trade, business, and workers. The forest industry, traditionally the largest resource industry in the province, is the *Key Topic* in this chapter.

What makes British Columbia so attractive for industry and people? While much can be attributed to its mild climate and rich resources, British Columbia's strategic position on the Pacific coast has made it a trade centre. Much of its population increase is due to the arrival of immigrants from Asia who, in turn, created business and established trade with Asian countries. Canadians from other provinces have also been attracted to the mild climate and economic opportunities. Since its population has grown and since size is a measure of political power, British Columbia can now demand more political respect and attention from Ottawa. Indeed, BC's demands for more political power underscore the struggle between the province and Ottawa, which is described in this chapter as the centralist/decentralist faultline. The popularity of Cascadia, a regional alignment of Oregon, Washington, and British Columbia, signals both the appeal of a west coast solution to a new North American political arrangement and a sign of the level of disenchantment with Ottawa. Cascadia is also the product of geographic proximity to the American west coast. Within BC there is another faultline—the Aboriginal/non-Aboriginal faultline—whereby Aboriginal peoples are demanding more power through land-claim agreements.

British Columbia within Canada

British Columbia is an emerging giant within Canada's economic system. The region has undergone rapid economic growth and is ideally situated for trade with Pacific Rim countries. In fact, British Columbia's economy benefited from the rapidly expanding Asian economy during the 1980s and early 1990s, and this important trade already shows signs of recovery. A large number of Hong Kong immigrants to BC (especially to Vancouver) brought with them skills, capital, and Asian business connections, thereby stimulating the region's economy. In the past, this economy

was propelled by its vast array of natural wealth—fish, forests, and minerals. More recently, natural gas production has taken first place in the resource sector. Though exploitation of these resources remains very important, BC's economy is now powered by other sectors, especially high technology and its tourist industry.

BC's strong place in Canada is revealed by its share of Canada's GDP and population (Figure 7.1). The growing population of Aboriginal peoples, with ongoing land claims, makes them an important minority within British Columbia. On the other hand, the percentage of French-speaking people living in British Columbia is declining.

Since 1998, BC's economy has sputtered, allowing other provinces but especially Alberta and Ontario to outperform it. In 2002, economic recovery seemed just around the corner until Washington imposed a 27 per cent duty on softwood lumber from Canada. Since British Columbia is the major exporting province of softwood lumber, this economic blow hurt its major industry and caused layoffs in many single-industry towns. The forest industry was dealt another blow in the summer of 2003 when fierce forest fires ravaged many timber stands in the interior (Vignette 7.1). The fires spread into inhabited areas, including the city of Kelowna. On the bright side, Vancouver is the site of the 2010 Winter Olympics and that event will greatly stimulate the tourist industry and may have other economic spinoffs.

As in Ontario and Québec, there is a core/periphery arrangement of the population in BC, though it is not as prominent in this province. This geographic dichotomy is best expressed by British Columbia's population distribution. There is a population core in the southwest corner of the province, where over

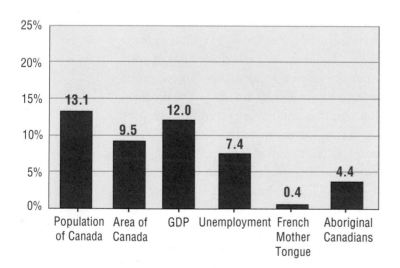

■ **Figure 7.1 British Columbia, 2001.** BC's strong place in Canada is revealed by its share of Canada's GDP and population. The Aboriginal population and French by mother tongue show the weak position of Aboriginal peoples and French-speaking Canadians in the province. The small percentage of Aboriginal peoples, however, does not reflect the extent of First Nations' potential ownership of land in BC through ongoing land-claim negotiations.

Sources: Tables 1.1, 1.5, and 4.19.

60 per cent of BC's residents are concentrated. The major urban centres, including Vancouver and Victoria, are within this core. Beyond this population core is more than 90 per cent of the province's territory. Much of this rugged terrain is sparsely populated, though there are growing population clusters in the Okanagan and Thompson valleys and on Vancouver Island north of Victoria. As well, population centres are located along the north coast around Prince Rupert, in the central interior at Prince George, and in the Peace River country. Finally, east of the Okanagan Valley, a few towns, such as Trail, are located near the US border.

In British Columbia, natural resources remain important but no longer dominate the province's economy. For instance, 50 years ago forestry alone accounted for 50 per cent of the provincial economy and for most of the employment. By the late 1990s, forestry comprised only 17 per cent of BC's economy and 14 per cent of employment (Barnes and Hayter, 1997: 5). Still, forestry, mining, and fishing generate most of the manufacturing jobs in BC. However, competition for land and water by a growing recreation industry and an expanding urban community is increasing. The fixed size of BC's natural resources is thus

■ **Figure 7.2 Vancouver wins Winter Olympics bid**. On 2 July 2003, British Columbia received a much needed boost when the International Olympic Committee declared Vancouver the Host City for the 2010 Olympic and Paralympic Winter Games. These games are projected to cost $1.3 billion but could generate up to $10 billion in direct economic activity. Vancouver will provide the facilities for hockey, figure skating, curling, and speed skating while alpine skiing and Nordic sports will take place at Whistler. Located 115 km north of Vancouver, Whistler boasts the largest ski area in North America. (© Adrian Raeside)

Vignette 7.1   BC's Precipitation: Too Much or Too Little?

British Columbia receives the greatest amount of precipitation along its Pacific coast. Inland, the annual precipitation decreases sharply. In simple terms, there are two precipitation areas in British Columbia—one in the Pacific Climatic Zone, which receives from 800 to 2,000 millimetres per year, the other in the Cordillera Climatic Zone, where less than 800 millimetres fall each year. These figures are average amounts of rain and snow. Fluctuations do occur. The interior of British Columbia has seen five consecutive years of weather drier than usual and this has resulted in extensive forest fires and affected production of hydroelectric power. For example, the power station at Kemano is dependent on its water supply from the Nechako Reservoir located east of the Coast Mountains. Low levels at this reservoir in 2000 forced Alcan to reduce its production at its smelter in Kitimat and to suspend power deliveries to two customers, BC Hydro and Europcan Pulp and Paper. After an extremely dry summer in 2003, forests in the interior were tinder dry. The Okanagan Valley has three characteristics that can lead to fires spreading rapidly—strong winds, steep slopes, and a combination of dry underbrush, grass, and lodgepole pine trees. On 23 August 2003, such a fire reached Kelowna. As a forest fire raged out of control and swept into its suburbs, 10,000 residents had to abandon their homes and more than 250 houses were destroyed.

Sources: Hasselback (2000: D7); Boei and O'Brian (2003: A1).

under pressure and is faced with environmental degradation from industrial, recreational, and urban users. Herein lies a dilemma: economic growth depends on the resource industries, yet BC's outdoor lifestyle and tourist industry rely on its pristine wilderness. The challenge will be to balance the needs of these two sectors in the coming years, to exert better control over the misuse of the natural environment, and to reduce air and water pollution. Outstanding land claims pose a second dilemma, for failure to negotiate with Indian bands could stall BC's economy and keep Indians from participating in BC's economy.

Environmental Challenges

British Columbia faces several environmental challenges. As outlined above, the seemingly limitless natural riches found in this Pacific province have been subjected to past mismanagement and wasteful practices that have led to resource loss, environmental degradation, and conflicts. Clear-cutting of the forest remains one such controversial practice. Clear-cutting is the harvesting method most widely used in BC. The clear-cutting method is to cut down every tree within a large area. In the past, huge areas were stripped of their trees. Now provincial regulation restricts such logging to 40–60 hectares, but clear-cutting still takes place on steep slopes where soil erosion is greatly increased. The alternate to clear-cutting is selective cutting. However, logging companies claim that selective logging is too expensive and would not allow them to compete in world markets. In the past, clear-cutting extended right to the banks of streams and rivers, leaving surrounding land vulnerable to rapid soil erosion and stream sedimentation. Under such conditions, fish habitat is damaged and spawning grounds may be destroyed. The Nisga'a claim (Nisga'a, c. 2000) that salmon spawning grounds along the Nass River have been damaged by such logging practices, thus significantly reducing the salmon run.

Land Claims and the Aboriginal/Non-Aboriginal Faultline

Who owns British Columbia? This question goes to the heart of the land-claims issue. Indian bands are pressing to settle their claims to lands they occupied prior to the arrival of Europeans. The overlapping land claims of BC's 197 Indian bands equal 110 per cent of the province (Howard, 1996: A5), but land-claim settlements rarely exceed 10 per cent of the claimed territory. Nevertheless, the stakes are very high. Until 1990, the government of British Columbia was dragging its feet on land claims, but now BC recognizes its obligation to settle outstanding Aboriginal land claims. Following the results of the Treaty Negotiations Referendum in May 2002, all parties are willing to push forward. Once land claims have been addressed, the tension underlying the Aboriginal/non-Aboriginal faultline in BC will be reduced, and First Nations will be able to focus their energies on improving the economic and social well-being of their members.

Almost 200,000 Aboriginal people live in British Columbia. A small number of Indian bands have signed treaties: some, in northeastern BC, fall under Treaty No. 8, and others under the Douglas and Vancouver Island treaties. These treaties extinguished Aboriginal rights to the land in exchange for various benefits for the Indians. Treaty No. 8, propelled by the Klondike gold rush of 1898, had two pur-

poses: (1) to open access to miners and prospectors heading to Yukon across the Canadian Northwest, and (2) to allow the Indians to continue their traditional way of life and receive annual payments. The 14 Douglas treaties, on the other hand, focused on providing land reserves for Indians along with their right to hunt and fish on Crown land.

Most of the First Nations in BC have still not negotiated a settlement with Ottawa and Victoria. However, a combination of blockades, legal action, and public protests has made this a central political issue in the province. In 1990, Victoria made a commitment to partici-

Totem poles not only represent Indian family kinships and stories, but support their land claims, which are based on their ancestors' occupancy of British Columbia. This well-known totem pole is located at Prospect Point in Stanley Park. In the background are the North Shore mountains. (Search4Stock Inc.)

Vignette 7.2   Facts about the Nisga'a Agreement

Location:

The Nass Valley extends from the Pacific Ocean across and beyond the Coast Mountains approximately 150 km inland. This valley is the homeland of the Nisga'a. Prince Rupert is the closest large city and it lies some 100 km southwest of the headwaters of the Nass Valley.

Population:

The total population of the Nisga'a is almost 10,000, with approximately 6,000 living in British Columbia and the rest in other parts of Canada and the United States. Some 2,400 Nisga'a reside in the Nass Valley in the villages of New Alyansh (Gitlakdamiks), Canyon City (Gitwinksihikw), and Kincolith (Gingolx).

Agreement:

The Nisga'a and the federal and BC governments approved and signed the agreement in 1999. In exchange for their traditional land claim, the Nisga'a receive title to almost 2,000 km² in the Nass Valley of British Columbia; access to forest and fishery resources; self-government powers, including a Native judicial system and policing; and nearly $500 million in cash, grants, and program funds from Victoria and Ottawa. British Columbia supplies the land and 30 per cent of the cash settlement with the federal government paying 70 per cent of the cash settlement.

pate in resolving land claims. Much has happened since then, including the establishment of a treaty commissioner, whose role is to coordinate negotiations between the Aboriginal leaders and the federal and provincial government representatives. By November 1994, 42 First Nations groups representing about 70 per cent of the Aboriginal population had entered the process of negotiations. By 1996, one claim, that of the Nisga'a, had been approved in principle by the Aboriginal leaders and the public officials representing Ottawa and Victoria. By 1998, the Nisga'a had approved this agreement, and in 1999 the Nisga'a Agreement was ratified by the provincial legislature and the federal Parliament (Vignette 7.2).

The Agreement sees the Nisga'a surrender their traditional claim to lands, remove themselves from the Indian Act, and agree to cede a portion of their tax-free status for a settlement worth $490 million, which includes 2,000 km² of land. The Nisga'a also gain governance

powers, access to resources, a share of the commercial fishing catch, and timber rights. This land-claim settlement, which is viewed as a precedent for the remaining land claims in British Columbia, transfers considerable resources, capital, and political power (self-government) to the Nisga'a in exchange for the surrender of much of their traditional homeland. The Nisga'a now have an opportunity to put politics behind them and to move forward on their economic and social agenda. The political dimensions of this agreement, which date back many decades and include the landmark 1973 *Calder* decision by the Supreme Court of Canada, have been so critical that Frank Cassidy (1992: 11) described it as 'a centrepiece in the historical development of the province of British Columbia'. Given the number of land claims and the slow nature of negotiations, however, BC's obligation to its Aboriginal citizens is expected to take a great deal of time to resolve.

British Columbia's Physical Geography

The physical geography of British Columbia is far more varied than that of any other region of Canada and even extends offshore to a number of islands. The climate of the west coast is unique in Canada. Winters are extremely mild and freezing temperatures are uncommon. Summer temperatures, while warm, are rarely as high as temperatures common in the more continental and dry climate of the Interior Plateau of British Columbia. Moderate temperatures, high rainfall, and mild but cloudy winters make the west coast of British Columbia an ideal place to live and a popular retirement centre for those Canadians wanting to escape long cold winters.

The Pacific Ocean has a powerful impact on BC's climate, resource base, and transportation system. Unlike in Atlantic Canada, the continental shelf in BC extends only a short distance from the coast. Within this narrow zone there are many islands, the largest being Vancouver Island followed by the Queen Charlotte Islands (Haida Gwaii). The riches of the sea include salmon, which return to the rivers, such as the Fraser and the Skeena, to complete their life cycle. Most of BC's natural wealth, however, is not in the sea but in the province's diversified physical geography, which provides valuable resources, particularly forests, minerals, and rivers.

Most of British Columbia lies in the physiographic region known as the Cordillera—a combination of mountains, plateaux, and valleys. A small portion in the northeast of British Columbia, known as the Peace River country, extends into the Interior Plains region (Figure 2.1). The Cordillera is a complex physiographic region. Extending from southern British Columbia to Yukon, this region encom-passes over 16 per cent of Canada's territory. As explained in Chapter 2, the Cordillera was formed by severe folding and faulting of sedimentary rocks. This coastal zone is subject to earthquakes because of tectonic movement.

The Cordillera has at least 10 mountain ranges, the most prominent of which are the Coast Mountains, which extend northward from Vancouver to the Alaskan Panhandle, and the Rocky Mountains, which stretch from the Canada–US border almost to Yukon (Figure 7.3). Other north–south mountain ranges include the Insular Mountains that rise above the sea to form the Queen Charlottes and Vancouver Island. Two more mountain ranges, which lie in the southern interior of BC, are the Cascade Mountains, in south-central BC along the US border, and the Columbia Mountains. The Columbia Mountains consist of three parallel, north–south mountain ranges—the Purcell, Selkirk, and Monashee—and a fourth range, the Cariboo Mountains, forms the Columbia's northern extension. Further north are four mountain ranges—the Hazelton, Skeena, Omineca, and Cassiar mountains.

The Interior Plateau separates the Coast Mountains from the mountains of the interior (Figure 7.3). North of the Interior Plateau is the Stikine Plateau. The topography of these plateaux consists of gently undulating land with occasional deeply trenched river valleys. The Fraser Canyon, one of BC's best-known landforms, drops about 300 to 600 m below the Interior Plateau and extends from just south of Quesnel to Hope. Just south of Lytton, the canyon walls rise about 1,000 m above the river. This rocky gorge is called Hell's Gate. The Fraser Canyon is a major fault zone that separates the Cascades from the Coast Mountains.

British Columbia has two climatic zones, the Pacific and the Cordillera. Because of the

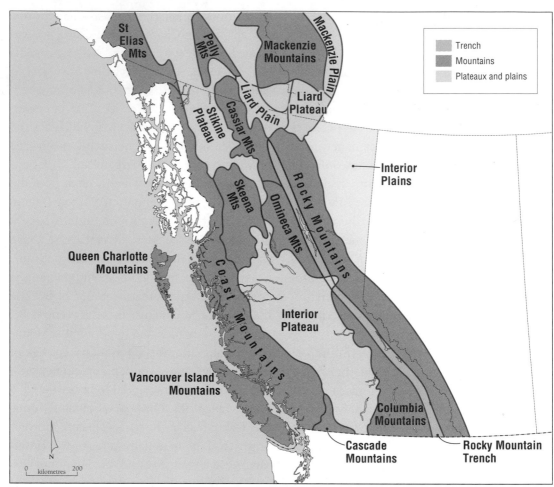

Figure 7.3 Physiography of the Cordillera. British Columbia's complex physiography is evident in the physiographic subregions in the Cordillera. The difficulty of constructing east-west transportation routes can be appreciated if one considers the number of north-south mountain ranges that must be traversed. For that reason, the importance of the Pacific Ocean for transportation prior to the completion of the CPR becomes clear. As well, the grain of the land makes north-south land transportation construction into the Pacific Northwest of the United States relatively simple. Herein lie the natural factors behind the political concept of Cascadia. (Further resources: Student Web site, National Atlas section, Map 25. Web site instructions are found on p. xxi.)

extremely high elevations in the Coast Mountains, few moist Pacific air masses reach the Interior Plateau. The spatial variation in precipitation is remarkable. Heavy orographic precipitation occurs along the western slopes of the Insular and Coast mountains where 2,000 mm of precipitation fall annually (Vignette 7.2). In sharp contrast, the Interior Plateau receives less than 400 mm per year. In the Thompson Valley and Okanagan Valley of

the Interior Plateau, hot, dry conditions result in an arid climate with sagebrush in the valleys and ponderosa pines on the valley slopes.

The Pacific coast of British Columbia has the most temperate climate in Canada, dominated by the constant flow of moist Pacific air masses across its terrain. The result is mild and often wet, cloudy weather. Because eastward-moving Pacific air masses must rise above the Vancouver Island Mountains and the Coast

Mountains, orographic precipitation frequently occurs (Vignette 2.9). This, combined with frontal precipitation, gives the west coast a high level of annual precipitation. Known locally as 'liquid sunshine', the annual precipitation along the west coast of Vancouver Island can exceed 3,000 mm. Further east, the annual precipitation declines. At Vancouver, located near the Coast Mountains, the annual precipitation declines to about 1,000 mm. Victoria, however, lies in the partial rain shadow of the Insular Mountains of Vancouver Island and thus avoids the heavy orographic rainfall that affects Vancouver. Consequently, the capital city receives nearly 40 per cent less rainfall than Vancouver.

As a mountain climate, the Cordillera climatic zone consists of a number of microclimates. Changes in elevation (height above sea level), latitude (from 49° N to 60° N), and distance from the Pacific Ocean to the Rocky Mountains (600 km) control each microclimate, which in turn affect micro-vegetation and micro-soil conditions. In general, these microclimates become drier as distance from the Pacific Ocean increases and cooler as either elevation or latitude increases. The range of natural vegetation and soil types is enormous. In the Thompson Valley near Kamloops, an arid microclimate results in a desert-like grassland vegetation with chernozemic soils. In sharp contrast, huge trees in the coastal rainforest grow in podzolic soils. Different still are the higher elevations of the Rocky Mountains extending beyond the treeline, where tundra consisting of grasses, mosses, and lichens grows on cryosolic soils. For much of northern British Columbia, the northern coniferous forest (evergreen, needle-leaf trees) is the most common natural vegetation.

Because the Cordillera stretches from southern British Columbia to Yukon, the climate turns into a Subarctic mountain climate north of 58° N. Higher elevations in these latitudes have Arctic climatic conditions. South of 58° N, however, summers are longer and warmer, partly because the eastward-moving Pacific air masses are more dominant and partly because of greater annual solar energy.

The natural vegetation of British Columbia is far from homogeneous (Figure 2.7). Along the west coast, the mild, wet climate encourages a rainforest. The Insular Mountains of Vancouver Island and the Coast Mountains along the mainland both rise abruptly from the Pacific Ocean and therefore receive much rainfall throughout the year. The vegetation cover responds with lush evergreen and deciduous trees. Characteristic tree species are western hemlock, Douglas fir, western red cedar, and Sitka spruce. Under this vegetation cover, podzolic soils are common. These soils are strongly acidic and low in plant nutrients, making it necessary for farmers to use fertilizers. Fertilizers are used because nutrients are washed away by heavy rainfall, and the needles of the coniferous trees, when dropped, add to the ground cover and produce an acidic soil. This coastal region contrasts with the sagebrush and yellow grasses in the lower elevations of the southern Interior Plateau, where higher temperatures occur. At higher elevations and in more northerly areas, where cooler temperatures prevail, forest covers the Interior Plateau. The series of mountain ranges, aligned north to south, are also forested.

Because of the rugged nature of the Cordillera, there is little arable land. Only about 2 per cent of the province's land is classified as arable. British Columbia's largest area of cropland lies outside of the Cordillera physiographic region in the Peace River country. Within the Cordillera, most arable land is in the Fraser Valley, while a smaller amount can be found in the interior, especially the Okana-gan and Thompson valleys. This shortage of arable land poses a serious problem for British Columbia. With urban developments spreading onto agricultural land, British Columbia is losing some of its most productive farmland.

British Columbia's Historical Geography

Indians lived along the Pacific coast of British Columbia for over 10,000 years before European explorers reached the northern Pacific coast in the mid-eighteenth century. The Spanish had already sailed northward from Mexico to California, but the Russians were the first to reach Alaska and establish fur-trading posts along its coast. In 1778, Captain James Cook established Britain's interest in this region by sailing into Vancouver Island's Nootka Sound, where he and his sailors found the Nootka village of Yuquot. The Nootka, now known as Nuu-Chah-Nulth, fished for salmon and hunted the sea otter. Upon landing, Cook engaged in trade for sea otter pelts, then sailed to China

where he and his men sold the pelts for a handsome profit. After the Royal Navy published Cook's record of his voyage, British and American traders came to the Pacific Northwest to seek the highly valued sea otters. Russian fur traders, based in Alaska, also harvested sea otters. Spain, which considered these lands Spanish territory, was disturbed by these interlopers and sent a fleet northward from Mexico in 1789. At Nootka Sound on the west coast of Vancouver Island, the Spanish seized several ships and built a fort to defend their claim to these lands. In 1792, Captain George Vancouver of Britain's Royal Navy sailed around Vancouver Island. In the following year, Alexander Mackenzie travelled overland from Fort Chipewyan to just south of Prince George and then to the Pacific coast near Bella Coola, which is just over 400 km north of Vancouver. Under the Nootka Convention (1794), the Spanish surrendered their claim to the Pacific coast north of 42° N, leaving the British and Russians in control.

In the early nineteenth century, the North West Company established a series of fur-trad-

Captain James Cook's ships were moored in Nootka Sound, as depicted in a watercolour by M.B. Messer. Four years earlier, the Spanish explorer Juan Hernandez sailed along the BC coast. However, there is the possibility that Francis Drake, while on a secret mission to find a western entrance to the Northwest Passage, reached these waters in 1579 (Hume, 2000: B1). (*National Archives of Canada C11201*).

ing posts along the Columbia River. From 1805 to 1808, Simon Fraser, a fur trader and explorer, explored the interior of British Columbia on behalf of the North West Company. He travelled by canoe from the Peace River to the mouth of the Fraser River. As elsewhere, the strategy of the North West Company was to develop a working relationship with local Indian tribes based on bartering manufactured goods for furs. After 1821, the North West Company merged with its rival, the Hudson's Bay Company. The Hudson's Bay Company then took charge of the Oregon Territories, which extended from the mouth of the Columbia River to Russia's Alaska.

In 1843, American settlers began to arrive on the coast from the eastern part of the United States. In the same year, the Hudson's Bay Company relocated its main trading post from Fort Vancouver at the mouth of the Columbia River to Fort Victoria at the southern tip of Vancouver Island. The increasing number of American settlers who came west along the Oregon Trail represented a challenge to the authority of the Hudson's Bay Company. A few years later, the United States claimed all of the Pacific coast northward to Alaska, where Russian fur-trading posts existed. In 1846, Britain and the United States agreed to place the boundary between the two nations at 49° N and then to follow the channel that separates Vancouver Island from the mainland of the United States. While the loss of the Oregon Territories was substantial, Britain was fortunate to hold onto the remaining lands administered by the Hudson's Bay Company. Britain recognized that its hold on these lands through the Hudson's Bay Company was tenuous and could not withstand the political weight of the growing number of American settlers. Indeed, without the presence of the Hudson's Bay Company and Britain's negotiat-

ing skills, Canada might well have lost its entire Pacific coastline.

The gold rush of 1858 brought about 25,000 prospectors from California to the Fraser River. Prospectors walked upstream along the sandbanks and sandbars of the Fraser River and its tributaries, panning for placer gold (small particles of gold in sand and gravel deposits). The major finds were made in the BC interior, where the town of Barkerville was built near the town of Quesnel. By 1863, Barkerville had a population of about 10,000, making it the largest town in British Columbia. To ensure British sovereignty over territory north of 49° N, the British government established the mainland colony of British Columbia in 1858 under the authority of Sir James Douglas, who was also governor of Vancouver Island. In 1866, the two colonies were united.

Confederation

By the 1860s, the British government was actively encouraging its colonies in North America to unite into one country. Once the first four colonies were united in 1867, Ottawa adopted the British strategy to create a transcontinental nation. An important part of that strategy was to lure British Columbia into the 'national fabric'. The Canadian Pacific Railway was the first expression of this national policy.[1] Ottawa promised to build a railway to the Pacific Ocean within 10 years after British Columbia joined Confederation. In Fort Victoria, however, some wanted to join the United States. By the middle of the nineteenth century, British Columbia had developed significant commercial ties with Americans along the Pacific coast. San Francisco was the closest metropolis and, with its railway to New York, offered the simplest and quickest route to London. In 1859, Oregon became a state and

Washington was soon to follow. Commercial links with the United States were growing stronger. But the majority of people in Fort Victoria wanted to remain British. In 1871, British Columbia chose to become a province of Canada (Figure 3.5). British Columbians, however, had to wait 14 years for the Canadian Pacific Railway to reach the Pacific coast at Port Moody, near Vancouver, on Burrard Inlet. Later, the railway was extended 20 km westward to the small sawmill town of Vancouver, where there was a better harbour and terminal site for the railway (Figure 7.4).

When British Columbia joined Confederation in 1871, its official population was 36,247 (McVey and Kalbach, 1995: 35). This figure likely underestimated the number of Indians and prospectors who were living in the more remote areas of the province.[2] At the time of the first comprehensive census of British Columbia in 1881, there were approximately 4,200 Chinese, 19,000 white settlers (American, Canadian, and British), and about 30,000 Indians. Most British settlers lived around Fort Victoria. Beyond Fort Victoria, the vast majority of the inhabitants were Aboriginal peoples.

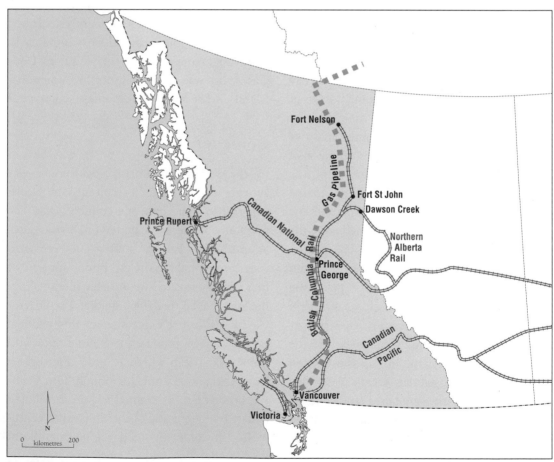

Figure 7.4 Railways in British Columbia. The first railway to cross the mountains of the Cordillera was the Canadian Pacific Railway in 1885. To open the northern interior of the province, Victoria built the Pacific Great Eastern Railway (now known as British Columbia Rail). After several extensions, British Columbia Rail joined North Vancouver to Fort Nelson by 1971. An eastern link from this railway joins Dawson Creek to the Northern Alberta Railway.

In the previous decade, about 25,000 prospectors, mainly Americans, had been scattered in small camps along the Fraser River or at Barkerville. Most Americans had left at the end of the gold rush.

Post-Confederation Growth

At first, Confederation had little effect on British Columbia. The province was isolated from the rest of Canada, Canada's fledgling factories, and even the halls of power in Ottawa. Goods still had to come by ship from San Francisco or London. When the Canadian Pacific Railway was completed in 1885, British Columbia truly became part of the Dominion, and BC's role as a gateway to the world began.

The main line of the Canadian Pacific Railway and its many branch lines were responsible for the formation of many of the province's towns and cities and for providing access to its forest and mineral wealth. The Esquimalt and Nanaimo Railway, the Canadian National Railways, and British Columbia Rail (BC Rail) added to the rail network in British Columbia. In 1886, the Esquimalt and Nanaimo Railway was built, connecting the coalfields of Nanaimo with the capital city of Victoria, and stimulated logging and sawmilling along its rail route. With the closure of the coal mines in the early 1950s, the railway lost its main function and now plays a minor role in the transportation system of Vancouver Island. By 1914, the Grand Trunk Pacific Railway (later the Canadian National Railways) provided a trans-Canada rail route to Prince Rupert and an alternative rail service to Vancouver. The Pacific Great Eastern Railway (later BC Rail) was incorporated in 1912 but laid few rails until the early 1950s, when the provincial government made a commitment to complete the railway in order to facilitate resource development in the interi-

or of British Columbia. By 1956, this rail line extended from North Vancouver to Prince George. Two years later, BC Rail extended its rail system to Dawson Creek and Fort St John in the Peace River country, and by 1971 it reached Fort Nelson in BC's forested northeast.

After the completion of the CPR, Vancouver grew quickly and soon became the major centre on the west coast. By 1901, Vancouver had a population of 27,000 compared to Victoria's 24,000. As the terminus of the transcontinental railway, Vancouver became the transshipment point for goods produced in the interior of BC and Western Canada. As coal, lumber, and grain were transported by rail from the interior of British Columbia and the Canadian Prairies, the port of Vancouver spearheaded economic growth in the southwest part of the province. It was then possible to tap the vast natural resources of BC and ship them to world markets through the new port of Vancouver. By the twentieth century, Vancouver had become one of Canada's major ports. Vancouver, located on Burrard Inlet, has an excellent harbour. Unlike Montréal, Vancouver has an ice-free harbour, thanks to the warm Pacific Ocean. As Canada's major Pacific port, Vancouver became the natural transportation link to Pacific nations. With the opening of the Panama Canal in 1914, British Columbia's resources were more accessible to the markets of the United Kingdom and Western Europe.

Between the completion of the Canadian Pacific Railway in 1885 and the end of World War I in 1918, British Columbia underwent a demographic explosion. By 1921, BC had over half a million inhabitants. The province had been gradually transformed from a fragile political entity in 1871 into a self-confident political and social entity. During this time, the populations of both European and Asian origin increased about tenfold. The European

Vancouver has a magnificent harbour that has facilitated trade with Pacific Rim countries. With Vancouver in the background, a cruise ship sails through Burrard Inlet and passes under Lions Gate Bridge. Stanley Park is located on the south side of Lions Gate Bridge while West Vancouver is on the north side. (Al Harvey/The Slide Farm)

population had reached nearly half a million, while the Asian population was about 40,000. At the same time, a combination of disease and social dislocation caused the number of Aboriginal people to decline sharply, perhaps from 40,000 to 20,000. This magnitude of demographic decline was matched in other regions of Canada. While a number of factors were at work, smallpox, tuberculosis, and other communicable diseases caused the greatest losses. (See Chapter 4 for more details.)

While BC's economy and population continued to grow in the 1920s, the Great Depression of the 1930s caused the province's economy to stall and unemployment to rise sharply. Economic disaster struck British Columbia in 1929. Exports of Canadian products from BC, so necessary for its economic well-being, slowed and prices dropped. In the Prairies, the collapse in agricultural prices was accompanied by prolonged drought, turning the land into a dust bowl. Many Prairie farmers abandoned their farms and fled to British Columbia, adding to the burden of unemployment in that province.

World War II and the Post-War Economic Boom

World War II called for full production in Canada, thereby pulling British Columbia's depressed resource economy out of the doldrums. Military production, including aircraft manufacturing, greatly expanded BC's industrial output. As well, resource industries based on forestry and mining (especially coal and copper) were producing at full capacity.

When the war ended in 1945, BC's resource boom continued. With world demand for forest and mineral products remaining high, the provincial government focused its efforts on developing the resources of its hinterland, the central interior of British Columbia. The first step was to create a transportation system from

Vancouver to Prince George, the major city in the central interior. The highway system was improved and extended from Prince George to Dawson Creek in 1952. More important for economic development was the completion of the Pacific Great Eastern Railway to Prince George in 1956 and then to Dawson Creek in 1958. With rail access to Vancouver, forestry, as well as other resource industries, expanded rapidly, thereby leading to the integration of this hinterland into the BC and global industrial core.

Rapid Economic Growth in the 1990s

From 1990 to 1997, BC's increasing economic strength, partly driven by the Asian economy, outpaced that of all other regions in Canada. In 1994, British Columbia's robust economy accounted for over 13 per cent of Canada's GDP, up several percentage points from a decade earlier. Several factors account for the province's rapid economic ascent. The principal force behind BC's growth is a rich and varied resource base accompanied by high commodity prices. Like the Prairie economy, this Pacific region relies heavily on its natural resources for its economic well-being. While mining and fishing are important industries, forestry remains the centrepiece of British Columbia's vibrant resource economy.

Although British Columbia has only 14.5 per cent of the forest lands in Canada, its share of the area of productive forest land amounts to 21 per cent (Statistics Canada, 1998a: 19). Moreover, of Canada's productive forest land, British Columbia has the most valuable timber stands because of the size, type, and density of the trees in its rainforest. Along the coastal communities and in the interior centres, logging and wood processing have provided employment for workers, wealth for its forest product companies, and valuable primary products for domestic and foreign customers. Just under 2 per cent of British Columbia's labour force is employed in logging and related primary forestry jobs (British Columbia, 2000c: 1). Another 10 per cent are involved in wood-processing industries, including sawmilling, plywood manufacturing, and paper-producing activities.

Trade is another dynamic force propelling British Columbia's economy. Vancouver and, to a lesser degree, Prince Rupert and other Pacific ports serve as trade outlets for coal, lumber, potash, and grain from the interior of British Columbia and the Prairie provinces to reach world markets. In BC's export-oriented economy, it is usually much less expensive in transportation costs to ship the raw material than the finished product. This has led to trade focused on resources in BC. Also, BC has higher labour costs and more stringent environmental regulations than most countries importing BC's products. Those factors make BC a high-cost processing region, which has inhibited the development of manufacturing.

During the last 20 years, China and other Pacific Rim countries have led the world in economic growth, creating new markets for Canadian products shipped through Vancouver. One measure of the potential for trade with Pacific Rim countries is the 18-member Asia-Pacific Economic Co-operation forum. Trade opportunities are almost endless between British Columbia and the population of 2.2 billion people in the Pacific nations. However, in 1998, Asia's economy faltered. Asia's economic problems quickly led to a drop in demand for primary products, including BC's forest products. This decrease in demand and a fall in commodity prices caused BC's economy to slide into a recession.

British Columbia Today

Over the years, the economy of British Columbia has diversified. This economic trend is expected to continue, creating a more balanced economy. Already, its dependency on the resource industry has diminished and the so-called new economy has expanded. Three components of the new BC economy are tourism, filmmaking, and high technology. British Columbia is recognized as a world-class tourist destination. With the Winter Olympics in Vancouver and Whistler in 2010, tourism is anticipated to expand rapidly and the economic fallout is expected to spill over into other sectors of BC's economy. Filmmaking has taken hold in the Vancouver area, with American film companies making the Vancouver area a major film production centre. In turn, these activities have created a new sector of employment and business opportunities. The high-technology industry has made the greatest impact on BC's economy. It has two components, manufacturing and service. Manufacturing involves those processes that require a substantial amount of technology. Three examples are telecommunications, pharmaceuticals, and scientific instruments. The service sector includes the application of high technology to customers through computer, engineering, and medical services.

Since Confederation, BC's population has grown. This population increase has caused a shift of the population centre of Canada towards the West and to BC in particular. In turn, BC has gained more seats in the House of Commons. This has changed relationships between central and eastern Canada and British Columbia. In short, economic, demographic, and political power has shifted to the west.

Population and the Centralist/ Decentralist Faultline

One measure of BC's expanding power lies in its rapid population growth. In 1969, Premier W.A.C. Bennett boasted that 'With the population of British Columbia growing at twice the rate of the rest of Canada, the presence of British Columbia as an economic region of its own is more obvious as each day passes' (Francis et al., 1996: 428). In this case, the numbers support the politician's statement. In 1901, the population of this Pacific province amounted to only 3 per cent of the Canadian population. By 1996, British Columbia's 3.7 million residents constituted 12.7 per cent of Canada's population. By the start of the twenty-first century, the population of British Columbia surpassed 4 million, forming 13.2 per cent of the national population.

British Columbia is the only province that has made substantial population gains in each decade since Confederation. In comparison to Ontario, British Columbia has had a higher rate of population increase from 1951 to 2000, reaching an average annual rate of population increase of 4 per cent compared to 3.6 per cent for Ontario. From 1991 to 1996, BC's annual rate of increase was 2.7 per cent, well above Ontario's rate of 1.3 per cent and almost two and a half times greater than the national average of 1.1 per cent. From 1996 to 2000, this trend continued.

Most of BC's population increase comes from immigration and interprovincial migration. Since the 1970s, more Canadians have resettled in British Columbia than in any other province. Many immigrants also select British Columbia (after Ontario and Québec) as their new home. From 1990 to 1996, many Hong Kong immigrants came to British Columbia before Britain turned its colony over to China.

This significant influx of immigrants to BC has changed the cultural makeup of the province. In 1996, for instance, 8.5 per cent of all British Columbians declared Chinese as their ethnic origin, while another 4.5 per cent had South Asian origins (Statistics Canada, 1998c: 19). In comparison, 9.2 per cent said that they were of French ancestry, though only 1.5 per cent of British Columbia's residents declared that French was their mother tongue.

Many immigrants have brought considerable wealth and economic connections to invest in British Columbia. Asian investment, particularly from Japan, has had an important impact on British Columbia's economy (Edginton, 1994: 32). Since the 1960s, Japanese multinational firms have invested heavily in mining, forestry, and tourism in BC. Japanese companies made a massive investment in the Northeast Coal Project in Tumbler Ridge. This project was designed to supply coal to the steel plants in Japan (Bone, 1992: 139–40). As well, Japanese investors have made many real estate purchases in the Vancouver area rather than in larger centres like Toronto and Montréal because of the large influx of Japanese tourists to Vancouver (Edginton, 1996). In the early 1990s, Japanese investors (Nippon Cable) developed a ski resort just north of Kamloops, turning it into the largest ski resort in British Columbia after the Whistler ski resort just north of Vancouver (Jim Miller, personal communication, 28 Apr. 1998).[3]

BC's growing population has not only affected its economic status and cultural diversity; it has also bolstered its political might. The political importance of British Columbia's rapid population increase lies in the increased number of seats held by BC members in the House of Commons. However, since the redistribution of the number of seats assigned to each province occurs after the last census of Canada's population, there is a lag in translating BC's increased population size into more seats in the House of Commons. In 1872, British Columbia had six representatives in the House of Commons, which was 3 per cent of the members. By 1996, BC had 32 members, forming nearly 11 per cent of the membership of the House of Commons. In 1997, British Columbia held only 11.3 per cent of the seats in the House of Commons, although its population size warranted nearly 12.7 per cent. This lag poses a serious irritant in the province's relations with Ottawa, which reinforces the centralist/decentralist faultline.

Industrial Structure

Is British Columbia a core or a periphery? No doubt this Pacific province is an emerging powerhouse within Canada, but its economy still has a relatively small manufacturing base. Approximately one-third of BC's manufacturing activities involve the processing of wood or wood fibre. In 2001, forestry and related activities, such as wood manufacturing, made up only 4.4 per cent of the total employed labour force while the high-technology sector contributed 2.9 per cent (British Columbia, 2002: 46). BC exhibits one of the classic characteristics of a resource hinterland: close to 90 per cent of its resource exports are either unprocessed or semi-processed goods. Natural gas and forest products are two leading exports and they fall into that category.

According to employment figures by industrial sector, 3.5 per cent of BC's workers are employed in the primary sector, 16.7 per cent in the secondary sector, and nearly 80 per cent in the tertiary sector (Table 7.1). In comparison with Ontario, BC's primary sector is proportionally larger than that of Ontario, while BC's secondary (manufacturing) sector is much smaller and its tertiary sector is much larger than Ontario's service sector (Table 7.2). Relatively few workers are actually employed in the primary and secondary sectors; rather, most are involved in the tertiary sector. For example, the forest industry, which supports cities like Prince George, is no longer expanding, and this has potentially dire consequences for single-industry towns and even for cities like Prince George. Almost all towns that lost population from 1996 to 2001 were heavily dependent on the forest industry (Table 7.10).

High-technology companies and, to a lesser degree, an expanding tourist industry are playing a key role in diversifying BC's economy. With a core of high-technology industries, such as advertising, computer, engineering, and management firms, BC is well on the road to a more mature economy, but, unlike Ontario, it lacks a solid manufacturing base. Most high-technology companies are concentrated in Vancouver and Victoria (as they are in major cities in Ontario and Québec). Often these industries are located near research centres and universities, where highly skilled labour is available and innovative research takes place. On the national scene, BC's high-technology industry is tied with Alberta for third place. In comparison to Ontario, BC's high-technology industry is about one-fifth the size.

Clearly, British Columbia is in a state of change, the leading edge of which is in the

Table 7.1	Employment by Industrial Sector in British Columbia, 2002			
Industrial Sector	BC Workers (000s)	BC Workers (%)	Ontario Workers (%)	Percentage Difference
Primary	69.0	3.5	1.8	1.7
Agriculture	30.1	1.5	1.3	0.2
Secondary	329.9	16.7	25.1	(8.4)
Manufacturing	196.9	10.0	18.5	(8.5)
Tertiary	1,574	79.8	73.1	6.7
Trade	319.0	16.1	15.7	0.4
Transport	110.1	5.6	4.7	0.9
Finance	118.9	6.0	6.6	(0.6)
Professional services	136.7	6.9	7.2	(0.3)
Management	74.7	3.8	4.2	(0.4)
Education	139.7	7.1	6.2	0.9
Health care	217.3	11.0	9.3	1.7
Culture	105.6	5.4	4.8	0.6
Accommodation	171.0	8.7	6.0	2.7
Other services	96.8	4.9	4.2	0.7
Public administration	84.7	4.3	4.8	0.5
Total	1,973.4	100.0		

Source: Statistics Canada (2003).

Table 7.2	Comparison of Ontario and British Columbia Industrial Structures, 1995 to 2002					
Sector	Ontario % 1995	Ontario % 2002	Percentage Difference	BC % 1995	BC % 2002	Percentage Difference
Primary	3.0	1.8	(1.2)	4.7	3.5	(1.2)
Secondary	23.6	25.1	1.5	18.1	16.7	(1.4)
Tertiary	73.4	73.1	(0.3)	77.2	79.8	2.6
Workers (000s)	5,231	6,068		1,762	1,973.4	

Sources: Adapted from Statistics Canada (1996b, 2003).

Vancouver/Victoria urban cluster, where high-technology and tourism services are concentrated. Beyond that cluster, the resource economy remains strong. Paradoxically, the industrial structure of this resource-rich province is more oriented to service activities than to primary employment. Over the years, the percentage of workers in the primary sector has declined, not because resource extraction efforts were decreasing but because firms increased labour productivity by substituting machines for labour, and because other sectors of the economy were expanding.

In 2002, BC's secondary sector comprised nearly 330,000 workers, or 16.7 per cent of the total workforce. The bulk of these workers are employed by manufacturing firms (Table 7.1). Manufacturing is an important but small element of British Columbia's economy; it lags far behind manufacturing in Ontario. Manufacturing in British Columbia is based on the processing of forest and other resources. By value, just over half of all exports from British Columbia are forest products, while food items contribute another 11 per cent. Coal, grain, and lumber exports have transformed the port of Vancouver into Canada's major shipping centre. Most resources, however, are exported in a semi-processed form, and the

key to expanding the manufacturing sector in British Columbia lies in greater processing of its resources, especially to the final product stage. Such diversification would greatly increase both the value of production and the size of the labour force. Yet, the simple fact is that the value of manufacturing has decreased. In 1999, the value of manufacturing was $37 billion while in 2001 it had dropped to $34 billion worth of goods (Table 7.3).

In 2002, the tertiary sector accounted for 79.8 per cent of employment in BC (Table 7.1). Not only does service employment in BC exceed the Ontario average by 6.7 percentage points but most subsections within this sector outstrip their Ontario counterparts. Employment in transport, trade, finance, and service indicates the relative strength of the service industry. The service sector owes its strength to several factors, including British Columbia's growing domestic market, strong participation in world trade, a vibrant tourist industry, and its producer services firms. Tourism is a growing industry with many overseas visitors coming to Vancouver. In 2001, $9.2 billion was spent by 28 million overnight visitors to the province (British Columbia, 2003a). **Producer services** are also playing a greater role in BC's economy. These firms have been pivotal in

Table 7.3	Manufacturing Shipments by Industry, British Columbia, 2001		
Industry	$Millions in 1999	$Millions in 2001	Per cent 2001
Wood	12,211	10,327	30.3
Paper	6,402	5,421	15.9
Food	3,502	4,149	12.2
Metal	1,949	1,577	4.6
Machinery	1,193	1,467	4.3
Transportation equipment	1,851	1,341	3.9
Non-metallic mineral products	961	1,160	3.4
Plastics	776	1,039	3.0
Chemicals	949	988	2.9
Beverages	791	905	2.7
Petroleum and coal	1,391	792	2.3
Printing	1,270	681	2.0
Computer products	1,231	511	1.5
Furniture	323	458	1.3
Clothing	333	313	0.9
Textiles	179	147	0.4
Other	681	1,751	5.1
Total	37,008	34,087	100.0

Sources: British Columbia (2000a, 2003a).

transforming Vancouver from 'its traditional role as a provincial higher-order service centre to a place in a network of increasingly interdependent cities of the Pacific Rim' (Davis and Hutton, 1994: 18).

The evidence is clear—British Columbia is changing, but it is still not an important manufacturing region. In 1982, a leading Canadian geographer, John Bradbury, saw BC as a resource hinterland that also had several characteristics common to a core region. In the past two decades, however, BC has experienced growth in high technology and in its tourist industry—the tertiary sector, not the secondary sector, is leading BC to a new level of economic development. The high-technology sec-

tor focuses on the manufacturing of electronics, telecommunications equipment, and pharmaceuticals. For example, Ballard Power Systems Inc. has achieved global markets through its development of fuel cells that produce electrical power with virtually no pollution. By 2001, BC's high-technology industry ranked third behind forestry and construction. In the same year, its 5,500 firms employed nearly 46,000 employees with total revenues of $3.3 billion and exports valued at $1 billion.[4]

British Columbia's Wealth

The key to British Columbia's past economic prosperity has been the province's natural resources. The industries that have historically led the province's growth and continue to contribute to its economic well-being include: fishing, mining, hydroelectric power, and especially forestry. These activities have not only fuelled the primary sector, but also the secondary sector. Moreover, a portion of the tertiary sector, the tourism industry, relies almost entirely on the province's natural beauty. Like Ontario and Quebec, BC can be divided into two parts—a core and a periphery—though the core, centred in Vancouver, is much smaller than the industrial cores found in Ontario and Québec. Furthermore, hinterland activities continue to play a pivotal role outside of Vancouver and Victoria, and there is less of a sense of a declining resource hinterland than is found in Ontario and Québec.

Fishing Industry

British Columbia, like Atlantic Canada, has an important fishery. In BC more than 80 species of fish and marine animals are harvested from the ocean, freshwater bodies, and aquaculture areas. Salmon is the most valuable species, followed by herring, shellfish, groundfish, and halibut. Most salmon today comes from fish farms. In 2001, the wholesale value of farm salmon was $292 million compared to $160 million for wild salmon (Table 7.4). In recent years, excessive harvesting has reduced fish stocks and, in turn, the number of landings (fish caught). At the same time, production from fish farms has made fresh salmon available to the consumer throughout the year, thus popularizing the product. Environmentalists, however, have long expressed concern about fish farming. In January 2004, fears that salmon from fish farms contain high levels of PCBs and other pollutants that could increase the risk of cancer surfaced in the media and these fears may affect sales of such salmon (Mittelstaedt, 2004).

Unfortunately, BC shares another similarity with Atlantic Canada—overexploitation of fish stocks, particularly the very valuable salmon stocks. The results of years of overfishing are: overexploitation by excessively large fishing fleets; lower returns per fisher; and longer idle periods for fishers, fishing vessels, and handling/processing facilities.

Ottawa, which is responsible for managing salmon stocks within Canada and for international fishing agreements, must deal with four perplexing and interrelated issues: (1) the nat-

Table 7.4	Fishery Statistics, British Columbia, 2001	
Species	**Wholesale Value ($millions)**	**Percentage**
Farm salmon	292	28.9
Shellfish	213	21.0
Groundfish	177	17.5
Wild salmon	160	15.9
Herring	119	11.8
Halibut	50	4.9
Total	1,010	100.0

Source: British Columbia (2003a).

ural cycle (up to five years) of salmon to spawn in rivers, migrate to the sea, and then return to the rivers to spawn again; (2) the harmful effects of the forestry and hydroelectric industries on salmon spawning grounds; (3) the division of the salmon catch among commercial, Native, and sports fishers; and (4) the harvesting of 'Canadian' salmon in international waters.

Management of fish resources is based on the estimated size of the fish stock. The problem of determining the size of the stock, and therefore the quota to be set for each year's catch, is complicated by the natural population cycle of the fish. The natural fluctuation is illustrated by pink salmon catches in the past. The size of catch has varied from a low in 1975 of 38,000 tonnes to a high of 108,000 tonnes in 1985. In 1994, an estimated 2.3 million fish did not return to BC rivers to spawn. What caused this decline is unknown, but a similar drop in 1995 in the number of salmon returning to the Fraser River forced Ottawa to close fishing on the Fraser. One of the possible reasons for this decline is pollution of fish habitat. Forestry companies, for example, have affected salmon habitat through their logging practices, which have blocked streams and thereby interfered with migration routes to spawning areas. Pulp and paper plants also discharged toxic wastes into rivers and the ocean. Another possibility is a rise in ocean temperature above levels suitable for salmon—the so-called El Niño effect. El Niño's warming of the Pacific waters could provide a natural explanation for the declining salmon stocks, but so far the El Niño effect on ocean temperatures and then on salmon is unsubstantiated by scientific evidence.

A more likely explanation for the apparent decline in salmon stocks is a combination of a larger fishing fleet and the application of new technology (radar and sonar equipment), both of which lead to huge catches initially but then, inevitably, result in the annual catch exceeding the replacement rate for fish. Overfishing is generally regarded as the leading cause of declining fish stocks. Fishers have simply taken more fish than the fish populations can sustain, and overly optimistic stock estimates established by the federal Department of Fisheries and Oceans (DFO) have meant that quotas, historically, have been higher than they should have been. The DFO, which is responsible for fish management, announced a decline in the number of salmon returning to the Fraser River to lay eggs in 1994 and 1995 (DFO, 1996: 4). The fear is that future runs will decline even more.

Salmon are migratory fish. Like other fish, they are common property until caught. This principle is based on the 'rule of capture'. Fishers therefore try to maximize their share of a harvest so no one else will take 'their' fish. The problem is complicated further because the Canadian government cannot regulate the 'Canadian' salmon stocks because they migrate to American waters, where the American fishing fleet harvests them.

The net result is that the salmon stocks are threatened. This problem is commonly referred to as the **tragedy of the commons**. (See Chapter 9 for an account of the overexploitation of the northern cod.) However, Ottawa is taking action to alleviate the pressure on salmon stocks. It plans to reduce the 4,500 fishing vessels by about one-third. At the same time, Ottawa is allowing Aboriginal fishers, who have treaty rights to harvest fish for subsistence purposes, a share of the commercial stock. Ottawa is also negotiating with Washington to reduce the American catch of the Fraser River salmon. Ottawa is able to exert some management of fish stocks in the Pacific Ocean because of its 200-mile fishing zone (Figure 7.5) and because of its role in the Pacific Salmon Commission. Salmon fishing on the Pacific coast is regulated, based on the 1985 Pacific Salmon Treaty between Canada and the United States (the treaty determines the size of the catch taken by each nation).[5] Unfortunately, Canada has been unable to convince the US to reduce each nation's salmon take. If each country does not lower its catch through a new treaty, salmon stocks may decline to dangerous levels similar to those experienced by the northern cod in Atlantic Canada.

Salmon catches by fishing fleets decreased in the 1990s, not because of conservation measures that restrict the allowable amount of catches but simply because the fish were not present in large numbers. Salmon stocks may be failing. Such a collapse spells disaster not only for the fish but also for the people who rely on the fisheries for their livelihood, including the many who work in the fish-processing plants. In fact, several large companies, including BC Packers and the Canadian Fishing Company, control the processing of fish. They have strategically placed their canneries at the mouths of the Skeena and Fraser rivers. Such large companies process about 80 per cent of all the salmon caught each year. The repercussions of declining fish stocks for the 25,000 full-time and part-time employees of these fish-processing plants are most serious.

Mining Industry

The mineral wealth of British Columbia is found in both of the province's physiographic regions—the Cordillera and the Interior Plains. In 2001, the value of total mineral production was $9.2 billion, nearly double the 1999 figure of $4.7 billion (Table 7.5). The explanation for this remarkable rise in value is due to the higher prices for natural gas coupled with increased output of natural gas in the northeast corner of British Columbia. By value, natural gas accounted for nearly 50 per cent, followed by coal at 11 per cent, and other fuels at 12 per cent. Mining and mineral pro-

West coast fish canneries, like this one at Sechelt, have long played a key role in British Columbia's economy. However, given the presence of American and Canadian fishers as well as Aboriginal fishers who have special rights, the salmon fishery could suffer the same fate as the Atlantic cod fishery. (©Natalie Forbes/Corbis/Magma)

cessing employ about 3 per cent of the labour force, but yield nearly 25 per cent of the value of primary production. Because nearly all of the mineral production is consumed outside of the province, the fortunes of the mining industry are largely determined by external markets and global prices. The outlook for the natural gas industry is bright, while coal faces a bleak future. The closure of the Quintette mine in northeastern BC is due to the weak demand from Japanese steel mills. This pattern of development is common to all resource development in British Columbia and again emphasizes both the strength and the vulnerability of BC's hinterland economy.

The importance of the export market for mineral products has shaped BC's transporta-

tion system—first, by requiring a network that would allow mining operations to reach more isolated areas and ship ore back to its markets; second, by adopting cost-reducing transportation methods. For instance, competition from other resource hinterlands, such as Australia, has forced Canadian producers to reduce their costs of operation and marketing. Transportation costs have therefore been reduced by unit trains and bulk-loading facilities. Unit trains consist of a large number of ore cars, sometimes over 100, pulled by one or more locomotives. The coal-loading terminal at Robert's Point is a large bulk-loading facility. It is able to dump each ore car of coal and then move the coal by conveyor belts to the ship's hold. These ships must be moored in deep water,

Table 7.5	Mineral Production in British Columbia, 2001	
Mineral product	**Value in $millions**	**Per cent**
Metals		
Copper	679	9.7
Gold	322	4.4
Zinc	155	3.3
Silver	132	1.8
Molybdenum	77	0.8
Lead	27	0.4
Other	20	0.3
Total metals	1,412	20.7
Fuels		
Coal	1,068	10.6
Natural gas	5,180	49.6
Other fuels	978	12.5
Total fuels	7,226	72.7
Structural materials	475	5.9
Industrial minerals	41	0.7
Total mineral production	9,154	100.0

Source: British Columbia (2003a).

which necessitated building a long causeway to reach the ships.

As with other resource development, the capital investment in mining is enormous. For that reason, only very large companies, often multinational firms, can venture into such developments. The Northeast Coal Project is one such example. In the late 1970s Denison Mines and Teck Corporation were the major mining companies behind this $4.5 billion construction project (Bone, 2003: 140–1). With a 14-year contract to sell coal to Japanese steel companies, banks and other financial institutions were happy to supply much of the capital (about $2.5 billion) to Denison and Teck. The federal and provincial governments provided $1.5 billion of the $4.5 billion to upgrade the CNR line to Prince Rupert, to extend a BC Rail line to Tumbler Ridge, and to build a coal terminal to load ships that would carry the coal to Japan. The two mining companies contributed about $500 million. Both governments provided another $1 billion to build the coal town of Tumbler Ridge. These agreements were signed in the late 1970s, when the world economy was expanding and the demand for coal was increasing. The Northeast Coal Project comprised Quintette Coal Ltd at Tumbler Ridge, Teck Corporation at Bullmoose, and Gregg River Coal Ltd at Gregg River.

Like all large-scale construction projects, several years were required to complete the undertaking. Production began in 1984. By then, the world demand and price for coal had peaked and a slow but steady decline set in as the global economy slipped into a recession. At that point, the Japanese steel mills no longer needed so much coal and the price set with the Northeast Coal Project mines was no longer attractive. By 1984, when coal was first produced at the Northeast Coal Project, the world spot price (a price charged for goods available for immediate delivery) for coal was 10 per cent lower than the agreed-upon, contract price! Eventually, the price for coal was renegotiated, but Denison Mines went bankrupt during the process. Since then, Teck Corporation has taken over the coal mining operation. Declining demand and prices forced the company to close its Quintette mine in 2000, leaving the community of Tumbler Ridge with no economic base. Efforts to convert the community into a retirement and 'second-home' town face an uphill battle because of the remote location of Tumbler Ridge (Halseth and Sullivan, 2002).

The example of the Northeast Coal Project illustrates the vulnerability of the mining industry, indeed, of all resource industries. With a heavy reliance on export markets and the fluctuations in world prices, BC's mining industry faces ongoing challenges. The economy will also have to adapt to the new economic order by increasing secondary activities and relying less on such primary industries.

Hydroelectric Power

British Columbia has many natural hydroelectric sites. Within the Cordillera, a combination of elevation, steep-sided valleys, and steady-flowing rivers provides ideal conditions for the construction of hydroelectric dams. The Columbia, Fraser, and Peace rivers offer many excellent hydroelectric sites. In addition, heavy precipitation and meltwater from the mountain snowpack ensure a regular and abundant supply of water. Major hydroelectric developments have taken place on the Columbia and Peace rivers as well as on the Nechako River, a tributary of the Fraser River. Minor developments have occurred on Vancouver Island.

In the early days of hydroelectric development, small-scale hydroelectric projects were established near Vancouver and Victoria. These two cities are the major markets in the province for electricity. Hydroelectric dams were also built near smelters that required vast amounts of power to operate. For example, early in this century, power was generated from privately owned dams on the Kootenay River for the lead-zinc smelter at Trail.

After World War II, public and private companies began to undertake megaprojects. Among these giant industrial construction efforts, three—one on the Columbia River, another on the Peace River, and the third on the Nechako River—had enormous impacts on the economy. They all involved the harnessing of water power from the province's rivers to generate low-cost electrical power, but they also flooded valuable farmland and Indian lands and led to the loss of salmon spawning grounds. Such construction projects were considered major engineering feats. In 1951, Alcan completed the construction of the Kenney Dam across the Nechako River, impounding the water draining from an area more than twice the size of Prince Edward Island. In 1968, BC Hydro built the W.A.C. Bennett hydroelectric dam on the Peace River,

creating Williston Lake, the largest freshwater body in the province. From this giant reservoir, vast quantities of water flow through the turbines at the power station, generating much of British Columbia's electrical power and creating a surplus of power for sale in the US.

The most ambitious of these early megaprojects was the one developed on the Columbia River. A combination of rising demand for power in the state of Washington and innovations in the transmission of electricity over long distances enabled the development of water power along the Columbia River in British Columbia. In the 1960s, dams were constructed on the Columbia River to increase the flow of water to the hydroelectric power plant at Grand Coulee in the United States (Vignette 7.3). Initially, BC Hydro sold its share of the electrical power generated at Grand Coulee to its counterpart in the state of Washington. In 1997, this power was no longer needed in the US and British Columbia was faced with a huge surplus of electrical power. Similar to the Québec government's efforts to stimulate industrial development in that province, Victoria has now begun offering this surplus power at a reduced price to firms that locate in British Columbia. So far, no large

Vignette 7.3   The Columbia River Treaty

In 1961, Canada and the United States agreed to co-operate in the development of the Columbia River for the production of hydroelectric power. Canada undertook to construct three dams for water storage in the Canadian portion of the Columbia River Basin and to operate them to produce maximum flood control and power downstream. In return, the United States would return half the power generated to Canada and pay for half the value of the flood protection for prop-

erty in the United States. The three dams constructed as a result of this agreement are the Duncan (1967) north of Kootenay Lake, the Hugh Keenleyside (1968) on the Columbia River, and the Mica (1973) north of Revelstoke. The major hydroelectric power station is located in the state of Washington at Grand Coulee. Before the construction of the Grand Coulee Dam, the Columbia River was one of the world's great spawning grounds for salmon.

firms have expressed an interest, but Victoria has used this surplus as part of a settlement it reached with Alcan over the cancellation of one of its hydroelectric projects (see below).

Alcan in Northwest British Columbia

The Alcan hydroelectric development in British Columbia illustrates the advantages and pitfalls of megaprojects. The Aluminum Company of Canada (Alcan) is one of the giants in the world aluminum industry. To keep its production costs low, Alcan is attracted to sites where low-cost hydroelectric power can be developed and then used by its smelting plants to reduce bauxite (a clay-like mineral) into alumina (the chief source of aluminum) and then aluminum. In Canada, water resources are under provincial jurisdiction, so Alcan seeks long-term arrangements with provincial governments to develop the water power for its smelting plant. In turn, provincial governments are attracted to megaprojects because they can develop a region and employ large numbers of people.

In the late 1940s, Alcan proposed to build an aluminum complex in northwest British Columbia. After obtaining the rights to the waters of the Nechako River for 50 years from the provincial government, Alcan began construction of the first phase of this complex in 1951.[6] Three years later, Alcan built a new town called Kitimat, a power plant at Kemano, and a dam on the Nechako River. To produce hydroelectric power, Alcan dammed the Nechako River, thereby reversing its flow westward to the Pacific Ocean near Kemano. To accomplish this task, a tunnel had to be drilled through the Coast Mountains. At Mount DuBose, a 16-km tunnel allowed waters from the Nechako reservoir to flow to Kemano where the power station is located. These waters plunge 860 m from the western

end of the tunnel to drive the turbines at Kemano. Electricity is transmitted about 80 km to Alcan's aluminum smelter at Kitimat. Until 1999, Alcan maintained a permanent workforce at Kemano. In that year, Alcan announced that its Kemano powerhouse will be operated and maintained by crews from Kitimat working in Kemano on rotating shifts.

In the late 1940s, concerns about environmental and social impacts of industrial projects were not given much attention. Both the provincial and federal governments bent over backwards to assist Alcan in what was considered one of the great engineering and construction feats of the twentieth century. The flooding of the Cheslatta Indian Reserve, which resulted from the Alcan project, was not considered a significant social impact and the Department of Indian Affairs quickly arranged for a relocation of this tribal settlement. As for the sockeye salmon spawning grounds that were affected, no protest was raised. Three reasons account for the lack of concern: (1) the general public was convinced that industrial growth was 'good' for the province; (2) the public and the fishing industry were not aware that the destruction of the Nechako spawning grounds could have a major impact on salmon stocks; (3) the environmental movement and Aboriginal groups had not yet emerged as forces capable of challenging such projects.

In 1989, Alcan began the second phase of its hydroelectric complex—the Kemano Completion Project. Estimated costs were $1.3 billion. The plan was to enlarge its hydroelectric facility by boring a second tunnel through Mount DuBose, thereby diverting more waters of the Nechako reservoir westward. This time, both the Carrier-Sekani Tribal Council (whose members claim the Nechako River as part of their land claim) and environmental groups spearheaded by the Rivers Defense Coalition openly challenged the project, claiming that the social and environmental costs were too great. Even so, the project was approved by the federal government without an environmental review. By 1991, the project was half-completed at a cost of about half a billion dollars, but opposition from environmental groups and Aboriginal leaders had grown so strong that the federal government was forced to order an environmental inquiry. Alcan ceased its construction efforts, hoping to restart work after federal approval.

In 1995, the provincial government cancelled its approval for Alcan's Kemano Completion Project. Alcan threatened to sue. Later, the British Columbia government agreed to compensate Alcan for the money spent on construction by selling the company electricity at a very low price (BC was able to make use of the surplus power it had). Alcan claimed to have spent $500 million towards the construction of the $1.3 billion project. In the arrangement reached in 1997, Alcan will receive electrical power for its proposed smelter in two ways: (1) 115 MW a year from BC Hydro at a price pegged to the price of aluminum on the London Metal Exchange until 1

January 2024; and (2) an additional 60 MW a year from BC Hydro, at 1997 prices, until 1 January 2024 (Cernetig, 1997: 1). At 1997 prices for power, the difference between the cost to Alcan and the market value of the 175 MW a year results in a savings of approximately $9 million a year, or $243 million over the course of the 27 years.

The prospects for the construction of new hydroelectric projects in BC in the near future are poor. Environmental and social impacts may well be too great, as was the case with the Kemano Completion Project. Other adverse factors include: BC has a surplus of power; world demand for aluminum is very weak; and Alcan has just completed a huge expansion of its aluminum operation in Québec, where the government is 'more friendly'.

Key Topic: Forestry

The forest is British Columbia's greatest natural asset. Forest covers 62 per cent of the province. Almost all of this is coniferous **softwood forest** and constitutes about half of the Canadian softwood timber stock. Within Canada, British Columbia harvests just over

Table 7.6	Timber Harvest by Species, British Columbia, 2001	
Species	**Volume (million cubic metres)**	**Per cent**
Lodgepole pine	22.9	31.7
Spruce	12.4	17.2
Douglas Fir	10.8	15.0
Hemlock	8.3	11.5
Balsam	7.1	9.8
Cedar	6.2	8.6
Other	4.5	6.2
Total	72.0	100.0

Source: British Columbia (2003a).

Vignette 7.4	Making Pulp

Wood is reduced to fibres called pulp by mechanical means (grinding wood), by chemical means (cooking wood in a chemical solution), or by a combination of these two methods. Each of the approaches produces a different type and quality of pulp. They are: groundwood pulp, thermomechanical pulp, and chemical pulp. **Groundwood pulp**, made by grinding wood chips or sawdust into fine fibres, is used to make high-quality newsprint. **Thermomechanical pulp** is a combination of the groundwood and chemical pulp processes. It produces pulp with many of the qualities of groundwood pulp, but with the addition of greater strength such as found in chemical pulps. In both the groundwood and thermomechanical processes, grinding the wood produces very soft pulp and makes effective use of about 95 per cent of the cellulose content of the wood. However, grinding the wood breaks the fibres and shorter fibres produce a pulp with little strength. **Chemical pulp** mills change wood into sulphate pulp by chemical means, thereby retaining the wood's long fibres. However, only about 50 per cent of the cellulose fibre of the wood is used. The chemical pulp is used to make industrial paper, cardboard, and a wide range of quality paper products.

half of the total logging output. The main species harvested are lodgepole pine, spruce, hemlock, balsam, Douglas fir, and cedar (Table 7.6). The processing of timber into lumber, pulp, newsprint, paper products, shingles, and shakes supports a major manufacturing industry in British Columbia. Within Canada, British Columbia leads in the production of wood products like lumber and plywood, while Ontario and Québec produce a greater proportion of pulp and paper products.

In 2001, British Columbia accounted for the following share of Canadian forest output (British Columbia, 2003a):

- 77 per cent of softwood plywood;
- 46 per cent of softwood lumber;
- 45 per cent of **market pulp**;

Table 7.7	Forest Product Exports from British Columbia, 2001	
Commodity	**Value in $millions**	**Per cent**
Softwood lumber	6,522	45.1
Pulp	3,068	21.2
Paper and paperboard	1,220	9.2
Wood products	956	6.6
Newsprint	623	4.3
Cedar shakes and shingles	334	2.3
Other	1,407	9.7
Total exports	14,130	100.0

Source: British Columbia (2003a).

- 15 per cent of **newsprint**;
- 15 per cent of paper and paperboard.

The timber harvest in 2001 dropped to 72 million m³, a decrease of 5 per cent over the 1999 harvest. The value of this harvest, including processing, was $17 billion. Led by lumber and pulp, forest exports are valued at over $14 billion (Table 7.7). Wood processing consisted of softwood lumber, pulp, paper, and plywood. The forest industry accounts for nearly 100,000 jobs and another 200,000 indirect

jobs. Almost 30,000 people are employed in logging operations, 26,000 in wood processing, and 11,000 in pulp and paper making. This does not include the many workers who rely indirectly on forestry, such as tree planters, public officials, and furniture makers.

Forest Regions

British Columbia's forest is far from homogeneous, largely because of its climatic zone and the varied relief in the Cordillera. The varia-

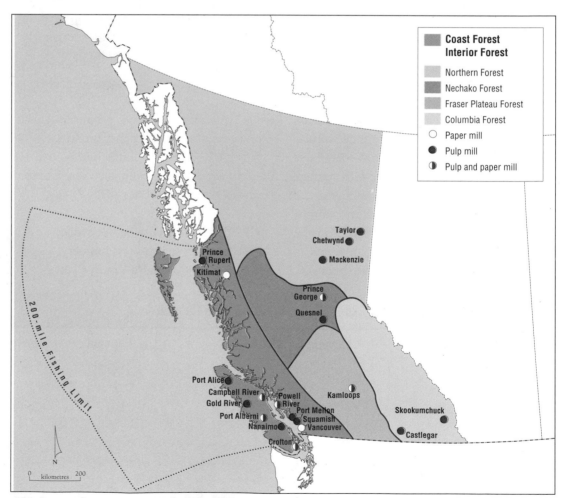

Figure 7.5 Forest regions in British Columbia. The two principal regions are the Coast Forest and the Interior Forest. The Interior Forest is subdivided into four areas that reflect the variations in growing conditions. (Further resources: Student Web site, National Atlas section, Maps 21, 22, and 23. Web site instructions are found on p. xxi.)

tions in the type of forest have strong implications for forestry. For example, trees along BC's moist west coast grow much more quickly than those in the dry interior. BC's forest lands are divided into two major regions: the Coast Forest and the Interior Forest. Within the Interior Forest, there are four subregions: the Northern Forest, the Nechako Forest, the Fraser Plateau Forest, and the Columbia Forest (Figure 7.5). Within each of these subregions there is great variation in the forests due to the differences in local growing conditions, which are affected by precipitation and length of growing season. Other factors are elevation, soil conditions, and topography.

The Coast Forest is the most luxuriant coniferous forest in Canada. With its wet and mild marine temperatures and abundant rainfall, this is one of the most densely forested areas of North America. The key species are Douglas fir, western red cedar, and western red hemlock. Under ideal conditions, mature cedar and hemlock trees reach 45 m in height and about 1 m in diameter. The Douglas fir is an even larger tree. Mature stands may average 60 m in height and over 1.5 m in diameter. Logs from the mature stands (known as old growth) are highly valued for lumber and plywood, while logs from immature stands (known as secondary growth) are used for pulp and paper.

Inland from the Coast Forest, the climate changes from a wet marine climate to a semi-arid one. At this point, the Interior Forest begins. However, differences exist within the Interior Forest. The Fraser Plateau subregion is an open woodland with much smaller trees than those in the Coast Forest. The controlling factor is the dry climate. Trees in the Fraser Plateau subregion must be able to cope with drought conditions. Ponderosa pine and lodgepole pine are the most common species.

They often attain heights of 25 m and a diameter of 1 m. These trees are usually converted into lumber.

The Nechako Forest subregion, which lies to the north of the Fraser Plateau, receives more precipitation and experiences lower summer temperatures. The net result is more moisture for tree growth. Consequently, the forest cover is denser than that in the Fraser Plateau. A common species in the Nechako Forest is the Engelmann spruce, which can attain heights of 40 m and diameters of 60 cm. Timber from this region is processed by pulp and paper mills at Prince George.

The Columbia Forest subregion lies in the easternmost area of southern British Columbia. Again, a dry climate limits tree growth. Because of widely varying terrain, the forest cover is very heterogeneous. Western red cedar and western hemlock are common species, though stands of ponderosa pine and

While logging is both highly efficient and mechanized, the practice of clear-cut logging makes the goal of a sustainable industry problematic. According to a report, *Clear-cutting Canada's Rainforests*, produced by the David Suzuki Foundation, clear-cutting remains the dominant system of harvesting the old-growth trees in British Columbia's rainforest. Two impacts of clear-cutting are blowouts (strong winds knock over isolated trees) and landslides (moderate to steep slopes are subject to mud and land slides during periods of heavy rain). (Al Harvey/The Slide Farm)

even Douglas fir are also found in this region. Large timber is sent to sawmills, while the smaller logs are used for making pulpwood.

The Northern Forest subregion is the most remote of BC's forests. High transportation costs to ship wood to world markets hinder logging. Tree growth is hampered by cool growing conditions and poorly drained land. Large blocks of land that contain muskeg are either devoid of trees or have trees with little commercial value.

Forest Industry

Of all its natural resources, the exploitation of the forest was the mainstay of the British Columbia economy. While other sectors have become equally important, the forest industry remains a cornerstone of BC's economy. British Columbia has always been oriented to **staples production** (or primary production) to supply its resources to markets in Canada, the United States, and other foreign countries. Such a pattern of development is the mark of a hinterland economy. As with other resource hinterlands, attempts to diversify its resource-based industry by increasing the amount of processed resource products has been slow. Most processing has not reached a final product stage, e.g., paper. Manufacturing increases the value of a product, such as a log, by changing it into a final product, such as a window frame. This process is called **value-added production**. Besides increasing the value of production in British Columbia, this strategy would also provide more employment. In 1997, the British Columbia government announced the Jobs and Timber Accord, an agreement it reached with the forest industry. One aspect of the Accord has increased the value added to forest products by requiring lumber producers to sell one-sixth of their output to local manufacturing firms, such as a furniture factory. In this way, the government has increased the size of forestry's manufacturing sector and thereby generated more jobs.

At first, the forest industry consisted of small-scale logging and sawmilling operations. Prior to World War II, the forest industry was concentrated along the Pacific coast. After the war, logging, sawmilling, and pulp mills began to locate in the interior. Sawmills in the interior account for about three-quarters of the lumber production. Pulp and paper mills are mainly along the coast and in the Prince George area of the interior. Wood manufacturing plants are heavily concentrated in the Lower Mainland. With the increased demand for forest products, logging intensified during the 1970s. By the 1980s, almost all of British Columbia's vast commercial timber lands were leased to forestry companies.

Today, 10 forestry companies account for over half of the timber harvest and three-quarters of the processing of wood products. The pulp and paper companies are controlled by a few multinational corporations. Since the 1970s, Japanese firms have invested heavily in the forest industry, often as minority partners. This shift towards multinational corporations is common in resource hinterlands. The rise in multinational ownership is due to two factors: (1) the huge capital investment required to undertake megaprojects, and (2) resource hinterlands' integration into the global economy with foreign owners securing their source of raw materials.

In 2002, enormous forest reserves supported a major industry with an output valued at over $17 billion. Just over 95 per cent of the 60 million hectares of forest lands were owned by the British Columbia government and allocated to forestry companies on long-term leases. In 1999 Weyerhaeuser Canada purchased

MacMillan Bloedel, thus becoming the largest forestry company in BC. It holds about 15 per cent of the leased forest area and, like other firms, has its headquarters in Vancouver. Despite the forest industry's success, through the 1990s it became more and more difficult to obtain raw material. Two casualties were the pulp mills at Gold River and Port Alice. In 1999, the closure of the Bowater pulp mill at Gold River on Vancouver Island occurred for three reasons: limited wood fibre supply, loss of sales to Asia, and inability to compete with larger mills in a shrinking market Port Alice suffered a similar fate. Prince Rupert's pulp mill has been struggling to survive for a number of years. Twice the provincial government has rushed to the rescue (once in the 1970s and again in the 1990s) after the owners declared bankruptcy. As an old, less efficient mill, its chances of survival are not encouraging.

The Forest Myth

At the time of Confederation, the vast timber lands of British Columbia seemed endless. Today, there is a wood fibre crisis. Few forest leases for mature timber stands are available, and what there is usually includes inaccessible stands. In 2003, forest fires reduced the amount of timber stands in the interior of British Columbia. What was thought to be a limitless supply of wood fibre was a myth.

The rapid depletion of wood is revealed in the changes, over a five-year period, in the volume of British Columbia's exports of wood chips. In 1989, nearly 2 million tonnes were exported. By 1994, this amount had fallen by more than half, to 630,000 tonnes (Statistics Canada, 1998a: 36). Imports of wood fibre from Alberta and Alaska, on the other hand, increased during the same period. Several factors account for this change from a surplus of wood chips in BC to a shortage. BC forestry companies' past logging practices were one factor. Driven by profit, companies harvest the more accessible (and more cheaply harvested) stands first. A second factor is the scarcity of accessible timber stands. In the past, logging companies simply obtained another timber lease, which provided them with easy access to commercial timber stands. Now, with the depletion of forest stands, fewer accessible ones are available. The third factor is the intervention of the provincial government. Environmental activists forced the provincial government to 'reassess' its forestry position over the issue of clear-cutting the ancient rainforest on Vancouver Island's Clayoquot Sound. In 1995, the provincial government introduced new restrictions on logging and, at the same time, announced the establishment of wilderness parks in areas of mature forests. The combination of these two policies has reduced the amount of Coast Forest available to forestry companies by about 10 per cent and reduced the **annual allowable cut** by 6 per cent (British Columbia Ministry of Forests, 1996: 5).

Forest Management

In the late 1970s, the provincial government had a system of long-term forest leases that encouraged companies to manage these lands to allow another 'crop' of trees to mature to a commercial state. The assumption was that such a management system would lead to a sustainable harvesting of the forest. Typically, these licences extend for 25 years but are subject to renewal. Companies responded by cutting the best and most accessible trees. The failure to build additional logging roads to reach new areas often led to overcutting in the more accessible timber areas. Private companies' attempts to plant seedlings have met with mixed success. In 1994, the Forest Renewal Act was passed, nearly doubling the stumpage fees the companies paid for each tree cut. This revenue is designed to finance tree planting through a government corporation, Forest Renewal BC.

Until 1995, neither the province nor companies had approached forestry from a 'serious' sustainable perspective. Now forest management practices are enforceable by law (British Columbia Ministry of Forests, 1996: 1). The assumption of the 1960s that 'old' timber stands would last into the next century, by which time 'natural' and 'artificial' reforestation of earlier logged areas would develop into **commercial forest** stands, no longer has any credence. The primary reason is that the increase in logging over the last 25 years was much greater than anticipated in the 1960s. This increase has reduced the mature forest stands much more quickly and has left less time for the logged areas to regenerate. By the mid-1990s, the wood fibre shortage pushed timber prices so high that it became economical to transport logs from Alberta's foothills to mills in the southern interior of British Columbia and from southern Yukon to mills in northern areas of the province. At some point, however, additional transportation costs will force the closure of such mills. It is hoped that forest management measures will alleviate problems related to the depletion of forest stands.

Free Trade and Forest Exports

The major markets for BC forest products lie outside of the province. Like most Canadian resource industries, the economic well-being of BC's forest industry depends on export markets. BC's principal markets are the United States, which absorbs about half of its production, with Japan and the European Union accounting for 25 and 15 per cent, respectively. Given the dominance of the American mar-

ket, one of the challenges facing the forest industry in BC extends beyond the province's borders and relates to trade relations between Canada and the United States. The annual value of Canada–US trade is nearly $400 billion, which is the largest in the world between two countries. The vast majority of this trade takes place without incident, but there are 'hot points' and lumber is one of them.

NAFTA has reduced trade barriers between the United States and Canada, but it has not prevented trade disputes involving lumber from erupting. Forest exports from British Columbia have already been subjected to the heavy hand of Washington. Most Canadians do not understand why NAFTA has not given Canadian products 'free' access to US markets. After all, was not 'free' access to each other's market the purpose of the original Free Trade Agreement and the subsequent NAFTA? The answer to that question is 'freer' but not 'free' trade. For example, when an American producer is adversely affected by imported products, the US company complains to the American government about the lower-priced products, claiming that such lower prices are a form of unfair trade. The company expects Washington to protect it by creating a trade barrier, which it has done on occasion. The trade agreement has a dispute-settlement mechanism, but resolving trade disputes takes time, during which the Canadian exporter loses sales and profits.

Problems over free and fair trade have been particularly frequent in the forest industry. American lumber production, which operates mainly in the Pacific Northwest and Georgia, can produce a maximum of 15 billion board feet per year. Canadian lumber production nearly doubles the American figure. In fact, lumber production from British Columbia (Table 7.8) is roughly equal to that of the

Figure 7.6 Traditional US ball games. Back in 2002, the United States invoked a punitive countervailing duty of 27.2 per cent on Canadian softwood lumber, claiming that stumpage fees charged to logging companies are too low and therefore represent a subsidy. So far, the trade war has taken a toll on Canadian companies and jobs. By April 2004, companies had paid $2 billion in duties and thousands in the industry lost their jobs, including about 15,000 forestry workers who were laid off in British Columbia. Canada took the dispute to the World Trade Organization. Even though in December 2003 the WTO ruled in Canada's favour by rejecting American claims, Washington has not removed this trade barrier. (Victoria-Times Colonist, Copyright © Adrian Raeside)

entire United States. The main difference is that US-produced lumber serves its domestic market, while most Canadian-produced lumber is exported to the United States, Japan, and other foreign customers. Trade disputes over lumber exports can therefore significantly affect BC's forest industry.

When American lumber producers lose market share to their Canadian counterparts, they turn to their lobby organization, the US Coalition for Fair Lumber Imports. The US political system is extremely sensitive to lobby efforts because individual members of the House of Representatives and the Senate have the power to intercede on trade matters. In addition, these elected officials rely on lobby organizations for support at election time. Since 1982, the US Coalition for Fair Lumber Imports has managed to convince the American government to reduce Canadian forest imports into the United States. In 1986, for example, the US launched a trade action, claiming that Canadian softwood lumber production was subsidized by low provincial stumpage fees. In the following year, the US Department of Commerce ruled that Canadian

Table 7.8	Lumber Production by Provinces/Territories, 2002	
Province	**Production (000 cubic metres)**	**Production (per cent)**
Yukon/Northwest Territories	>100	>0.1
Newfoundland/Labrador	>100	>0.1
Prince Edward Island	>100	>0.1
Manitoba	630	0.9
Saskatchewan	630	0.9
Nova Scotia	1,950	2.7
New Brunswick	3,378	4.7
Alberta	6,599	9.2
Ontario	7,345	10.2
Québec	17,667	24.5
British Columbia	33,555	46.7
Canada	71,982	100.0

Source: Natural Resources Canada (2003).

stumpage rates constituted a countervailing subsidy. Ottawa agreed to impose a 15 per cent export tax on lumber exported to the US. British Columbia and several other provinces raised their stumpage rates to counter the American claim of low stumpage rates and to replace the federal export tax. In 1991, Canada terminated the 15 per cent export tax. In response, the US Coalition for Fair Lumber Imports called for another trade action to restrict the flow of Canadian lumber into the United States. In 1992, the United States government imposed a countervailing duty on Canadian lumber. Over a two-year period, the US government collected more than $800 million from Canadian exporters. In 1994, a bilateral trade panel (the dispute mechanism created by NAFTA) declared the tax invalid under NAFTA and ordered the US government to reimburse the Canadian companies. Not to be outdone, the US Coalition for Fair Lumber Imports called for new restrictions.

In 1996, the US government insisted on a ceiling for Canadian lumber shipments to the US, and Ottawa and Washington reached a new agreement: a limit of 14.7 billion board feet would be allowed into the US in exchange for five years of non-interference with the lumber trade by the US government. The agreed-to figure was an average of the volume of Canadian lumber exported to the United States over the previous three years. The first 650 million board feet in excess of the figure faced a US tax of $50 per 1,000 board feet. Even with this additional cost, Canadian lumber will still be competitive in the US market. Once the 650 million board feet figure is exceeded, the tax will jump to $100 per 1,000 board feet. At that tax level, Canadian lumber still could be competitive in the US market. All of these efforts by the US lumber lobby and the US government are attempts to bypass NAFTA in order to protect US lumber interests. This US–Canada agreement ended 31 March 2001. In May 2002, Washington announced a duty of 27 per cent on Canadian softwood lumber entering

Vignette 7.5   The Ongoing Softwood Lumber Dispute

Trade friction between Canada and the United States over softwood lumber has existed for many years. Washington has imposed a number of countervailing duties against Canadian softwood lumber in 1982, 1986, 1991, and 2002. In each case, Washington's action has been driven by pressure from the American lumber lobby, the Coalition for Fair Lumber Imports. In the May 2002 duty, Washington claims that provincial stumpage fees are too low and constitute a form of public subsidy. From a Canadian perspective, the softwood lumber trade dispute is not about subsidy but rather a means by which Washington is protecting the US lumber producers. The American position becomes more transparent when we see that the 1996 Softwood Lumber Agreement committed Canada to limit its lumber exports to 14.7 billion board feet per year. Beyond that point, a duty of US $50 per thousand board feet was placed on the first 650 million board feet and US $100 per thousand board feet for quantities in excess of 650 million board feet.

the United States. By May 2004, a new agreement had not yet been reached, mainly because American lumber producers want to restrict the volume of softwood imports from Canada and thus allow the US companies to maintain market share (Vignette 7.5).

In the coming years, the forest industry in BC will continue to face a number of challenges. In addition to trade disputes, BC will have to contend with environmental concerns, dwindling forest resources, and value-added production. Industry analysts say that the province has too many pulp mills facing a dwindling supply of wood fibre. The forest industry remains an important part of BC's economy, but in order to protect it, and the economy in general, such issues will have to be addressed.

British Columbia's Urban Geography

The urban geography of British Columbia is dominated by its largest city, Vancouver, and

Table 7.9   Major Urban Centres in British Columbia, 1996–2001

Centre	Population 1996	Population 2001	Percentage Change
Chilliwack	66,254	69,776	5.3
Prince George	87,731	85,035	(3.1)
Nanaimo	82,691	85,664	3.6
Kamloops	85,407	86,491	1.3
Abbotsford	136,480	147,370	8.0
Kelowna	136,541	147,739	8.2
Victoria	304,287	311,902	2.5
Vancouver	1,831,665	1,986,965	8.5

Source: Statistics Canada (2002).

by Victoria (Table 7.9 and Figure 7.7). Like the province as a whole, Vancouver's population has increased at a rate well above the national average. Until 1996 it was the fastest-growing large metropolitan area in Canada. Since then, Vancouver's spectacular growth rate has continued, but from 1996 to 2001, Calgary was the fastest-growing city in Canada.

Another striking aspect of the population geography of British Columbia is the concentration of people in the southwest corner of the province and the dominant position of Greater Vancouver within that population cluster. Greater Vancouver encompasses a number of cities and towns, including Burnaby, Coquitlam, Delta, Langley, New Westminster, North Vancouver, Richmond, Surrey, and West Vancouver. Besides Greater Vancouver, four other metropolitan centres, Victoria, Abbotsford, Nanaimo, and Chilliwack, are located in this densely populated area of British Columbia. The Okanagan Valley cities of Kelowna, Vernon, and Penticton constitute a modest secondary cluster.

Besides the rapid growth of Vancouver, smaller urban centres in BC are also growing

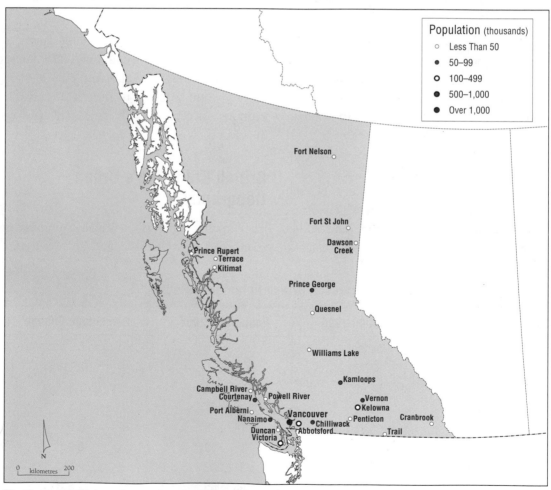

Figure 7.7 Major urban centres in British Columbia. Vancouver dominates BC's urban geography, followed by Victoria and Kelowna. BC's population is clustered in the southwest corner of the province, with smaller centres dotting the northern region.

Victoria is not only the capital of British Columbia, but it is also a major tourist destination. Its strong British heritage, as seen in this English-style pub, makes Victoria appealing to many visitors. (Al Harvey/The Slide Farm)

quickly. Since 1981, Kelowna, Abbotsford, Chilliwack, and Nanaimo have exhibited the greatest rates of increase of all cities in British Columbia (Table 7.9). From 1996 to 2001, these four cities recorded rates of population increase of 8.2 per cent, 8.0 per cent, 5.3 per cent, and 3.6 per cent respectively. In comparison, Vancouver and Victoria had increases of 8.5 per cent and 2.5 per cent.

Greater Vancouver is the dominant urban centre of British Columbia. Unlike Ontario and Québec, which each have more than one population cluster, BC can only claim the Vancouver area as a veritable population cluster. Defined in terms of a census metropolitan area, Vancouver's 2001 population was 2 million. Victoria, the second largest urban centre, has a much smaller population at just under 312,000 inhabitants (Table 7.9). As the capital of the province, Victoria is a 'government' town, as well as an important tourist and serv-

ice centre. Its mild climate has also attracted retired people, especially from the Prairie provinces. Kelowna and Abbotsford, with populations in 2001 of 148,000 and 147,000 respectively, fall into a distant third and fourth place. After Kelowna, there are four cities with populations over 65,000 but under 86,000: Nanaimo, Kamloops, Prince George, and Chilliwack. Below the 60,000 mark are 17 smaller cities with populations of 10,000 or more (Table 7.10).

Vancouver

Vancouver has a majestic physical setting. The city, located on the shores of Burrard Inlet, lies across the water from the snowcapped peaks of the North Shore mountains. To the west lies the island-studded Strait of Georgia, while the Fraser River and its deltaic islands (flat, low islands composed of silt and clay near the

Table 7.10	Smaller Urban Centres in British Columbia, 1996–2001		
Centre	**Population 1996**	**Population 2001**	**Percentage Change**
Kitimat	11,136	10,285	(7.6)
Squamish	14,236	14,435	1.4
Prince Rupert	17,414	15,302	(12.1)
Fort St John	15,021	16,034	6.7
Dawson Creek	18,039	17,444	(3.3)
Powell River	18,402	18,269	(0.7)
Terrace	20,941	19,980	(4.6)
Cranbrook	24,151	24,275	0.5
Parksville	22,629	24,285	7.3
Quesnel	25,074	24,426	(2.6)
Williams Lake	24,992	25,122	0.5
Port Alberni	26,893	25,396	(5.6)
Campbell River	33,849	33,872	0.1
Duncan	38,464	38,813	0.9
Penticton	41,276	41,574	0.7
Courtenay	46,297	47,051	1.6
Vernon	49,701	51,530	3.7

Source: Statistics Canada (2002).

mouth of a river) mark Vancouver's southern edge. Vancouver has a mild, marine climate, though some find the frequent rain and overcast skies unappealing. Vancouverites are rightfully proud of their city, which has one of the most beautiful settings in the world.

Vancouver emerged from its forest wilderness in a remarkably short time to become one of the great cities of North America. Few cities have grown as quickly. In 1870, the frontier settlement of Granville on Burrard Inlet had 50 residents. Little changed until the arrival of the Canadian Pacific Railway in 1886. At that time, the population of Granville had soared to about 1,000 and the residents incorporated their village as the City of Vancouver. By 1891, the newly incorporated city had reached a population of 15,000. The spectacular growth continued, partly as a result of amalgamation with neighbouring municipalities. By 1931, Vancouver's population had reached a quarter of a million. Twenty years later, Vancouver's population had doubled. By 2001, Vancouver's population was 2 million.

Like Montréal, much of Vancouver's commercial strength stems from its role as a trade centre. Most of the world's population is located along the Pacific Rim and Vancouver's economic well-being is closely tied to these countries. Most of BC's exports go to the United States and Japan. In 2002, nearly 80 per cent of exports from British Columbia were shipped to those two countries (British Columbia, 2003b: 29). Forest products made up approximately half of all 2002 exports, of which softwood lumber comprised 44 per cent, pulp 20 per cent, newsprint 5 per cent, and paper and paperboard 10 per cent. The remaining 21 per cent consisted of manufactured wood products such as doors and window frames (ibid., 28).

Vancouver is also a transshipment point for resource products from the interior of British Columbia, Alberta, Saskatchewan, and Manitoba; a service centre; and a tourist destination. Head offices of private companies (especially fishing, forestry, and mining companies) and federal and provincial public offices are located in Vancouver. While Vancouver (known in 1867 as Gastown) began as a sawmill centre, these air-polluting mills relocated to other centres or were dismantled. False Creek, in downtown Vancouver, had been the prime site for processing logs, but in 1986, False Creek became the site of Expo '86 and then the site of an upscale residential and farmers' market complex known as Granville Island (Vignette 7.6).

Much like British Columbia itself, Vancouver lies at a crossroads. It continues to rely on the business and trade generated by resource industries for its prosperity. However, in recent years it has developed into a major financial,

The face of Vancouverites is changing. Nearly 60 per cent of new arrivals in the census metropolitan area of Vancouver speak Chinese. The impact on Vancouver's schools is profound. In 2001, nearly one in five school-age children were new arrivals while in the City of Richmond, which is part of Vancouver's census metropolitan area, the figure jumped to one in three. (Al Harvey/The Slide Farm)

tourist, and cultural centre, much like the major cities of Central Canada. The staging of the 2010 Winter Olympics in Vancouver and the ski resort of Whistler will focus world attention on these two sites and may expand its role as a world destination for tourists.

Vignette 7.6   Granville Island

Granville Island, in the heart of downtown Vancouver, is strategically located between two residential areas, the West End and Kitsilano. In 1915, Granville Island was converted from a nondescript sandbar into an industrial core for the growing forest industry. In the 1970s, Granville Island began to change into an upscale residential and specialized commercial area centred on the Granville Island Public Market. Other enterprises have widened its appeal—artists' studios and shops, a wide variety of restaurants, and features like the Kids Market, Maritime Market, and the Coast Salish Houseposts, a joint endeavour between the Emily Carr College and First Nations. Granville Island is unique to Vancouver and has added another dimension to the wide-ranging tourist attractions in the Greater Vancouver region.

British Columbia's Future

The future poses several challenges for British Columbia. The first will be to regain the high rates of economic growth and population increase experienced prior to 1998. The US and Asian economies have shown the first signs of recovery. With the activity preparatory to hosting the 2010 Winter Olympics, the BC economy may spring back into action. The second challenge will be to avoid environmental deterioration of BC's major cities, parks, and wilderness areas while pursuing economic growth. A third challenge will be to reduce the mounting pressure placed on renewable resources in the province, particularly its agricultural lands, old-growth forest, and salmon stocks. Urban sprawl has already spread into the best agricultural land in the Lower Mainland. Related to BC's resources is the question of industrial diversification and this resource hinterland's position in the Canadian, North American, and global economies. Unless BC's resource industries are able to shift to a more value-added production, additional economic growth in manufacturing is unlikely, and, with continued automation in the resource industries, fewer jobs will exist. The final and most contentious challenge will be settling Aboriginal land claims. Land-claim agreements will benefit both First Nations and BC's economy as First Nations buy land, set up businesses, and purchase goods and services. The negotiations of these agreements, however, will come up against considerable opposition because they call for the sharing of BC's resources. The idea of sharing resources, particularly fish and forest resources, is not popular among other resource users.

British Columbia's past has been based on resource development; its future may well lie in the expansion of the high-technology, tourism, and manufacturing sectors. In the nineteenth century, the sheer extent of BC's natural resources suggested limitless wealth. By the end of World War II, harvesting of these resources had increased exponentially, forcing governments and private companies to rethink their approach to resource extraction. Now public policy emphasizes sustainable resource development—a policy that industries seem to support. But do they support it in practice? Now that the fish and forest resources are dwindling, private companies can no longer count on expanding their production without threatening the capacity of these renewable resources to replenish. Instead, fishers and loggers will pay the price for a declining fish stock and a shrinking volume of old-growth timber. Perhaps the recognition of these resource problems suggests that it will become easier to address them. At the 1998 shareholders' meeting of MacMillan Bloedel, the president and chief executive officer, Tom Stephens, announced that the company may halt logging of old-growth forests (Lush, 1998: B1), but when Weyerhaeuser purchased MacMillan Bloedel in 2000, the matter of harvesting old-growth timber was no longer on the table.

How can British Columbia adjust to these new resource constraints? Hayter (1996: 101) argues that the forest industry must change its outlook from that of a staple producer to that of a processor. More specifically, forestry companies must shift their efforts from low-value processing of timber, such as lumber, to high-value processing, such as furniture. But can forestry companies make this adjustment? If private forestry companies must maintain their profits, why would they change the nature of their operations from the profitable practice of exporting lumber, pulp, and other semi-processed forest products to the uncertainties of wood manufacturing? Given that scenario, Barnes and Hayter (1997: 7) pose two possibilities:

> The pessimistic interpretation is of an industry on the decline, fizzling out as the resource base itself disappears. The optimistic interpretation, though, is of a kind of industrial renaissance, of a newly fashioned forest products industry that emphasizes high-value products, skilled labour, and leading edge technology.

Herein lies the key to BC's complete transformation from a resource hinterland to an industrial core.

Summary

Since World War II, British Columbia's population has grown rapidly and its economy has diversified. Except for the late 1990s, when the downturn in the Asian economy dampened BC's exports and stalled its economic growth, British Columbia's economy outperformed the economies of other provinces. In 2001, BC constituted 13.1 per cent of Canada's population but accounted for 12 per cent of Canada's GDP.[7] These factors indicate that British Columbia is an emerging economic powerhouse within Canada. High-technology companies, producer services firms, and resource industries combine to drive BC's economy. Trade with Asia, coupled with Asian investments in BC, has further stimulated British Columbia's economic growth. Together these economic forces have propelled BC's economy into Canada's third economic force behind Ontario and Québec.

Before BC becomes an industrial core, however, several important issues must be resolved:

- The resource economy must be based on sustainable harvesting practices.
- In order to break its dependency on the primary sector and its inevitable boom-and-bust cycles, BC must increase the processing of its natural resources to expand its manufacturing base and increase employment opportunities.
- Land-claim agreements must be negotiated with BC's First Nations to settle past obligations and remove uncertainty over landownership.

Notes

1. Besides the issue of luring British Columbia into Confederation, there were several other political reasons for constructing a transcontinental railway. First, there was the urgent need to exert political control over the newly acquired but sparsely settled lands in Western Canada. As in British Columbia, the perceived threat to this territory was from the United States. Second, there was the need to create a larger market for manufactured goods produced by the firms in southern Ontario and Québec.

2. The exact number of people in British Columbia in 1871 is not known. How many Aboriginal peoples is a guess because their numbers declined sharply as they came into contact with European diseases. Similarly, the number of Americans and people from other countries who remained in the country after the gold rush is unknown. Certainly most moved on to the next gold rush, but some stayed. The gold rush of 1858 may have attracted about 25,000 Americans who sailed from San Francisco to New Westminster at the mouth of the Fraser River.

3. Professor Jim Miller was Chair of the Depart-

ment of Geography, University College of the Cariboo, Kamloops, BC. He has since moved to the Maritimes.

4. The provincial government created the Technical University of British Columbia in 1997 and this virtual university began to provide on-line courses in 2000. The objective is to prepare students for the high-tech work industry. The university offers certificate programs in electronic commerce and software development. High-technology companies are encouraged to locate offices and research laboratories near the campus. Students will undertake internships and co-operative work sessions with high-technology firms. Its theoretical basis lies in the concept of high-tech clusters around a university.

5. The Pacific Salmon Commission is an outcome of the 1985 Pacific Salmon Treaty. However, the Commission does not regulate the salmon fisheries but provides regulatory advice and recommendations to the two countries. It has responsibility for all salmon originating in the waters of one country that are subject to interception by the other, affect management of the other country's salmon,

or affect biologically the stocks of the other country. In addition, the Pacific Salmon Commission is charged with taking into account the conservation of steelhead trout while fulfilling its other functions. On 27 June 1997, negotiations between Canada and the United States over the size of the salmon catch in the Pacific Northwest collapsed. Efforts to resume negotiations have failed, leaving the fish quotas at the levels established in the 1985 Pacific Salmon Treaty.

6. In 1950, Alcan secured the rights to the water in the Nechako River system until 1999. When the British Columbia government signed this long-term agreement, it saw this industrial development as the key to opening up the province's Pacific Northwest. At that time, Victoria imagined that Kitimat would become an industrial complex with a population quickly reaching 20,000.

7. The negative impact of the decline of Asian exports accounts for much of BC's economic slowdown. In 1996, for example, BC's share of Canada's GDP was 13.3 per cent, but by 2000 its share had dropped to 12.4 per cent.

Key Terms

annual allowable cut
The amount of timber that can be cut under a publicly controlled timber lease; the amount of allowable cut is based on sustainable harvesting principles. The main factor in calculating the allowable harvest is the growth rate of the forest.

chemical pulp
A soft, shapeless mass of fine, long wood fibres produced from wood chips boiled in a chemical solution placed under pressure.

commercial forest
Forest land that is able to grow commercial coniferous (softwoods), deciduous (hardwoods), and mixed woods timber within an acceptable time frame.

groundwood pulp
A soft, shapeless mass of fine but short wood fibres produced from wood chips that are ground by mechanical means.

market pulp
Pulp sold in competition with other suppliers.

newsprint
A general term used to describe very thin paper used primarily in the publication of newspapers.

producer services
Services that have enabled firms and regions to maintain their specialized roles in marketing, advertising, administration, finance, and insurance industries. Producer services are one of several parts of the growing service sec-

tor of the economy.

softwood forest

The predominant forest in Canada. Softwood forests consist mainly of coniferous trees, characterized by needle-like foliage.

staples production

Resource development results in primary or staples production.

thermomechanical pulp

A soft, shapeless mass of fine, long, wood fibres produced by grinding wood chips into a soft mass and then boiling the fibrous mass.

tragedy of the commons

The destruction of renewable resources because of the absence of collective control over these resources, which are available to all. The problem lies in individuals maximizing the use of such resources for personal gain; such usage, in total, overwhelms the capacity of renewable resources to maintain and regenerate themselves.

value-added production

Manufacturing that increases the value of primary (staple) goods.

Bibliography

Barman, Jean. 1996. *The West Beyond the West: A History of British Columbia*, 2nd edn. Toronto: University of Toronto Press.

Barnes, Trevor, and Roger Hayter. 1997. *Trouble in the Rainforest: British Columbia's Forest Economy in Transition*. Canadian Western Geographical Series, vol. 33. Victoria: Western Geographical Press.

————, ————, and E. Grass. 1990. 'MacMillan Bloedel: Corporate Restructuring and Employment Change', in M. De Smidt and E. Wever, eds, *The Corporate Firm in a Changing World Economy*. London: Routledge, 145–65.

Boei, Bill, and Amy O'Brian. 2003. '10,000 flee Kelowna, B.C., Blaze', *National Post*, 23 Aug., A1.

Bone, Robert M. 2003. *The Geography of the Canadian North: Issues and Challenges*, 2nd edn. Toronto: Oxford University Press.

Bradbury, John. 1982. 'British Columbia: Metropolis and Hinterland in Microcosm', in L.D. McCann, ed., *Heartland and Hinterland: A Geography of Canada*. Scarborough, Ont.: Prentice-Hall, ch. 10.

British Columbia. 2000a. British Columbia Home Page: Quick Facts—the Economy [on-line database], Victoria. Searched 31 Aug. 1998 and 21 Dec. 2000: <http://www.bcstats.gov.bc.ca/data/QF_econo.HTM#for>.

————. 2000b. Quick Facts—Statistical Appendix [on-line database], Victoria. Searched 31 Aug. 1998: <http://www.bcstats.gov.bc.ca/data/QF_stats.HTM>.

————. 2000c. Quick Facts—People [on-line database], Victoria. Searched 31 Aug. 1998: <http://www.bcstats.gov.bc.ca/data/QF_peopl.HTM>.

————. 2002. Profile of the British Columbia High Technology Sector, 2002 Edition [on-line database], Victoria. Searched 31 Aug. 2003: <http://www.bcstats.gov.bc.ca/data/bus_stat/hi_tech.htm>.

————. 2003a. British Columbia Home Page: Quick Facts—the Economy [on-line database], Victoria. Searched 31 Aug. 2003: <http://www.bcstats.gov.bc.ca/data/QF_econo.HTM#for>.

————. 2003b. British Columbia Financial and Economic Review, 63rd edn [on-line database], Victoria. Searched 31 Aug. 2003: <http://www.bcstats.gov.bc.ca/tbs/F&Review03.pdf>.

British Columbia Ministry of Forests. 1996. *Forest Practice Code: Timber Supply Analysis*. Victoria: Ministry of Forests.

Cassidy, Frank. 1992. 'Aboriginal Land Claims in British Columbia: A Regional Perspective', in Ken Coates, ed., *Aboriginal Land Claims*. Toronto: Copp Clark, 10–43.

Cernetig, Miro. 1997. 'BC Makes Peace with

Alcan', *Globe and Mail*, 6 Aug., A1, A5.

Christensen, Bev. 1995. *Too Good to Be True: Alcan's Kemano Completion Project*. Vancouver: Talonbooks.

Davis, H. Craig, and Thomas A. Hutton. 1994. 'Marketing Vancouver's Services to the Asia Pacific', *The Canadian Geographer* 38, 1: 18–28.

Department of Fisheries and Oceans (DFO). 1994. *Canadian Fisheries Statistical Highlights 1992*. Ottawa: DFO.

———. 1996. *Fraser River Pink: Report of the Fraser River Action Plan Fisheries Management Group*. Ottawa: DFO.

Edginton, David W. 1994. 'The New Wave: Patterns of Japanese Direct Foreign Investment in Canada During the 1980s', *The Canadian Geographer* 38, 1: 28–36.

———. 1996. 'Japanese Real Estate Investment in Canadian Cities and Regions, 1985–1993', *The Canadian Geographer* 40, 4: 292–305.

——— and T.A. Hutton. 2000–1. 'Multiculturalism and Local Government in Vancouver', *Western Geography* 10–11: 1–29.

Farley, A.L. 1979. *Atlas of British Columbia*. Vancouver: University of British Columbia Press.

Forward, Charles N., ed. 1987. *British Columbia: Its Resources and People*. Victoria: Department of Geography, University of Victoria.

Francis, R. Douglas, Richard Jones, and Donald B. Smith. 1996. *Destinies: Canadian History Since Confederation*, 3rd edn. Toronto: Harcourt Brace.

Gourley, Catherine. 1988. *Island in the Creek: The Granville Island Story*. Maderia Park, BC: Harbour Publishing.

Halseth, Greg. 1999. '"We came for the work": Situating Employment Migration in B.C.'s Small, Resource-based Communities', *The Canadian Geographer* 43, 4: 363–81.

——— and Lana Sullivan. 2002. *Building Community in an Instant Town: A Social Geography of Mackenzie and Tumbler Ridge, British Columbia*. Prince George: University of Northern British Columbia Press.

Harris, Cole. 1997. *The Resettlement of British Columbia: Essays on Colonialism and Geographical Change*. Vancouver: University of British Columbia Press.

Hasselback, Drew. 2000. 'Water shortage leaves Alcan high and dry', *National Post*, 9 Dec., D7.

Hayter, Roger. 1992. 'The Little Town That Did: Flexible Accumulation and Community Response in Chemainus, British Columbia', *Regional Studies* 26: 647–63.

———. 1996. 'Technological Imperatives in Resource Sectors: Forest Products', in John N.H. Britton, ed., *Canada and the Global Economy: The Geography of Structural and Technological Change*. Montréal and Kingston: McGill-Queen's University Press, ch. 6.

———. 1997. 'High-Performance Organizations and Employment Flexibility: A Case Study of In Situ Change at the Powell River Paper Mill, 1980–1994', *The Canadian Geographer* 41, 1: 26–40.

Howard, Ross. 1996. 'BC Land-Claim Stakes Highest Yet', *Globe and Mail*, 16 Jan., A5.

Hume, Stephen. 2000. 'Did Francis Drake discover B.C.?', *National Post*, 5 Aug., B1–B2.

Hutchison, Bruce. 1942. *The Unknown Country: Canada and Her People*. Toronto: Longmans, Green and Company.

Hutton, Thomas A. 1998. *The Transformation of Canada's Pacific Metropolis: A Study of Vancouver*. Montréal: Institute for Research on Public Policy.

Koroscil, Paul, ed. 1991. *British Columbia: Geographical Essays in Honour of A. McPherson*. Burnaby, BC: Department of Geography, Simon Fraser University.

Liu, Xiao-Feng, and Glen Norcliffe. 1996. 'Closed Windows, Open Doors: Geopolitics and Post-1949 Mainland Chinese Immigration to Canada', *The Canadian Geographer* 40, 4: 306–19.

Lush, Patricia. 1998. 'MacBlo Posts Profit, Sells MB Paper', *Globe and Mail*, 24 Apr., B1.

McCullough, Michael. 1998. *Granville Island: An Urban Oasis*. Vancouver: CMHC.

McKee, Christopher. 1996. *Treaty Talks in British*

Columbia: Negotiating a Mutually Beneficial Future. Vancouver: University of British Columbia Press.

McVey, Wayne W., and W.E. Kalbach. 1995. Canadian Population. Toronto: Nelson Canada.

Marchak, M. Patricia. 1995. Logging the Globe. Montréal and Kingston: McGill-Queen's University Press.

Marsh, James H., ed. 1988. The Canadian Encyclopedia, 2nd edn. Edmonton: Hurtig.

May, Elizabeth. 1998. At the Cutting Edge: The Crisis in Canada's Forest. Toronto: Key Porter Books.

Miller, Jim. 1998. Personal communication, 28 Apr., University College of the Cariboo, Kamloops, British Columbia.

Mittelstaedt, Martin. 2004. 'Farmed salmon laced with toxins, study finds', Globe and Mail, 9 Jan., A1.

Natural Resources Canada. 2003. Selected Forestry Statistics, 2003 [on-line database], Victoria. Searched 17 Oct. 2003: <http://mmsd1.mms.nrcan.gc.ca/forest/section1New/section1New_e.asp>.

Nisga'a. c. 2000. Nisga'a History [on-line database], Vancouver. Searched 17 Nov. 2003: <http://www.schoolnet.ca/aboriginal/nisga1/hist-e.html>.

Prudham, W. Scott. 2001. 'Looking to Oregon: Comparative Challenges to Forest Policy Reform and Sustainability in British Columbia and the US Pacific Northwest', BC Studies 130: 5–40.

Rajala, Richard A. 1998. Clearcutting the Pacific Rain Forest: Population, Science, and Regulation. Vancouver: University of British Columbia Press.

Reed, Maureen G. 1997. 'Seeing Trees: Engendering Environmental and Land Use Planning', The Canadian Geographer 41, 4: 398–414.

Robinson, J. Lewis, ed. 1972. British Columbia. Studies in Canadian Geography. Toronto: University of Toronto Press.

——— and W.G. Hardwick. 1973. British Columbia: 100 Years of Geographical Change. Vancou-

ver: Talonbooks.

Saku, James C., Robert M. Bone, and Gérard Duhaime. 1998. 'Toward an Institutional Understanding of Comprehensive Land Claim Agreements in Canada', Études/Inuit/Studies 22, 1: 109–21.

Shidelar, Janet. 2000. The Geography of Canada Bibliography Series: Vol. 5, British Columbia. Plattsburgh, NY: Plattsburgh State University, Center for the Study of Canada.

Sparke, Matthew. 1998. 'A Map That Roared and an Original Atlas: Canada, Cartography, and the Narration of Nation', Annals of the Association of American Geographers 88: 463–95.

Stanford, Quentin H., ed. 1998. Canadian Oxford World Atlas, 4th edn. Toronto: Oxford University Press.

Statistics Canada. 1982. Census Metropolitan Areas and Census Agglomerations with Components. Catalogue no. 95–903. Ottawa: Ministry of Supply and Services Canada.

———. 1988a. Canadian Statistics—Labour Force, Employed and Unemployed [on-line database], Ottawa. Searched 2 Sept. 1998: <http://www.statcan.ca/english/Pgdb/People/Labour/labor07a.htm>.

———. 1988b. Labour Force Characteristics for Both Sexes, Aged 15 and Over [on-line database], Ottawa. Searched 12 Aug. 1998: <http://www.statcan.ca/english/econoind/lfsadj.htm>.

———. 1988c. 1981–1996 Census: Labour Force Activity [on-line database], Ottawa. Searched 2 Sept. 1998: <http:www.statcan.ca/english/census96/mar17/labour/table6/t6p59s.htm>.

———. 1992. Mother Tongue. Catalogue no. 93–313. Ottawa: Industry Canada.

———. 1996a. Labour Force Annual Averages 1995. Catalogue no. 71–220–XPB. Ottawa: Statistics Canada.

———. 1996b. Canadian Statistics: Distribution of Employed People, by Industry, by Province [on-line database], Ottawa. Searched 20 June 1996: <http://www.statcan.ca/english/Pgdb/

labor21b.htm>.

———. 1997a. *A National Overview: Population and Dwelling Counts*. Catalogue no. 93–357–XPB. Ottawa: Industry Canada.

———. 1997b. 1996 Census: Nation Tables—Population by Mother Tongue, Showing Age Groups, for Canada, Provinces and Territories, 1996 Census—20% Sample Data, 2 Dec. 1997 [on-line database], Ottawa. Searched 15 July 1998: <http://www.statcan.ca/english/census96/>.

———. 1997c. *The Daily*—1996 Census: Mother Tongue, Home Language and Knowledge of Languages, 2 Dec. 1997 [on-line database], Ottawa. Searched 14 July 1998: <http://www.statcan.ca/Daily/English/>.

———. 1998a. *Canadian Forestry Statistics 1995*. Catalogue no. 25–202–XPB. Ottawa: Ministry of Industry, Science and Technology.

———. 1998b. *The Daily*—1996 Census: Aboriginal Data, 13 Jan. 1998 [on-line database], Ottawa. Searched 14 July 1998: <http://www.statcan.ca./Daily/English/>.

———. 1998c. The Daily—1996 Census: Ethnic Origin, Visible Minorities, 17 Feb. 1998 [on-line database], Ottawa. Searched 16 July 1998; <http://www.statcan.ca/Daily/English/>.

———. 1998d. *Canadian Economic Observer*. Catalogue no. 11–010–XPB. Ottawa: Statistics Canada.

———. 2002. 2001 Census: Population and Dwelling Counts, for Census Metropolitan Areas and Census Agglomerations, 2001 and 1996 Censuses. 16 July 2002 [on-line database], Ottawa: <http://www.statcan.ca/english/IPS/Data/93F0050XCB2001013.htm>.

———. 2003. Canadian Statistics: Distribution of Employed People, by Industry, by Province [on-line database], Ottawa. Searched 20 May 2003: <http://www.statcan.ca/english/Pgdb/labor21b.htm>.

Tollerson, Christopher. 1998. *The Wealth of Forests: Markets, Regulations, and Sustainable Forestry*. Vancouver: University of British Columbia Press.

Wilkinson, Bruce W. 1997. 'Globalization of Canada's Resource Sector: An Innisian Perspective', in Barnes and Hayter (1997: 131–47).

Willems-Braun, Bruce. 1997. 'Buried Epistemologies: The Politics of Nature in (Post)colonial British Columbia', *Annals of the Association of American Geographers* 87: 3–31.

Wynn, Graeme, and Timothy Oke, eds. 1992. *Vancouver and Its Region*. Vancouver: University of British Columbia Press.

Yamamoto, Daisaku. 2002. 'Issues of Globalization and Reflexivity in the Japanese Tourism Production System: The Case of Whistler, British Columbia', *The Professional Geographer* 54, 1: 83–93.

Further Reading

Molloy, Tom. 2000. *The World Is Our Witness: The Historic Journey of the Nisga'a into Canada*. Calgary: Fifth House.

On 11 May 2000, the Nisga'a treaty passed into law, marking a historic agreement between this small group of Indians and the rest of Canadian society. In *The World Is Our Witness*, Molloy, who was the chief federal negotiator, describes how this agreement ends the centuries-old colonization by the British and, later, Canadians of the Nisga'a lands and people. Furthermore, this treaty begins the search for a place within Canadian society by the Nisga'a. For the other Indian tribes of British Columbia, this treaty has far-reaching implications. Molloy not only provides an insider's view of the struggle to achieve an agreement, but also explains its significance to Canadians. The Nisga'a treaty represents a compromise between the Nisga'a and other Canadians on how to share the lands and resources found in the traditionally occupied lands of the Nisga'a. Originally, there was no sharing. The early English settlers who established British colonies along the east coast of the United States nearly 400 years ago claimed the land by the right of 'discovery' and acknowledged no Aboriginal rights to land. Recognition of Aboriginal rights began with the Royal Proclamation of King George III. In 1763, the Royal Proclamation declared that lands to the west of the Appalachian Mountains were Indian lands. While these lands were lost to the Americans in 1783, the legal precedent laid the foundation for treaty-making across British North America and later Canada. The Nisga'a treaty is not only the most recent agreement but it has expanded the scope of treaties into the area of resource-sharing.

Overview and Objectives

There are two distinct subregions within Western Canada: a northern resource hinterland and a more densely populated agricultural/industrial core in the south. Because of low world prices for agricultural products, restricted access to US markets, and the removal of the transportation subsidy for grain, this rural economy is in a state of change. At the same time, rapid growth has taken place in the major cities, especially Calgary and Edmonton. Alberta leads this transformation, powered by its oil and gas industry, the diversification of its industrial base, and the rapid expansion of its high-technology industry. Like BC, Western Canada has moved from a resource frontier to an upward transitional region. But like BC's forest industry, Western Canada's natural resource base—agriculture—is struggling. The transformation of its agricultural economy is the *Key Topic* in this chapter.

- Describe Western Canada's physical and historical geography.
- Present the basic characteristics of its population and economy.
- Examine Western Canada's population and economy within the context of the region's physical setting and dry continental climate.
- Explore Western Canada's changing position within the core/periphery model—from a resource frontier to an upward transitional region.
- Focus on the transition in Western Canada's agricultural sector, and to a lesser degree, in its resource and manufacturing sectors.

Chapter 8

Western Canada

■ Introduction

Western Canada is at a turning point. Behind lies its history as Canada's chief exporter of grain, oil, and other primary products. Ahead is the prospect of playing a different economic role, one that is much more involved in the processing of Western Canada's agricultural and resource products. This chapter examines the factors underlying this radical departure for Western Canada from its traditional role as an agricultural hinterland. The *Key Topic*, 'Transition', looks at the economic and political forces propelling Western Canada's economy to diversify, shifting more and more into an urban industrial economy and forcing rural land-use changes in the Canadian Prairies. The exodus from rural villages and towns represents a major internal migration. In a broad context, this transition is illustrated in Friedmann's core/periphery model as a shift from a resource (agricultural) frontier to an upward (industrial) transitional region (see Chapter 1).

Western Canada within Canada

Western Canada is part of the vast western interior of North America (Figure 1.2). This huge area, comprising the three western provinces—Alberta, Saskatchewan, and Manitoba—contains many natural resources, but shipping these products to world markets has proven to be a major stumbling block to the region's economic development. These resources range from the fertile soils of the Canadian Prairies to the commercial timber stands in the northern coniferous forest. As well, there is considerable mineral wealth in the Interior Plains, the Canadian Shield, and the Cordillera.

In terms of economic output, Western Canada falls behind Ontario and Québec. Overall, Western Canada's economy accounted for just over 20 per cent of Canada's GDP in 2001 (Figure 8.1). Western Canada's economic performance translates into a $120 billion economy. An important measure of its economic well-being is Western Canada's low unemployment rate. In 2001, Western Canada had, at 4.9 per cent, the lowest unemployment rate of the six regions, well below the national average of 7.2 per cent (Figure 8.1).

Since World War II, most of the region's economic and demographic growth has occurred in oil-rich Alberta. While oil and gas are the region's most valuable non-renewable assets, agriculture remains its key renewable resource. A common problem confronting farmers and resource companies in Western Canada is the high cost of transportation, particularly the long rail distance their products must travel to reach ocean ports, from where they are then shipped to world markets. Once products are aboard ocean vessels, the transportation cost per kilometre drops drastically. Overcoming the cost of shipping products to ocean ports has been a persistent struggle in Western Canada's development of its agricultural, forest, and

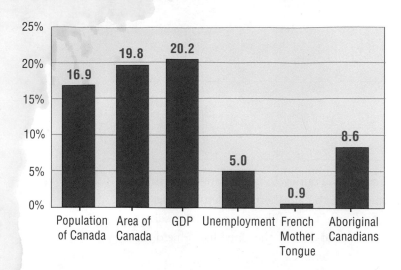

Figure 8.1 Western Canada, 2001. Although Western Canada covers a large area of the country, the region's population strength is more modest. The region has fewer French-speaking Canadians than other parts of the country, but a significant number of Aboriginal Canadians.

Sources: Tables 1.1, 1.5, and 4.19.

mineral resources. The Hudson Bay Railway was one such effort. In the 1920s, Prairie farmers thought they could reduce their railway costs to Atlantic ports by constructing a rail line from The Pas to Churchill, Manitoba (York Factory) on Hudson Bay and then shipping their grain to their major market in the United Kingdom. The Hudson's Bay Company had used this route for centuries so why not Prairie farmers? Unfortunately, the high cost of marine insurance made this rail/sea route from Churchill to London unattractive.

Farmers are also confronted with low world prices for their products. Grain prices, for example, have been declining for several decades while costs of production have been increasing. Prices for non-grain crops such as canola have encouraged farmers to switch to these more profitable crops. Unfortunately, non-grain crops also suffer from fluctuating prices and, as more farmers grow such crops, prices tend to decline.

The population of Western Canada totals over 5.1 million. The vast majority live in the **regional core**, an area of major cities, towns, and villages located in the southern half of the region. Agricultural, industrial, and service activities are well established within this population core. Calgary, Edmonton, Winnipeg, Saskatoon, and Regina are the leading cities. In the northern half of Western Canada lies its regional hinterland, where the larger centres are linked to resource developments. Western Canada's northern hinterland has less than 5 per cent of the region's population. While most live in resource towns, such as Fort McMurray and Thompson, many Aboriginal settlements exist near former fur trading posts.

Western Canada's Physical Geography

Western Canada, which extends over nearly 20 per cent of Canada, has two major physiographic regions—the Interior Plains and the Canadian Shield—and two minor ones—the Hudson Bay Lowland and the Cordillera (Figure 2.1). A thin portion of the Cordillera, the Rocky Mountains, forms a natural and political border between southern Alberta and British Columbia. Each physiographic region has a particular set of geological conditions, physical landscapes, and natural resources. (See also 'The Interior Plains', 'The Canadian Shield', 'The Hudson Bay Lowland', and 'The Cordillera' in Chapter 2.)

The tiny section of the Cordillera, along the eastern flank of the Rocky Mountains in Alberta, provides logging and mining opportunities for Western Canada. However, the main attraction of this slice of the Cordillera is its spectacular mountain landscape. This mountainous terrain has two internationally acclaimed parks, Banff National Park and

Jasper National Park, which attract visitors from around the world. Calgarians are especially fortunate in having easy access to the Kananaskis Country Provincial Park, where a number of mountain recreational activities—camping, hiking, and skiing—are available.

In the Interior Plains region the sedimentary rocks contain valuable deposits of fossil fuels. By value, the four leading mineral resources are oil, gas, coal, and potash. Most petroleum production occurs in a geological structure known as the **Western Sedimentary Basin**, which underlies most of Alberta and portions of British Columbia, Saskatchewan, and Manitoba. However, not all mineral extraction takes place deep in these sedimentary rocks. A few deposits are exposed at the surface of the ground. In southeastern

Saskatchewan, brown coal is extracted through open-pit mining and then burned to produce thermal electricity. In northeastern Alberta, the huge petroleum reserves in the Athabasca tar sands are exploited by surface mining techniques, such as hydro-transport in which the oil sands are mixed with extremely hot water and transported to the upgrader plant by pipeline.

In the Canadian Shield, which extends into Manitoba and Saskatchewan, rocky terrain makes cultivation virtually impossible. While forestry does take place along the southern edge of the Canadian Shield, particularly in southeastern Manitoba, large-scale commercial enterprises are generally limited to mining and the production of hydroelectricity. In northern Saskatchewan, uranium companies

Lake Louise is one of Banff National Park's most stunning natural features. The lake's famous turquoise colour is caused by fine rock particles, called 'glacial flour', that are contained in the stream water from alpine glaciers. (© Royalty-Free/Corbis/Magma)

produce most of Canada's uranium from open-pit and underground mines. Large-scale hydroelectric dams and generators are located on Manitoba's northern rivers, particularly the Nelson River.

All of Western Canada has a continental climate characterized by extreme daily and seasonal fluctuations in temperature and low annual precipitation (Figure 2.5). The region is dry not only because of its distance from the ocean, but also because the mountain ranges in the Cordillera block the eastward movement of mild, moist Pacific air masses. During the winter, Arctic air masses often dominate weather conditions in the Prairies, placing the region in an Arctic 'deep-freeze' (Figure 2.3). A combination of strong winds and sub-zero temperatures can produce blizzard-like weather. (See 'Prairies Zone' in Chapter 2 for further details.) In southern Alberta, strong winds that become warm and dry as they flow down a mountain slope are known as chinooks.

Precipitation is a critical climatic element in Western Canada. While southern Manitoba, which receives rainfall from the moist Gulf of Mexico air masses, is rarely troubled by droughty conditions, the remaining prairie lands, but especially the dry lands of southern Saskatchewan and Alberta, are not so fortunately. The annual precipitation in Western Canada is low, varying from a high of 550 mm in southern Manitoba to less than 400 mm in the dry lands of southern Alberta and Saskatchewan (the so-called Palliser Triangle described in Vignette 8.2). In addition to this spatial variation, precipitation varies from year to year, causing so-called wet and dry years. This annual variation has its greatest effect in the dry lands of Palliser's Triangle. For instance, in the 1930s a series of dry years resulted in the disastrous Dust Bowl, a period of severe drought that devastated Prairie farming and led to extensive wind erosion of the topsoil. The Dust Bowl drove thousands of homesteaders off the land in the short-grass Prairies. Annual variations in precipitation have led Prairie farmers to coin the term 'Next Year Country', referring to the bad harvest years that prompt farmers to hope for a better crop the following year. (See 'Next Year Country' later in this chapter for further information.)

In Western Canada the natural vegetation and the soil, and therefore the success of commercial crops, are largely determined by the

Too dry and too wet: dust storm southwest of Lakenheath near Assiniboia, Saskatchewan, in 1934. The drought that struck Western Canada in the 1930s was so severe that some grain farmers abandoned their homesteads. Yet within this brown soil zone of southern Saskatchewan, sudden rainstorms can account for much of the local area's annual precipitation. For example, on 30 May 1961, a severe thunderstorm struck Buffalo Gap, some 150 km south of Regina near the United States border. In less than an hour, more than 250 mm of rain fell, ending a long dry spell (Phillips, 1993: 88). (*Saskatchewan Archives Board R-A4665*).

evapotranspiration rate, which is a combination of precipitation and temperature (Vignette 8.1). The evapotranspiration rate varies within Western Canada and therefore creates differences in the region's soil and vegetation. In the northern parts of Western Canada, in the Canadian Shield, cooler weather leads to a lower evaporation rate, resulting in sufficient moisture for tree growth. In this area the boreal forest grows in a podzolic soil (Figures 2.7 and 2.8). The only exception is in the Peace River country, a gently rolling plain where black chernozemic soils are found beneath an aspen parkland. Further south, where the

Vignette 8.1　Water Deficit and Evapotranspiration

A dry, continental climate prevails over the interior of Western Canada. Due to a combination of low annual precipitation and high summer temperatures, this zone has a 'water deficit'. This zone's low precipitation is primarily the result of the eastward-moving Pacific air masses losing most of their moisture as they rise over the Cordillera's mountain chains. By the time they descend the eastern slopes of the Rocky Mountains, these moist air masses have turned into dry ones. As a result, the Prairies receive little rain or snow from the Pacific Ocean. This water deficit is often measured in terms of potential evapotranspiration, which is the amount of water vapour that can potentially be released from an area of the earth's surface through evaporation and transpiration (the loss of moisture through the leaves of plants). There is a water deficit if the evapotranspiration rate is greater than the average annual precipitation. For example, most of the Prairies zone usually receives less than 400 mm annually, while its evapotranspiration rate exceeds 500 mm. The difference indicates a water deficit of over 100 mm.

Rich, dark brown soils represent one of the major chernozemic soils in the Canadian Prairies. In this photograph near Bigger, Saskatchewan, the practice of crop rotation is illustrated. (Barrett and Mackay Photography, Inc.)

evaporation rate is much higher, the podzolic soil gives way to chernozemic soils, which are more favourable for agriculture.

Indeed, the soil and natural vegetation of the southern part of Western Canada have had a significant impact on the region's role as an agricultural heartland. This southern portion, in the Interior Plains, has two natural vegetation zones (parkland and grassland) and three chernozemic soil zones (black, dark brown, and brown) (Figures 2.7 and 8.4). Just south of the boreal forest, podzolic soil gives way to the black chernozemic soil that supports a parkland vegetation. This parkland is a transition zone between the boreal forest and the grassland natural vegetation zone, which is located a little further south and is supported by dark brown and brown chernozemic soils. Within the grassland, the evaporation rate increases towards the American border, causing this mid-latitude grassland to change from tall grass to short grass as one moves southward. Tall-grass natural vegetation occurs in a wide arc from southwestern Manitoba to Saskatoon to Edmonton. Here there are dark-brown chernozemic soils, except in southern Manitoba, where black soils occur because of higher annual precipitation. This tall-grass natural vegetation zone and the parkland to its north comprise the area known as the **Fertile Belt**, where farming has traditionally been more successful. South of this fertile arc is the area of short-grass natural vegetation and brown chernozemic soils known as the **Dry Belt** (Vignette 8.2). This semi-arid area has presented the highest risk for grain farming, so much of the land is now worked in fallow rotation (a crop is planted every second year) or used for pasture.

Environmental Challenges

Western Canada faces several environmental challenges. One is the urgent need to remove the radioactive wastes surrounding abandoned uranium mines along the north shore of Lake Athabasca. Like a ticking time bomb, these radioactive tailings are slowly seeping into the groundwater and the waters of Lake Athabasca (Bone, 2003: 165). The cost of this cleanup is estimated to be at least $1 billion (Saskatchewan Environment, 2002). Another

| Vignette 8.2 | Palliser's Triangle |

In 1857, the Palliser Expedition set out from England to assess the potential of Western Canada for settlement. This expedition spanned three years (1857–60). In his report to the British government, John Palliser identified two natural zones in the Canadian Prairies. The first zone was described as a sub-humid area of tall grasses, while the second, located further south, was described as a semi-arid area with short-grass vegetation. Palliser considered the area of tall grasses to be suitable for agricultural settlement. He named this area the Fertile Belt. In Manitoba this belt is south of the Canadian Shield and stretches to the border with the United States. Palliser believed that the semi-arid zone, located in southern Alberta and Saskatchewan, was a northern extension of the Great American Desert. According to Palliser, these semi-arid lands were unsuitable for agricultural settlement. The area described by Captain Palliser became known as Palliser's Triangle—its area overlaps with, but is slightly larger than, the agriculture zone known as the Dry Belt (see Figure 8.5).

problem is the challenge to reduce the release of carbon dioxide and sulphur dioxide gases from the mining and processing operations associated with the Alberta tar sands developments. While the release of carbon dioxide increases the greenhouse gases in the atmosphere and therefore furthers the prospect of global warming, sulphur dioxide emissions cause acid rain, thus making thousands of lakes lying in the Canadian Shield in northern Saskatchewan vulnerable to increased acidity. Problems exist in the agricultural zone where the growth of the livestock and hog industries has led to intensive livestock operations that see cattle and hogs contained in fenced or enclosed areas. Alberta leads in this shift from the small traditional farms with small herds to large, high-technology operations with thousands of animals. This process of large-scale and highly specialized livestock factory-like operations is part of the trend towards an industrial agricultural economy in Alberta and the rest of North America.

Large-scale hog farms are the fasting-growing sector of the livestock industry and the waste products from hogs provide the greatest threat to the water resources. The heavy concentration of manure in a small area poses a danger to surface water through runoff and to groundwater through infiltration. Untreated manure may also contain bacteria such as E. coli and salmonella, viruses, and parasites. The potential of fecal bacteria seeping into the limited water supplies in this dry zone of Western Canada is great and yet provincial governments have been slow to react, preferring to encourage agricultural diversification. While world demand for pork is increasing, many European countries, several American states, and Québec have refused to permit additional large-scale hog operations because existing hog farms have caused the contamination of

some of their water resources. In the late 1990s, water pollution was so serious that Taiwan banned the construction of factory-like hog barns, forcing Taiwan Sugar Corporation, a major hog operator, to seek permission to build hog barns in Alberta. Strong opposition from concerned citizens in Alberta took the Taiwan hog barn development plan to the Alberta Court of Appeal, where a ruling revoked the company's development permit in January 2001.

Public awareness and pressure from environmental organizations have prompted governments and industry to address the environmental risks associated with intensive livestock operations (Price, 2003: 38). Even so, the so-called 'beneficial management practices' designed to minimize the impacts of animal waste products on water quality are just in the research stage. Even if such research is successful, the likelihood of the livestock industry causing a serious health problem from the contamination of a local water supply along the lines of Walkerton, Ontario, or North Battleford, Saskatchewan, is too great not to invoke the precautionary principle, whereby risk to the environment mandates that certain practices should not be pursued.

Western Canada's Historical Geography

Western Canada's history began long before the three Prairie provinces became part of Canada. In fact, Western Canada's recorded history goes back to the fur-trading days. Beginning in 1670, the Hudson's Bay Company administered for 200 years much of Canada's western interior. This area was part of Rupert's Land (all the land draining into Hudson Bay). In 1821, when the company merged with its rival, the North West Company, the

Hudson's Bay Company acquired control over more land, known as the North-Western Territory (lands draining into the Arctic Ocean). Before 1870, when Canada was ceded these lands by the British government (Figure 3.4), the Hudson's Bay Company used these lands exclusively for the fur trade.

The land in Western Canada began to be used for purposes other than fur trading at the beginning of the nineteenth century. In 1810, Lord Selkirk, a Scots nobleman who was concerned with the plight of poor Scottish crofters (peasants) evicted from their small holdings, acquired land in the Red River Valley from the Hudson's Bay Company. The first Scottish settlers arrived in 1812 to form an agricultural settlement near Fort Garry, the principal Hudson's Bay trading post in the region. This settlement became known as the Red River Colony. Selkirk's settlers, however, faced an unfamiliar and harsh environment and had great trouble establishing an agricultural colony. Over the years, many gave up and left for Upper Canada and the United States.

At the same time, many former officers and servants of the Hudson's Bay Company, along with their Indian wives and children, settled at Fort Garry. In addition to these English-speaking people were the French-speaking Métis who had worked for the North West Company. Because the Métis were Catholic and spoke French, they formed a separate cultural group within the Red River Colony. After the consolidation of the Hudson's Bay Company and the North West Company in 1821, many who worked for the North West Company were no longer employed by the new company. Many Métis, particularly those who were French-speaking, settled in the Red River Colony, where they turned their attention to subsistence farming, freighting, and buffalo hunting.

Agricultural Potential

Despite the modest settlement near Fort Garry and the thriving fur trade, little was known about the geography of the Canadian West in the mid-nineteenth century. To the south, in the United States, American settlers had begun to occupy land west of the Mississippi River. By 1854, a railway stretched across the United States from New York through Chicago to St Paul on the Mississippi River. In 1858, Minnesota had a sufficiently large population to warrant statehood. By then, American homesteaders began moving beyond the Mississippi into the northern Great Plains, including the Red River Valley. West of the Red River Valley, however, were the dry lands of the Great Plains, which limited further settlement. In fact, about 30 years earlier American explorers had dubbed Montana part of 'the Great American Desert'. Blocked to the west by these dry lands of North Dakota and Montana, settlers began to look northward to the unoccupied lands of British North America.

In 1857, the British government and the Royal Geographical Society sponsored an expedition into the Canadian West. Their central task was to determine the suitability of the Canadian West for agricultural settlement. John Palliser, an explorer, led the British North American Exploring Expedition. After his party arrived from England, they quickly travelled by rail from New York to St Paul and then by steamboat to the Red River Colony. They travelled by horseback from Fort Garry across the Canadian Prairies to the Rocky Mountains. Palliser reported that there was fertile land in much of the western interior, but that the land in southern Alberta and Saskatchewan near the border with Montana was far too dry for farming (Vignette 8.2). Palliser believed the Great American Desert that American explorers had already identified extended into the grasslands of southern Alberta and Saskatchewan. Named after him, this semi-arid area is now known as **Palliser's Triangle**. Another expedition in 1858, led by Henry Hind, a geologist and naturalist, confirmed that the parkland (the natural vegetation zone between the grasslands and the boreal forest) offered the best land for agricultural settlement.

Opening the West

By the middle of the nineteenth century, arable land in Upper Canada was in short supply. Some settlers began to look to the American West for new land. During the negotiations with Britain over Confederation, the subject of the annexation of Rupert's Land into the Dominion of Canada arose, and provision was made in the British North America Act for its admission into Canada. In 1869, the Hudson's Bay Company signed the deed of transfer, surrendering to Great Britain its chartered territory for £300,000—with the exception of the lands surrounding its posts and about 1,133,160 ha of farmland. In 1870, Great Britain transferred Rupert's Land to Canada.

In 1869, Canada's government sent surveyors into the Red River Colony to prepare a land registry system for the expected influx of settlers. The surveyors employed a grid system known as a township survey. A township consisted of 93 km², or 36 sections. Each section was subdivided into four quarter sections of 65 ha each. Each settler would receive a quarter section and would be required to till the land and build a house on that section. The opening of Western Canada for agricultural settlement officially began in 1870 and continued until 1914, when World War I halted the influx of European immigrants into Western Canada. Following World War I, veterans

were encouraged to establish homesteads, especially in the Peace River country, where there remained some arable land.

Original Inhabitants

When the Canadian government sent surveyors to the West in 1869, it was not uninhabited. Indians and Métis lived there, but the new agrarian economy would quickly marginalize them and overtake their lands.[1] The territories of the Dene and Woodland Cree were outside of the settled areas and these people remained tied to a hunting and trapping economy (Figure 3.2). The Plains Indians (Sarcee, Blood, Peigan, Stoney, Plains Cree, Nakota, Lakota, Blackfoot, and Saulteaux) lost control of the Plains and were driven to sign treaties and live on Indian reserves (Figure 3.9). In 1867, the Red River Colony had a population of nearly 12,000 people, mostly Métis. The arrival of land surveyors and settlers led the Métis, under the leadership of Louis Riel, to mount the Red River Rebellion in 1869. (See 'The Red River Rebellion' in Chapter 3 for more details.) The Métis wanted to negotiate the terms of entry into Canada from a position of strength—that is, as a government. The Métis obtained major concessions from Ottawa—they were guaranteed ownership of land, recognition of the French language, and permission to maintain Roman Catholic schools. In 1870, the fur-trading district of Assiniboia became the province of Manitoba. But the rebels' victory was hollow. As the newcomers poured into Manitoba, the Métis society was overwhelmed.

Many Métis left the colony to search for a new place to settle in the Canadian West. Such a place was Batoche, just north of the site where the city of Saskatoon now stands. Within 15 years, however, settlers would again encroach upon the Métis agricultural settlement. In 1885, as before, the Métis staged a

The Métis search for a place within Confederation began with rebellions in 1869–70 and 1885. Both rebellions were led by Louis Riel. After losing the Battle of Batoche, Riel surrendered to the Canadian forces. He was tried and convicted of high treason. On 16 November 1885, Riel was hanged as a traitor, though he remains a hero to the Métis to this very day. (National Archives of Canada)

rebellion led by Louis Riel. This time, the Canadian militia defeated the Métis at the Battle of Batoche and Louis Riel was captured, found guilty of treason, and hanged. To this day, Riel remains a controversial figure in Canadian history—a traitor to some, he remains a hero to the Métis.

The experiences of Indian tribes during the early period of western settlement were somewhat different. Indian tribes, such as the Blackfoot, had roamed across the Canadian

Prairies and the northern Great Plains of the United States long before the arrival of European explorers, fur traders, and settlers. The tribes were nomadic and hunted buffalo. By the 1870s, the buffalo had virtually disappeared from the Prairies, leaving the Prairie Indians destitute. By then, life for the Blackfoot, Blood, Plains Cree, Peigan, and Saulteaux tribes became almost unbearable. They had little choice but to sign treaties with the federal government. Between 1873 and 1876, all the tribes (except for three Cree chiefs—Big Bear, Little Pine, and Lucky Man—and their followers) signed numbered treaties in exchange for reservations, cash gratuities, annual payments in perpetuity, the promise of educational and agricultural assistance, and the right to hunt and fish on Crown land until such land was required for other purposes. In 1882, impending starvation for his people also forced the Cree leader, Big Bear, to accept Treaty No. 6. Over the next few years, however, the Cree sought other concessions from the federal government. When these efforts failed, Cree warriors supported the doomed Métis rebellion in 1885 by attacking several settlements, including Fort Pitt, the Hudson's Bay post on the North Saskatchewan River near the Alberta–Saskatchewan border.[2]

Treaties with Ottawa offered the Indians prospects for survival and time to find a place in a new economy, but the treaties also made them wards of the Crown. Living on reserves, Indians were isolated from the evolving Canadian society and became increasingly dependent on the federal government. Further north, the Woodland Cree and Dene (Chipewyan) tribes who lived in the boreal forest were not as affected by the encroachment of western settlers. Although they, too, signed treaties, these northern Indians continued their migratory hunting and trapping lifestyle well into the next century. In the 1950s, their dependency on Ottawa grew with the demise of the fur trade and their subsequent relocation to settlements.[3]

Canadian Pacific Railway

Once treaties ensured the availability of land for homesteading, the Canadian government needed to make this land more accessible for new settlers. The next step in the plan to open Western Canada to agricultural settlement was a transcontinental railway. Macdonald's vision of Canada extending from the Atlantic to the Pacific hinged on a transcontinental railway. There were already three transcontinental railways in the United States. Without a Canadian counterpart, Ottawa feared the worst, namely, that the West would be lost to the Americans. Even if Canada could retain its western territories in the absence of a Canadian transcontinental railway, the north–south transportation pull exerted by the American railways would prevent Ontario's fledgling industrial core from reaching the market in Western Canada, and western settlers would be unable to ship their products to eastern markets. For instance, in 1870, the new province of Manitoba was linked by steamboat to the rail centre of Fargo, North Dakota. Only from Fargo could passengers and freight from Manitoba quickly reach Toronto, Ottawa, and Montréal.

British and Canadian companies were not interested in a risky railway construction project across Canada unless they could obtain substantial financial assistance from Ottawa. Two reasons accounted for their lack of interest: the Canadian Shield and the Cordillera were two formidable (and therefore costly) barriers to overcome in building a railroad. Indeed, physical geography posed a much greater challenge to Canadian railway builders than to their American counterparts. In 1881,

Ottawa announced generous terms: the Canadian Pacific Railway Company was awarded a charter, whereby the company received $25 million from the federal government, 1,000 km of existing railway lines in eastern Canada owned by the federal government, and over 10 million ha of Prairie land in alternate square-mile sections on both sides of the railway to a maximum depth of 39 km. The terms were successful—the Canadian Pacific Railway was completed in 1885.

Settlement of the Land

The settling of Western Canada marks one of the world's great migrations and the transformation of the region into an agricultural resource frontier. Under the Dominion Land Act that Ottawa had passed in 1872, homesteaders were promised 'cheap' land in Manitoba—by building a house and cultivating some of the land, they could obtain 65 ha of land for only $10. Following 1872, an influx of prospective homesteaders began arriving, most coming from Ontario and, to a lesser degree, the Maritimes, Québec, and the United States. When the Canadian Pacific Railway was completed, the settlement of Saskatchewan and Alberta began. Many homesteaders now came from Great Britain.

By 1896, the federal government sought to increase immigration by promoting Western Canada in Great Britain and Europe as the last agricultural frontier in North America. The Canadian government initiated an aggressive campaign, administered by Clifford Sifton, Minister of the Interior, to lure more settlers to the Canadian West. Thousands of posters, pamphlets, and advertisements were sent to and distributed in Europe and the United States to promote free homesteads and assisted passages. Prior to 1896, most immigrants came from the British Isles or the United

States—these were 'desirable' immigrants. Sifton's campaign, however, cast a wider net to areas of Central and Eastern Europe that were not English-speaking and therefore provided 'less desirable' immigrants. The strategy generated considerable controversy among some English-speaking Canadians who believed in the racial superiority of British people.

Nevertheless, Clifford Sifton's efforts paid off. At the end of the 1880s, the Canadian Prairies had few settlers beyond Manitoba, and most of them had taken land near the Canadian Pacific Railway. Following the recruitment campaign, a flood of settlers arrived and the land was quickly occupied. Thus began the great migration to Western Canada. After 1896, the majority of settlers—about 2 million—were Central or Eastern Europeans from Germany, Russia, and the Ukraine. This large influx of primarily non-English-speaking immigrants led to a quite different cultural makeup in Western Canada from that in Central Canada, where the French and English dominated. Within a remarkably short span of time, cultural and linguistic acculturation had forged a non-British but English-speaking society from the sons and daughters of these immigrants. Some, however, kept separate. For instance, Doukhobor settlers were Russian-speaking peasants who adhered to a sect-like religion and who preferred to live in a communal setting (see 'The Doukhobors' in Chapter 3). By 1905, Alberta and Saskatchewan had sufficient numbers to warrant provincial status. By the outbreak of World War I, the region of Western Canada was settled.

The decision to build the CPR along a southern route (from Winnipeg to Regina to Calgary) meant that much of the land opened to homesteaders lay in the driest part of Western Canada.

Life was not easy for homesteaders. Many were ill-prepared for farming, let alone farming in a dry continental environment. Securing supplies of wood and water often posed a problem. Those settlers who could not afford to import lumber were forced to live in sod houses and burn buffalo chips and cow dung for heat. While many members of ethnic and religious groups settled together in the same area, forming communities, isolation still posed a problem for many. The land survey system encouraged a dispersed rural population, with individual farmsteads rather than rural villages. As a result, farm families sometimes did not visit the town or see their neighbours for weeks or even months. Such isolation was particularly hard on farmwives. In spite of these difficulties, the land was settled, towns sprang up, and institutions were created to meet the local and regional needs. In short, a new society was in the making.

By 1921, there were over 250,000 farms in Western Canada. Homesteaders now had to turn to the Peace River country for arable land. The Peace River country, part of the high Alberta Plain, is much further north and its short growing season makes agriculture risky. Even so, the Peace River country has a climate and soils that allow for mixed farming (a combination of grain and hay crops with livestock). The section of the Rocky Mountains located to the west of the Peace River country is considerably lower and thus allows more rainfall from Pacific air masses to reach the area. The final settlement of the Peace River country took place after World War I, when returning soldiers were encouraged to settle there.

Emergence of an Agricultural Economy

By the early twentieth century a Prairie agricultural economy had emerged. Based on a single staple—grain—this economy had to contend with fundamental geographic weaknesses. The first of these was its geographic position within the interior of Canada. Western Canada's geographic isolation from the rest of the world was partly overcome by the building of a railway system across the Prairies. The purpose of the CPR was twofold: (1) to bring settlers to the West, and (2) to allow these settlers to export their farm products to markets in Central Canada, Britain, and other European countries. However, once the railway system was in place, farmers faced a new problem—the high cost of shipping grain long distances by rail. (See Vignette 8.4 and 'Grain Transportation Subsidy' later in this chapter.)

Railway expansion within the Prairies solved the second geographic problem, namely, the difficulty of transporting grain from the farm to the grain elevator. In the late nineteenth century most farmers transported their grain by horse-drawn wagons to loading points (grain elevators) along the two east–west railway lines. Under the best conditions, farmers operating 15 km from a grain elevator were fortunate if they could haul their grain to an elevator within one day. For that reason, farmers did not cultivate much land beyond 15 km of a grain elevator. By building many branch lines to the main east–west railway, railway companies expanded their rail systems and made grain farming commercially viable in almost all areas of the Prairies.

A third geographic challenge was the region's short growing season. In 1910, a new strain of wheat, Marquis wheat, was developed. Its shorter maturation period overcame the threat of frost.[4] Marquis wheat also extended the growing area for wheat in the Prairies. By 1920, Marquis wheat was the most popular spring wheat in Western Canada and the adjoining Great Plains states.

Drought posed the fourth geographic problem for farmers, especially those farming in the semi-arid Dry Belt of southern Alberta and Saskatchewan. While drought remained a threat, the technique of dry-land farming reduced the risk of crop failure. In dry-land farming, part of the land is left in **summer fallow** each year. In this way, sufficient soil moisture is accumulated over several years, allowing for the seeding of the land every other year.[5]

From Intensive to Extensive Agriculture

Over the last century, agriculture in Western Canada underwent significant changes. The first change was the shift in the farm economy from a labour-intensive operation to a capital-intensive one. At the end of the nineteenth century, many hands were required to successfully deal with the sowing, growing, and harvesting of grain. The introduction of machinery, such as self-propelled steam tractors and threshing machines, changed the way farms were run and reduced the need for labour. Further technological changes continued to affect the size of the farm labour force. For instance, the development of the combine harvester, which can cut and harvest a swath of grain as wide as 15 metres, allowed farmers to harvest hundreds of hectares in a single day.

After the mechanization of the farm economy, a trend of consolidating farms into larger and larger units occurred. The number of farms declined, while the size of farms increased (Tables 8.1 and 8.2). In the Canadian West, this shift began after World War II but was most readily apparent in the 30-year period from 1971 to 2001. Over this period, the number of farms declined by 28 per cent, while the average farm size increased by about

44 per cent. Nearly 50,000 farms disappeared, with Saskatchewan alone losing 26,000 farms. Some lost heart and sold their lands while others were less efficient farmers or lacked sufficient capital. For the surviving grain farmers, larger farms increased productivity and thereby lowered per unit costs.

The move from intensive to extensive agriculture transformed the grain economy of Western Canada. With fewer people engaged in agriculture, the rural landscape lost much of its farm population to towns and cities. The ripple effect was that many villages, which had been small service centres, were abandoned and disappeared from the landscape.

Economic Diversification

While Western Canada remains an important agricultural region, development of its forests, minerals, and petroleum deposits has diversified the region's economy, particularly in Alberta. Following World War II, rising world prices for primary products led to a resource boom in Western Canada. American demand for oil and gas from Alberta rose sharply, while Saskatchewan saw major resource developments in potash and uranium. At the same time, Manitoba became a major producer of nickel and hydroelectric power. Furthermore, the forest industry in all three provinces

Table 8.1	Number of Farms in Western Canada, 1971–2001			
Year	**Alberta**	**Saskatchewan**	**Manitoba**	**Western Canada**
1971	62,702	76,970	34,981	174,653
1981	58,056	67,318	29,442	154,816
1991	57,245	60,840	25,706	143,791
2001	53,652	50,598	21,031	125,281
Change 1971–2001	(9,050)	(26,372)	(13,950)	(49,372)
Percentage Change	(14.4)	(34.3)	(39.9)	(28.3)

Sources: Statistics Canada (1992b, 2002a, 2003a).

Table 8.2	Average Size of Farms in Western Canada, 1971–2001 (acres)			
Year	**Alberta**	**Saskatchewan**	**Manitoba**	**Western Canada**
1971	790	845	543	726
1981	813	952	639	801
1991	898	1,091	743	911
2001	970	1,283	891	1,048
Change 1971–2001	180	438	348	322
Percentage Change	23	52	64	44

Sources: Statistics Canada (1992b, 2003a).

expanded due to a rising demand for lumber, pulp, and paper in the United States and other industrial countries.

While these resource developments helped diversify the economy of Western Canada, rising oil prices in the 1970s triggered a major resource boom that had a most dramatic impact on the economy of Alberta. This boom had important spinoffs for the Albertan economy, including jobs and royalties, technological advances that allowed for the mining of the vast tar sands in northern Alberta, pipeline construction projects to supply the large markets of Ontario and the United States, and the emergence of Calgary as the headquarters for the offices of major oil companies.

By the 1980s, Alberta's economy had not only grown rapidly, it had also diversified, transforming Alberta into a have province. While Alberta's economy and population expanded rapidly during the 1970s and 1980s, Manitoba and Saskatchewan followed at a much slower pace. In fact, during the decade that followed, low prices for primary products had a greater impact on Manitoba's and Saskatchewan's economies because they continued to rely heavily on agriculture and resources. Even today, Manitoba and Saskatchewan remain more vulnerable to a resource-driven economic downturn than Alberta.

The Prairie Psyche and Western Alienation

Western Canada's geography and its role as a resource hinterland forged a 'Prairie psyche'. The Prairie psyche originated in the days of homesteading. Farmers, confronted with a harsh environment and long shipping distances to ocean ports, developed a sense of collectivity and a feeling of alienation caused by the lack of control over their environment and economy. Unlike southern Ontario and even southern Quebec, arable land in Western Canada lies much further north. Windsor at 42° N and Québec City at 47° N represent Central Canada's latitudinal range of arable land while Winnipeg, 50° N, and Edmonton, 54° N, represent the western latitudinal range. Hence, climatic conditions in the Canadian Prairies are much less favourable for a successful harvest. In fact, drought, grasshoppers, and untimely frosts or rain can turn a bumper crop into a crop failure.

Before World War I, farmers felt frustrated by external forces—world prices for grain remained low, while prices for agricultural

Over half the wheat grown in Canada comes from Saskatchewan. While depressed world prices have encouraged farmers to try other crops, wheat remains the principal crop in Saskatchewan because of its ability to grow with little soil moisture. However, low precipitation in 2001, 2002, and 2003 resulted in below-average yields, pushing farmers to the edge of bankruptcy. (Al Harvey/The Slide Farm)

machinery made in Ontario increased. Fur- thermore, dealers at the Winnipeg Grain Exchange bought low from the farmers and sold high to the grain buyers, and operators of private grain elevators along the railways assigned the farmers a low grade for their wheat (resulting therefore in a low price). These conditions led farmers to form new institutions or movements. For example, in 1913 farmers banded together to create the United Grain Growers, to provide an alterna- tive means of selling their grain and thus pro- tect themselves from the low grades and prices offered by private grain companies.

The deep sense of alienation expressed by farmers has been an ongoing theme in the his- tory of Western Canada and can now be found in all sectors of society in Western Canada. This negative feeling stems from the peripher- al position of Western Canada within Canada and the global economy. Ottawa is often seen as either an uncaring government that ignores western grievances or a manipulative state power that places the interests of Central Canada over those of Western Canada. For instance, at the time that Alberta and Saskatchewan joined Confederation in 1905, the two western provinces were denied control over natural resources while Ontario, Québec, Nova Scotia, and New Brunswick had obtained this control (and taxing power) when they united to form the Dominion of Canada in 1867. A more recent example is the federal initiative of the early 1980s known as the National Energy Program. In this case, Ottawa exerted its control over oil prices and levied taxes on oil production. In Alberta's eyes, the National Energy Program was both a 'tax grab' and a political means by which Ottawa favoured energy-deficient Ontario over Alber- ta, securing low oil prices for Central Canada's manufacturing industry while interfering with

Western Canada's resource revenue.

Sometimes western alienation has led to the formation of new political movements, particularly political parties, as a means of combatting the apparent inequities of central- ist governments. One such example was the Co-operative Commonwealth Federation (CCF), a political party formed in 1932 by a coalition of labour and farming interests in an effort to combat the destitution people were experiencing during the Great Depression. (The CCF later became the NDP.) It met with considerable success, and is attributed with leading the way for the creation of a variety of social programs in Canada. Social Credit, another Western-based political party, came into existence at about the same time. The Reform Party of Canada, which became the Canadian Reform Conservative Alliance in 1999 and then amalgamated with the old-line Progressive Conservative Party to form the Conservative Party in 2003, provides the most recent example of a western protest party.

The political fragmentation of Canada's federal parties is strongest in the West (British Columbia and Western Canada) and Québec. In the 1960s, Howard Richards, who founded the Department of Geography at the Universi- ty of Saskatchewan in 1960, argued that isola- tion from other populated regions and from the political decision-making in Ottawa is at the root of western alienation. This isolation is worsened by the physical barriers of the Cordillera and the Canadian Shield. As a con- sequence, westerners feel exploited by Central Canada's businesses and politicians, who wield the economic and political power that ulti- mately affects those living in the western hin- terland. For farmers, the targets of their resent- ment were the banks and railways.[6] Later, for oil producers and provincial governments, the target would become Ottawa's centralist poli-

Vignette 8.3   The Kyoto Accord Rankles Alberta

Alberta's heavy oil sands have attracted some $90 billion in actual and planned investments. The construction boom that has already propelled Alberta's economy forward would continue for at least another 10 years. Fort McMurray is at the centre of the construction effort and is 'enjoying' its role as a boom town. (Since 1 April 1995, Fort McMurray became part of a new regional municipality known as Wood Buffalo. As the largest regional municipality in Canada, its lands stretch from north-central Alberta to the borders of Saskatchewan and the Northwest Territories, including the vast oil sand deposits.) Unfortunately, the separation of the oil from the sand takes a great deal of energy, resulting in the release of high levels of carbon dioxide into the atmosphere. The purpose of the Kyoto Accord is to reduce the release of greenhouse gases into the atmosphere and, in this way, reduce the

threat of global warming. Fearing that some companies would not proceed with their plans to mine and process the oil sands, Alberta urged Ottawa not to sign the Accord. At the same time, companies sought assurances from Ottawa that the cost of implementing the Accord's obligations would maintain the competitive nature of their investment in the Alberta oil sands compared with similar opportunities in other countries that have not signed the Kyoto Accord. According to Pierre Alvarez, president of the Canadian Association of Petroleum Producers, Petro-Canada's threat to shift its investment to Venezuela 'underscores that large, successful companies have an opportunity to invest anywhere in the world and they will pick and choose where the best opportunities are, where there is the greatest amount of certainty, and where there are the highest returns' (Cattaneo, 2002: FP1).

Figure 8.2 Planning to implement Kyoto Accord. The Accord is one of Jean Chrétiens legacies to Canadians. Albertans, however, are concerned about its impact on their heavy oil industry.

cies. Western alienation remains rooted in the centralist/decentralist faultline where western interests are seemingly ignored by Ottawa in favour of national (i.e., Central Canadian) interests. Following the Liberal re-election in November 2000, Prime Minister Jean Chrétien attributed Western alienation to voters focusing too much on local issues. The Prime Minister went on to say: 'It's possible [that we will try to have a better approach] but we have to ask them [Westerners] to look at the national scene from a national perspective' (Bellavance and Fife, 2000: A1). In 2002, the Prime Minister signed the Kyoto Protocol, which he believed was in the national interest. Alberta, however, saw the Kyoto Protocol as a threat to its heavy oil industry (Vignette 8.3).

Western Canada Today

Western Canada has evolved from a narrowly based agrarian economy to a more diversified one. Increased exploitation of its natural resources, a growing trend towards processing these resources, and a fundamental shift in its agricultural sector are the factors contributing to the region's transformation. Alberta's economy, driven by the oil and gas industry, has become more diversified. Saskatchewan and Manitoba have also diversified but, lacking huge petroleum reserves and still relying heavily on agriculture, they lag behind Alberta's economy. Of particular concern to the agricultural sector of the Prairie provinces is the 1995 cancellation of the Crow Benefit, a transportation subsidy that allowed farmers to ship their products by rail at a reduced cost (Vignette 8.4). This cancellation initiated massive changes in the agricultural sector. In 1995, prices for grain were relatively high and farmers could manage the transportation charges. But in the following years, grain prices dropped, making it much more difficult for farmers. The idea of shifting to greater processing of agricultural products within Western Canada still requires a market and that

Vignette 8.4   The Origin and End of the Crow Benefit

When the Canadian Pacific Railway was built, the railway company required public financial assistance to cover the construction costs. In exchange for that assistance, the CPR agreed to lower its shipping rates to Thunder Bay, thus reducing the cost of marketing grain in Canada's major market, Great Britain. Signed in 1897, the Crow's Nest Pass Agreement between the Canadian Pacific Railway and the federal government called for the CPR to lower the rate of moving grain to Thunder Bay by 3 cents per hundredweight in return for a $3 million federal subsidy to extend the rail line from Lethbridge to the Kootenay Valley in British Columbia. The purpose of this subsidy was to ensure that the rail rates for grain were low, that is, below actual shipping costs. This would allow the agricultural sector in Western Canada to thrive despite geographic distance from markets.

Over time, the transportation subsidy for Canada's two national railways increased, reaching $550 million in 1994. It increased for two reasons: shipping costs increased, and the subsidy was extended to include west coast ports. On 1 August 1995 the Western Grain Transportation Act ended and so did the railway subsidy known as the Crow Benefit. Now farmers would bear the full burden of shipping grain by rail to Canadian ports.

Table 8.3	Basic Statistics for Western Canada by Province, 2001				
Province	Population (000s)	Unemployment Rate (%)	GDP (%)	Aboriginal Population (%)	Canada's Cropland (%)
Alberta	2,975	4.6	14.9	16.0	27.3
Saskatchewan	979	5.8	3.1	13.3	41.2
Manitoba	1,120	5.0	3.2	15.4	13.5
Western Canada	5,074	4.9	20.2	44.7	82.0
Canada	30,007	7.2	100.0	100.0	100.0

Sources: Statistics Canada (2002a, 2002c, 2003c).

market is the United States. Washington has, from time to time, placed trade barriers at the border with the objective of limiting Canadian exports.

Alberta is the economic giant of the three provinces. It has over half of the population in Western Canada and produces about 63 per cent of the region's GDP (Table 8.3). In Alberta the 2000 GDP per capita amounted to just over $40,000. Both Saskatchewan and Manitoba fall well below this level. What accounts for this variation? Each province has much natural wealth. Saskatchewan, for example, has most of the cropland and is the leading producer of potash and uranium. In addition to having the richest agricultural land in the West, Manitoba produces vast amounts of hydroelectric power from the Nelson River. Even so, Alberta holds the trump resource card—oil and gas.[7]

Back in 1972, when a barrel of oil was worth $2.00 on the world spot market, Alberta oil was not valuable enough to dominate the western economy. Now that oil prices have increased by about 20 times, as of May 2004, the same oil reserves have greatly appreciated in value and thus in economic importance. The annual value of natural gas and oil production in Alberta was over $40 billion in

2002. In comparison, Saskatchewan's annual value of petroleum production was $5.6 billion, while Manitoba's was just above $150 million (Table 8.7).

Though Western Canada exports both primary and processed products to other countries—recently increasing its exports to Pacific Rim countries—the petroleum industry is more closely tied to the North American market. Most of the natural gas and oil goes to markets in Ontario and the United States.

Alberta's economic success stems from its huge petroleum deposits and high world prices for oil. With the continued development of the huge oil sands deposits near Fort McMurray, Alberta's place as the leading oil producer in Canada seems assured. (Al Harvey/The Slide Farm)

After World War II, a network of oil and gas pipelines was constructed to serve both the Canadian and American markets. Energy, unlike manufactured goods, had easy access to American markets before the Free Trade Agreement.

World prices hold the key to economic growth for the resource industries. Imagine the impact on the Prairie economy if other **primary prices** (prices for primary resources) followed the upward path of oil prices! Imagine, too, if Europeans and Americans dropped their wheat subsidies to Canadian levels! In Europe, 56 cents of every dollar of a farmer's wheat sales are a public subsidy, while in the United States the effective subsidy is 38 cents. In Canada, it is 9 cents (National Farmers' Union, 1999: C6). Unfortunately, Prairie farmers, miners, and loggers are all too well acquainted with a long-term downward trend in their commodity prices, as well as sudden fluctuations in the prices for their primary resources—all caused by changes in world demand. While a rise in spring wheat prices is highly unlikely, dramatic land-use changes that might turn around the agricultural sector are taking place in Western Canada.

Population

Western Canada's population has undergone significant changes since settlers first came to this region, and these changes have affected and been affected by many aspects of Prairie life. In 1921, Western Canada had a population of nearly 2 million. At that time, it comprised 22 per cent of Canada's population. By 2001, its population had increased to 5.1 mil-

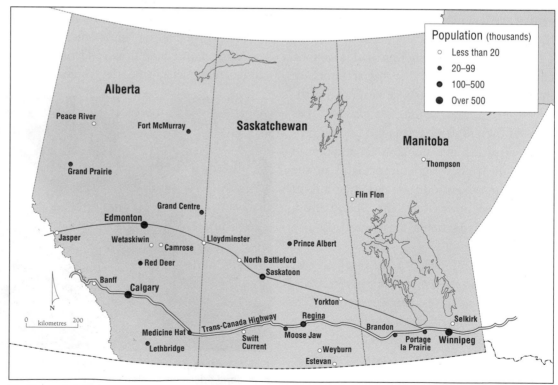

■ **Figure 8.3 Major urban centres in Western Canada**. Edmonton, Calgary, Winnipeg, Saskatoon, and Regina are the major population centres in the region.

lion but accounted for less than 17 per cent of the national population.

Two migrations transformed Western Canada. Just 100 years ago, land-hungry homesteaders poured into the Prairies, creating a rural landscape of small farms and villages. No one could have foreseen that this rural landscape would lose most of its population in a remarkably short time. In the second migration, the so-called rural-to-urban migration, rural people moved from the countryside to towns and cities in Western Canada and to urban centres in British Columbia, Ontario, and the United States. The push factors of this migration were the mechanization of the agri-

cultural sector, the consolidation of farms, and the shrinking need for farm labour that resulted. Pull factors included the employment opportunities and greater amenities that existed in towns and cities. The rural-to-urban migration resulted in a redistribution of people within Western Canada. Today, most people live in the five major cities of the region: Calgary, Edmonton, Winnipeg, Saskatoon, and Regina (Figure 8.3).

Within Western Canada, Alberta has the largest population with 3 million residents. Manitoba has just over 1 million and Saskatchewan just under 1 million. Over the last 20 years, the rate of population increase

Table 8.4   Employment by Industrial Sector in Western Canada, 2002

Industrial Sector	WC Workers (000s)	WC Workers (%)	Ontario Workers (%)	Difference Percentage points
Primary	**260.2**	**9.6**	**1.8**	**7.8**
Agriculture	144.0	5.3	1.3	4.0
Secondary	**463.6**	**17.0**	**25.1**	**(8.1)**
Manufacturing	245.4	9.0	18.5	(9.5)
Tertiary	**1,998.8**	**73.4**	**73.1**	**0.3**
Trade	418.6	15.4	15.7	(.03)
Transport	158.2	5.8	4.7	1.1
Finance	115.7	4.2	6.6	(2.4)
Professional services	163.0	6.0	7.2	(1.2)
Management	87.4	3.2	4.2	(1.0)
Education	182.9	6.7	6.2	0.5
Health care	290.2	10.7	9.3	1.4
Culture	111.3	4.1	4.8	(0.7)
Accommodation	188.0	6.9	6.0	0.9
Other services	130.5	4.8	4.2	0.8
Public administration	128.0	4.7	4.8	(0.1)
Total	2,722.8	100.0		

Source: Statistics Canada (2003c).

has been highest in Alberta, followed by Manitoba. Recent population changes between the 1996 and 2001 censuses confirm that this trend continues, with Alberta growing at a rate of 10.3 per cent over this five-year period followed by Manitoba at 0.5 per cent. Saskatchewan's population decreased by 1.1 per cent (Statistics Canada, 2002b).

Another demographic feature of Western Canada is the size of the Aboriginal population. By 2001, approximately 440,000 Aboriginal peoples lived in Western Canada, forming nearly 9 per cent of Western Canada's population. A growing number reside in urban centres, largely because Indian reserves and Métis communities have insufficient employment opportunities and limited urban amenities.

Industrial Structure

Employment by industrial sector in Western Canada reveals the prominence of primary economic activities. While Table 8.4 is only a generalized picture of the western economy, it illustrates two important aspects. One is the importance of the primary sector. The percentage of people employed in this sector (9.6 per cent) is over five times the figure for Canada's principal industrial core region—Ontario. The second is the small size of the secondary sector. Employment in the secondary sector is well below that of Ontario (Table 8.4).

Manufacturing

Though mechanization of Western Canada's resource industries has created a more balanced economy, the manufacturing sector is still relatively small. In fact, in 1984 Blackbourn and Putnam (1984: 160) were surprised to learn that such a vast area as Western Canada accounted for only 8 per cent of Canada's manufacturing employment. Their explana-

tion for the weak state of manufacturing in the Prairies was that manufactured goods could be produced in Ontario and Québec and sold in Western Canada at less cost than manufacturing the same goods in Western Canada; in other words, market size and economies of scale are necessary for manufacturing to flourish. From 1981 to 2001, however, there was a significant increase in Western Canada's domestic market as the region's population went from 4.2 million to 5.1 million. Western Canada's portion of total Canadian manufacturing remained around 9 per cent. While much of this increase stems from the petrochemical industry, small-scale manufacturing aimed at particular market niches in Western Canada is expanding through the production of specialized products for the agricultural, forestry, and mining industries. In a few cases, these specialized products are being purchased by customers in the United States. Perhaps the highly specialized manufacturing, the increase in the domestic market, and access to the United States market explain this modest rise in Western Canada's share of the country's manufacturing industry over the past 15 years.

Agriculture

Agriculture was the driving force behind the settlement and development of Western Canada in the late nineteenth and early twentieth centuries. At that time, most homesteaders grew spring wheat for export to Great Britain. While spring wheat remains the principal crop grown in Western Canada, other crops, but particularly durum wheat, oilseed, and specialty crops, are strengthening their place in Prairie agriculture. Cattle and hog production is on the rise, too. These changes are due primarily to low prices for grain, rising costs of shipping grain by rail, the loss of the grain rail subsidy (the Crow Benefit), and much higher grain subsidies in the European Union and the United States.[8] The combination of these four factors has forced some farmers into bankruptcy while others have sought alternative crops or have turned to livestock and hog production. As well, approximately 30 per cent of farmers seek off-farm employment to supplement their farm incomes. In the process of change, the number of farmers has decreased sharply, leaving few people on the land and in the rural communities. Signs of decline, such as decaying communities, abandoned branch rail lines, and the closure of many rural grain elevators, are everywhere in the rural landscape. Today, agriculture has lost its central place in the economy of Western Canada and is struggling to find its footing.

Western Canada is blessed with rich chernozemic soils that are well-suited for growing cereal crops (Figure 8.4). Other crops, such as peas and lentils, require more moisture and are therefore restricted to the black soil zone. Unlike canola, durum wheat thrives in hot, dry weather commonly occurring in the brown soil zone. It also requires a longer growing season than spring wheat and therefore is usually grown south of the fifty-first parallel. Consequently, the agricultural land use varies, forming three distinct agricultural regions. They are: (1) the Fertile Belt (black soil associated with parkland and long-grass natural vegetation, (2) the Dry Belt (brown soils with short-grass natural vegetation), and (3) the agricultural fringe (southern end of the boreal forest) and the Peace River country (Figure 8.5). These subregions each have very different growing conditions. The major factors controlling those conditions are the number of frost-

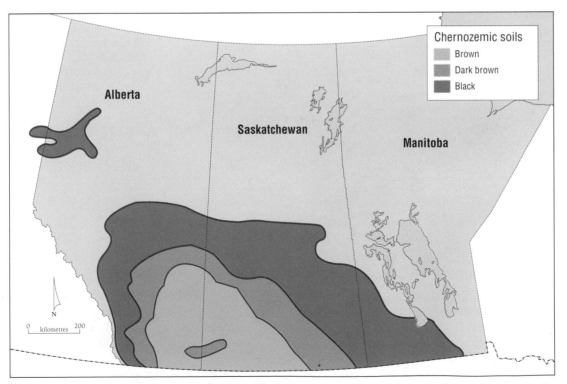

Figure 8.4 Chernozemic soils in Western Canada. There are three types of chernozemic soils in the Canadian Prairies: black, dark brown, and brown. The differences in colour are due to varying amounts of humus in the soil. The Peace River country's soil, formed under an aspen forest, is 'degraded' black soil.

free days and the soil moisture. The Fertile Belt provides the best environment for crop agriculture. The Dry Belt occupies semi-arid lands and has become a grain/livestock area. The **agricultural fringe** is along the southern edge of the boreal forest, and the Peace River country forms a pocket in the northwestern area of this forest. In the agricultural fringe, the short growing season encourages farmers to grow feed grains and raise livestock, while in the Peace River country, farmers grow both grain and feed grain for raising livestock.

The Fertile Belt

The Fertile Belt extends from southern Manitoba to the foothills of the Rocky Mountains west of Edmonton (Figure 8.5). The higher levels of soil moisture, an adequate frost-free period, and rich soils make this belt ideal for a variety of crops and livestock. The most popular crop, since farmers first arrived in the West, has been wheat. In recent years, however, the acreage in grain has declined, while the planting of oilseed and specialty crops, such as beans, field peas, and sunflower plants, has increased. This change was fuelled by rising prices for these crops and declining prices for wheat. Between 1996 and 2001, farmers moved away from traditional crops to alternatives that would reduce their input costs or increase their per acre revenues. Wheat is still the largest single crop. In Saskatchewan, wheat makes up close to 40 per cent of the area under field crops, but just five years earlier the figure was over 50 per cent.

The predominance of wheat in the Fertile

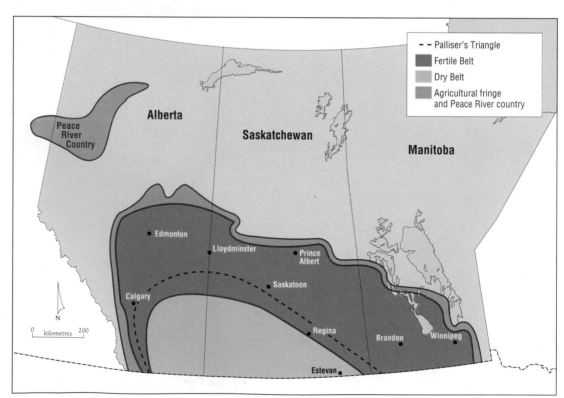

■ **Figure 8.5 Agricultural regions in Western Canada**. Farming in the Prairies can be divided into three areas: the Fertile Belt, the Dry Belt, and the agricultural fringe and Peace River country. Each region has a different type of agriculture because of variations in physical geography.

Belt is most evident in western Saskatchewan and eastern Alberta. In southern Manitoba and the adjacent parts of eastern Saskatchewan, there is more annual precipitation so farmers there can grow a wider variety of crops as well as keep livestock. As a result, mixed farming is common. Grain and specialty crops (canola, flax, sunflowers, and lentils) are combined with beef, pork, and poultry production. Near Winnipeg, for instance, a livestock industry has developed where cattle are fattened before shipment to meat-packing plants in Winnipeg and Ontario. Feed lots and nearby meat-pack-

ing plants are located in other parts of Western Canada, particularly in Brandon, Calgary, Edmonton, Lethbridge, and Saskatoon. Since NAFTA, meat production has increased, especially pork, and most meat products are being exported to markets in the United States.

As the urban markets grow, market gardens, dairy farms, and other specialized forms of intensive agriculture are developing near major cities in the Fertile Belt. The demand for specialty-crop products has spawned a number of smaller but very intense production units, including nurseries and greenhouses, to

Vignette 8.5 The Death of a Grain Elevator

The destruction of Saskatchewan Wheat Pool No. 8 near Melfort, Saskatchewan, in November 2000 reflects the economic transition occurring in the Prairie provinces. Until recently, the rural landscape of Western Canada was dominated by local grain elevators such as the one operated by United Grain Growers shown in the photograph. Each small rural community had at least one grain elevator, which was connected by a branch line to the main railway line to Vancouver and Thunder Bay. These elevators were the economic basis of rural communities but are no longer economically viable. The grain companies are replacing them with a smaller number of very large cement terminals at strategic locations along mainline railroads. The loss of these smaller elevators not only symbolizes the restructuring of Western Canada's agriculture but also signals the demise of

many rural communities. The consolidation of the grain transportation network calls for the destruction of over 2,000 wooden grain elevators and the abandonment of many branch lines. Farmers are now faced with hauling their grain by truck to large grain terminals located on the CPR and CN main lines. As well, the heavy truck traffic on rural roads has caused their rapid deterioration, forcing provincial governments to repair and upgrade these country roads.

A grain elevator is demolished near Melfort, Saskatchewan. (*Peter Wilson*, Saskatoon Star Phoenix, *9 December 2000, E1.*)

produce flowers and vegetables for local sale. For that reason, farm sizes are much smaller around the major cities.

The Dry Belt

The Dry Belt contains both cattle ranches and large grain farms. It extends from the Saskatchewan–Manitoba boundary to the southern foothills of the Rockies and north nearly to Saskatoon (Figure 8.4). However, the driest area, or heart of the Dry Belt, occupies a much smaller area, stretching southward from the South Saskatchewan River to the US border. The arid nature of the Dry Belt is not due to low annual precipitation (it is comparable to other areas of Western Canada) but to longer summers and higher evaporation rates. Within the Dry Belt, feed grain and hay crops are grown to supply winter feed for the cattle ranching that dominates in this area. Prior to settlement, the Dry Belt's short-grass vegetation provided a natural grazing area for buffalo. Later, cattle replaced the buffalo. Cattle ranch-

ing began in this area in the 1880s. Today, ranches are large, often many times the size of grain farms, because of the lower productivity of the dry land and the need for huge grazing areas to support a rotational grazing system.

Along the northern edge of the Dry Belt, grain farming is pursued, but the risk of crop failure is high. To conserve soil moisture, summer fallowing is practised—crops are only planted on parcels of land in alternate years. Until the 1980s, farmers planted wheat in narrow rows separated by a band of fallow land. This method of farming is called **strip farming**. Strip farming is not really practised now but summer fallowing remains common. The hope is that two years of precipitation will accumulate sufficient soil moisture to germinate the seed and sustain the young wheat plant. The crop will still require summer rainfall to reach maturity. On average, half of the arable land in the Dry Belt is kept in summer fallow each year.

However, land left to fallow has been declining in favour of the growing trend towards continuous cropping. Though summer fallow was practised partly to conserve moisture and partly to control weeds, a drawback of this technique was that the ploughed (fallow) land was subject to water and wind erosion. In a dry spring, windy weather could result in extensive loss of topsoil with huge clouds of dust stretching for many kilometres across the Prairies. Advances in technology have allowed for 'one-pass' seeding, spraying, and fertilizing. The expense and time not spent on repeated tilling of a field reduces a farmer's costs and conserves moisture, which in turn allows for continuous cropping.

Irrigation on these semi-arid lands has provided another solution to dry conditions. The most extensive irrigation systems are in southern Alberta. In fact, nearly two-thirds of the

The export of grain to foreign markets requires an elaborate transportation and handling system. Grain elevators play a key role by serving as a collection point along railway lines where the grain is graded, cleaned, and weighed before being transported by railway cars to Canada's major ports. (Al Harvey/The Slide Farm)

The Great Sand Hills, situated approximately 100 km northwest of Swift Current, Saskatchewan, is a fragile ecosystem that is unique within the larger environment of the short grass vegetation zone. Located in the centre of the Palliser Triangle, some sand dunes remain void of natural vegetation making them subject to wind erosion while others have been stabilized by a covering of native prairie grasses. Cacti, creeping juniper, and small shrubs like rose, saskatoon, chokecherry, and silver sagebrush grow in the Great Sand Hills. (Al Harvey/The Slide Farm)

750,000 ha of irrigated land in Canada are located in Alberta. In the 1950s and 1960s, two major irrigation projects were developed in the dry lands of Alberta and Saskatchewan: the St Mary River Irrigation District is based on the internal storage reservoirs of the St Mary and Waterton dams in southern Alberta, and Lake Diefenbaker serves as a massive reservoir on the South Saskatchewan River. In 1992, the Oldman River Dam was completed. At first, development of irrigated land in Saskatchewan was hampered by the problem of finding crops that would provide sufficient revenue to offset the high cost of irrigation. Farmers grew cereals, alfalfa, hay, oilseeds, sugar beets, and canning vegetables. Until the location of three large french-fry-processing plants near Lethbridge, few potatoes were grown. The demand for potatoes by Lamb-Weston, McCain Foods, and Maple Leaf Foods resulted in potato acreage jumping from 28,000 acres in 1998 to nearly 40,000 in 1999 (Robertson, 1999: D3).

Another problem, faced by Saskatchewan farmers in particular, is that with a shorter growing season their selection of crops is more limited compared to the corn, sugar beets, and other specialty crops that farmers in southern Alberta are able to grow. However, the major problem is distance to major markets.

The Agricultural Fringe and Peace River Country

After World War I, much new land was brought under cultivation in the Peace River country in northern Alberta and along the southern edge of the boreal forest that constitutes the agricultural fringe of the three Prairie provinces. Settlers moved into the agricultural fringe after the more promising lands of Western Canada were occupied. Unfortunately, much of the agricultural fringe was ill-suited for crops. In these higher latitudes, the threat of frost increased and, in the boreal forest, soil quality decreased. These two natural factors,

The shortage of water in the arid lands of southern Alberta has been partly solved by damming rivers and using the water for irrigating sugar beet, hay, and potato crops. Fodder crops, including hay, supply the feeder cattle and hog farms with necessary animal feed. The Oldman River Dam, shown above, added significant amounts of water for irrigation. Alberta accounts for 60 per cent of Canada's irrigated farmland. (Bayne Stanley/Viewpoints West)

plus the increased cost of shipping agricultural products to local and world markets at greater distances, made farming in these areas even more risky and marginal than in the rest of Western Canada.

During World War II, agriculture in these areas declined—some farmers joined the armed forces, while others migrated to the cities to find jobs. Farmers in the Peace River country fared better—particularly grain farmers—but they had to contend with a circuitous rail route in getting their products to markets. Abandoned farms reverted to bush while remaining farmers turned to mixed farming. In 1955, however, farmers in the Peace River country benefited from the construction of a rail line from Prince George to Dawson Creek. This shorter link to grain terminals in North Vancouver and then to Pacific Rim customers reduced shipping costs for these farmers.

Next Year Country

The dry continental climate in Western Canada makes farming a risky business. The threat of crop failure is greatest along the margins—the northern agricultural fringe is subject to frost, while grain farmers in the Dry Belt constantly face the possibility of drought. Prairie farmers often describe the land as 'Next Year Country'. The meaning behind these words is simple: Our crops did poorly this year, but we hope that they will do better next year. Often the reason for a bad year with low yields is insufficient precipitation during the growing period. However, farmers face a whole range of natural hazards including: summer frosts, which occur when a cold air mass slides unimpeded from the Canadian Arctic to Western Canada; hail and early snowfall; pests, such as grasshoppers; and diseases, such as stem rust. All these hazards can have devastating effects—from delaying a harvest or lowering the grade of wheat, to reducing a yield or destroying an entire crop. Nevertheless, when compared to the days of the pioneer farmers, who did not have the advantages of modern farm technology and improved strains of wheat, farmers now have a better chance of dealing with adverse weather conditions.

Since 2000, crops have suffered from a lack of moisture. If this dry spell continues, then grain farmers and ranchers in the Dry Belt of Alberta and Saskatchewan will face several years of poor yields, thereby threatening their survival. Some fear that global warming could create such arid conditions again. Others argue that since weather records indicate that Prairie weather follows a cycle of wet and dry years, it is only a matter of time before another dry cycle occurs (Akinremi et al., 2001). Successive hot, dry summers can quickly dry up soil moisture, bringing back the spectre of drought, dust storms, and crop failure in the semi-arid areas. David Jones's *Empire of Dust* (1987) describes the settling and abandonment of homesteads in the dry belt of Saskatchewan and Alberta. Perhaps Palliser's original assessment of the limited agricultural potential of the semi-arid areas of Western Canada was correct. In any case, a rise in temperature, without a corresponding increase in precipitation, would threaten grain-growing in the Dry Belt and increase the risk of crop failure in the Fertile Belt.

The Canadian Wheat Board

Besides the constant struggle against natural elements, grain farmers have no control over the prices for their commodities. Unlike dairy and poultry farmers, whose output and prices are set by marketing boards, grain farmers live with prices that are set by the market forces of supply and demand. In order to assist Western Canadian farmers to market their grains and to obtain the best possible prices, the federal government established the Canadian Wheat Board (CWB) in 1935. Since 1943, Canadian wheat farmers have been compelled by law to sell their crops only to the Board. While the CWB ceased to be a Crown corporation in 1998, it remains a quasi-federal government agency. The CWB remains the grain-handling and marketing agency for wheat, barley, and oats destined for export or use by the food industry in Canada. The twin goals of the CWB are to sell as much grain as possible at the best possible prices and to ensure that each producer gets a 'fair' share of the market.

To achieve these goals, the CWB issues delivery quotas. These quotas tell farmers that the Board is willing to accept a certain volume of a specified grain from each farmer at their local elevators in a defined geographic region. Quotas, based on the Board's sales commitments and stocks on hand in the elevator system, are issued to maintain a relatively even flow of grain from farms through the primary (local) elevator system. The CWB may sell directly to foreign governments' buying agencies, to commercial interests in foreign countries, or to private grain-trading companies, which then resell to foreign buyers.

Over the years, farmers have had mixed reactions towards the Canadian Wheat Board. At times, it has provided much-needed protection from an unstable market and low prices. At other times, farmers have resented having to sell their grain through the Board at a Canadian price that is lower than the American price. In recent years, some prairie farmers living near the border have challenged the authority of the Canadian Wheat Board by trucking their grain across the border to the US market, where prices have been higher for durum. In 2000, the Canadian Wheat Board retaliated, taking the issue to court, and in 2002 the court ruled in favour of the CWB. Grain farmers in Saskatchewan paid a $500 fine while a handful of Alberta farmers refused to pay the fine and, as a protest against this monopoly, went to jail (CBC, 1999, 2002).

The Canadian Wheat Board is also under attack from Washington. In 2003, the United States served notice that it intends to try to dismantle all state trading enterprises because Washington believes such enterprises subsidize exports. For that reason, Washington has gradually raised the duties on durum and spring wheat. By August 2003, the duties were raised again, reaching 13.55 per cent duty on high-quality durum wheat and a 14.6 per cent duty on spring wheat. The Wheat Board sells about $400 million worth of wheat a year in the US market, representing about 5 per cent of the American market. Millers and bakeries in the United States prefer Canadian wheat because of its high quality, which is ideally suited for pasta.

Key Topic: Transition

In the 1990s, global markets and national policies were unleashed that began the process of reshaping Western Canada's economy. While this transformation affects all sectors of the western economy, the changes are most visible in the agricultural sector. For the most part, these changes are being driven by three factors: global forces such as trade restrictions and fluctuating world prices; national policies such as government programs, including the reduction of agricultural subsidies; and regional conditions such as the high costs of transporting products to distant markets, low yields due to dry growing conditions, and the livestock crisis due to mad cow disease.

The end of the grain transportation subsidy, the continued use of agricultural subsidies by the United States and Europe, and the persistent use of trade barriers by Washington to slow Canadian grain and livestock products from entering the US market have made this transition to a more open market difficult. Prices for the major crop, spring wheat, remained low over the past five years. Coupled with these challenges, the twenty-first century began with Prairie farmers confronted with drought, grasshoppers, and mad cow disease or BSE (bovine spongiform encephalopathy). Since 2000, Western Canada has experienced unusually warm and dry weather. Drought affected both farmers and ranchers—farmers saw their yields drop and ranchers had little hay for winter feed. Warm, dry conditions are perfect breeding grounds for grasshoppers, which have the capacity to consume whole fields of hay and grain. In 2001 and 2002, crop and hay production fell, almost reaching the 1988 level when drought ravaged prairie crops. In May 2003, the discovery of BSE in one animal in Alberta had devastating impli-

cations for the livestock industry, as did the appearance at the end of 2003 of a BSE-infected dairy cow across the border in Washington state that had been born in Canada. While these cattle were identified and removed from the food-processing chain, exports of live and processed beef ceased immediately. The Canadian livestock industry is concentrated in Western Canada, with Alberta having almost half of Canada's 13.4 million cattle. Beef cattle comprise Alberta's largest agricultural sector, providing about half of farm revenue. Exports are extremely important, making up around 40 per cent of total sales. The largest customer is the United States, followed by Japan and Mexico. With the halt of exports, Canadian prices to the ranchers dropped dramatically, though retail prices did not follow.

The liberalization of international trade has opened up foreign markets for Western Canada, but this has made resource sectors more vulnerable to downturns in these foreign markets. For Western Canadian farmers, trade liberalization is a double-edged sword. This new world of so-called open markets has provided export opportunities, but as exports to the United States increased, duties were imposed. Trade liberalization has also led to the elimination of farm subsidies, such as the transportation subsidy for grain, and it threatens farm marketing boards and farm organizations, such as the Canadian Wheat Board.[9] Canada, being so dependent on exports to the United States, is vulnerable to US trade barriers. While Washington aims at protecting its domestic producers, the cost of this protection is borne by Canadian farmers and ranchers. Perhaps the worst aspect of trade with the United States is the uncertainty—if Canadian producers expand production to service the US market, then their operations are extremely vulnerable to barriers to that market.

Grain Transportation Subsidy

The greatest change affecting agriculture in Western Canada is the loss of the Crow Benefit in 1995. Since then, farmers have had to pay the full cost of rail transportation to ship their grain to port. The magnitude of the new transportation costs is illustrated by the size of the former subsidy. In 1994, the Crow Benefit subsidy amounted to approximately half the cost of shipping grain. Once this subsidy was cancelled, most agricultural economists expected that the loss of the annual $550 million rail subsidies would slash farm income by a similar amount. For farmers in Saskatchewan, their share of the rail cost of shipping spring wheat to port doubled. For instance, a Swift Current farmer's grain shipping costs to Vancouver were $13.82 per tonne under the Crow Benefit, but rose to $29.39 per tonne immediately after August 1995 (Budhia, 1995b: 43). Manitoba farmers face the largest increases because they are the farthest from ocean ports. Spring wheat shipments from Winnipeg to Montréal via Thunder Bay could triple rail costs for Manitoba farmers, causing them to lose money growing spring wheat. At the same time, however, the cost of shipping durum wheat by rail from Winnipeg to Montréal is about half the cost of shipping spring wheat (Budhia, 1995a: 38–9). What accounts for the variation? The rail freight rates for durum were lower in the mid-1990s because large quantities of durum were sold in the US market, where the price for durum was high. Consequently, Canadian railways were forced to lower their rates to remain competitive.

The changes in rail transportation costs for grain farmers are altering the agricultural land-use pattern in Western Canada.[10] The speed with which these changes occur will depend on the world price for spring wheat. High prices in the 1994–5 crop year did not last, forcing farmers to grow alternative crops and in some cases to raise cattle or hogs. A combination of crop and livestock farming creates diversity and allows flexibility, i.e., if grain prices are extremely low and cattle/hog prices are high, then the farmer can convert his low-priced grain to beef or pork. In 2003, spring wheat was still the dominant crop in the Prairie provinces, but the land seeded to spring wheat declined sharply between 1996 and 2001. In Saskatchewan, farmers reduced their acreage of spring wheat by 23 per cent from 1996 to 2001; Alberta farmers by 10 per cent; and Manitoba farmers by 8 per cent (Statistics Canada, 2003a).

Two long-term changes to agriculture should lead to a new pattern of land use and create a more profitable agriculture system in Western Canada.

(1) Farmers in eastern Saskatchewan and in Manitoba may grow less grain but more feed grains and specialty crops, as the climate here is both wetter and warmer than in most other areas of the Canadian West, allowing for a wider variety of crops. Feed grains can be sold to new beef and hog producers in these provinces, while specialty crops, such as canary seed, can be grown under contract to larger firms. The expansion of livestock slaughter plants is expected if access to the US market remains open.

(2) Farmers in western Saskatchewan and in Alberta may not decrease their wheat acreage for two reasons. First, wheat is an ideal crop for semi-arid growing conditions. Second, western Saskatchewan and Alberta have the lowest rail costs for shipping spring wheat to Vancouver. Ranchers, too, benefit from short distances to major markets. The short rail haul to Vancouver to reach the Asian market and the short trucking route for live cattle

and hogs to slaughter plants in Montana provide strong incentive for an expansion of the livestock industry and the growth of more hay and fodder crops for winter feed.

The Prairie Staple

Grain production, particularly spring wheat, has been the prairie staple for nearly 100 years. Recent developments, however, such as low world prices for spring wheat and the loss of the rail transportation subsidy for grain, have led grain farmers to seek alternative crops.[11] The North American Free Trade Agreement and the World Trade Organization (formerly GATT) have opened American and other foreign markets to farmers, thereby further encouraging a diversification in crops (see Table 8.5). Access to the American and global markets has encouraged:

- growing more high-priced durum wheat for sale in the US market;
- producing and processing more beef and pork in Western Canada, thereby providing a local market for feed grains;
- growing and locally crushing seed crops, such as canola, and shipping the oil product to the US, Japan, and other Asian countries;
- growing specialty crops, such as sunflowers and field peas, under contract to US firms;
- producing alfalfa pellets from hay for shipment to the US and Japan for feed in their dairy industries.

Faced with increasing rail rates for grain products, farmers will be discouraged from shipping grain to foreign markets. They have several alternatives. One is to grow more

Table 8.5	Shift in Acreage: Wheat, Durum, and Canola, 1986–7, 1995–6, and 2001–2			
Crop	1986–7 (million acres)	1995–6 (million acres)	2001–2 (million acres)	% Change 1995–6 to 2001–2
Spring wheat	17.8	11.5	10.6	(8)
Canola	2.5	6.1	4.7	(23)
Durum	3.5	4.4	4.2	(5)

Sources: Saskatchewan Agriculture and Food (1997); Saskatchewan Agriculture, Food, and Rural Revitalization (2002).

Table 8.6	Shift in Price: Wheat, Durum, and Canola, 1986–7, 1994–5, and 2001–2			
Crop	1986–7 ($/tonne)	1994–5 ($/tonne)	2001–2 ($/tonne)	% Change 1994–5 to 2001–2
Spring wheat	105	167	140	(16)
Canola	199	348	244	(30)
Durum	122	237	193	(19)

Sources: Saskatchewan Agriculture and Food (1997); Saskatchewan Agriculture, Food, and Rural Revitalization (2002).

durum wheat, which is ideal for making pasta products. As a more specialized wheat, durum commands a higher price and therefore can better withstand the higher rail transportation costs. For instance, in the 1994–5 crop year, the price of durum was 42 per cent higher than that of spring wheat (Table 8.6). Another attractive feature is the durum market in the neighbouring states of Montana and North Dakota, where the distance to the final delivery point is much shorter than to other export markets.

Farmers are also growing more oilseed crops, such as canola and flax, and more specialty crops, such as peas, lentils, and canary seed. The primary reason for the viability of these alternative crops is that their prices, although fluctuating, have risen over the last 20 years, while the price for spring wheat has not changed appreciably. In fact, the price of spring wheat dropped so low that from the crop year 1998–9 to 2001–2 many farmers were losing money. However, price fluctuations in alternate crops do occur. The price for canola, for example, dropped by 30 per cent from 1994–5 to 2001–2 (Table 8.6). During that seven-year period, however, the price for canola did not decline steadily but varied widely from year to year, making choices about what crops to seed difficult and risky. Another factor is the lower transportation costs for the farmer because oilseed and specialty crops only need to be trucked to local processing plants. Major food companies have built large canola-crushing plants in Western Canada over the past 10 years, encouraging farmers to grow more canola. By shipping their products to local plants by truck, farmers' transportation costs are minimized. However, canola has its limitations. Canola cannot be grown in the more arid areas of Western Canada, and the crop is prone to disease and weed infestations if cultivated repeatedly.

With the end of the Crow Benefit in 1995, livestock producers in southern Ontario can no longer afford to import feed grains from Western Canada. However, a growing livestock industry in Western Canada now supplies beef and pork products to Ontario and Québec. As well, the demand for pork is increasing, opening more markets for Western Canadian producers in both the United States and Pacific Rim countries. To meet the demand, huge hog farms, similar to those in the United States, have sprung up across Western Canada that take advantage of modern agro-technology, mass production techniques, and economies of scale. Because of the smell and the risk of contaminating groundwater, large hog barns are located in rural settings far from settlements. Here, the hog barns are providing employment for residents of rural communities, though their waste poses a threat of fecal bacteria seeping into groundwater.

The livestock industry is also undergoing change—a combination of restructuring and consolidation of its processing plants. As a result of the Free Trade Agreement in 1989, competition within the North America market has become fierce, forcing Canadian operators to build larger hog-processing plants, to specialize in a single product in each plant, and to demand lower wages from employees. In Western Canada, new hog-slaughtering plants have been built at Brandon, Red Deer, and Lethbridge and another one is planned for Winnipeg. Plant specialization is driven by cost of production—larger plants to gain economies of scale, and specialized production lines to gain higher productivity. While it does cost to ship meat from the slaughterhouse to a number of processing plants, the saving gained from efficiencies in processing plants dedicated to a single product offsets the shipping costs. For example, Maple Leaf Foods' new hog-slaughtering operation at Brandon supplies carcasses to its Winnipeg 'ham' plant and its North Battleford 'bacon' plant. Michael McCain, president of Maple Leaf Foods, describes competing for a place in the North American market for pork products as 'akin to dancing with elephants. We have to be a little more nimble, a little more agile than our dancing partners' (Bell, 1999: 60).

Western Canada now accounts for 40 per cent of hog production in Canada. Manitoba leads with 3 million pigs, Alberta has 2.1 million, and Saskatchewan follows with 1.2 million. Large-scale hog barns have led the way to this remarkable expansion. Hog barns, like cattle feeding lots, produce a great deal of waste that could threaten the local water sup-

ply. At a community meeting in Foam Lake, Saskatchewan, the issue of economic benefits from a proposed large hog barn was challenged by the possible environment costs. The proposal includes a breeder farrow barn for 5,000 sows, a nursery barn, and three finisher barns. Pig excretion would be placed in lagoons and, in time, would be removed as manure that could be placed on fields as fertilizer. Grain farmer Harry Abtosway had heard enough. He complained that 'Pollution, environment, smell, you name it. I think we are going to ruin our land, our water supply' (Hall, 2003: C8). Large hog barns are not popular with local people, but then, as Terry Markusson, the proponent of the proposed hog barn, put it: 'There is nothing that smells as bad as a dying community.'

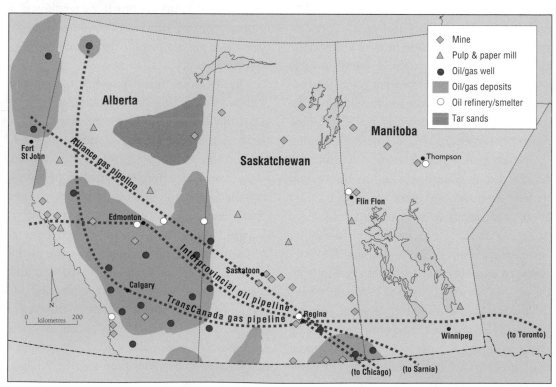

■ Figure 8.6 Western Canada's resource base, 2003. The number of mines, pulp and paper mills, oil and gas wells, and oil refineries and smelters that dot Western Canada's landscape attest to the region's range of natural wealth. The Alliance Pipeline, completed in 2000, is a natural gas pipeline extending 3,000 kilometres from just north of Fort St John, BC, to Chicago, Illinois.

Western Canada's Resource Base

Western Canada's resource economy began to diversify in the 1970s, when energy developments began in Alberta. Since then, some of the same factors that have been affecting the region's agricultural sector have also been fuelling further growth and changes in its resource sectors. The significance of resource exports to the economy of Western Canada is evident in the 2002 export figures. In that year, the leading exports from Western Canada were grain, petroleum, potash, and uranium. By value, these exports amounted to $60 billion, about 15 per cent of Canada's total exports (Statistics Canada, 2003b). Besides its fertile soils, Western Canada's main assets are its minerals, fuels, and forests (Figure 8.6).

Mining Industry

The mining industry has helped to diversify the economy of Western Canada. The variety and value of mineral production in Western Canada are enormous. Mining companies produce a full range of mineral products, including metals, non-metals, structural materials, and fuels. In 2002, the three Prairie provinces produced $50.5 billion of the energy and mineral production in Canada (Table 8.7). This production amounted to 67 per cent of the national output (Statistics Canada, 2003b).

In Western Canada the geology of each province differs sufficiently to produce three distinct types of mining. Alberta contains rich coal reserves along the eastern slopes of the Rocky Mountains. Coal mining began in 1872 just west of Lethbridge. By the 1990s, the province of Alberta produced more coal than any other province, which contributed to Canada's position as the fourth-largest coal exporter in the world. Today, coal mining takes place in the East Kootenay and Peace River coalfields. Like other resource industries, these Alberta mines depend on world demand and prices. In 1998, the Asian economic slump resulted in a drop in demand and price for Alberta coal.

Potash and uranium are the major mineral deposits in Saskatchewan. The potash deposit lies approximately 1 km below the surface of the earth, reaching its thickest extent around Saskatoon, where six of the nine potash mines are located. Canada is the world's largest producer and exporter of potash. Saskatchewan has the largest and highest-quality deposit in the world, and except for a potash mine in New Brunswick, all of Canada's potash production comes from Saskatchewan. Potash is used to produce potassium-based fertilizers

Table 8.7	Mining Production by Value in Western Canada, 2002 ($000s)			
Mineral Product	Alberta	Saskatchewan	Manitoba	Western Canada
Metals	600	662,754	734,005	1,397,559
Non-metals	593,562	1,640,886	25,387	2,259,835
Fuels	41,021,762	5,642,656	152,639	46,817,057
Total	42,615,924	7,946,295	982,031	50,474,451

Source: Natural Resources Canada (2003a).

and is sold to firms in the United States and Asia, especially China and Japan. Even though China and Japan have been suffering from the Asian economic downturn, they continue to import Canadian potash to ensure high yields for their crops. Still, world demand varies, forcing potash mines to adjust the volume of their production.

Uranium mining also takes place in Saskatchewan. Production first took place on the northern shores of Lake Athabasca near Uranium City. Since the late 1970s, uranium mining has shifted south to the geological area of the Canadian Shield known as the Athabasca Basin. Three mines are currently operating (Cluff Lake, Key Lake, and Rabbit Lake). These older open-pit mines will soon cease production and, subject to the approval of the federal environmental agency, four new mines (Cigar Lake, McArthur River, McClean Lake, and Midwest) are expected to begin production.

Saskatoon is the major supply centre for both potash and uranium mines as well as their corporate headquarters.

Manitoba has two major mineral deposits—copper-zinc and nickel—both located in the Canadian Shield. Both the copper and nickel mining in Manitoba are relatively expensive operations. As well, the high cost of shipping the processed product to distant markets adds to their economic disadvantage. Copper and zinc ore bodies are found near Flin Flon, a mining and smelter town in northern Manitoba that began production in 1930, shortly after a rail link to The Pas was completed in 1928. At one time, Flin Flon produced most of Canada's copper and zinc but today it is an aging resource town with a declining population. Thompson, located some 740 km north of Winnipeg, is a nickel-mining town. In 1957, after a rail link to the Hudson Bay Railway was completed, the mine facility, smelter, and town

Vignette 8.6   Westward Expansion of the Oil Industry

While oil was first discovered in the southwestern corner of southern Ontario in 1858, the search for petroleum soon switched to southern Alberta. In 1883, Canadian Pacific Railway workers found gas while drilling for water near Medicine Hat. In 1890, gas was again found in the area by miners drilling for coal. Natural gas was soon used as a fuel for heating and lighting the homes and offices of Medicine Hat. The streets of the young Prairie city were lighted by gas lamps; natural gas was so cheap that the lamps were never turned off. Soon major natural gas deposits were found throughout southern Alberta. In 1912, a gas pipeline was built, linking Calgary to the Bow Island gas fields near Medicine Hat, 275 km away.

In 1902 oil was discovered in southwest-

ern Alberta. However, Alberta did not experience a true oil boom until 10 years later, when oil was discovered at Turner Valley. Drilling companies also discovered natural gas, but because it was so abundant it was considered a nuisance and burned at the site of the well. Turner Valley was Canada's largest and most important oil field in the years before World War II. In 1947, when Imperial Oil discovered a major oil field near the small town of Leduc, just outside of Edmonton, the Alberta oil boom began in earnest. This oil field lies in a geological formation of the Western Sedimentary Basin. Since such basin formations have potential for oil and natural gas deposits, they were the focus of more oil and gas exploration.

were constructed. Unlike Flin Flon, Thompson was a specially designed resource town that had a complete array of urban amenities. The first nickel was produced in 1961. During the 1960s, Thompson's population soared to 20,000. Since then, two events have reduced the population size of Thompson and many other resource towns: (1) productivity of nickel mining, enhanced by greater mechanization, has increased, resulting in a smaller labour force, and (2) the service sector of Thompson has contracted. In 1991, Thompson had a population of 14,977, but by 2001 it had fallen to 13,256. For similar reasons, Flin Flon's population dropped from 7,119 in 1991 to 6,000 in 2001.

Fuels: Oil and Natural Gas

What has proven to be Western Canada's most valuable mineral deposit is its vast reserves of oil and natural gas (Vignette 8.6). In 2002, Western Canada's production of fuels (including coal, oil, and natural gas) was valued at $47 billion. Alberta, with 87 per cent, and Saskatchewan, with 12 per cent, accounted for 99 per cent of this production (Table 8.7). Western Canada contains over 70 per cent of Canada's oil reserves. These oil deposits are located in the southern half of the Western Sedimentary Basin. Alberta has the largest oil reserves (55 per cent of Canada's reserves), followed by Saskatchewan (17 per cent). Manitoba has less than 1 per cent.

In addition to oil and natural gas deposits, vast amounts of oil are contained in the tar sands—oil mixed with sand, known as **bitumen**—of northern Alberta. The two major oil sands mining sites are situated near Fort McMurray and Cold Lake. Alberta has the major oil and gas industry, so it leads the three Western provinces in the value of mineral pro-

duction. Major oil sands projects planned or underway in the Fort McMurray area are (1) Syncrude expansion; (2) Suncor Energy expansion; (3) Canadian Natural Resources planned project; (4) Mobile Canada planned project; (5) Petro-Canada planned project; (6) Gulf Canada planned project; (7) PanCanadian Petroleum planned project; (8) Shell Canada underway; and (9) True North Energy planned project.

The economic impact of petroleum on Western Canada, especially Alberta, was caused by the sudden rise in oil prices in the 1970s. During the 1970s, the sale of large quantities of oil and natural gas to Ontario at much higher prices than in the previous decade resulted in a transfer of enormous wealth and economic power from the consuming provinces (mainly Ontario) to the producing provinces (mainly Alberta). Alberta was transformed into a 'rich' province. Oil was a source not only of economic power but also of political power. In terms of the core/periphery model, the export of oil reversed a century-old economic relationship—Central Canada was dependent on the western provinces. Ottawa interfered with this new relationship by introducing the National Energy Program. The federal government was seen by Alberta as intruding on provincial rights and attempting to prevent the transfer of wealth. The results of this federal program were to widen the political gap between Alberta and Ottawa and to deepen the sense of western alienation and mistrust towards Ottawa and Central Canada.

Forest Industry

The boreal forest stretches across the northern part of Western Canada. Within the boreal forest, a few coniferous species—spruce, fir, pine, and tamarack—predominate. The commercial forest zone lies in the Interior Plains where soil conditions are more favourable for tree growth than in the thin, rocky soils of the Canadian Shield. In 2001, logging in Western Canada amounted to 15 per cent of Canada's total timber harvest (Table 8.8). Approximately 75 per cent of this production takes place in Alberta.[12] Alberta's advantage comes from its large northern forest stands. The forest-covered Interior Plains region, which extends beyond the North Saskatchewan River to the border with the Northwest Territories, is the largest forest stand in Western Canada. This, together with the timber in the foothills of the Rocky Mountains, means that Alberta has by far the largest commercial forest stands of the three Prairie provinces.

Each of Western Canada's three provinces has a diversified timber operation, ranging from sawmills to pulp and paper plants. Most pulp and paper mills are located on the southern edge of the forest. In Saskatchewan these mills are located at Prince Albert and Meadow Lake. In Alberta they are in Athabasca, Grande Prairie, Hinton, and Peace River. In Manitoba The Pas is the site of a major pulp and paper mill.

By the end of the 1980s, virtually all the commercial forest stands had been leased, sig-nalling the end of available virgin timber. In fact, several mills were unable to obtain additional Crown timber leases and have suffered from a shortage of wood fibre. Now that the original boreal forest has been harvested, the second growth is best suited for the pulp and paper industry. These trees, small and large, known as pulpwood, are cut and then trucked to a pulp mill. Each mill has a forest lease, giving it the right to log on Crown land. While a small amount of lumber production does occur, pulp and paper products predominate. Distance from markets also affects the focus of the forest industry. Pulp and paper products, which have a much higher value per tonne than lumber, are better able to withstand rail transportation costs from Western Canada to distant markets.

In the 1980s, it was discovered that aspen forests in Alberta, Saskatchewan, and Manitoba provided an excellent source of pulpwood for high-quality paper production. With considerable provincial financial support, mills were established at Hinton, Grande Prairie, Prince Albert, and The Pas. The Alberta government, seeking to diversify its economy by encouraging more processing of resources, tried to lure pulp and paper companies to northern Alberta by offering them vast north-

Table 8.8	Forest Area and Production in Western Canada, 1999		
Province	Area (millions of ha)	Productive Forest Area (millions of ha)	Industrial Roundwood (millions of m³)
Manitoba	26.3	13.2	2.2
Saskatchewan	28.8	10.6	4.5
Alberta	38.2	20.9	21.9
Western Canada	93.3	44.7	28.6
Per cent Canada	22.3	20.9	15.0

Source: Natural Resources Canada (2003b).

ern hardwood timber leases. Two Japanese firms, Daishowa Canada Company and Alberta-Pacific Forest Industries, established mills at Peace River and Athabasca in northern Alberta. By the end of the 1980s, virtually all Alberta timber had been leased to five pulp and paper companies.

Similar developments occurred in Saskatchewan and Manitoba. One Alberta-based firm, Millar-Western, turned to Saskatchewan where it arranged to purchase pulp logs from Norsask, a local company that holds the timber lease to lands north of Meadow Lake. With an assured supply of pulp wood, Millar-Western built a mill at Meadow Lake. The Millar-Western pulp mill is a unique industrial development because: (1) the mill's internal circulating system means that no toxic wastes are released into the local rivers and lakes, and (2) it has a business arrangement with Norsask Forest Products, which is jointly owned by the Meadow Lake Tribal Council and Techfor Services, a company owned by the employees of the local sawmill (Anderson and Bone, 1995: 127). This business venture represents a new approach for resource development, combining the financing and expertise of an established pulp and paper company with the commitment by Aboriginal peoples to supply the raw materials.

Western Canada's Urban Geography

The process of urbanization in Western Canada has lagged behind that of Ontario, British Columbia, and Québec. Even so, nearly three-quarters of Western Canada's residents live in

Table 8.9	Major Urban Centres in Western Canada, 1996–2001		
Centre	Population 1996	Population 2001	% Change
Moose Jaw	34,829	33,519	(3.8)
Grande Prairie	31,353	36,983	18
Brandon	40,581	41,037	1.1
Prince Albert	41,706	41,460	(0.6)
Wood Buffalo (includes Fort McMurray)	36,124	42,602	17.9
Medicine Hat	56,570	61,735	9.1
Lethbridge	63,053	67,374	6.9
Red Deer	60,080	67,707	12.7
Regina	193,652	192,800	(0.4)
Saskatoon	219,056	225,927	3.1
Winnipeg	667,093	671,274	0.6
Edmonton	862,597	937,845	8.7
Calgary	821,628	951,395	15.8
Total	3,096,169	3,371,658	8.9

Source: Statistics Canada (2002a).

urban centres, and most of these reside in the five major cities—Regina, Saskatoon, Winnipeg, Edmonton, and Calgary. A second order of urban centres includes Red Deer, Lethbridge, Medicine Hat, Wood Buffalo (Fort McMurray), Prince Albert, Brandon, and Moose Jaw. In 2001, these cities had a total population of 3.4 million (Table 8.9).

An important urban corridor is emerging within Alberta, linking Edmonton with Calgary. Some 2 million people live and work within this corridor, so it not only forms a major market within Western Canada but also has the potential for the establishment of high-tech industries. While the three Prairie provinces have not drawn high-tech industry in the same concentration as Ontario, Québec, and British Columbia, all five major cities have a cluster of high-tech firms. Alberta leads with

Saskatoon, situated on the South Saskatchewan River, is Saskatchewan's largest city with a population of nearly 226,000 (2001). Known as the 'Bridge City', Saskatoon has witnessed rapid population growth over the last 20 years. During that period, many Indians and Métis have settled in Saskatoon. By 2001, the Aboriginal population comprised 9.1 per cent of Saskatoon's residents. (Al Harvey/The Slide Farm)

66,000 high-tech workers, followed by Manitoba with 16,000 and Saskatchewan with 12,000 (Howes and McKinnon, 2000: C7). Two anchors—Calgary and Edmonton—each contribute special functions to the corridor. Calgary (951,000 in 2001) is the headquarters of the oil and gas industry, while Edmonton (937,000 in 2001), besides being the provincial capital, is known as the 'Gateway to the North'. Over the past five years, this urban corridor has had a high rate of population growth, exceeding 10 per cent.

From 1996 to 2001, the rate of urban growth varied considerably for the towns and cities of Western Canada. This variation reflects differences in local economic growth and in the pace of consolidating populations into regional centres. From 1996 to 2001, the fastest-growing cities in Western Canada were in Alberta: Grande Prairie, Wood Buffalo (Fort McMurray), Calgary, and Red Deer. The slowest-growing cities were in Saskatchewan and Manitoba—Saskatoon had the highest rate at 3.1 per cent while Winnipeg was at 0.6 per cent. Regina, Moose Jaw, and Prince Albert saw their populations decline (Table 8.9).

Western Canada has, in effect, two types of urban settlement pattern. These two urban patterns reveal a north–south split based on a regional core and a hinterland. In the south, centres are aligned close to the southern border with the US, whereas an oasis-like pattern exists in the northern hinterland with settlements far apart. Such a pattern is common in resource hinterlands where a single-industry town is located near an ore body. Three examples are: Flin Flon (copper), Fort McMurray (heavy oil), and Thompson (nickel).

Across Western Canada, the trend for over 60 years has been a rural-to-urban migration with many of the smallest centres disappearing. At the turn of the twentieth century, these small communities formed the heart of the local farming area, but today they have lost their function. Professors Stabler and Olfert, who have studied the migration from rural areas in Western Canada for many years, see the rural decline as inexorable, causing many small centres to disappear. They also predict that this migration in Western Canada, but in Saskatchewan in particular, will escalate (Stabler and Olfert, 2002). The main reasons why these centres dwindled are complex and interconnected:

- Farms consolidated, thereby reducing the size of the rural population.
- Highways were built, facilitating shopping in larger centres, where the variety and prices of goods were superior to those in smaller centres; eventually, stores in smaller communities closed.
- Schools, hospitals, and other public services were concentrated in larger centres, attracting the people who use these services.
- Railways abandoned branch lines, causing grain elevator companies to close their operations; for smaller towns, the closure of the grain elevator and the loss of rail service meant the loss of the town's last function.
- Grain companies built larger grain elevators on main lines, expecting that the lower freight rates offered at the larger grain elevators would attract farmers, drawing business away from elevators in small centres.

Vignette 8.7   Winnipeg

Winnipeg, the capital and the largest city in Manitoba, is located at the confluence of the Red River and the Assiniboine River. With the building of the Canadian Pacific Railway, Winnipeg became known as the 'Gateway to the West'. Because of its strategic location, Winnipeg was and remains a transportation centre. By the turn of the twentieth century, Winnipeg was the principal city in Western Canada, controlling the grain trade and also acting as the administrative, financial, and wholesale hub for Western Canada.

Winnipeg's role in Western Canada has gradually weakened over time. After the completion of the Panama Canal in 1914, Alberta farmers and some Saskatchewan farmers began to ship their wheat through Vancouver instead of through Winnipeg. Next, service industries began to emerge in Edmonton, Calgary, Saskatoon, and Regina. Each city captured some of the trade previously held by Winnipeg merchants.

Winnipeg's stranglehold over the sale and distribution of agricultural machinery and products to the farmers of Western Canada was broken.

After World War II, Winnipeg still remained the largest city in Western Canada. In 1951 Winnipeg's population was 357,000, while Edmonton and Calgary were much smaller: 177,000 and 142,000, respectively. However, Calgary and Edmonton grew rapidly in the following years, partly because of developments in the oil and gas industries. By 1981, both of these cities surpassed Winnipeg in population size. At that time, Winnipeg had a population of 592,061, while Edmonton's population was 740,882 and Calgary's was 625,966. Today, Winnipeg dominates the economy of Manitoba, but its role within Western Canada has been considerably diminished by Calgary and Edmonton and, to a lesser degree, by Saskatoon and Regina.

- Farmers moved to larger centres, where more urban amenities were available, and commuted to their farms.

Western Canada's Future

Western Canada began the twentieth century as an agricultural hinterland. This hinterland slowly began to diversify into other forms of resource development. After World War II, the process of change accelerated, particularly in Alberta, where petroleum diversified the resource economy and provided the basis for a strong petrochemical industry. Two events, trade liberalization and the termination of the Crow Benefit, launched further economic changes in the 1990s, especially in agriculture.

In Western Canada, agriculture is undergoing most of the economic restructuring caused by free trade and the loss of the transportation subsidy. As farmers shift to crops of higher value and crops that can be sold locally, the importance of spring wheat will diminish. The structural changes to transportation rates will also encourage local processing of agricultural products, thereby expanding the meat-processing industry and stimulating the development of new food-processing industries. Agricultural land use is also changing across Western Canada, with an expansion in the growing of specialty crops. For instance, farmers in Manitoba pay more for shipping their wheat to Thunder Bay than Alberta farmers pay for sending their grain to Vancouver—this spatial price differential is encouraging a shift in agricultural land use with more spring wheat being grown in Alberta than in Manitoba. Manitoba farmers now sell more feed grain to the province's expanding livestock industry, which then ships its meat products to markets in Ontario, Québec, and the United States.

The greatest impact of economic diversifi-cation within Western Canada has been in Alberta. Since the 1970s, soaring oil and gas prices transformed Alberta into a wealthy province. Yet resource industries, while they add greatly to the value of provincial production, required only a small labour force and therefore had a limited effect on the workforce. The indirect effects of oil, however, were substantial in the construction, pipeline, and service sectors of Alberta's economy. Saskatchewan and Manitoba, while also diversifying but without the benefit of a large oil and gas industry, have proceeded at a much slower rate. With the basic changes now underway in agriculture, these two provinces may hasten their pace towards more diversified economies.

Summary

Western Canada occupies the western interior of Canada. The region has a range of resources, including arable land, vast forests, and mineral wealth. The geography of Western Canada has posed several challenges to economic development. Two major challenges are distance to ocean ports and world markets, and a dry climate for agriculture. Initial attempts to overcome distance took the form of railway construction and a subsidy for shipping grain by rail. To overcome the dry climate, farmers grew wheat, which can withstand the semi-arid conditions and thus became the staple crop.

In the twenty-first century, the role of Western Canada as a resource hinterland is changing through economic diversification and continued population growth, especially in Alberta. The expansion of the petroleum industry, the liberalization of world trade, and the loss of the Crow Benefit have accelerated this trend. While the region is still far from becoming a major industrial core, Western Canada has become an upward transitional region as described in Friedmann's core/periphery model. The process of diversification has included strong growth in the manufacturing sector, agricultural transition, and the greater processing of non-renewable resources and of forest and agricultural products.

Alberta leads the three Prairie provinces in economic growth as a consequence of a sharp increase in natural gas prices, the continuing expansion of its heavy oil sands in the Fort McMurray area, and the rapid growth of its livestock and hog industry. Manitoba and Saskatchewan, with a less rich resource base, are lagging behind Alberta's economic performance. Cities, but particularly Calgary, Edmonton, Winnipeg, Saskatoon, and Regina, serve as 'engines of economic and population growth'. Yet, the region's economy was dampened by several unanticipated events in the early twenty-first century: the continuation of

low precipitation for three years hurt the agricultural sector; the economic downturn in the United States economy (though it recovered by late 2003) slowed industrial expansion; and American duties on softwood lumber and grain had a devastating impact on those two economic sectors. In 2003, the worldwide ban on Canadian beef due to the fear of mad cow disease or BSE (bovine spongiform encephalopathy) hurt the livestock industry, and this ban may not be lifted for another year or even longer. Nonetheless, the overall economy, especially that of Alberta, continued to grow because it has become more broadly based. The direction of future change in Western Canada's economy is clear—more processing, a larger service sector, and more diversification—but the final shape of this new economy will not be revealed until well into the twenty-first century.

Notes

1. Two novels illustrate the powerful impact of settlement on the Indians and Métis. Rudy Wiebe's *The Temptations of Big Bear* (1973) focuses on the Cree; Guy Vanderhaegh's *The Englishman's Boy* (1996) looks at the Cypress Hills Massacre.
2. The Cree attacked the outpost at Frog Lake, laying siege to Fort Battleford and defeating the North-West Mounted Police at Fort Pitt and Cut Knife Hill. With the arrival of the Canadian militia from eastern Canada, the Cree were defeated.
3. It was only in 1960 that the federal government extended the rights and privileges accorded to Canadian citizens to its Aboriginal peoples. Until then, Indians could not vote in provincial and federal elections without losing their status. The right to vote marked the start of a long journey to find a place in Canadian society. This journey is

far from over.
4. The search for a quicker-maturing wheat began in 1892 when a cross was made between Red Fife, a popular wheat grown on the Prairies, and an earlier maturing wheat. After a decade of trials, Marquis wheat was tested at the Dominion Experimental Farms at Indian Head, Saskatchewan. By 1910, this variety of wheat made Canada famous for producing an exceptionally high-quality, hardy spring wheat.
5. Summer fallowing accomplished two goals: it conserved moisture and controlled weeds. In the 1990s, farmers began to abandon this technique because advances in technology enabled them to accomplish these goals without having to resort to summer fallowing.
6. Western alienation is based on past experi-

ence of real or perceived 'abuse' by big business, such as the CPR, and big government, meaning government controlled by eastern interests. Such feelings were deeply felt by homesteaders who settled the West. By early in the twentieth century, western alienation was particularly strong and the subsequent resentment was often aimed at the Canadian Pacific Railway. In fact, farmers regarded this railway company as the most rapacious agent of eastern Canadian interests. Not only had the company obtained millions of acres of fertile land, but its real estate offices often manipulated station sites to ensure their location on Canadian Pacific Railway property. Since farmland increased in value with proximity to rail-loading sites, such Canadian Pacific Railway land increased in value and could be sold to settlers at a higher price. Even more galling to westerners was the fact that CPR landholdings could not be taxed for 20 years. To extend this tax-free period, the CPR sometimes delayed selecting land, which meant that large areas were not available for homesteading because the railway could opt to select such lands as part of its grant.

7. Alberta, Manitoba, and Saskatchewan did not gain full control over Crown lands, and therefore over natural resources, until 1930. At that time, Ottawa transferred federal property to these provinces, giving them access to lucrative sources of taxation associated with natural resource developments. Until then, taxes from these lands went to Ottawa, supposedly to pay for railway building in the West. At first, the revenue from natural resources was small, but with the discovery of oil at Leduc, Alberta, in 1947, royalties from the production of petroleum provided most of the revenue for Alberta. Equally important, the exploitation of the vast oil and gas deposits in Alberta grew rapidly and eventually resulted in a variety of petroleum-processing plants and pipeline-construction firms in Alberta. Later, more modest energy and mineral developments were found in Saskatchewan, British Columbia, and Manitoba.

8. In June 1999, the Organization for Economic Co-operation and Development announced that 1998 wheat subsidies in American dollars were $141 a tonne in the European Union, $61 a tonne in the US, and $8 a tonne in Canada (Hursh, 1999: C9).

9. Changes in the structure of farming are also anticipated. With expansion of the livestock and meat-processing industries, the number of farm workers may also increase. Large-scale hog farms, for instance, require farm workers. Also, the increase in specialty crops is causing more farmers to undertake contract farming; for example, they grow a certain quantity of canary seed for a set price. Some of these crops are now processed locally, a trend that is expected to intensify.

10. Farmers are becoming less sheltered from economic forces. Through contract farming, agribusiness—the sector of the economy that provides inputs to farms, such as chemical fertilizers, and procures agricultural products from farms for processing and distribution to consumers—is increasingly affecting the daily lives of farmers. Some fear that in the next severe economic downturn or drought, family farming operations will be replaced by corporate farms. In fact, the trend towards contract farming by agribusiness may be the first sign of such an ownership shift. While not yet apparent in the grain industry, agribusiness has established itself in specialty crops, poultry, and hog farming.

11. Farmers' decisions are affected by a complex set of variables, but ultimately the cost of doing business is the determining factor. For instance, the increased production of canola may affect the price of livestock feed.

A protein by-product derived from the crushing of canola seeds is suitable for livestock. Instead of importing a similar protein supplement and paying for the built-in freight costs, using a local canola-based product to feed livestock may reduce costs.

12. Since the early 1990s, sawmills in the interior of British Columbia have exhausted local supplies of timber. They have had to purchase logs from Alberta ranchers who own timber lands in the foothills around Calgary.

Key Terms

agricultural fringe
Agriculture at its physical limits. Along the southern edge of the boreal forest and in the Peace River country, farmers clear the land, but the short growing season prevents most crops from maturing, so many farmers turn to cattle.

bitumen
A tar-like mixture of sand and oil.

Dry Belt
An agricultural area in the semi-arid parts of Alberta and Saskatchewan that is primarily devoted to grain farms and cattle ranches. Crop failures due to drought are more common.

Fertile Belt
A mixed farming area where crop failures due to drought are less common. This area of long-grass and parkland natural vegetation is associated with black and dark-brown chernozemic soils.

Palliser's Triangle
Captain John Palliser conducted a survey of the Canadian West in 1857–60. He concluded that this short-grass natural vegetation area in southern Alberta and Saskatchewan was a northern extension of the Great American Desert and was therefore unsuitable for agricultural settlement.

primary prices
The prices for commodities such as foodstuffs, raw materials, and other primary products.

regional core
Within the core/periphery model, cores can occur at different geographic levels. A regional core is an area (often a large city) that dominates trade and stimulates economic growth in the region.

strip farming
An old farming practice in which alternating strips of land were cultivated and left fallow. The purpose was to conserve soil moisture and reduce wind and water erosion. Due to advances in agricultural technology, farmers can now maximize soil moisture conservation while practising continuous cultivation.

summer fallow
The farming practice of leaving land idle for a year or more to accumulate sufficient soil moisture to produce a crop; also practised to restore soil fertility; being replaced by continuous cropping.

Western Sedimentary Basin
Within the geological structure of the Interior Plains, the normally flat sedimentary strata are bent into a basin-like shape. These basins often contain petroleum deposits.

Bibliography

Akinremi, O.O., S.M. McGinn, and H.W. Cutforth. 2001. 'Seasonal and Spatial Patterns of Rainfall on the Canadian Prairies', *Journal of Climate* 14, 9: 2177–82.

Anderson, Robert B., and Robert M. Bone. 1995.

'First Nations Economic Development: A Contingency Perspective', *The Canadian Geographer* 39, 2: 120–30.

Artibise, Alan. 1975. *Winnipeg: A Social History of Urban Growth, 1874–1914*. Montréal and

Kingston: McGill-Queen's University Press.

Barr, Brenton M., and John C. Lehr. 1982. 'The Western Interior: Transformation of a Hinterland Region', in L.D. McCann, ed., *A Geography of Canada: Heartland and Hinterland*. Scarborough, Ont.: Prentice-Hall, ch. 8.

Barron, F. Laurie, and Joseph Garcea, eds. 1999. *Urban Indian Reserves: Forging New Relationships in Saskatchewan*. Saskatoon: Purich Publishing.

Bell, Ian. 1999. 'Dancing with Elephants', *Western Producer*, 2 Sept., 60–1.

Bellavance, Joël-Denis, and Robert Fife. 2000. 'PM may try "tough love" on West', *National Post*, 23 Dec., A1.

Blackbourn, Anthony, and Robert G. Putnam. 1984. *The Industrial Geography of Canada*. London: Croom Helm.

Bone, Robert M. 2003. *The Geography of the Canadian North: Issues and Challenges*, 2nd edn. Toronto: Oxford University Press.

Bonsal, B.R., X. Zhang, and W.D. Hogg. 1999. 'Canadian Prairie Growing Season Precipitation Variability and Associated Atmospheric Circulation', *Climate Research* 11: 191–208.

Broadway, Michael J. 1998. 'Where's the Beef? The Integration of the Canadian and American Beefpacking Industries', *Prairie Forum* 23, 1: 19–30.

Budhia, Narendru. 1995a. 'Manitoba—Outlook for 1995 Improves', *Provincial Outlook: Economic Forecast* 10, 2 (Spring): 38–41.

———. 1995b. 'Saskatchewan—As the Crow Flies', *Provincial Outlook: Economic Forecast* 10, 2 (Spring): 42–5.

Cattaneo, Claudia. 2002. 'Hedging over Kyoto begins in oilpatch', *National Post*, 11 Dec., FP1.

CBC. 1999. 'Farmers fined $500 for each illegal border crossing', 22 Nov.: <http://www.cbc.ca/news/features/wheat_board.html>.

———. 2002. 'Prairie farmers would rather go to jail', 1 Oct.: <http://www.cbc.ca/news/features/wheat_board.html>.

Evans, Simon M. 1979. 'American Cattlemen on the Canadian Range, 1874–1914', *Prairie Forum* 4: 121–35.

Eyton, J. Ronald. 1997. 'Spatial Externalities and Edmonton Dwelling Values', *The Canadian Geographer* 41, 2: 202–6.

Found, William C. 1996. 'Agriculture in a World of Subsidies', in John N.H. Britton, ed., *Canada and the Global Economy: The Geography of Structural and Technological Change*. Montréal and Kingston: McGill-Queen's University Press, ch. 9.

Garrod, Stan. 1984. *Economics in Society: Canadian Case Studies*. Don Mills, Ont.: Addison-Wesley.

Hall, Angela. 2003. 'Hogs touchy issue in rural Sask.', *Saskatoon Star-Phoenix*, 25 Feb., C8.

Hamburg, Karen T., C. Emdad Haque, and John C. Everitt. 1997. 'Municipal Waste Recycling in Brandon: Determinants of Participatory Behavior', *The Canadian Geographer* 41, 2: 149–65.

Haque, C.E. 1998. 'Coping Responses to the 1997 Red River Valley Flood: Research Issues and Agenda', *Prairie Perspectives* 1: 47–62.

Hildebrandt, Walter, and Brian Hubner. 1994. *The Cypress Hills: The Land and Its People*. Saskatoon: Purich Publishing.

Howes, Carol, and Ian McKinnon. 2000. 'Calgary's Critical Mass', *National Post*, 15 Nov., C7.

Hursh, Kevin. 1999. 'Subsidy Complaints Don't Hold Water', *Saskatoon Star-Phoenix*, 28 July, C9.

Jones, David, C. 1987. *Empire of Dust: Settling and Abandoning the Prairie Dry Belt*. Edmonton: University of Alberta Press.

Kariel, H.G. 1996. 'Land Use in Albert's Foothill Country', *Western Geography* 7: 20–46.

MacLachlan, Ian. 2001. *Kill and Chill: Restructuring Canada's Beef Commodity Chain*. Toronto: University of Toronto Press.

McNicol, B.J. 1997. 'Views about Industrial Tourist Pressures in Canmore, Alberta', *Western Geography* 7: 47–72.

Marsh, James H., ed. 1988. *The Canadian Encyclopedia*, 2nd edn. Edmonton: Hurtig.

National Farmers' Union. 1999. 'Farm Subsidies', *National Post*, 27 Oct., C6.

National Forestry Database Program. 2000. Forest Products: Industrial Roundwood by Category, 1990–8 [on-line database], Ottawa. Searched 30 Dec. 2000: <http://nfdp.ccfm.org/frames2 _e.htm>.

Natural Resources Canada. 2003a. Mineral Production of Canada, by Provinces and Territory [on-line database]. Searched 30 Aug. 2003: <http://mmsd1.mms.nrcan.gc.ca/mmsd/production/production_e.asp>.

———. 2003b. The State of Canada's Forests 2001/02 [on-line database]. Searched 31 Aug. 2003: <http://www.nrcan.gc.ca/cfs-scf/national/what-quoi/sof/sof02/profiles_e.html>

Neatby, L.H. 1979. *Chronicle of a Pioneer Prairie Family*. Saskatoon: Western Producer Prairie Books.

Paul, Alec H. 1997. 'Shortlines, Mainlines, Branchlines, Dead Lines: Rural Railways in Southwestern Saskatchewan in the 1990s', in John Welsted and John Everitt, eds, *The Yorkton Papers: Research by Prairie Geographers*. Brandon Geographical Studies No. 2. Brandon, Man.: University of Brandon.

Phillips, David. 1993. *The Day Niagara Falls Ran Dry!* Toronto: Canadian Geographic and Key Porter Books.

Potyondi, Barry. 1995. *In Palliser's Triangle: Living in the Grasslands, 1850–1930*. Saskatoon: Purich Publishing.

Price, Jacqueline D. 2003. 'Are Factory Farms Fouling Our Water?', *Alberta Views* (May/June): 34–9.

Rannie, W.F. 1998. 'The Red River Flood in Manitoba, Canada', *Prairie Perspectives* 1: 1–24.

Rice, Murray D. 1996. 'Functional Dynamics and a Peripheral Quaternary Place: The Case of Calgary', *Canadian Journal of Regional Science* 19, 1: 65–82.

Richards, J. Howard. 1968. 'The Prairie Region', in John Warkentin, ed., *A Geographical Interpretation*. Toronto: Methuen, ch. 12.

Robertson, Grant. 1999. 'Big Business Harvests Potato Profit', *Calgary Herald*, 20 July, D1, D3.

Rumney, Thomas. 1999. *The Geography of Canada Bibliography Series: Vol. 2, The Prairies*. Plattsburgh, NY: Plattsburgh State University, Center for the Study of Canada.

Russell, Bob. 1999. *More with Less: Work Reorganization in the Canadian Mining Industry*. Toronto: University of Toronto Press.

Saskatchewan Agriculture and Food. 1997. *Agricultural Statistics 1996*. Regina: Statistical Branch.

Saskatchewan Agriculture, Food, and Rural Revitalization. 2002. Agricultural Statistics 2001 [on-line database]. Searched 2 Sept. 2003: <http://www.agr.gov.sk.ca/DOCS/statistics/finance/other/Handbook01.pdf>.

Saskatchewan Environment. 2003. *An Assessment of Abandoned Mines in Northern Saskatchewan*. A report prepared by Clifton Associates Ltd. File R3160. Regina: Clifton Associates Ltd.

Sauchyn, David J., ed. 1993. *Quaternary and Late Tertiary Landscapes of Southwestern Saskatchewan and Adjacent Areas*. Canadian Plains Research Centre, University of Regina. Winnipeg: Hignell Printing.

Schmitz, A., and H. Furtan. 2000. *The Canadian Wheat Board*. Regina: Canadian Plains Research Centre.

Selwood, J., and S. Kohm. 1988. 'Location, Location, Location: Selling Sex in the Suburbs', *Prairie Perspectives* 1: 161–71.

Semple, R.K. 1994. 'The Western Canadian Quaternary Place System', *Prairie Forum* 12: 81–100.

Simpson, J., and S. Hathout. 1998. 'Optimum Route Location Model for an All-Weather Road on the East Side of Lake Winnipeg', *Prairie Perspectives* 1: 113–24.

Smith, P.J., ed. 1972. *The Prairie Provinces: Studies in Canadian Geography*. Toronto: University of Toronto Press.

Spry, Irene M. 1963. *The Palliser Expedition: An Account of John Palliser's British North American Expedition 1857–1860*. Toronto: Macmillan.

———, M.R. Olfert, and Murray Fulton. 1992.

The Changing Role of Rural Communities in an Urbanizing World: Saskatchewan 1961–1990. Canadian Plains Report no. 8. Regina: Canadian Plains Institute.

Stabler, Jack C. 2002. *Saskatchewan Communities in the 21st Century: From Places to Regions*. Regina: Canadian Plains Research Centre.

———— and M.R. Olfert. 1992. *Restructuring Rural Saskatchewan: The Challenge of the 1990s*. Regina: Canadian Plains Institute.

Stanford, Quentin H., ed. 1998. *Canadian Oxford World Atlas*, 4th edn. Toronto: Oxford University Press.

Statistics Canada. 1982. *Census Metropolitan Areas and Census Agglomerations with Components*. Catalogue no. 95–903. Ottawa: Minister of Supply and Services Canada.

————. 1992a. *Census Divisions and Subdivisions*. Catalogue no. 93–304. Ottawa: Minister of Supply and Services.

————. 1992b. *Census Overview of Canadian Agriculture 1971–1991*. Catalogue no. 93–348. Ottawa: Ministry of Industry, Science and Technology.

————. 1993. *Canada Year Book 1994*. Ottawa: Ministry of Industry, Science and Technology.

————. 1994a. *Canada's Aboriginal Population by Census Subdivision and Census Metropolitan Area*. Catalogue no. 94–326. Ottawa: Ministry of Industry, Science and Technology.

————. 1994b. *Canadian Forestry Statistics 1993*. Catalogue no. 25–202–XPB. Ottawa: Ministry of Industry, Science and Technology.

————. 1996a. *Canada's Mineral Production*. Catalogue no. 26–202–XPB. Ottawa: Mining Sector, Natural Resources Canada.

————. 1996b. *Labour Force Annual Averages 1995*. Catalogue no. 71–220–XPB. Ottawa: Statistics Canada.

————. 1997a. *A National Overview: Population and Dwelling Counts*. Catalogue no. 93–357–XPB. Ottawa: Industry Canada.

————. 1997b. 1996 Census: Nation Tables—Population by Mother Tongue, Showing Age Groups, for Canada, Provinces and Territories,

1996 Census—20% Sample Data, 2 Dec. 1997 [on-line database], Ottawa. Searched 15 July 1998: <http://www.statcan.ca/english/census96/>.

————. 1997c. *The Daily*—1996 Census: Mother Tongue, Home Language and Knowledge of Languages, 2 Dec. 1997 [on-line database], Ottawa. Searched 14 July 1998: <http://www.statcan.ca/Daily/English/>.

————. 1998a. *Canadian Economic Observer*. Catalogue no. 11–010–XPB. Ottawa: Statistics Canada.

————. 1998b. *The Daily*—1996 Census: Aboriginal Data, 13 Jan. 1998 [on-line database], Ottawa. Searched 14 July 1998: <http://www.statcan.ca/Daily/English/>.

————. 1998c. *The Daily*—1996 Census: Ethnic Origin, Visible Minorities, 17 Feb. 1998 [on-line database], Ottawa. Searched 16 July 1998: <http://www.statcan.Daily/English/>.

————. 1998d. Census of Agriculture—Canada Highlights [on-line database], Ottawa. Searched 2 Sept. 1998: <http://www.statcan.ca/english/censusag/can.htm>.

————. 1998e. *Canadian Forestry Statistics 1995*. Catalogue no. 25–202–XPB. Ottawa: Ministry of Industry, Science and Technology.

————. 2000a. Table 2: Revised Statistics of the Mineral Production of Canada, by Province, 1996 [on-line database], Ottawa. Searched 30 Dec. 2000: <http://www.nrcan.gc.ca/mms/efab/mmsd/production>.

————. 2000b. Canadian Statistics—Export of Goods on a Balance-of-Payments Basis [on-line database], Ottawa. Searched 30 Dec. 2000: <http://www.statcan.ca:80/english/Pgdb/Economy/International/>.

————. 2002a. Canadian Statistics: Census of Agriculture Farms: Land tenure, provinces. [on-line database], Ottawa. Searched 30 Aug. 2003: <http://www.statcan.ca/english/Pgdb/econ113a.htm>.

————. 2002b. 2001 Census: Population and Dwelling Counts, for Census Metropolitan Areas and Census Agglomerations, 2001 and

1996 Censuses, 16 July 2002 [on-line database], Ottawa: <http://www.statcan.ca/english/IPS/Data/93F0050XCB2001013.htm>.

———. 2002c. Census of Canada 2001—Census Geography. Highlights and Analysis: Canada's 2001 Population: <http://www12.statcan.ca/English/census01/>.

———. 2003a. Agriculture 2001 Census Farm Operations: provincial/regional trends. [on-line database], Ottawa. Searched 30 Aug. 2003: <http://www.statcan.ca/english/agcensus2001/first/regions/contents.htm>.

———. 2003b Canadian Statistics—Export of Goods on a Balance-of-Payments Basis [on-line database], Ottawa. Searched 2 Sept. 2003: <http://www.statcan.ca/english/Pgdb/gblec04.htm>.

———. 2003c. Canadian Statistics: Distribution of Employed People, by Industry, by Province [on-line database], Ottawa. Searched 20 May 2003: <http://www.statcan.ca/english/Pgdb/labor21b.htm>.

Thraves, B.D., and G. Barriault. 1998. 'Change in the Size and Functions of Regina's Central Business District, 1964–1997', *Prairie Perspectives* 1: 141–60.

Vanderhaeghe, Guy. 1996. *The Englishman's Boy*. Toronto: McClelland & Stewart.

Wiebe, Rudy. 1973. *The Temptations of Big Bear*. Toronto: McClelland & Stewart.

Zakreski, Dan. 2000. 'Change in the Wind: destruction of elevators signals symbolic shift in Saskatchewan agriculture', *Saskatoon Star-Phoenix*, 9 Dec., E1–E2.

Further Reading

Fung, Ka-iu, ed. 1999. *Atlas of Saskatchewan*. Saskatoon: PrintWest. (Also available as a CD.)

The *Atlas of Saskatchewan* adds greatly to our understanding of Western Canada. The section on agriculture (pp. 217–58) is particularly pertinent to the *Key Topic* in this chapter. Maps, graphs, and text provide the reader with a detailed account of agriculture in Saskatchewan. For example, approximately half of Saskatchewan is classified as farmland. Of this land, natural pasture forms some 20 per cent while another 20 per cent is placed in summer fallow. Cropland, including that for hay and alfalfa, comprises the remaining 60 per cent. From 1966 to 1996, cropland increased from 18.6 million ha to 20 million ha. Spring wheat remains the dominant crop, grown on 30 per cent of the cropland in 1996, followed by barley at 9 per cent, durum wheat at 8 per cent, and canola at 7 per cent. Trends in agricultural production provide the reader with a clear picture of the agricultural shift underway in Saskatchewan. From 1966 to 1996, six major trends were identified: (1) spring wheat cropland declined 25.4 per cent, from 7.5 million ha to 5.6 million ha; (2) durum wheat cropland increased nearly five times, from 366,000 ha to 1.7 million ha; (3) barley nearly doubled in acreage, from 913,000 ha to 1.8 million ha; (4) canola increased threefold, from 296,000 ha to 1.6 million ha; (5) specialty crops, such as mustard, lentils, peas, and canary seed, jumped from 36,000 ha to 1 million ha; and (6) the livestock industry is expanding. The removal of the freight subsidy on grains in 1995 was a major impetus to expand hog production. Large-scale hog producers within Saskatchewan purchase feed grain from local farmers. Saskatchewan farmers account for 38 per cent of Canada's primary agricultural exports, 44 per cent of its cultivated land, and 21 per cent of its farms.

Overview and Objectives

Diamond Cove, Newfoundland (Victor Last, Geographical Visual Aids D060045)

Atlantic Canada, located on the eastern margins of the country, is far from the centres of power. Its dominant role in transatlantic trade faded long ago and, in the age of super-tankers, seems unlikely to return. As an 'old' resource hinterland, Atlantic Canada lags far behind the other geographic regions in economic growth. As well, it suffers from high out-migration, which slows its population growth. While the North American Free Trade Agreement has provided access to the larger New England market, Atlantic Canada has few resources suitable for export apart from oil, natural gas, hydroelectricity, and iron ore. The demise of the northern cod stocks has seriously hurt the region's econo-my, especially in Newfoundland, which has the highest unemployment rate in Canada. Although offshore oil and gas developments have quickened the region's economic pace and are having a ripple effect, other economic sectors remain mired in the 'old' resource economy.

- Describe Atlantic Canada's physical and historical geography.
- Present the basic characteristics of Atlantic Canada's population and economy from the perspective of an 'old' resource hinterland.
- Examine these two elements within the context of Atlantic Canada's physical setting, especially its fragmented geography.
- Explore Atlantic Canada's role as a resource hinterland within Canada and North America.
- Focus on the cod fishery as an example of the 'tragedy of the commons'.

Chapter 9 Atlantic Canada

■ Introduction

Stretching along the country's eastern coast, the region of Atlantic Canada consists of two parts: the Maritimes (Nova Scotia, Prince Edward Island, and New Brunswick) and Newfoundland/Labrador (Figure 1.2).[1] Despite the differences this physical separation has created between the Maritimes and Newfoundland/Labrador, Atlantic Canada is united by a rich sense of place that has grown out of the region's history and geographic location. The Atlantic Ocean has dominated the history of this region of Canada from the early days of the fishery to the 'Golden Age of Wooden Ships and Iron Men'. Today, Atlantic Canadians continue to rely heavily on the sea, fishing on the continental shelf and conducting energy exploitation below it. Most important of all, the sea provides a link between Atlantic Canada and countries on the Atlantic Rim. As in the region's early history, the Atlantic Ocean asserts itself in the affairs of Atlantic Canada. In this chapter, the *Key Topic* is the fishing industry.

Atlantic Canada within Canada

As Canada's oldest hinterland, Atlantic Canada has experienced both growth and decline over the years. Today, Atlantic Canada is, in Friedmann's regional version of the core/periphery model, a downward transitional region. It is a region troubled by past exploitation of its renewable resources and the exhaustion of its most accessible and richest non-renewable resources. Yet, Atlantic Canada received a second chance with the discovery of vast hydrocarbon deposits beneath its continental shelf. Newfoundland is benefiting from oil while Nova Scotia has natural gas.

As a resource hinterland Atlantic Canada has not been faring well. As the region enters the twenty-first century, the performance of the Atlantic economy falls well below the national average. Residents of Atlantic Canada have a strong sense of place and are proud of their land and heritage. At the same time, people are increasingly determined to overcome their position as poor cousins within the Canadian federation. In each province of Atlantic Canada, people are struggling to find better ways to manage the region's resources in a more sustainable fashion; to achieve a higher degree of processing of its non-renewable resources; and, at the same time, to expand its base of high-technology industries. Some changes are already taking place. Offshore oil and gas developments have stimulated the growth of highly specialized manufacturing firms in both Newfoundland and Nova Scotia, while high-technology firms are locating in the major cities of Atlantic Canada. Information technology is also growing in the Maritimes. For instance, Bathurst, Fredericton, Halifax, Moncton, Saint John, and St Stephen are telephone call centres for firms operating in Canada, such as Air Canada, AT&T Canada, Canadian Pacific, IBM, Northern Telecom, Purolator Courier, Royal Bank of Canada, and United Parcel Service. (Call centres enable customers to phone a 1–800 number to obtain the services that companies provide.) However, these information service jobs are at the low end of the labour market in terms of pay and skills and, unlike the innovative high-technology industries, provide limited spinoffs.

Atlantic Canada's economy continues to struggle. While Halifax leads the region in economic growth, most other communities are not faring so well and many rural communities are on the edge of collapse. This urban/rural split in economic well-being is due to several factors. First, jobs in the primary sector, such as logging and fishing, are disappearing as firms replace workers with more advanced machinery. Second, the collapse of the northern cod fishery has resulted in great hardships in fishing communities, especially the outports of Newfound-

land. Third, the need to share the fishing resource (and possibly other natural resources) with the Mi'kmaq will give the Indian communities a much needed boost but, as a zero-sum game, non-Indian communities will see some of their fishers leave this industry. Economic decline has created another problem—out-migration. With few prospects in sight, many in Atlantic Canada have been leaving the region to seek opportunities in other parts of the country. This makes the task of rebuilding the region's economy even more difficult.

One measure of Atlantic Canada's overall economic performance is reflected in the region's per capita gross domestic production figures and its level of unemployment. In 2001, Atlantic Canada's GDP per capita was the lowest in Canada, while the region's 2001 unemployment rate was the highest (Figure 9.1 and Table 9.1). The primary reasons for Atlantic Canada's weak economic performance include the following:

- The geographic division of Atlantic Canada into subregions adds to companies' transportation costs and discourages the emergence of an integrated economy in which economies of scale might occur.
- Atlantic Canada has been exploiting its resources for a long time and some of these, such as coal and iron, have been exhausted, while its renewable resources, such as the northern cod, have been overexploited.
- Atlantic Canada has a series of small internal markets.
- Atlantic Canada, but especially Newfoundland/Labrador, is far from Canada's major markets.

All these factors have made it extremely difficult for economic development to flourish

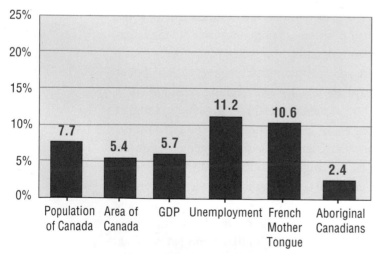

Figure 9.1 Atlantic Canada, 2001. Atlantic Canada has a weak economy. The region has 7.7 per cent of Canada's population but produces only 5.7 per cent of the country's GDP. The high rate of unemployment is another indication of its poor economic performance—it is the highest in the country. Acadians account for the high number of French-speaking Canadians in Atlantic Canada, which has the second highest percentage by region in Canada. Aboriginal peoples, however, form only 2.4 per cent of Atlantic Canada's population.

Sources: Tables 1.1, 1.5, and 4.19.

Table 9.1	Basic Statistics for Atlantic Canada by Province, 2001				
Province	Population (000s)	Population Density (per km2)	% GDP	Unemployment Rate	% Aboriginal
Prince Edward Island	135	23.8	0.3	11.9	0.1
Newfoundland/Labrador	513	1.4	1.9	16.1	1.9
New Brunswick	730	10.2	1.7	11.2	1.7
Nova Scotia	908	17.2	2.1	9.7	1.7
Atlantic Canada	2,286	4.9	6.0	11.2	5.4
Canada	30,007	3.3	100.0	7.7	100.0

Sources: Statistics Canada (2000a, 2000b, 2003c).

in the region. Furthermore, Atlantic Canada has, over time, become heavily dependent on Ottawa for economic support through equalization payments and social programs. For both economic and political reasons, this cycle of dependence seems unlikely to end.

Atlantic Canada's Physical Geography

Atlantic Canada is not a homogeneous region but 'a region of geographical fragmentation' (Macpherson, 1972: xi). It consists of three subregions: the Maritimes, Newfoundland, and Labrador. Macpherson argued that this physical division has impacted on all aspects of life in Atlantic Canada, hindering the emergence of a common political will, a common regional consciousness, and an economic union among the four Atlantic provinces. Whether or not an economic union is a desirable political arrangement for Atlantic Canada is uncertain, though the topic of union has long been debated.

Each subregion has its distinct set of physical characteristics. In this section, climate and physiography are emphasized. The climate in

the Atlantic Canada is far from homogeneous. As shown in Figure 2.6, Atlantic Canada has three climatic zones—Atlantic, Subarctic, and Arctic zones. Why do three zones exist in Atlantic Canada? The great north–south extent of this region is one reason. For example, the distance from the southern tip of Nova Scotia (44° N) to the northern extremity of Newfoundland/Labrador (60° N) is over 2,000 km. In addition, the Labrador Current chills the Labrador coastline, creating an arctic climate, while inland a subarctic climate is found. Milder air masses affect the Atlantic climatic zone, including the hot, moist air masses from the Gulf of Mexico.

The general atmospheric circulation system brings several air masses to the Maritimes and Newfoundland, where they interact and result in unsettled weather conditions. The main air masses originate in the interior of North America, the Gulf of Mexico, and the North Atlantic Ocean. Consequently, summers are usually cool and wet, while winters are short and mild but often associated with heavy snow and rainfall. Most precipitation falls in the winter. Temperature differences between inland and coastal locations are striking.

Coastal locations are usually several degrees warmer in the winter than inland locations. During the summer, the reverse is true, with coastal areas usually several degrees cooler than inland areas.

Along the narrow coastal zone of Atlantic Canada, the climate is strongly influenced by the Atlantic Ocean. The summer temperatures of coastal settlements along the shores of Newfoundland are markedly cooled by the very cold water of the Labrador Current (a cold ocean current in the North Atlantic Ocean), which brings with it icebergs. Another effect of the sea occurs in the spring and summer—fog and mist result when the warm Gulf Stream waters (which originate in the Gulf of Mexico) mix with the cold Labrador Current. In the winter, the clash of warm and cold air masses sometimes results in severe winter storms characterized by heavy snowfall. For example, St John's had a record 648.2 cm of snow (almost 22 feet) in the winter of 2000–1 (see Vignette 9.1). Such storms are usually short-lived and more normal weather conditions return quickly. Annual precipitation is abundant throughout Atlantic Canada. It averages around 100 cm in New Brunswick and 140 cm in Newfoundland.

Atlantic Canada is in two of Canada's physiographic regions: the Maritimes and Newfoundland are part of the Appalachian Uplands, and Labrador is part of the Canadian

Vignette 9.1 Weather in St John's

St John's has one of the most varied but mild climates in Canada. Weather is characterized by foggy days, overcast skies, and stormy, wet weather. Throughout the year, a salty smell is in the air. While most of the year has mild temperatures, strong northwest winds result in heavy winter snowfalls. In the winter of 2000–1, St John's had a record amount of snow—just under 6.5 metres! Jutting into the Atlantic Ocean at the eastern tip of the island of Newfoundland, the marine climate of St John's is affected by the westerly circulation of the atmosphere and by the waters of the North Atlantic. As David Phillips (1993: 155), Canada's well-known weather pundit, put it, 'Of all major Canadian cities, St John's is the foggiest, snowiest, wettest, windiest, and cloudiest.' These weather characteristics are due to four major climatic controls:

- The proximity of St John's to the Atlantic Ocean, which keeps summer temperatures cool and winter temperatures mild. Its marine location also brings abundant precipitation, with just over 1,500 mm falling each year.

- The westerly flow of continental air masses brings warm weather in the summer and cold weather in the winter. These continental air masses intersect with marine air masses, resulting in highly unstable weather.

- The cold waters of the Labrador Current bring a taste of Arctic weather by filling St John's harbour with pack ice each winter. Icebergs are often sighted just off shore sometimes as late as June.

- The warm, moisture-laden air masses from the Gulf of Mexico make their way to St John's and, by mixing with the cold air just above the waters of North Atlantic, result in thick, cool fog. Fog is common from April to September.

Newfoundland still has hundreds of small outports along its rocky and indented coastline. Since the loss of the cod fishery, most outports no longer have an economic base and their future is in doubt. (Al Harvey/The Slide Farm)

Shield. Within the Appalachian Uplands, the region's landforms are dominated by a rugged coastline and an upland interior consisting of hilly to mountainous terrain. In the Canadian Shield, where arable land is even more scarce than in the Appalachian Uplands, the geography consists of a coastline with fjords, a rugged interior, and the lofty Torngat Mountains.

Atlantic Canada's physiography lends itself to a resource hinterland. The region has little arable land, except on Prince Edward Island, in the Saint John River Valley, and in the Annapolis Valley (Vignette 9.1). Beyond these few favoured lowlands, the land is too rough and rugged for agricultural settlement. As a consequence, most settlements in Atlantic Canada are clustered along the coastline, where for years fishers have made their living harvesting the riches of the ocean. Further inland, settlements evolved where mineral and forest resources were profitable. For instance, in the Appalachian Uplands, a dense forest

bolstered the economies of the Maritime provinces, especially New Brunswick. Similarly, mining has long been a way of life for some Atlantic Canadians—from the famous coal mines of Cape Breton Island, Nova Scotia, to the recent discovery of a rich nickel-copper-cobalt deposit at Voisey's Bay, Labrador. Furthermore, the region has long benefited from its extensive continental shelf, where fishers pursued codfish and recent discoveries of petroleum and natural gas have led to the development of major oil-field sites, such as Hibernia on the Grand Banks.

The natural division of Atlantic Canada has led to different settlement patterns and a diverse cultural geography. The settlement of Atlantic Canada began the process of cultural diversity. For instance, Nova Scotia was settled first by the Acadians, who were later expelled and replaced by the New England Planters; German Protestants arrived at Lunenburg; the Loyalists, including about 3,000 black refugees,

Vignette 9.2   The Annapolis valley

The Annapolis Valley is a low-lying area in Nova Scotia. At its western and eastern edges the land is at sea level, but it rises to about 35 m in the centre. The area is surrounded by a rugged, rocky upland that reaches heights of 200 m and more. Beneath the valley's fertile soils are soft sedimentary rocks. The sandy soils of the Annapolis Lowlands originate from marine deposits that settled there about 13,000 years ago. After glacial ice retreated from the area, sea waters flooded the land, depositing marine sediments that consisted of minute sand and clay particles. Isostatic rebound then caused the land to lift and slowly these lowlands emerged from the sea. In the seventeenth century, the favourable soil of the Annapolis Valley attracted early French settlers, who later became known as Acadians. Today, the Annapolis Valley's stone-free, well-drained soils and its gently rolling landforms provide the best agricultural lands in Nova Scotia.

settled in Halifax and other parts of Nova Scotia; and Gaelic-speaking Highlands Scots, both Catholic and Protestant, made Cape Breton and the Northumberland shore their home. The British-run merchant fishery saw many English and Irish workers settle along the coast of Newfoundland. In addition to these different settlement patterns, the physical obstacles between the subregions and the separation of the region from the rest of Canada fostered and maintained cultural differences:

- The Maritimes, though a relatively cohesive economic unit, has closer ties to New England than to the rest of Canada.
- The island of Newfoundland stands alone, separated from the rest of Atlantic Canada by the Cabot Strait and from Labrador by the Strait of Belle Isle. These water barriers hamper Newfoundland from having close economic links with the Maritimes and Labrador.
- Labrador, while part of the province of Newfoundland, has been drawn into Québec's economic orbit because of the land connection between Québec and Labrador. Examples include: (1) the trans-mission of hydroelectric energy to markets across southern Québec; (2) the shipment of Labrador's iron ore to the port of Sept-Îles, Québec; and (3) the Labrador–Québec highway that runs from Labrador's largest town, Happy Valley–Goose Bay, to Baie-Comeau and Québec's provincial highway system.

Despite the differences between Atlantic Canada's provinces, many similarities exist. In addition to a common geographic location—the waters of the Atlantic Ocean wash up on the shores of all four provinces—the Atlantic provinces have all undergone a similar economic process of growth, decline, and dependence. As the region of Canada where Europeans first settled, Atlantic Canada has undergone a very long period of resource exploitation. Its role as a resource hinterland brought early prosperity, which was followed by decline as its non-renewable resources were exhausted and some of its renewable resources were overexploited. This economic malaise drew Ottawa into the affairs of Atlantic Canada, often at the request of the Atlantic provinces, which turned to Ottawa for financial assistance in times of crisis. Ottawa continues to have a strong presence in the region and, some say, an unduly strong influence. Ottawa's role in the management of the Atlantic fisheries, for example, is particularly resented. In 2003, the Newfoundland/Labrador government sought a role with Ottawa in the management of the Atlantic fishery. Yet, Atlantic Canada relies on the support provided by Ottawa and therefore its demands are often ignored by the federal government. The challenge is to break the cycle of dependence by revitalizing Atlantic Canada's economy.

Environmental Challenges

Without a doubt, the major environmental disaster to strike Atlantic Canada was the collapse of the cod fishery. But this disaster was due more to technological advances that enable much larger catches, coupled with federal mismanagement, than to natural factors such as the warming of the Atlantic waters or the expanded seal population consuming vast quantities of cod. Clearly a case of a 'tragedy of the commons', this ecological collapse of one of the world's greatest fish stocks is documented in the latter pages of this chapter. Here, our attention is focused on an unwanted remnant of Cape Breton's glory days as a major iron and steel centre in Canada—the so-called tar ponds.

By the end of the twentieth century, Sydney, Nova Scotia, not only lost its industrial sector but the community was saddled with an environmental disaster—the Sydney Tar Ponds. But what are the tar ponds and how did this industrial disaster occur? The tar ponds are composed of tar and a wide variety of chemicals that are dangerous to human beings. Arsenic, lead, and other chemicals are found in the tar ponds. These toxic wastes came from the operation of the Sydney Steel Co., or Sysco, especially its coke ovens. A coke oven is a large chamber where coal is heated at extremely high temperatures. Waste products from coke ovens include tar and toxic gases. These toxic wastes included benzene, kerosene, and naphthalene.

In the case of Sysco, this effluence was discharged into a nearby stream and gradually seeped into Muggah Creek. Since the 1980s, residents living near the tar ponds complained of an orange goo seeping into their basements, while others found that dust from the tar ponds was blowing into their yards and houses. When Health Canada stated that those living closest to the tar ponds were in greater danger than those living further away, Ottawa and Halifax began to pay attention. Action began after a cancer specialist concluded that there was a higher risk of dying from cancer in Whitney Pier and Ashby, communities closest to the tar ponds, than the national average. At that point, the federal and provincial governments joined forces with the local government to fund a cleanup of the tar ponds.

The Sydney Tar Ponds are the site of the biggest environmental cleanup project in Canadian history. In 1998, Ottawa and Halifax agreed to a $62 million cost-sharing agreement to clean up the tar ponds and the site of the coke ovens. Since 2001, the tar ponds, coke ovens site, and other areas within the Muggah Creek watershed have been the scene of intense activity to repair the environmental damage. Since the cleanup began, the community is seeing improvements on a number of fronts: the construction of a sewer interceptor that will divert the tonnes of raw sewage that now flows daily into the tar ponds; the demolition and removal of derelict structures on the coke ovens site; and the closure and capping of the old Sydney landfill. Still, much remains to be done.

Atlantic Canada's Historical Geography

Atlantic Canada was the first part of North America to be discovered by Europeans. While the Vikings arrived earlier, the first documented discovery of land in North America was made by John Cabot. The Italian navigator, employed by the king of England, reached the rocky shores of Atlantic Canada (the exact location, either Cape Bonavista, Newfoundland, or Cape Breton Island, Nova Scotia, is in dispute) on 24 June 1497. Like Columbus, Cabot was searching for a sea route to Asia. In England, Cabot's report of the abundance of groundfish—cod, grey sole, flounder, redfish, and turbot—in the waters off Newfoundland lured European fishers to make the perilous voyage across the Atlantic to these rich fishing grounds. Canada's Atlantic coast quickly became a popular area for European fishers and though landings on shore took place—for drying the fish and establishing temporary habitation during the fishing season—permanent settlements were slow to take hold in this part of North America.

The first permanent settlement in North America north of Florida was a cluster of French settlers at Port Royal on the Bay of Fundy in 1605. French settlement then spread into the Annapolis Valley and other lowlands in the Maritimes. The expanse of French-settled lands that evolved during the seventeenth and part of the eighteenth centuries became known as Acadia. Acadia consisted of a collection of French settlements united by culture, language, and a common economy.

While French settlers were populating the Maritimes, English settlers began arriving in 1610 at Conception Bay, Newfoundland. Over the next 100 years, English, French, and Irish settlements developed because fishers pre-

ferred to remain in Newfoundland at the close of the fishing season rather than make the long journey back to Britain. By the 1750s, there were over 7,000 permanent residents living in hundreds of small fishing communities along the shores of the island of Newfoundland. The French Shore, for example, is a coastal area of southwest Newfoundland where French fishers enjoyed treaty rights to fish. These rights were granted by the British in 1713 and did not end until 1904. Today, many communities along the French Shore contain descendants of these French settlers.

During the first half of the eighteenth century, war between the two European colonial powers in North America—England and France—was almost continuous. During that time, the French forged an alliance with the Mi'kmaq and Maliseet, drawing them into the conflict with the English and their Iroquois allies.[2] Under the terms of the 1713 Treaty of Utrecht, France surrendered Acadia to the British. However, many French-speaking settlers (known as Acadians) remained in this newly won British territory, which was renamed Nova Scotia. During the previous century, the Acadians had established a strong presence in the Maritimes with settlements and forts. Most Acadians lived along the Bay of Fundy coast, tilling the soil. Until the mid-1700s, Britain made little effort to colonize these lands, leaving the Acadians to till the land in this British-held territory. By 1750, French-speaking Acadians numbered over 12,000.

British settlement began to spread to the Maritimes in the mid-eighteenth century, with the first serious attempt occurring in 1749 when the British government began to recruit settlers for its newly acquired lands in Nova Scotia. Halifax, founded in 1749, would become the centre of British influence and military power in Nova Scotia and would provide

Vignette 9.3   Strangers in Their Own Land

Four historic events changed the Mi'kmaq into strangers in their own land. First, the defeat in 1713 of their European ally, the French, made the Mi'kmaq vulnerable to the British. Second, the expulsion of the Acadians in 1755 eliminated another ally. Third, following the founding of Halifax in 1749, relations between the British and the Mi'kmaq deteriorated. British rangers were unleashed to harass the Mi'kmaq, to destroy their villages, and to drive them far beyond the British settlements. Fourth, by 1783 the British Loyalists from the new republic of the United States relocated to Nova Scotia and New Brunswick, where they occupied the fertile lands of the recently departed Acadians and the prime hunting lands and fishing areas of the Mi'kmaq. No longer regarded as a military threat, the Mi'kmaq were pushed into more marginal lands and were surrounded by a growing British population. By then, the ancient world of the Mi'kmaq had collapsed. Denied a place in the new economy, the Mi'kmaq were forced into a state of dependence in which they soon became outcasts in their own land.

In 1999, the Supreme Court of Canada ruled that the Mi'kmaq should have a share of the commercial fishery (See *Further Readings* for a summary of the *Marshall* decision). Five years later, the Mi'kmaq had gained access to about 3 per cent of the federal lobster allocation by receiving about 320 commercial fishing licences. When operating at full capacity, these licences should allow the Mi'kmaq fishers to generate around $25 million worth of lobster each year. (CP/Jaques Boissinot)

a counterbalance to the well-fortified French military base of Louisbourg on Cape Breton Island. Through a series of wars culminating with the Seven Years' War (1756–63), the British would gradually gain control of Atlantic Canada.

In 1756, the final struggle for North America began. Britain and France engaged in an all-out battle for supremacy around the world. In North America, the British commanders began preparing for war in 1755. When the Acadians refused to swear an oath of unconditional loyalty to the British Crown, the British deported 6,000 Acadians to more secure parts

of the British colonies in North America. After the British defeated the French, all French territories in North America were under British control, ending the dream of a French North America. Under the terms of the Treaty of Paris in 1763, France ceded all its territories to Britain, except for the islands of Saint-Pierre and Miquelon near the southern shore of Newfoundland. Today, most Acadians live in northern New Brunswick, where they constitute just under 40 per cent of the province's population.

The next event to influence the evolution of Atlantic Canada was the American Revolution.

Following victory by the American colonies, approximately 40,000 Loyalists made their way to Nova Scotia and New Brunswick. (See 'The Loyalists' in Chapter 3 for more details.) Most settled along the shores of the Bay of Fundy, in the Annapolis Valley, on Cape Breton Island, and around the British stronghold of Halifax. Halifax became known as the 'Warden of the North' because it provided a superb port for the ships of the British Navy. Over the next 50 years, the arable land was cleared and occupied. During that time, more British settlers came to the Maritimes. Nova Scotia alone received 55,000 Scots, Irish, English, and Welsh. Most Scots went to Cape Breton and the Northumberland shore. While many newcomers tried their hand at farming, the fishery still dominated life in the Maritimes and Newfoundland. Through the years, until the end of the nineteenth century, Atlantic Canada changed rapidly: the population increased, towns grew, and the economy evolved from one based on subsistence production to one emphasizing commercial production.

In the early nineteenth century, the harvesting of Atlantic Canada's natural wealth increased. This frontier hinterland of the British Empire exploited its rich natural resources—the cod off the Newfoundland coast and the virgin forests in the Maritimes—and became heavily dependent on transatlantic trade of these resources. Overseas trade with other British colonies in the Caribbean, as well as with Britain, formed the cornerstone of Atlantic Canada's trade and prosperous economy. In the first half of the nineteenth century, trade with New England was relatively limited because both the Maritimes and New England had almost identical resource-oriented economies. However, after the American Civil War, New England would industrialize, leading to greater trade between the two regions.

In addition, Britain's move to free trade in 1849 would mean the loss of Atlantic Canada's protected markets for its primary products, resulting in even greater interest in the American market, especially in the Maritimes.

Historic Head Start

With the fishery and the timber trade well underway in the early nineteenth century, Atlantic Canada had a head start in developing a commercial economy. Furthermore, the availability of timber and the region's favourable seaside location provided the ideal conditions for shipbuilding in Atlantic Canada. By 1840, Atlantic Canada entered the 'Golden Age of Sail', with Nova Scotia and New Brunswick becoming the leading shipbuilding centres in the British Empire. Exports from Atlantic Canada were primarily cod and timber, while imports were manufactured goods from England and sugar and rum from the British West Indies. Several world events in the mid-nineteenth century added strength to this economic boom in the Maritimes. Demand for its exports rose, especially for timber, which was used to build British merchant ships and warships. The Crimean War (1853–6), the Reciprocity Treaty with the United States (1854–66), and the American Civil War (1861–5) opened new markets for fish, minerals, and lumber from the Maritimes and cod from Newfoundland. International trade was increasing and Atlantic Canada was looking outward to the rest of the world, while the Province of Canada was more consumed with internal developmental issues at that time.

Just before Confederation, however, external events dampened Atlantic Canada's resource-based economy. Iron was replacing wood in shipbuilding; the three-way trade with Britain and its Caribbean possessions collapsed, largely because a world glut of sugar caused prices to drop; and the end of the Reciprocity Treaty in 1866 cut off access to the Maritimes' natural trading partner, New England. These events resulted in the deterioration of Atlantic Canada's economic position among the Atlantic trading nations and colonies.

Confederation

The provinces of Atlantic Canada joined Canada at different times. Nova Scotia and New Brunswick joined at the time of Confederation (1867); Prince Edward Island followed in 1873; while Newfoundland did not join Canada until 1949. At the time of Confederation, Newfoundlanders, being so isolated, saw little advantage to joining Canada. Then, too, Newfoundlanders cherished their independence and they feared that Newfoundland would be dominated by the larger provinces. Besides, Newfoundland's trade relations at the time were tied to Britain, not Canada. For the Maritimes, the decision to join Canada was not an easy one. In spite of the Maritime provinces' historic head start and their excellent access to sea and world markets, they were on the margins of the new country and their once-booming economy lost its momentum.

Confederation favoured economic growth in Central Canada. The federal government's National Policy of 1879 reinforced Central Canada's advantage, placing the Maritimes' fledgling manufacturing base in danger. In order to stimulate Canadian manufacturing, Canada imposed duties on imported goods. In retaliation, Americans increased their trade barriers on Canadian manufactured goods. Canada even placed a tariff on foreign-produced coal to protect the high-cost coal mines of Nova Scotia. In effect, Canada had created a 'closed' economy that facilitated the industrial-

ization of Central Canada by delegating the rest of Canada to the dual roles of domestic market for Canadian-manufactured goods and resource hinterland for Central Canada, the United States, and the rest of the world (see Chapters 1, 3, and 4).

At Confederation, the Maritimes did have a fledgling manufacturing base, but local markets were too small and the Maritimes' natural market, New England, was now less accessible because of high US tariffs. What Atlantic Canada needed was access to the market of Central Canada. Ottawa's answer was the Intercolonial Railway (completed in 1876). Built to fulfill the terms of Confederation, the Intercolonial provided rail access to Central Canada. The railway was never a commercial success, but it did stimulate economic growth in the Maritimes. The Intercolonial was operated by the federal government: freight rates were kept low in order to promote trade, and substantial annual deficits were paid by Ottawa. With low-cost transportation to the national market and the possibility of firms achieving economies of scale, a general economic surge occurred in the Maritimes. Cotton mills, sugar refineries, rope works, and iron and steel manufacturing plants were established or expanded to serve the much larger national market and to grab a share of the expanding western market. The railway boom in the Canadian Prairies had a positive impact on Nova Scotia in the form of heavy investment in the province's steel mills. With the construction of new railway lines, steel production was focused on manufacturing steel rails and locomotives. By taking advantage of Cape Breton's coalfields and iron ore from Bell Island, Newfoundland, the steel industry in Sydney prospered, accelerating Nova Scotia's economic growth well above the national average.

The Maritime economy, however, suffered a deadly blow when, in 1919, freight rates were increased to levels common in Central Canada. Immediate access to the national market became more difficult and, with the loss of sales, many firms had to lay off workers, while others were forced to shut down their operations. Even before these troubled times, the Maritimes' economy was unable to absorb its entire workforce, leading many to migrate to the industrial towns of New England and southern Ontario. After World War I, the out-migration increased.

After World War II, the Maritimes grew but at a slower pace than other parts of Canada. Newfoundland joined Canada in 1949, mainly for economic reasons. Newfoundland stood to gain from the social programs available to Canadians—for instance, unemployment insurance assisted seasonal workers like fishers and family allowance payments provided a monthly cheque to each mother. Above all, Newfoundlanders hoped that Canada's booming economy would pull Newfoundland from its poverty and its dependence on the fishery. Even during periods of national affluence—for instance, the 1950s and 1960s—Atlantic Canada experienced limited economic growth; even numerous federal loans and grants to firms in Atlantic Canada made little difference. No doubt this region languished because of its scattered and relatively small population, its narrow resource base, and the high cost of transporting goods to the national market. In 1968, Professor David Erskine at the University of Ottawa drew a rather dismal picture of Atlantic Canada as a hinterland:

> The region is, in the Canadian context, one of 'effort' rather than of 'increment'. Small scale resources once encouraged small scale development, but only large scale resources encourage modernization. The small scale and lack of concentration of its resources makes the region one in which government investment is easily dispersed without bringing about growth. Low levels of professional services and low levels of education result from the high taxes from low incomes; thus, still further retardation of economic growth occurs. (Erskine, 1968: 233)

Until the Free Trade Agreement, Atlantic Canada's limited access to external markets hindered its economic development. Manufacturing firms in Atlantic Canada have had to overcome high transportation costs to deliver their products to Canada's major markets in Central Canada. Competition from firms in Central Canada often drove Atlantic-based firms from the marketplace for two reasons:

1. Manufacturing firms in Ontario and Québec could achieve economies of scale, making their production costs lower and resulting in lower prices for consumers.
2. Manufacturing firms in Ontario and Québec had much lower transportation costs in reaching their customers than similar firms in Atlantic Canada.

Unable to compete within the national marketplace, most manufacturers in Atlantic Canada had to limit their production to the local market and keep wages low in order to compete with imported products from Central Canada.

Atlantic Canada Today

Speaking of Atlantic Canada as a whole belies the geographic differences among its four provinces: New Brunswick, Nova Scotia, Prince Edward Island, and Newfoundland.

One difference lies in history. For example, Newfoundland joined Confederation over 80 years later than the Maritime provinces.

Another difference is related to physical geography. Newfoundland and Labrador, for instance, were not well endowed by nature. Both have climates less conducive to agriculture than the Maritimes, partly because both Newfoundland and Labrador are located much farther north and because of the chilling effect of the Labrador Current. Added to this climatic disadvantage, neither Newfoundland nor Labrador has much arable land. Instead, they consist mainly of rocky landforms. Also, Newfoundland, an irregularly shaped island jutting into the North Atlantic, makes access to and from Canada more difficult. With little fertile land and limited timber stands, Newfoundland and Labrador's inhabitants remained dependent on the sea for their livelihood.

In contrast, the Maritime provinces are a comparatively more compact unit with rail and road links to Central Canada and New England. The Maritimes also rely less on the sea than Newfoundland because they have a milder climate that allows for agriculture and a more varied resource base. Prince Edward Island, blessed with gently rolling farmland and red, fertile soils, has an agricultural base as well as good fishing grounds. In Nova Scotia and New Brunswick, agriculture, forestry, and tourism play a strong role in their economies. The Maritimes' proximity to New England also attracts American tourists more than does the more complicated automobile route and ferry crossing to Newfoundland. In fact, the importance of land links for tourism is made evident by the completion of Confederation Bridge from New Brunswick to Prince Edward Island, which resulted in a sharp increase in the number of out-of-province vehicles and visitors to this island.

Economic variations also exist in Atlantic Canada, and many of these are related to the

The Confederation Bridge, completed in 1997, connects Prince Edward Island with New Brunswick. As an integral part of the Trans-Canada Highway network, the 12.9 km Confederation Bridge is the longest bridge over ice-covered waters in the world. (Ron Garnett/AirScapes)

provinces' different geographies and resource bases. Nova Scotia and New Brunswick, with their more diversified resources, have the strongest economies, while Newfoundland, with its heavy reliance on the seasonal and declining fisheries, has by far the weakest economy (though the offshore oil boom has sparked Newfoundland's economy). Prince Edward Island trails behind New Brunswick. These economic differences are illustrated in the region's unemployment rates, with Newfoundland having the highest rate at 16.1 per cent in 2001 and Nova Scotia having the lowest rate at 9.7 per cent (Table 9.1). Nova Scotia is poised to expand its economic base with natural gas developments and the growth of high-technology firms in the Halifax area.

Population

Since Confederation, Atlantic Canada's population has increased very slowly, well below the national average. In recent years, its population growth has stalled and then declined. From 1996 to 2001, Atlantic Canada experienced a population decline of 2.1 per cent, which amounted to a loss of 48,000 residents. The population decline varied among the four provinces, largely reflecting differences in economic conditions. From 1996 to 2001, only

Prince Edward Island did not suffer a population decline. Newfoundland had the largest drop—7 per cent—followed by New Brunswick at 1.2 per cent and Nova Scotia at 0.1 per cent (Table 9.2). The implication of this population loss on the regional economy is quite serious. For one reason, the loss leads to smaller markets within Atlantic Canada and smaller markets will affect local businesses. A second reason is that the large number of people leaving Atlantic Canada to search for jobs in more prosperous regions of Canada, especially southern Ontario, Alberta, and British Columbia, represents a loss of the younger, more ambitious members of the labour force. A third reason lies in the psychological cost associated with a declining population. Dark humour touched with a degree of fatalism epitomizes the situation as 'Will the last person to leave, please turn off the lights.'

Since the 1920s, the economic circumstances in Atlantic Canada have been on a downward cycle, driving hundreds of thousands to seek their fortunes elsewhere. Over the years, Atlantic Canada has lost more people through out-migration than it has gained through in-migration. In fact, few immigrants to Canada choose to settle in Atlantic Canada because they see few opportunities. The resulting loss of people, while boosting economies

Table 9.2	Population Change in Atlantic Canada, 1996–2001			
Province	Population 1996	Population 2001	Difference	% Difference
Prince Edward Island	134,557	135,294	737	0.5
Newfoundland/Labrador	551,792	512,930	(38,862)	(7.0)
New Brunswick	738,133	729,498	(8,635)	(1.2)
Nova Scotia	909,282	908,007	(1,275)	(0.1)
Atlantic Canada	2,333,764	2,285,729	(48,035)	(2.1)

Source: Statistics Canada (2002b).

elsewhere, has hindered economic growth in Atlantic Canada. An indication of the serious nature of this net outflow is shown in the proportion of Canadians living in Atlantic Canada. In 1949, when Newfoundland joined Canada, Atlantic Canada constituted 11.6 per cent of the nation's population. By 2001, the region's population had fallen to 7.7 per cent of the national total.

Until the United States placed restrictions on the movement of Canadians into the United States in the 1930s, New England was the favourite destination of migrants from the Maritimes. Many went to Boston. After the 1930s, most Maritime migrants settled in Ontario. When Newfoundland joined Confederation in 1949, its migrants went mainly to Ontario. From 1986 to 2001, about 200,000 people left Atlantic Canada for Ontario, Alberta, and British Columbia. During that 15-year period, Ontario was the destination of just over half of the migrants from Newfoundland, nearly half from Nova Scotia, about 40 per cent from New Brunswick, and one-third from Prince Edward Island. Within Ontario, Toronto attracts most of the migrants from Atlantic Canada.

Dispersed Population

Atlantic Canada's population of over 2 million is highly dispersed along its coastline in both small and large centres. Geography explains why this settlement pattern developed. The original selection of sites for fishing villages, particularly in Newfoundland and Nova Scotia, was largely determined by small, sheltered harbours suitable for mooring boats and affording quick access to good fishing grounds. While many of these coastal villages still remain, they are anachronisms in the twenty-first century. Delivering basic public services, such as education and health care, in

these coastal places is expensive because populations are small and are located in isolated, hard-to-reach areas.

From an economic perspective, this dispersed pattern of population distribution is a major weakness for three reasons:

1. The highways and rail lines required to connect the urban centres in Atlantic Canada are more extensive than in regions with a more compact population geography, and are therefore more expensive to build and maintain.
2. High construction and maintenance costs for transportation networks result in higher transportation costs for firms operating in Atlantic Canada, thus discouraging companies from investing and locating in the region.
3. Because of the absence of a large population/market, firms are hampered from achieving economies of scale and **economies of association** (having parts and service firms close to the manufacturing plant).

The rural-to-urban migration, common to other regions of Canada, is occurring in Atlantic Canada. Migration of younger members of outport families is accelerating, while older members may prefer to live in the outport where the cost of living is much lower than that in a regional centre. For the province, the cost of supplying public services is extremely high. Yet 'planned' relocation is fraught with two dangers: social disruption and economic transfer of rural poverty to urban poverty. The so-called remedy for isolated communities in Newfoundland was the 'planned' resettlement program.

How did Atlantic Canada develop such a dispersed coastal pattern of settlement? Back in the seventeenth and eighteenth centuries, when the first British and French families arrived in Newfoundland, the newcomers formed isolated coastal fishing outports (often with only three or four families each). As isolated self-sufficient outports, these communities were ideal for a small-scale coastal fishery. By 1945, nearly 1,400 tiny outports with populations under 200 had been established along the coast of Newfoundland (Staveley, 1987: 257). But by 2001, there were less than 400 outports with populations under 200. This decline was the result of two forms of relocation—voluntary and government-assisted. In short, people, particularly young people, no longer wanted to live in these tiny places where life revolved around a marginal fishery. Outport living could not survive a depressed fishing industry and the growing desire/need for urban amenities. The abandonment of outport villages began in 1946 and by 1954, 49 small communities were abandoned without government assistance.

Vignette 9.4   The End of Great Harbour Deep

For 400 years, people fished cod in the waters off Great Harbour Deep. Located on the Northern Peninsula, Great Harbour Deep was one of the most isolated communities in Newfoundland. Access to the outside world was by a three-hour boat ride in the summer and by a ski-equipped aircraft in the winter. While residents cherished their independence, the need for outside amenities, including health services, had become more and more pressing. In the 1990s, the loss of the inshore fishery and the closure of the town's fish plant led to the economic collapse of this outport. Some left, while others worked on the large trawlers for months at a time. By 2003, the population had dropped to 180 residents. After a series of public meetings, the community decided to accept the government's $5 million package and leave their community for good. In the summer of 2003, the 50 families of Great Harbour Deep left their community, leaving behind memories and empty homes but hoping to start a new life in a new community.

(CP/Mike McDonald)

After Confederation, both Ottawa and St John's became concerned about this unusual rural-to-urban migration. First, the Newfoundland government made financial help available under its Centralization Program. Under this scheme, all households in the community had to agree to relocate for public assistance to be made available. From 1955 to 1966, 110 communities were abandoned and the residents resettled in larger centres of their choice. In 1967, the federal government joined forces with Newfoundland. The federal-provincial Newfoundland Resettlement Program added new conditions to relocation. Migrants had to relocate to one of 77 designated 'growth centres'—selected communities that offer more social and economic opportunities. Another stipulation was that at least 75 per cent of the households must agree to move for them to receive public relocation funds. Under the joint program, another 150 communities were abandoned from 1967 to 1975. After this program ended in 1976, few communities were abandoned until the late 1990s when the cod fishery was closed. Following the 1992 cod moratorium, many communities saw their fish plants close, causing many to seek work elsewhere. Such was the fate of Great Harbour Deep, a tiny outport of 180 inhabitants on the Northern Peninsula that was founded some 400 years ago. In April 2003, the residents had accepted the provincial government's offer of financial help if all agreed to leave their community. Each household received up to $100,000 to help cover the costs of their relocating (Vignette 9.4).

Access to External Markets

The FTA and freer trade of the GATT/WTO regime greatly improved Atlantic Canada's access to foreign markets. Sales of fish, pulp, natural gas, and potatoes to consumers in New England rose sharply. In addition, a low Canadian dollar and an expanding American economy during the 1990s created ideal conditions for an increase in exports to the United States and in the number of US tourists visiting Atlantic Canada. At the same time, New England could export its manufactured goods to Atlantic Canada, thereby undercutting local manufacturers. By 2004, the US economy had recovered from a slump, but the rising Canadian dollar made exporting difficult.

Provincial Barriers to Trade

In addition to the obstacles that Atlantic Canada has faced in trading with external markets—Canadian and foreign—the region has also grappled with trade barriers between its provinces. Provincial barriers divide the region's population into four provincial markets, often impeding interprovincial trade and labour mobility. The purchasing policies of provincial governments and their various agencies frequently favour local producers and act as non-tariff barriers to producers from other provinces. Other barriers take the form of provincial marketing boards, provincially controlled liquor sales, and provincial-preference hiring practices in the construction industry. Such internal barriers are a major impediment to stimulating economic growth within Atlantic Canada. The very division of the region into four provinces results in higher administrative costs and thus higher tax rates, which exert an equally negative impact on economic development. Simply having four provincial governments for a relatively small population costs taxpayers in Atlantic Canada about $400 million annually, making them bear a higher tax burden than residents of other provinces.

From time to time, there has been talk of an Atlantic economic union or the creation of a single province to overcome this regional fragmentation. Such an amalgamation would not overcome the differences within the region's physical geography, but it might consolidate Atlantic Canada's economic and political power. It might even reduce the cost of government, but it would raise a host of issues including the thorny question of a capital city for Atlantic Canada. As early as 1949, Joey Smallwood, Premier of Newfoundland, proposed a union of the four eastern provinces to enable them to increase their bargaining power in Ottawa. In the 1970s, the Deutsch Commission on Maritime Union recommended a 'full political union as a definite goal'. However, there was no support within Atlantic Canada, and before the 1970s had ended Nova Scotia Premier Gerald Regan pronounced the proposal for a Maritime Union 'as dead as a doornail'. For the residents of Atlantic Canada, the advantages of a union are not readily apparent, and Atlantic premiers have not shown much enthusiasm for this proposal.

How much the taxpayers of Atlantic Canada would save with a single government is difficult to estimate, but political union could achieve considerable savings by eliminating

Table 9.3	Employment by Industrial Sector in Atlantic Canada, 2002			
Industrial Sector	AC Workers (000s)	AC Workers (%)	Ontario Workers (%)	Per cent Difference
Primary	**59.0**	**5.6**	**1.8**	**3.8**
Agriculture	17.1	1.6	1.3	0.3
Secondary	**178.8**	**16.9**	**25.1**	**(8.2)**
Manufacturing	108.8	10.3	18.5	(8.2)
Tertiary	**817.1**	**77.5**	**73.1**	**4.4**
Trade	178.2	16.9	15.7	1.2
Transport	53.5	5.1	4.7	0.4
Finance	52.2	4.9	6.6	(1.7)
Professional services	43.9	4.2	7.2	(3.0)
Management	53.1	5.0	4.2	0.8
Education	77.4	7.3	6.2	1.1
Health care	131.9	12.5	9.3	3.2
Culture	39.9	3.8	4.8	(1.0)
Accommodation	74.3	7.0	6.0	1.0
Other services	52.4	5.0	4.2	0.8
Public administration	66.7	6.3	4.8	1.5
Total	1,054.9	100.0		

Source: Statistics Canada (2003c).

overlapping public agencies and provincial trade barriers. Unfortunately, an Atlantic union would likely lead to massive layoffs in the public sector. For a region with high unemployment, such a prospect makes a union much less attractive.

Industrial Structure

Atlantic Canada's three primary resources are its fish, forests, and minerals, but the region's economic future lies in the development of its tertiary sector, especially high-technology industries. Employment provides an overall picture of the basic economic structure of Atlantic Canada and of each of its provinces. Employment in primary activities accounts for 5.6 per cent of the labour force in Atlantic Canada; secondary employment constitutes 16.9 per cent; and tertiary employment 77.5 per cent (Table 9.3).

Significant variations among the provinces' employment figures exist. Newfoundland, as expected, ranks first in employment in the fisheries, but it also ranks first in mining, primarily because of the large iron-mining operations in Labrador and the offshore oil extraction. Employment in manufacturing is much stronger in Nova Scotia and New Brunswick than in Newfoundland and Prince Edward Island, mostly because there is greater processing of local raw materials in these two provinces—from the manufacturing of agricultural, fish, and wood products to the processing of mineral products. However, manufacturing has also suffered its setbacks. For instance, heavy industry in Cape Breton used to employ many workers, but the closure of the coal mines and steel factory has hurt the economy of Nova Scotia and made Cape Breton's unemployment figures among the highest in Atlantic Canada.

Atlantic Canada's Economy

As a result of its long-term economic problems, Atlantic Canada has become heavily dependent on Ottawa, which sends money to the region through federal programs and transfer payments. The most important subsidies are equalization payments and employment insurance payments. Unemployment insurance in Atlantic Canada is extremely important, particularly to those engaged in seasonal industries, such as self-employed fishers and those employed in fish plants and logging operations. Ottawa first extended this insurance to self-employed fishers in 1947. Before the cod moratorium in 1992, inshore fishers caught fish during the summer and drew unemployment insurance payments in the winter. The combined loss of fish income and unemployment payments has contributed to the demise of small fishing communities.

Ottawa has tried to overcome the economic malaise in Atlantic Canada by providing funds for new businesses and for improving the region's transportation system. Such initiatives were meant to create jobs in areas of high unemployment. Since the 1960s, federal grants for businesses have totalled over $6 billion. Most federal funds have been administered by several key federal agencies. In 1962, the Atlantic Development Board was established to stimulate businesses and strengthen the region's public infrastructure. In 1969, the federal Department of Regional Economic Expansion (DREE) began to offer incentives to encourage companies to locate in Atlantic Canada and other less-favoured areas of the country. DREE also spent money on highway construction, schools, and municipal services. In the late 1980s, the Atlantic Canada Opportunities Agency was formed to develop the economy through business and job opportunities. But in the mid-1990s, Ottawa began to reduce its regional subsidies, including more stringent qualifications for employment insurance, in order to combat the federal debt, thereby threatening the financial aid upon which Atlantic Canada relied. The November 2000 election saw the Liberals reinstate regional subsidies and relax the qualifications for employment insurance. In July 2000, Jean Chrétien announced the Atlantic Investment Partnership, a new (but old) system of federal grants and loans to firms and business people in Atlantic Canada.

Ottawa's ongoing influence on Atlantic Canada's economy was evident after the closure of the cod fisheries in 1992. After it announced a moratorium on cod fishing because of the dismal state of the stocks, the federal government promised additional financial assistance for the short term. The Atlantic Groundfish Strategy (TAGS) began in May 1994. TAGS made nearly $2 billion in monthly payments available to about 40,000 fishers and employees displaced by the moratorium. While cod fishers are found in all parts of Atlantic Canada, the greatest impact of the cod moratorium occurs in the isolated fishing communities that are totally dependent on the fishing industry. TAGS was intended to shift fishers into other types of work through retraining and education programs, but most fishers considered the program to be another form of unemployment insurance. In their hearts, the fishers hoped that the cod would return before the payments from TAGS ran out. The success of TAGS was hindered by the following factors: alternative employment opportunities in remote fishing communities are virtually non-existent; few fishers choose to leave their homes and villages for an uncertain future in larger towns or cities; and given the age and education level of most fishers,

retraining becomes an almost impossible task.

By early 1998, TAGS was coming to a close but neither had the cod returned to their former numbers nor had the number of cod fishers declined. Ottawa was faced with a dilemma. Pressure from Atlantic Canada, particularly from Newfoundland, was intense. In August 1998, Ottawa announced that an additional $730 million would be available to cod fishers under the retraining program initiated by TAGS (MacAfee, 1998: A3). Such funding would extend payments to cod fishers for another year but Ottawa made it clear that this would be 'the end of the line' (Greenspon, 1998: A4). The cod fishery still has not reopened, as the depleted cod stocks have not returned to sustainable levels, and prospects for the foreseeable future are not encouraging.

As an older resource hinterland, Atlantic Canada is suffering from a problem that is all too common to such regions. Over the years, Atlantic Canada has depended heavily on the exploitation of its natural wealth. Except for the iron and steel industry based in Cape Breton Island, Atlantic Canada's natural wealth was exported to the United States, Britain, and other European countries where the resources underwent further processing. With the demise of the Cape Breton coal and steel complex, a strong industrial base no longer exists and the prospects for such a base in Atlantic Canada are dim. Other types of development are possible.

One of the promising alternatives is the vast wealth of energy lying below its continental shelf. However, the pattern of resource exploitation, processing, and export is the same as before. The problem is that little processing takes place. For example, the oil from Hibernia, the $6.2 billion offshore oil project east of St John's, is destined for refineries along the eastern seaboard of the United States, while a gas pipeline sends the natural gas from the Sable Island deposit off the Nova Scotia coast to New England. To counter the pattern of exploitation and export, provincial governments have tried to intervene in the marketplace to obtain more benefits from large-scale resource developments, but they have had only limited success. For example, the Newfoundland government has obtained some concessions from the oil industry—the key components for the Terra Nova oil-production vessel were manufactured within the province at Bull Arm, the construction site for the Hibernia platform (Vignette 9.5). With rising prices for oil but especially for natural gas, both Newfoundland and Nova Scotia should receive a sharp economic boost in coming decades. But do they? St John's and Halifax gain from the offshore oil royalties shared with Ottawa, but, because of this increased provincial revenue, they receive lower equalization payments.

As Atlantic Canada searches for its place in the highly competitive global economy, diversification of its resource hinterland role is essential. Economic diversification lies in two directions: greater emphasis on value-added activities (i.e., manufacturing and processing of its natural resources, especially fish stocks) in the primary sector, and an expansion of high technology in the tertiary sector.

Key Topic: The Fishing Industry

Nature has given Atlantic Canada a vast continental shelf that provides an excellent physical environment for fishing: the warm ocean currents from the Gulf of Mexico and the cold Labrador Current create ideal conditions for fish reproduction and growth. The continental shelf extends almost 400 km offshore (Figure 9.2). In some places, where the continental shelf is raised, the water is relatively shallow. Such areas are known as banks. The largest banks are the Grand Banks of Newfoundland and Georges Bank near Nova Scotia (Vignette 9.6).

The diversity that characterizes Atlantic Canada extends to the fishery as well. Although each province relies on the fishery,

| Vignette 9.5 | **Bull Arm Complex: Success or Failure?** |

In 1990–1 the Bull Arm complex, an elaborate and highly specialized construction site, was built by the provincial government at a cost of nearly $500 million. The complex was designed to build the Hibernia offshore oil-drilling platform, which was towed out to the Grand Banks and began production in 1997. The Newfoundland government hoped that the Bull Arm complex could continue to build offshore components for the oil and gas industry's future energy developments. While the Bull Arm complex did acquire work from two other offshore projects—Terra Nova and White Rose—this industrial site never regained the level of construction work achieved during the Hibernia project. Worse yet, the complex was idle for several years and, after its White Rose contract, the future of the complex appears uncertain. The spotty performance of the Bull Arm complex represents a classic problem faced by hinterlands—how to diversify the economy through public investment.

Source: Adapted from Jang (1998).

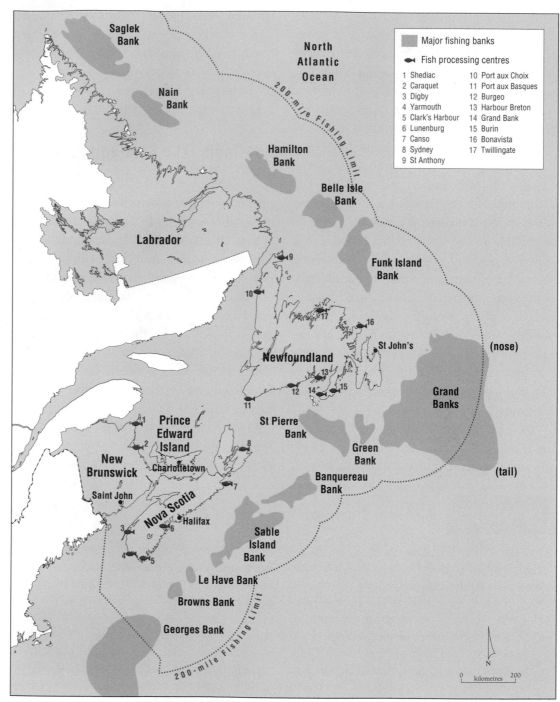

Figure 9.2 Major fishing banks in Atlantic Canada. The Atlantic coast fishery operates within a vast continental shelf that extends some 400 km eastward into the Atlantic Ocean, southward to Georges Bank, and northward to Saglek Bank. Within these waters, there are at least a dozen areas of shallow water known as 'banks'. The Grand Banks of Newfoundland is the most famous fishing ground while Georges Bank contains the widest variety of fish stocks. (Further resources: Student Web site, National Atlas section, Map 26. Web site instructions are found on p. xviii.)

Vignette 9.6   Georges Bank

As part of the Atlantic continental shelf, Georges Bank is a large, shallow-water area that extends over nearly 4,000 km². Water depth usually ranges from 50 to 80 m, but in some areas the water is 10 m or less. Georges Bank is one of the most biologically productive regions in the world's oceans because of the tidal mixing that occurs in its shallow waters. This brings to the surface a continuous supply of regenerated nutrients from the ocean sediments. These nutrients support vast quantities of minute sea life called plankton. In turn, large stocks of fish (such as herring and haddock), as well as scallops, feed on plankton. Cod also feed in these waters, but their numbers have dwindled due to overfishing, especially in international waters. Properly managed, Georges Bank can sustain annual fishery yields of about 400,000 tonnes.

there are striking differences between them (particularly Newfoundland and the Maritime provinces) in the type of catch and the location. Before the ban on cod fishing, Newfoundlanders depended mostly on cod fishing, while Maritimers harvest a variety of sea life, including cod, lobsters, and scallops. Fishers from the Maritimes ply the waters of Georges Bank and smaller banks just offshore of Prince Edward Island and Nova Scotia. In these waters, Maritimers find cod, grey sole, flounder, redfish, and shellfish. Newfoundlanders, on the other hand, primarily fish in the waters of the Grand Banks, around the island of Newfoundland, and along the shore of Labrador, where cod and other **groundfish** congregate in the shallow areas of the continental shelf.

The fishing industry consists of an inshore fishery, an offshore fishery, and fish-processing plants. Traditionally, Newfoundland and Maritime fishers used a flat-bottomed skiff called a dory and stayed close to the shore. Only the larger skiffs ventured to the Grand Banks and other offshore banks. Today, inshore fishers still use smaller boats to fish close to the shore. During the winter, rough waters and ice prevent the smaller craft from leaving port. Offshore fishers travel in steel-stern trawlers to the Grand Banks and other offshore fishing banks where they fish for weeks. Upon returning to port, they unload hundreds of tonnes of iced groundfish at a fish-processing plant. Offshore vessels, often much larger than 25 tonnes, are equipped with radar, radios, sonar detectors, and other sophisticated equipment that enable them to operate in winter as well as summer.

The modernization of the Canadian fishing industry, with its larger trawlers and more efficient nets and gear, was accompanied by huge investments by fish companies and individuals and an expansion of the fishing fleet in number and size. Atlantic Canada's fishery, particularly in Newfoundland and Nova Scotia, was able to increase its catch and its income. Now, fishing companies usually own the ships, so fishers no longer share in the value of the catch but receive a salary. However, modernization had many downsides:

- Fewer fishers were required by this more efficient fishing system.
- The new fishing system based on technological advances and huge dragger nets created the circumstances for overfishing.
- Overfishing and the collapse of the northern cod stocks forced inshore fishers to turn to government assistance to survive.

Management of Fish Stocks

The expansion and modernization of the world fishing fleets have threatened world fish stocks. With no management of fish stocks in international waters, pressure on selected fish stocks grew. By the 1970s, the world fishing fleets had greatly increased their take of groundfish in the Grand Banks. As is the case in British Columbia with the salmon, the issue of fishing limits in Atlantic Canada was crucial in trying to manage the fish stocks in a sustainable way. Management of the international fish stocks in Atlantic Canada was so important that, in 1977, Ottawa claimed the right to manage the fisheries within a 200-nautical-mile zone off the east coast of Atlantic Canada. Washington also extended its fishing zone to the 200-nautical-mile limit, making Georges Bank an area of dispute between the two countries.

Georges Bank is one of the richest fishing banks, where fish and shellfish, such as scallops and mussels, are found. Fishers from both the Maritimes and New England harvest the fish stocks on this bank, which is located at the edge of the Atlantic continental shelf between Cape Cod and Nova Scotia. In the 1970s, fish stocks, especially herring, dropped to dangerously low levels due to overfishing. A boundary decision made by the International Court at The Hague in October 1984 allocated five-sixths of Georges Bank to the United States. However, the easternmost one-sixth that was awarded to Canada is rich in groundfish and scallops. The Nova Scotia fishers have greatly benefited from this decision. With the scallops now under Canadian management, sustainable harvesting of the scallops and groundfish has brought a measure of stability to this sector of the fishing industry.

In spite of difficulties, fishing remains a way of life in Atlantic Canada. As shown in this photograph, inshore fishers operate close to the shore by using small boats and, because of their small-scale operations, pose little threat to the fishery. On the other hand, the Canadian and foreign fishing fleets with their drag nets endanger the Atlantic fishery. (Barrett and MacKay Photography, Inc.)

However, overexploitation of international fish stocks remains a worldwide problem. The root of the problem is that public resources are decimated by the selfish actions of individuals who have no regard for the well-being of the resource. Such mismanagement is known as the 'tragedy of the commons'. Such ecological disasters occur when no one is responsible for ensuring the proper management of resources. Northern cod stocks fell victim to this tragedy twice: in the 1970s when there was no management of fishing on the Grand Banks, and again in the 1990s when the resource was mismanaged (Vignette 9.7). Mismanagement was based on three factors: (1) the federal government's estimate of cod stocks was overly optimistic and Ottawa consequently set the quota too high; (2) strong pressure for a high cod quota came from Newfoundland politicians, various fisher groups, and outport communities; and (3) Canada had no control over foreign fishing along the nose and tail of the Grand Banks. The circumstances and consequences of the demise of cod stocks in Atlantic Canada can lead to a greater appreciation of the need for sustainable harvesting in the fishing industry.

Overfishing and the Cod Stocks

The Atlantic cod (*Gadus morhua*) is the major natural resource in Atlantic Canada. While the Atlantic cod (sometimes called the northern cod) is found as far north as the southern tip of Baffin Island and as far south as Cape Hatteras, cod stocks are concentrated off Newfoundland and along the Labrador coast. The largest single concentration of northern cod is at the Grand Banks.

Historically, the Atlantic cod has been one of the world's leading food fishes. Cod is sold fresh, frozen, salted, and smoked. Salted cod was very popular in the past, but now most of the cod sold is frozen. Cod has been fished on the Grand Banks for centuries. During the sixteenth century, the catch was small, probably averaging less than 25,000 metric tonnes a year. By the end of the seventeenth century, however, the annual catch may have reached 100,000 metric tonnes. Annual cod landings during the nineteenth century were about 150,000 to 400,000 metric tonnes. As fishing technology advanced, catches of cod rose to nearly 1 million metric tonnes in the 1950s. By the 1960s, the annual catch reached a peak of

---

Vignette 9.7   An Ecological Crisis

The decline in groundfish stocks in Atlantic Canada is an ecological crisis. The Canadian Atlantic Fisheries Scientific Advisory Committee revealed in mid-1992 that stocks had declined sharply since the 1960s, when the **biomass** (the volume of fish aged three years or older) was over 3 million tonnes. By 1991, the biomass had dropped to between 530,000 and 700,000 tonnes, possibly the lowest level ever observed. **Spawning biomass** (generally seven years or older) represents the reproductive part of the fish stock. In 1991 it amounted to between 50,000 and 110,000 tonnes. Thirty years before, the spawning biomass had been as high as 1.6 million tonnes, and even in 1987 it was estimated at 400,000 tonnes. Faced with such a serious crisis, the Canadian government announced in 1992 a moratorium on cod fishing in the waters of Atlantic Canada. By the summer of 2004, cod stocks had not yet recovered sufficiently to allow the reopening of the cod fishery.

almost 2 million metric tonnes. European trawlers accounted for most of this catch.

With the emergence of a modern fishing fleet, fish stocks around the world were overfished. Several fish stocks, including the California sardine, Peruvian anchovy, and Namibian pilchard, collapsed in the 1980s from overfishing—not so much from local fishing operations but from the highly modernized fishing fleets that scan the world's oceans for fish. The simple hook-and-line fishing system did not have the capacity to overfish, but foreign and Canadian fishing fleets that employ sophisticated fishing equipment and that drag huge, weighted nets across the ocean floor indiscriminately trapping all groundfish regardless of age, species, and value do have that capability. Known as draggers, these fleets create enormous waste because 'non-commercial' fish—the bycatch—are simply discarded.

In addition, the draggers destroy fragile ocean-floor ecosystems, including coral reefs and breeding habitat. In spite of the many excuses for the collapse of the northern cod stocks, including changing ocean temperatures or an increase in seal population, the primary cause lies at the feet of the owners who ran the technologically advanced fishing fleets and who ruthlessly exploited the cod and other fish of the sea.

In the case of cod, Ottawa's ability to manage the fish stocks was complicated because, as noted above, the nose and tail of the Grand Banks extend into international waters. This made it difficult for Canadian fisheries officials to monitor and enforce catches in these areas. An agreement was reached with the countries belonging to the European Common Market, but monitoring the actual catches proved difficult. Some European fishing vessels, particu-

Table 9.4	Atlantic Canada and Newfoundland Fisheries Landings, 1990–2001 (million tonnes)		
Year	Newfoundland	Atlantic Canada	Per cent Newfoundland
1990	245,896	395,266	62.2
1991	178,687	309,031	57.8
1992	75,138	187,804	40.0
1993	37,068	76,644	48.4
1994	2,292	22,719	9.8
1995	863	12,438	6.9
1996	1,147	15,541	7.4
1997	12,317	29,899	41.2
1998	22,764	37,894	60.0
1999	38,663	55,527	69.6
2000	30,216	46,177	65.4
2001	23,774	40,440	58.8

Source: Fisheries and Oceans Canada (2003).

larly those from Spanish ports, were exceeding their fish quotas and were catching undersized cod by using nets with a small mesh, which is illegal. During the 1980s, groundfish stocks, especially the northern cod stocks, dropped significantly—perhaps by as much as 60 per cent (Cashin, 1993: 19). This collapse translated into a decline in fish landings and processing. Cod landings were reduced by over half from 1991 to 1992. In 1992, Ottawa was forced to impose a moratorium on the cod fishery. The fear was that the northern cod stocks had collapsed and further fishing would make their recovery more difficult. While fisheries landings have remained small, the lowest point occurred in 1995 (Table 9.4). Nearly 10 years later, scientists are still unsure as to how long it will take for the northern cod stocks to recover, if indeed they ever will. Since cod must reach the age of seven before the fish can reproduce, recovery to a point to make the cod fishery marginally viable could take a decade, perhaps longer.

The impact of the cod moratorium on Atlantic Canada, particularly Newfoundland, was severe. The cod has been the mainstay of Newfoundland's economy. Quite justifiably, the Atlantic cod was called 'Newfoundland currency'. In the past, the total catch by Newfoundland fishers was close to four-fifths groundfish. By comparison, the groundfish would constitute only about half of the catch in Nova Scotia and about one-third of the catch in Prince Edward Island and New Brunswick.

Before the moratorium, most jobs were in the fish-processing plants. Often these jobs lasted for only a short period, sometimes less than a month. However, employment at fish-processing plants provided two highly valued items—much-needed cash and enough weeks of work to qualify for unemployment insur-ance. In May 1994, fish-processing employees became eligible for financial support from The Atlantic Groundfish Strategy and were included in the extension to TAGS announced in 1998. In Newfoundland, the fish-processing sector was dominated by two firms, Fisheries Products International and National Sea Products. Prior to the moratorium on cod fishing, these two firms accounted for about half of the fish-processing jobs or nearly 8 per cent of the total number of employed people in Newfoundland.

The fishing industry in Atlantic Canada has suffered under the July 1992 moratorium on cod fishing. Since 1995, however, Newfoundland's fishing industry has recorded higher and higher production values because of high prices for crab, shrimp, lobster, clams, turbot, and other stocks. In this highly managed industry, only a few fishers have licences for these special kinds of catches. For example, only about 1,200 licences to catch crab are issued to Newfoundland fishers each year. On the other hand, before the cod moratorium, approximately 7,000 licences were issued for groundfish. Without cod, wealth from the fishery is concentrated in fewer and fewer hands. By 1997, small catches of cod by inshore fishers were permitted, but, in spite of strong pressure from fishers, Ottawa is still not prepared to allow the cod fishery to reopen. In Atlantic Canada, fishing is not only an important economic industry, it is a way of life. By 2004, cod stocks show no signs of recovery and the inshore fishing way of life based on outports is disappearing.

Atlantic Canada's Land Wealth

Though Atlantic Canada's early settlement and subsequent development revolved very much around the harvesting of the sea, the land mass

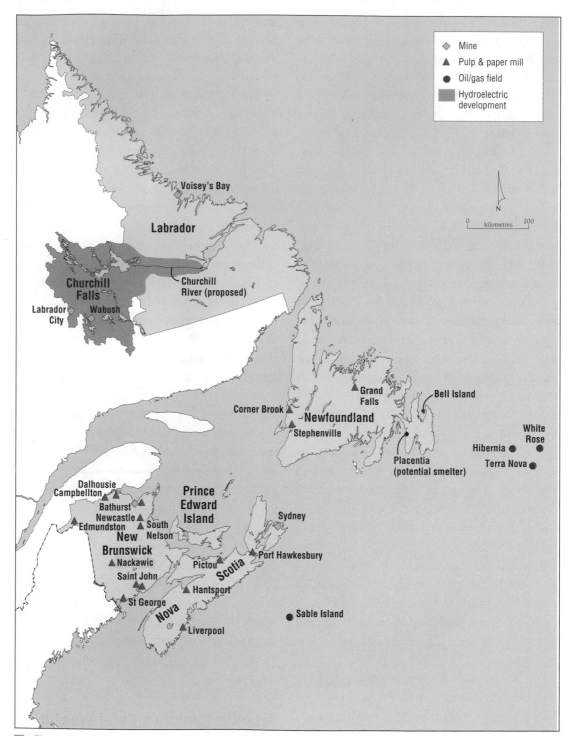

Figure 9.3 Natural resources in Atlantic Canada. Besides fish, Atlantic Canada contains important natural resources. Enormous iron deposits are located in Labrador, huge oil and gas reserves are found below the continental shelf, and one of Canada's largest hydroelectric facilities is situated on the Churchill River. Other natural resources, such as fertile soils and large forest reserves, are more limited.

of the four provinces has also provided the region's economy with important and profitable natural resources (Figure 9.3). The following section discusses the forest, mining, and agricultural industries within Atlantic Canada.

Forest Industry

Beyond the settled areas of Atlantic Canada, the land is covered by forest. Atlantic Canada has 22.9 million ha of forest land. Within the region, the forest industry plays an important role in the economy, particularly in the economy of New Brunswick. In 2001, the value of forest products was approximately $4 billion. Almost half of this production originated in New Brunswick. Most trees harvested, including logs from private woodlots and especially spruce, are sold to local pulp mills. There are nine pulp plants in New Brunswick, three in Nova Scotia, and three in Newfoundland. These mills account for most of the exports. With the exception of Saint John, where there are two pulp mills, mill towns have no other industrial activities and are therefore single-industry communities.

Employment in forestry is declining due to the introduction of efficient logging equipment and the popularity of clear-cutting operations. A single-grip timber harvester with one operator can cut as many trees in a day as 20 men using chainsaws.[3] Capital investment is also replacing high-priced labour in wood processing. The forestry companies' goal is to maximize profits by keeping their products competitive in the global marketplace. Unfortunately, this process adversely affects resource hinterlands by reducing the size of their labour force, leaving many unemployed. Mechanization has also made logging difficult for the small logging operator, who cannot afford the high cost of current machinery and who often owns a small woodlot where such machinery is unsuitable. Small operators, including those with woodlots, are finding it difficult to survive in the face of such stiff competition.

Because Atlantic Canada was settled first, its hardwood and softwood forests have been harvested for a very long time—over 300 years. In Nova Scotia and New Brunswick, some areas have been logged three or four times. Logging is an important occupation in Atlantic Canada. Unlike in the rest of Canada, where forest land is usually Crown land (public), the proportion of private timber lands to Crown lands in the Maritimes is extremely high. Private timber lands comprise 92 per cent of the commercial forest in Prince Edward Island, 70 per cent in Nova Scotia, and 50 per cent in New Brunswick. In Newfoundland, like the rest of Canada, private ownership makes up only 2 per cent of the forested area. On average, the rate of logging on private lands is very high, sometimes exceeding the annual allowable cut estimated by the province. Exceeding the allowable cut impedes sustainable harvesting practices in the region. The high rate of logging often takes place on farms where timber sales are an important source of income.

A new issue in the forest industry in Atlantic Canada relates to Aboriginal people's right to timber lands. Aboriginal Canadians are demanding greater access to timber. Most Crown land in Atlantic Canada has been leased to large companies, mainly to the Irving Forest Corporation. In 1998, a dispute erupted over cutting rights on Crown land. Mi'kmaq Indians in New Brunswick and the provincial government are trying to negotiate a compromise whereby some Crown land currently leased to large companies would be reallocated to Indians. Following a court ruling in

2003, the right of Mi'kmaq Indians to harvest timber on Crown lands that are also lands claimed by these Indians has been confirmed. Just how this ruling will translate into the assignment of logging to Indian loggers remains unclear at this time, but it does force the hand of the provincial government to do something quickly.

Mining Industry

Atlantic Canada is endowed with a number of minerals with commercial value. In 2002, mineral production in this region included metallics (such as iron), non-metallics (such as potash), and fuels (such as oil and coal). There are vast iron ore and nickel deposits in the Canadian Shield, and coal, lead, potash, and zinc deposits in Appalachia. Historically, coal mining on Cape Breton Island was extremely important to the local economy and, at its zenith, the coal industry was the major employer in Cape Breton.[4]

In 2002, the annual value of mineral production in Atlantic Canada was $7 billion (Statistics Canada, 2003). In 1999, the figure was $3.2 billion. This dramatic increase can be attributed to offshore petroleum developments. In 2002, Newfoundland/Labrador produced oil valued at $4 billion while natural gas

from Nova Scotia's Sable Island field contributed $1 billion (Table 9.5). The leading mining operations in Atlantic Canada in 2002 were based on the Newfoundland offshore oil deposits, lead-zinc deposits in New Brunswick, iron ore in Labrador, and coal in Nova Scotia. In the next few years, production of nickel from Voisey's Bay will add to Newfoundland/Labrador's mineral output.

Mining is an important employer in Atlantic Canada. For instance, the largest single employer in Labrador, the Iron Ore Company of Canada, employs nearly 400 workers. In 2002, these workers mined, processed, and shipped nearly $1 billion worth of iron in the form of pellets and concentrates to steel mills in Canada and the United States. Another 400 workers are employed at the lead-zinc mine operated by Brunswick Mining and Smelting Corp. near Bathurst, New Brunswick. Similar numbers of workers are employed in the oil and gas industry.

A Dangerous Occupation

Mining can be a very dangerous occupation—nowhere has this been more evident than in coal mining, where miners work underground in often volatile conditions. Coal has been very much a part of Nova Scotia's history. Although a coal miner's job may pay well, it is a dirty,

Table 9.5	Mineral Production in Atlantic Canada, 2002 ($000s)				
Mineral Product	Newfoundland/ Labrador	Prince Edward Island	Nova Scotia	New Brunswick	Atlantic Canada
Metals	918,429	0	0	440,923	1,359,352
Non-metals	48,156	3,521	247,443	189,274	488,394
Fuels	4,086,031	0	1,062,971	22,088	5,171,090
Total	5,052,616	3,521	1,310,414	652,285	7,018,836

Source: Natural Resources Canada (2003a).

physically debilitating, and dangerous occupation. As there were few other jobs available in Cape Breton and Pictou County, mining companies had little trouble recruiting miners. Danger goes with the job and, for some, the cost is their lives. In 1956, and again in 1958, Springhill Mine was the site of two underground explosions that took the lives of nearly 100 miners. In 1992, the Westray Mine in Nova Scotia was the site of another mine disaster. A buildup of methane gas caused an underground explosion. Twenty-six men lost their lives. The mine had failed safety inspections and yet continued to operate. No doubt, the marginal nature of this coal-mining operation contributed to the substandard conditions of the mine. Westray was under pressure to keep its costs low to remain competitive, so work stoppages to correct the problems identified in the safety inspections would not only have been costly but might have resulted in Westray's failure to deliver sufficient coal to Nova Scotia Power. Unfortunately, as is common in 'old' resource hinterlands, the choice between closing mines where the health and safety of miners are known risks and maintaining employment in an economically depressed region always favours the mines—until disaster and tragedy strike.

Steel, Iron, and Coal—The Rise and Fall

The iron and steel industry in Cape Breton Island near Sydney, Nova Scotia, was for a long time the heavy industrial heartland of Atlantic Canada. During that time, Sydney was the principal city on Cape Breton Island and the second-largest city in Nova Scotia.[5] Sydney's fate was closely tied to its major industrial firm, the Sydney steel mill, and local coal mines. As the mill and mines prospered in the early part of the twentieth century, the town of

Sydney expanded, but since the end of World War II, both the mines and mill suffered financial losses and the size of their labour forces were reduced. This process of deindustrialization, while delayed by federal and provincial subsidies, has taken its toll on the people of Sydney and Cape Breton. In 2001, the steel mill was closed, ending this long history of heavy industry. The future of Cape Breton, without the steel and coal industry to fuel its growth, is uncertain.

The rise and fall of heavy industry in Cape Breton spans a period of 100 years. In 1899, the Dominion Iron and Steel Company began to produce steel from Cape Breton coal and Newfoundland iron ore. Demand for its steel proved so great that plans were made to expand steel production. In 1901, a large, more integrated steel complex was built to supply steel rails to firms constructing new rail lines in Canada and other countries. Located at a fine harbour on the Atlantic coast, the iron and steel mill at Sydney on Cape Breton Island had ready ocean access to raw materials and world markets. It was able to use local coal and iron ore from Bell Island, Newfoundland. It supplied markets in Canada (via the St Lawrence River and the Intercolonial Railway) and in other Commonwealth countries. Now called the Dominion Steel and Coal Co. (Dosco), the firm benefited from the strong demand for steel during World War I. Under these prosperous conditions, Dosco became Canada's largest private employer.

In the 1920s, the Sydney steel mill began to fall on hard times. By 1925, Canada's railway boom had ended, weakening the demand for steel, and the demand for coal slumped as many coal customers turned to another source of energy, petroleum. A modest recovery for coal, however, occurred during World War II. After the war effort, it became evident that the

Sydney steel mill was no longer viable without public subsidies. Sydney simply could not compete with offshore steel producers in Brazil and South Korea. The basic problems facing the Sydney steel mill were threefold: little local demand in the Maritimes, stiff competition in Canada from low-cost producers in southern Ontario and even stiffer competition in external markets, and increasing costs for its raw materials and labour.

Even with the assistance provided by government subsidies, Dosco continued to lose money. As a result, in 1965, Dosco announced plans to close its coal mines, which would eliminate 6,500 jobs, and in 1966 the Bell Island iron mine, which had been a Dosco operation, was shut down. In 1967, the federal government took over the operation of the coal mines through a Crown corporation, the Cape Breton Development Corporation (Devco). The steel mill was on the brink of bankruptcy when it was sold to the Nova Scotia government in 1967. From that time on, a provincial corporation, the Sydney Steel Corporation (Sysco), not only ran the mill but, with help from the federal government, also paid for the annual shortfall. The two governments intervened because they did not wish to add to the unemployment problem in Cape Breton, an already economically depressed area with a very high unemployment rate. As Blackbourn and Putnam observed:

> The continued existence of an uneconomic and obsolete steel mill in an unsatisfactory location is a classic illustration of the importance of inertia in industrial location. The closure of the steel mill would create unemployment in a depressed province. Not only would jobs be lost in steel but the coal mines which supply the mill would probably close

leading to further unemployment. (Blackbourn and Putnam, 1984: 112)

In the late 1960s, neither the federal government nor the Nova Scotia government, for social and political reasons, could accept closure of the Sydney steel mill. At that time, the two governments' treasuries could absorb the annual losses and even advance more funds for plant improvements. However, by the early 1990s, governments had to deal with their enormous debts, and cutbacks were the order of the day. Even after the Nova Scotia and federal governments stepped forward with financial support, the industry continued to limp along. In 1999, the Crown corporation owned by Sydney Steel Corporation, which had received more than $2 billion in federal and provincial subsidies over the previous 20 years, tried to sell the company. By 1999, only 700 workers were employed at Sydney Steel Corporation, a far cry from the peak of 4,000 in the late 1960s. By 1987, Sydney's steel mill had been reduced to an electric-arc furnace using scrap metal to manufacture steel rails. Sales to the United States were crucial but the US imposed countervailing duties because the Nova Scotia government was subsidizing Sysco. Unable to find a buyer, the Nova Scotia government closed the plant and, in 2001, announced plans for its demolition.

Since the 1980s, demand for Cape Breton coal declined. With the switch at Sysco from a coal-burning furnace to an electric-arc furnace, a major customer for Cape Breton coal was lost. Cape Breton bituminous coal mines were operated by the federal agency, Cape Breton Development Corporation (Devco). Most coal was sold under contract to Nova Scotia Power. At first glance, coal from the Cape Breton coalfield was ideally suited for the nearby thermoelectrical stations operated by Nova

Scotia Power. While the cost of transporting coal to these generating stations is low, the cost of mining Cape Breton coal is high due to its underground mining operations, which extend several kilometres. As a result, coal from deep-tunnel mining at Glace Bay, Cape Breton, cost more to deliver to Nova Scotia Power than coal from Pennsylvania. With NAFTA, American coal is no longer barred from Canada by high tariffs. For that reason, Nova Scotia Power, now a privatized Crown corporation, purchased lower-priced American coal to fire its thermoelectrical-generating plants in order to keep its electricity rates at their current level. In February 1999, the government announced it was ending subsidization of Devco, which employed about 1,700 people at its two Cape Breton mines. One mine (Phalen) was closed in 2000 and the other one (Prince) closed in 2001.

Megaprojects

Large-scale resource projects do generate an immediate construction boom with large investments and high demand for workers, especially skilled tradesmen. Once operational, the economic benefits slow and the number of permanent workers is relatively small. In the first decade of the twenty-first century, the major megaprojects affecting Atlantic Canada are offshore oil and gas developments led by the Hibernia oil project and the Sable Island gas development. Terra Nova began to produce oil in 2002. These megaprojects have added a new dimension to Atlantic Canada's economy. Other potential projects—the White Rose oil field with production to commence in 2006, along with the Voisey's Bay nickel project, also scheduled to begin production in 2006—could make a fundamental difference to the economy of Newfoundland. Discovered in 1984, the White Rose offshore oil field is located in the Jeanne d'Arc Basin 350 km east of St John's. The field consists of both oil and gas pools, including the South White Rose oil pool. The oil pool covers approximately 40 km² and contains an estimated 200–50 million barrels of recoverable oil. Unlike the fixed platform used by Hibernia, the Terra Nova oil field uses a Floating Production Storage and Offloading vessel (FPSO). This facility is a ship-shaped vessel with integrated oil storage from which oil will be offloaded onto a shuttle tanker. Husky oil plans to use an FPSO vessel for its White Rose oil field.

The Hibernia oil project is located 315 km east of St John's, Newfoundland, on the Grand Banks above the site of a huge deposit of oil and natural gas. To tap the estimated 615 million barrels of oil from the Hibernia deposit, an innovative offshore stationary platform was needed (Vignette 9.8). About 4,000 workers built a specially designed offshore oil platform that can withstand the pounding storms of the North Atlantic and crushing blows from huge icebergs. The huge concrete and steel construction sits on the ocean floor, with 16 'teeth' in its exterior wall designed to absorb the impact of icebergs.

The Hibernia drilling site has an annual output of about 30 million barrels and adds greatly to Newfoundland's mineral output. Based on an average price of $30 per barrel over the next 20 years, the average annual value of production is estimated at $900 million. Oil production began in 1998. By 2000, Hibernia accounted for 12 per cent of Canadian oil production. Oil is exported to American and other foreign refineries. According to the owners, the Hibernia Consortium, production should continue for approximately 20 years.

A new megaproject is at Voisey's Bay. It is the site of a recently discovered huge nickel deposit, the Ovoid deposit, which lies close to

Vignette 9.8   The Hibernia Platform

The Hibernia platform is a gravity-based structure positioned on the ocean floor on the southeast corner of the Grand Banks. The 111-m-high oil platform and oil-storage units weigh over 650,000 tonnes. In the summer of 1997 the platform was placed on the ocean floor just above the oil deposits. The depth of the water at this point is about 80 m, leaving the oil platform approximately 30 m above the surface of the ocean. This structure is designed to be a platform for the oil derricks, to house pumping equipment and living quarters for the workers, as well as to provide storage for the crude oil. The rig will extract huge amounts of oil from the Avalon reservoir (2.4 km under the seabed) and from the Hibernia reservoir (3.7 km deep). The crude oil is then pumped from the Hibernia storage tanks to an underwater pumping station and then through loading hoses to three 900,000-barrel supertankers for shipment to foreign refineries.
Source: Adapted from Cox (1994).

The Hibernia platform has a massive concrete base which supports its drilling and production facilities as well as the workers' accommodations. Since the platform was positioned on the ocean floor in 1997, the province has joined the ranks of other oil-producing provinces. Unfortunately, the government of Newfoundland and Labrador receives only a portion of the royalties generated by these offshore developments. In 1984, the Supreme Court unanimously declared that Newfoundland's offshore resources belong to the federal government. (CP/Toronto Star/Peter Power)

the surface. It consists of 32 million tonnes of relatively rich ore bodies—2.8 per cent nickel and 1.7 per cent copper. Another deposit, Eastern Deeps, though larger (about 70 million tonnes), has a lower grade of nickel and copper (similar grade to that found at the Sudbury mines).

The Voisey's Bay deposit has three attractive features: high-grade nickel, a surface deposit that has the potential for open-pit mining, and proximity to ocean shipping. These characteristics may make it the lowest-cost nickel mine in the world. In fact, there is

concern that production from this mine might force higher-cost producers, such as the nickel mines in Sudbury, Ontario, to close their operations. For Newfoundland, the Voisey's Bay mining and smelting operation could employ over 1,000 workers. In late 1996, after Inco Ltd had purchased the Voisey's Bay deposit for $4.3 billion, the company announced that it would build a $1 billion nickel smelter and refinery at Argentia on the southeastern coast of Newfoundland with approximately 750 workers employed at the smelter and refinery. Later, Inco proposed to ship the nickel ore to its Sudbury smelter. Newfoundlanders were outraged and their government refused to sign the surface lease that would allow Inco to move forward with its plans for a mine. In 2002, the government of Newfoundland and Labrador concluded an agreement with Inco that included the possibility of some of the ore being processed in Newfoundland at a yet-to-be-built experimental hydromet smelter. In the meantime, the provincial government has approved the shipment of ore to Inco's Sudbury smelter.

Despite the lure of megaprojects, they present problems for regional development. First, they are capital-intensive undertakings. During the construction phase a large labour force is required, but in the operational phase relatively few employees are needed. Second, megaprojects in resource hinterlands lose much of their spinoff effects to industrial areas. As a consequence, economic benefits related to the manufacture of the essential parts of a

megaproject go outside the hinterland. Efforts by hinterland governments to address this classic problem of economic leakage from resource hinterlands to industrial cores have had mixed results, as is perhaps demonstrated by the indefinite nature of Inco's intent to build a smelter/refinery at Argentia, Newfoundland, and the difficulties faced by the Bull Arm complex (Vignette 9.5). Third, megaprojects sometimes take place on Crown lands where Aboriginal groups have a land claims based on traditional occupancy of the land. The Voisey's Bay nickel mine development is situated on such lands. Labrador, therefore, presents a special version of the Aboriginal/non-Aboriginal faultline.

Resource Development and the Aboriginal/Non-Aboriginal Faultline

The Voisey's Bay mining development has sparked a renewed interest in land-claims settlements in Labrador. With such agreements,

Discovered in 1993, the Voisey's Bay nickel deposit lies along the coast of Labrador approximately 350 km north of Happy Valley–Goose Bay. In 1996, Inco Ltd acquired the rights to the Voisey's Bay property and its subsidiary, Voisey's Bay Nickel Company, is responsible for constructing and operating a mine and concentrator. Once in operation, the mine and concentrator are expected to employ 400 people. Since the deposit lies within the land claims of both the Labrador Innu and Inuit, Inco was obliged to reach an Impacts and Benefits Agreement with each of the Aboriginal organizations. (CP/Andrew Vaughan)

resource developments are more assured of a trouble-free working environment. This point was made back in 1973 when the courts forced Hydro-Québec to reach an agreement with the Cree and Inuit of Québec who claimed that northern Québec was part of their traditional lands (Crown lands that they or their ancestors used for hunting and trapping). Such lands have given rise to a legal claim known as Aboriginal title. With an unresolved claim of Aboriginal title, there is an uncertainty over ownership of the land in dispute. The Québec government (as did all other provincial governments of the day) considered that Aboriginal people could use Crown lands for hunting and trapping at the pleasure of the Crown and therefore no special arrangement with Aboriginal residents was necessary when the provincial government granted a lease or sold the land to a developer. How wrong they were! Without such an agreement, construction of the James Bay hydroelectric project could have been halted. By the twenty-first century, however, companies can proceed with their developments before a land-claim agreement is achieved if they reach a special arrangement known as an Impact and Benefits Agreement with the concerned Aboriginal group(s).

By 2003, Inco had gained Impact and Benefits Agreements with both the Labrador Inuit and the Labrador Innu. Yet in all of the negotiations, the Innu—whose traditional homeland includes the site of the Voisey's Bay mining project, who are without treaty, who have never achieved a comprehensive land-claim agreement, and who continue to suffer from incredibly high suicide rates and substance abuse problems—are drawn into the vortex of a huge construction project and forced to answer the question, 'What do you want in exchange for development?' (Samson, 2003: 96–102). While the Innu, like the Labrador Inuit, have reached

an Impact and Benefits Agreement with Inco that provides specific benefits such as employment opportunities, some Innu (as well as many Inuit) have only a limited command of the English language, and most lack the basic education level (high school diploma) or employment experience to take advantage of opportunities provided by the construction of the Voisey's Bay nickel mine.

Then, too, the issue of a comprehensive land claim is more complicated for the Labrador Innu than for the Labrador Inuit because the Innu are considered Indians by the federal government and therefore they must be dealt with under the Indian Act. The Labrador Inuit began negotiations in the 1990s and, in August 2003, reached a land-claim agreement.[6] This agreement, which includes provision for self-government, is the first of its kind in Atlantic Canada. On the other hand, progress on the Innu land claim is proceeding very slowly because Ottawa insists that the Labrador Innu first register as Indians under the Indian Act, the benefits of which they had been denied for more than 50 years since Newfoundland joined Canada. This has meant that they must first negotiate for and be given small reserve lands, and only then seek a comprehensive land-claim settlement that would involve their traditional lands (lands that would extend far beyond their reserves) and that may then remove them from the Indian Act they had to adhere to before seeking a comprehensive agreement in the first place (Canadian Human Rights Commission, 2003).

Agriculture

Agriculture is limited by the physical geography in Atlantic Canada. Arable land constitutes less than 5 per cent of the Maritimes.

Arable land is even more rare in Newfoundland and Labrador, making up less than 0.1 per cent of its territory. Though limited in size, agricultural production significantly contributes to the economy of Atlantic Canada. In 2001, the value of agricultural production in the region was about $1 billion. Specialty crops, especially potatoes, contributed heavily to the value of this production.

Atlantic Canada has nearly 400,000 ha in cropland and pasture. Almost all of this farmland is concentrated in three main agricultural areas—Prince Edward Island, the Saint John River Valley in New Brunswick, and the Annapolis Valley in Nova Scotia. Specialty crops, especially potatoes and tree fruit, are extremely important cash crops. In all three agricultural areas, dairy cattle graze on pasture land. The dairy industry in Atlantic Canada has benefited from the orderly marketing of fluid-milk products through marketing boards.

Prince Edward Island is the leading agricultural area in Atlantic Canada. It has almost half of the arable land in the region. Most of Prince Edward Island's 155,000 ha of farmland is devoted to potatoes, hay, and pasture, with the principal cash crop being potatoes. Since the 1980s, most potato growers have had contracts with the island's major potato-processing plants—Irving's processing plant near Summerside and McCain's plant at Borden–Carleton now dominate the potato industry on the island. The second major agricultural area, the Saint John River Valley, is in New Brunswick. Its 120,000 ha of arable land comprise about one-third of Atlantic Canada's farmland. The Saint John River Valley has the best farmland in New Brunswick. Nova Scotia has nearly one-quarter of Atlantic Canada's farmland, with 105,000 ha. Nova Scotia's famous Annapolis Valley, the region's third agricultural area, is the site of fruit orchards and market gardens. The valley's close proximity to Halifax, the major urban market in Atlantic Canada, has encouraged vegetable gardening. In both New Brunswick and Nova

The rich, red soils of Prince Edward Island are famous for growing potatoes, which are the primary cash crop in the province. Prince Edward Island is Canada's leading potato province, responsible for almost one-third of Canadian production. Its potatoes are grown for three specific markets: seed, table potatoes, and processing. Seed potatoes are sold to commercial potato growers and home gardeners to produce next year's crop; table potatoes go to the retail and food service sectors; and processing potatoes are manufactured into french fries, potato chips, and other processed potato products. (Barrett and MacKay Photography, Inc.)

Scotia, potatoes are a major cash crop. Almost all potato farmers in these two provinces seed their potatoes under contract to McCain Foods, which is a multinational food-processing corporation based in New Brunswick. The company has benefited from NAFTA after the removal of tariffs on its food products, especially french fries, to the United States. Newfoundland has the least amount of farmland—just over 6,000 ha.

While agriculture plays a secondary role in much of Atlantic Canada, it dominates economic activity in Prince Edward Island. In 2001, the total net income for PEI farmers was $104 million. With nearly 2,000 farms, PEI farmers put most of their land into potatoes.

Atlantic Canada's Urban Geography

The urban geography of Atlantic Canada is characterized by few cities and a pattern of

When the British founded Halifax in 1749, the high hill overlooking the harbour offered a perfect place to construct a fortress to defend the new town and its naval base. Named the Halifax Citadel, this fortress is an impressive star-shaped masonry structure complete with defensive ditch, earthen ramparts, musketry gallery, powder magazine, garrison cells, guard room, barracks, and school room. The Citadel is now a National Historic Site. (Ron Garnett/AirScapes))

highly dispersed settlement. These features are attributed to the region's geography and natural resources. It is not surprising, then, that a relatively small percentage of Atlantic Canada's population lives in urban places. In 2001, Atlantic Canada was the least urbanized region of Canada with less than half of its population living in urban centres (Figure 9.4).

Atlantic Canada has a coastal settlement pattern. Major cities are situated along its coastline. Trade and fishing have played a key role in the development of this pattern. The region's interior is relatively 'empty'. The region's physical geography has shaped this pattern because interior lands have few resources to attract settlement. The main exception is mining towns. The iron-mining town of Labrador City is an example of a single-industry community in the interior of Labrador's mining hinterland. Labrador City's population has been declining due to cost reductions and the substitution of machinery for labour. Between 1996 and 2001, Labrador City's population declined from 10,473 to 9,638.

Halifax, the largest city in Atlantic Canada, ranks only thirteenth in size among Canada's major cities (Vignette 9.9). Halifax's relatively small size is due to three factors: (1) a less well-developed urban system and regional economy compared to other regions of Canada; (2) the fragmented nature of Atlantic Canada's population and resulting small markets; and (3) Atlantic Canada's relatively small economy. However, Halifax, with the best natural harbour on the east coast of North America, has the potential to become a superport for the eastern seaboard of North America. The need for such a port has arisen from the construction of container ships that are too large to pass through the Panama Canal. These ships, known as post-Panamax, are longer than three football fields. In 1999, Halifax's

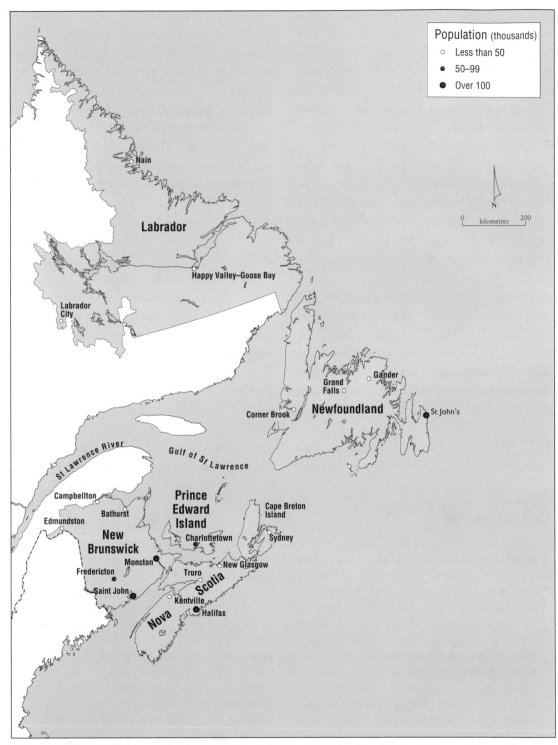

Population (thousands)
○ Less than 50
● 50–99
● Over 100

Nain

Labrador

Happy Valley–Goose Bay

Labrador
City

Gander

Grand
Falls

Corner Brook   Newfoundland

St John's

St Lawrence River

Gulf of St Lawrence

Campbellton

Prince
Edward
Island

Cape Breton
Island

Bathurst

Edmundston

New
Brunswick

Charlottetown

Sydney

Moncton

New Glasgow

Fredericton   Truro

Saint John   Scotia

Kentville

Nova   Halifax

Figure 9.4 Major urban centres in Atlantic Canada. Atlantic Canada has few large cities and most urban centres are scattered across the region. Only four cities—Halifax, St John's, Saint John, and Moncton—have populations exceeding 100,000.

dream of becoming a superport was dashed when two of the world's largest shipping companies selected New York City. The attraction of New York City was not its harbour but its huge market and its extensive rail/truck network that connects it to other major markets in North America. These two advantages of geographic location to major North American markets and excellent transportation infrastructure are exactly the factors that Halifax (and, by extension, Atlantic Canada) lacks in order to be a major player in the North American economic system. For the world's shipping companies, Halifax remains a secondary port for the time being.

Halifax, Saint John, and St John's, the three largest urban centres in Atlantic Canada, have experienced different population increases over the past 20 years. Saint John has seen virtually no change over 20 years with its population increasing from 121,000 to 123,000. The population of St John's increased by 18,000 from 155,000 in 1981 to 173,000 in 2001. Halifax witnessed the greatest increase. Over this 20-year period its population grew by 81,000, from 278,000 to 359,000 residents. Economic conditions have attracted few new immigrants to the cities of Atlantic Canada, and many of the region's residents have left for cities in other provinces. From 1996 to 2001, Halifax grew by 4.7 per cent while Saint John and St John's declined by 2.4 and 0.7 per cent respectively (Table 9.6). Moncton, located in eastern central New Brunswick, and Fredericton, capital of New Brunswick, fared much better. Their populations increased by 3.7 and 3 per cent respectively while Cape Breton's population declined by 7.2 per cent. Beyond these major urban cities, there are about a dozen smaller cities and towns that form a lower level in the urban hierarchy of Atlantic Canada. Most have populations between 10,000 and 50,000. Charlottetown, the capital of Prince Edward Island, is the largest of this group. With a population of nearly 60,000, Charlottetown has the highest rate of increase among these smaller cities. Smaller centres did not fare so well. Of the 10 towns, nine saw their populations decline from 1996 to 2001. The iron-mining town of

Vignette 9.9   Halifax

Halifax, the capital of Nova Scotia and the largest city in Atlantic Canada, was founded in 1749. On 1 April 1996 Halifax ceased to exist as an incorporated city and became the Halifax Regional Municipality, including the former cities of Halifax and Dartmouth. By 2001, Halifax had a population of 359,000. As in the past, its strategic location allowed Halifax to play a major role on the Atlantic coast as a naval centre and international port. Along the east coast of North America, its deep harbour is ideally suited for huge post-Panamax ships. Yet, because of its relative distance from the major markets in North America and its reliance on indirect transferring of goods between ships, trains, and trucks, Halifax cannot provide lower transportation costs than New York. The economic strength of Halifax, along with the neighbouring towns of Bedford and Dartmouth, rests on its defence and port functions, its service function for smaller cities and towns in Nova Scotia, and its provincial administrative functions. Halifax also has a small manufacturing base. However, the city's marginal location for manufacturing activities is illustrated by AB Volvo's announcement on 10 September 1998 that it would close its Halifax assembly plant (Keenan, 1998: B1, B4).

Table 9.6	Urban Centres in Atlantic Canada, 1996–2001		
Urban Centre	**Population 1996**	**Population 2001**	**Change (per cent)**
Labrador City	10,473	9,638	(8.0)
Gander	12,021	11,254	(6.4)
Campbellton	16,867	16,265	(3.6)
Grand Falls–Windsor	20,378	18,981	(6.9)
Edmundston	22,624	22,173	(2.0)
Bathurst	25,415	23,935	(5.8)
Kentville	25,090	25,172	0.3
Corner Brook	27,945	25,747	(7.9)
New Glasgow	38,065	36,735	(3.5)
Truro	44,102	44,276	0.4
Charlottetown	57,224	58,358	2.0
Fredericton	78,950	81,346	3.0
Cape Breton*	117,849	109,330	(7.2)
Moncton	113,491	117,727	3.7
Saint John	125,705	122,678	(2.4)
St John's	174,051	172,918	(0.7)
Halifax	342,966	359,183	4.7

*Cape Breton is a Statistics Canada creation that includes Sydney, North Sydney, Glace Bay, and other Cape Breton Island municipalities.
Source: Statistics Canada (2002a).

Labrador City suffered the largest percentage decline, at 8 per cent, followed by three other resource towns, Corner Brook, Grand Falls–Windsor, and Gander (Table 9.6).

Atlantic Canada's urban geography remains on the margins—highly dispersed, broken into provincial units, and lacking a dominant metropolitan centre. This has led to an absence of cohesion, which is symptomatic of the region's economy.

Atlantic Canada's Future

Atlantic Canada has a diverse geography and a long, rich history that combine to make the region a rewarding place to live. Yet geography and history have also made it a difficult place in which to survive. In the twenty-first century, the challenge for Atlantic Canada is to reverse current economic and demographic trends. For Newfoundland/Labrador and Nova Scotia, the discovery of offshore oil and gas could make them the Alberta of Atlantic Canada. No such hope yet exists for New Brunswick and Prince Edward Island. Still a nagging question persists—will energy production hold the key to economic rejuvenation? The problem is twofold. First, the two provinces receive only 30 per cent of the offshore oil and gas revenues compared to 100 per cent if the deposits were land-based. Second, the link between provincial revenues and the size of equalization payments reduces the actual gains by Newfoundland/Labrador and Nova Scotia. The reason is simple—as provincial oil and/or gas revenues increase, the amount of equalization payments declines. While not quite a zero-sum game, the impact of oil royalties is diminished by reduced equalization payments, thus slowing real economic gains in Newfoundland/Labrador and Nova Scotia. The Premier of Nova Scotia, Dr John Hamm, estimated that offshore gas will generate $36 billion in royalties by 2030 (Fife, 2001: A4). An earlier agreement between Ottawa and the provinces set the sharing of royalties at 70 per cent for Ottawa and 30 per cent for provinces. Prior to that agreement, the federal government would receive 100 per cent of offshore mineral developments. But Nova Scotia argues that it will receive less than 30 per cent ($10.8 billion) because equalization payments will be reduced by an estimated $4.8 billion.

Demography is working against Atlantic Canada. As its population shrinks, the prospects for economic growth diminish. The steady out-migration of its people is an indication of the region's economic malaise. In recent years, so many Newfoundlanders have gone to the heavy oil projects near Fort McMurray that this Alberta resource town boasts that it has more Newfoundlanders than most towns in Newfoundland/Labrador! The failure to keep its people and to attract new immigrants has serious implications for economic growth. Of the five major cities—Halifax, St John's, Saint John, Moncton, and Sydney—only two (Halifax and Moncton) recorded population increases from 1996 to 2001. Smaller centres fared even worst—only two (Fredericton and Charlottetown) out of 12 smaller cities posted population gains. With more and more economic growth taking place in large urban centres, Atlantic Canada finds itself at a distinct disadvantage. Halifax shows signs of economic growth and St John's, propelled by the oil industry, should do well, though its population declined slightly from 1996 to 2001. Most centres, however, are likely to continue to lose people, and this loss places more pressure on the small business sectors found in each community.

As a whole, Atlantic Canada's economy seems destined to lag behind that of other Canadian regions because it is so dispersed. Without a large population and a leading metropolitan centre, Atlantic Canada has been unable to build a strong industrial core—it therefore remains on the edge of the global economic system, making it doubly difficult to engage in new economic endeavours. Success has occurred in certain sectors—telecommunications in New Brunswick, software development in Nova Scotia, marine industries in Newfoundland, and tourism in Prince Edward Island. Overall, however, Atlantic Canada's economy remains weak.

Physical fragmentation is also working against a strong Atlantic Canada. Geography has divided Atlantic Canada into three subregions: the strongest economy and largest market are in the Maritimes, followed by the island of Newfoundland, and then Labrador. Within Atlantic Canada, future prospects seem brightest for Nova Scotia, which has the beginnings of a metropolitan centre in Halifax. Within other parts of the Maritimes, and in Newfoundland and Labrador, the opportunity to break out of a resource hinterland economy seems less assured.

Like other hinterlands, the relentless modernizing of the primary sector is working against a large labour force by reducing the number of workers. Primary activities are undergoing a technological revolution where capital is substituted for labour. Agriculture, aquaculture, fishing, forestry, and mining no longer require large workforces. As the cost of labour continues to rise, jobs will continue to be lost to machines. Consider the staple industry of Atlantic Canada, for example: even if the cod fishery were to regain its past productivity, the technological transformation of the fishing industry over the past 20 years means that large fishing vessels with smaller crews can catch most of the fish required by fish-processing plants. Furthermore, resource-led projects, like the iron mines in Labrador and the pulp mill at Corner Brook, Newfoundland, have resulted in single-industry towns, which, like the hydroelectric development at Churchill Falls, have a limited impact on the provincial economy and the province's unemployment rate.

Atlantic Canada faces the problem of dwindling resources and the promise of new discoveries. As an aging resource hinterland, many of its prime resources are gone. The major exception is offshore petroleum. On the negative side, in 1997, New Brunswick's potash mine closed because the mine flooded, and in 2001, the last coal mine in Cape Breton ceased production. For new discoveries, like the nickel deposit at Voisey's Bay, world prices for primary products will be a determining factor in the future of the primary sector. If prices rise sharply, these primary industries may generate significant spinoffs, triggering a diversification of Atlantic Canada's economy. But if commodity prices remain low, much of Atlantic Canada will stay trapped in a downward economic spiral. Oil, gas, and electricity prices are an exception to this rule. Prices are expected to increase, partly because of the demand forecast for New England, which suffered a major blackout in August 2003. By 2010, for example, the economic boost from the combined oil output of Hibernia, Terra Nova, and White Rose should boost Newfoundland/Labrador's economy and generate additional royalties for the provincial government. The search for another Sable natural gas deposit is continuing and such a discovery would greatly benefit Nova Scotia.

Ottawa's diminished capacity under international trade agreements to subsidize Atlantic Canada's economy adds to the region's economic dilemma. The federal government is less able to provide regional subsidies but is also less enthusiastic about them. Past efforts have had mixed results—usually the infusion of federal funds has a positive short-term impact but, since such efforts cannot overcome the basic structural problems facing Atlantic Canada, long-term impacts are negligible. Cape Breton Development Corporation's efforts to provide work for redundant coal miners by offering grants and subsidies to attract new industries to the region, as well as The Atlantic Groundfish Strategy's goal of assisting displaced fishers and fish-plant workers, have provided a temporary solution. But these measures have done little to resolve the fundamental weaknesses facing these two subregions and their people. Under these circumstances, the future of Cape Breton and of Newfoundland's outports is not bright.

The global economy forces local firms to reduce costs in order to survive. In Atlantic Canada, replacing workers with machines increases the number of unemployed workers. In turn, high unemployment rates trigger an out-migration from Atlantic Canada, thereby shrinking its population base. Such a decline affects market size, the age of the labour force, and the psychological outlook of the people of Atlantic Canada. In the core/periphery model, this 'old' resource hinterland would appear to be trapped in a downward spiral.

Atlantic Canada could break out of its role as a dependent hinterland, but for this to happen a combination of favourable circumstances, none of which is assured, would have to unfold. One of the most important would be the recovery of the codfish stocks. This, coupled with a harvesting system that ensures the well-being of the fish stocks and a sharing of the catch between fishing fleets and individual fishers, would achieve two goals—a supply of fish for the processing plants and a maximization of the number of fishers.

A second economic condition would be the expansion of the tourism industry. Atlantic Canada offers tourists from New England, Ontario, and Québec a 'down home' vacation experience. The north–south highway system attracts American tourists to the Maritimes. Atlantic Canada has a number of unique places that can draw selected tourists. For example, many Japanese tourists visit Cavendish in Prince Edward Island, the setting for Lucy Maud Montgomery's *Anne of Green Gables*. Others flock to Newfoundland for whale-watching.

A third economic necessity would be the development of high-technology industries. Other regions of Canada are enjoying an expansion in the high-tech sector due to Canada's highly skilled labour force. Often these innovative firms locate near universities. Halifax is particularly well-suited for such development; the success of the neighbouring American city of Boston provides an example. Such a development in Atlantic Canada would contribute greatly to the diversification of the region's economy.

Resource development is the last economic condition. Offshore oil and gas production at Hibernia, Terra Nova, and Sable Island is well underway. In the twenty-first century, three major resource projects are possible: a hydroelectric dam on the lower Churchill River; an oil well at the White Rose oil field near Hibernia; and a nickel mine at Voisey's Bay with the prospects of a smelter on the southeast coast of Newfoundland. These developments, if world prices for these resources remain high, could propel Atlantic Canada's economy into a more prosperous state—one in which unemployment rates decline, business opportunities increase, and out-migration slows. Such an economic state may be that elusive balance between economic growth and a 'down East' way of life.

Summary

Atlantic Canada lies on the eastern rim of Canada. The region is characterized by physical fragmentation, cultural diversity, and a slow-growing economy. Atlantic Canadians rely heavily on the export of their natural resources—especially fish and energy. With the Free Trade Agreement, Atlantic Canada, but especially the Maritimes, has increased its exports to the United States. This north–south trade is exemplified by the natural gas pipeline extended from the continental shelf near Nova Scotia to customers in New England. Still, Canada's oldest hinterland has been unable to generate enough jobs to support its people. As a result, each year thousands leave Atlantic Canada, especially from Newfoundland.

As Atlantic Canada enters the twenty-first century, the region's future is uncertain. A slowing of economic growth and a declining population do not lend themselves to an economic recovery. Somehow Atlantic Canada must break out of its role as a resource hinterland. Part of the answer lies in offshore energy developments, oil royalties, and a new arrangement over equalization payments.

Notes

1. On 30 April 1999, the Newfoundland House of Assembly gave unanimous consent to a constitutional amendment that would officially change the name of the province to Newfoundland and Labrador.

2. The Maritime region was a focal point for the struggle between France and Britain. Indian allies were critical for both European powers. 'Involvement with the French brought benefits—and immediate consequences. Before they realized the full scope of the French-British rivalry, the Mi'kmaq and the Maliseet discovered they had already chosen sides' (Coates, 2000: 31).

3. The FMG Timberjack 990 requires only one person—an operator—to fell a tree in a very short time. Priced at $600,000, this giant machine is one example of capital substitution for labour. The Timberjack grips the base of a tree, slices its steel blade straight across the trunk, and strips the tree of its branches. The operator then uses the Timberjack to cut the tree into measured lengths before utilizing a boom arm and clamp grip to load the cut logs onto a truck.

4. Coal, a combustible sedimentary rock formed from the remains of plant life during the Carboniferous Age (a geological period in the Paleozoic era), is classified into four types: anthracite, bituminous, sub-bituminous, and lignite. Anthracite is the highest grade of coal, while bituminous is used in the iron and steel industry and for generating thermoelectrical power. Most coal mined in Nova Scotia is bituminous coal.

5. As with Fort McMurray (Wood Buffalo), Alberta, Statistics Canada has created a geographically sprawling census metropolitan area—Cape Breton—out of Sydney and other Cape Breton Island municipalities.

6. Before coming into effect, the agreement, approved by the Labrador Inuit in May 2004, must be ratified by Newfoundland and Labrador and Canada. This agreement contains the provision that the Labrador Inuit will receive 25 per cent of the revenue from mining and petroleum production on their settlement land, as well as 3 per cent of provincial royalties from the Voisey's Bay project. The federal environmental review panel examining the possible environmental and social impacts of the Voisey's Bay Nickel Company's mine proposal expressed concern about the disposal of the 15,000 tonnes of mine tailings to be produced each day. The mining company proposes to deposit the toxic tailings in a pond and prevent this from draining into surrounding streams and rivers by building two dams. While federal and provincial environmental officials are satisfied with the company's solution to the tailings problem, local people, especially the Inuit and Innu, are skeptical and worried about the effects of these toxins on the wildlife they depend on for food.

Key Terms

biomass
The total quantity or weight of an organism (codfish) in a given area.

economies of association
Manufacturing plants that are located near their suppliers obtain lower prices because of the transportation savings associated with proximity to suppliers; localization economies.

groundfish
Fish that live on or near the bottom of the sea. The most valuable groundfish are cod, halibut, and sole.

spawning biomass
The total quantity or weight of a sexually mature organism that can reproduce. Cod, for example, reach sexual maturity around the age of seven.

Bibliography

Bailey, Alfred G. 1969. *The Conflict of European and Eastern Algonkian Cultures, 1504–1700.* Toronto: University of Toronto Press.

Bennett, Margaret. 1989. *The Last Stronghold: Scottish Gaelic Traditions in Newfoundland.* Edinburgh: Canongate.

Blackbourn, Anthony, and Robert G. Putnam. 1984. *The Industrial Geography of Canada.* London: Croom Helm.

Blades, Kent. 1995. *Net Destruction: The Death of Atlantic Canada's Fishery.* Halifax: Nimbus Publishing.

Bradfield, Michael. 1991. *Maritime Economic Union: Sounding Brass and Tinkling Symbolism.* Halifax: Canadian Centre for Policy Alternatives.

Canada. 1993. *Canada Year Book 1994.* Ottawa: Ministry of Industry, Science and Technology.

Canada, Natural Resources. 1994. *1993 Canadian Minerals Yearbook: Review and Outlook.* Ottawa: Minister of Supply and Services.

———. 1995. *The State of Canada's Forests: A Balancing Act.* Ottawa: Natural Resources.

Canadian Human Rights Commission. 2003. *Report to the Canadian Human Rights Commission on the Treatment of the Innu of Canada by the Government of Canada*, by Professors Constance Backhouse and Donald M. McRae. At: <http://www.chrc-ccdp.ca/publications/Rapport_Innu_Report/RapportInnuReport_Page3.asp?l=e>.

Cashin, Richard. 1993. *Charting a New Course: Towards the Fishery of the Future.* Ottawa: Department of Fisheries and Oceans.

Choyce, Lesley. 1996. *Nova Scotia: Shaped by the Sea.* Toronto: Viking.

Clapp, R.A. 1998. 'The Resource Cycle in Forestry and Fishing', *The Canadian Geographer* 42, 2: 129–44.

Conrad, Margaret, and James Hiller. 2001. *Atlantic Canada—A Region in the Making.* Toronto: Oxford University Press.

Coward, Harold, Rosemary Ommer, and Tony Pitcher, eds. 2000. *Just Fish: Ethics and Canadian Marine Fisheries.* St John's: ISER Books.

Cox, Kevin. 1994. 'How Hibernia Will Cast Off', *Globe and Mail*, 12 Nov., D8.

Department of Fisheries and Oceans. 1994. *Canadian Fisheries Statistical Highlights 1992.* Ottawa: Department of Fisheries and Oceans.

Doeringer, Peter B., and David G. Terkla. 1995. *Troubled Waters: Economic Structure, Regulatory Reform, and Fisheries Trade.* Toronto: University of Toronto Press.

DRI Canada. 1994. *Atlantic Canada: Facing the Challenge of Change.* Moncton: Atlantic Canada Opportunities Agency.

Eaton, Peter B., Alan G. Gray, Peter W. Johnson, and Eric Handert. 1994. *State of the Environment in the Atlantic Region.* Halifax: Environment Canada.

Erskine, David. 1968. 'The Atlantic Region', in John Warkentin, ed., *Canada: A Geographical Interpretation.* Toronto: Methuen, 231–80.

Fife, Robert. 2001. 'N.S. Recruits Albertans in cash grab', *National Post*, 7 June, A4.

Fisheries and Oceans Canada. 2003. Landings [on-line database]. Searched 3 Sept. 2003: <http://www.dfo-mpo.gc.ca/communic/statistics/landings/S1999aqe.htm>.

Forbes, E.R., and D.A. Muise, eds. 1993. *The Atlantic Provinces in Confederation.* Toronto: University of Toronto Press.

Foster, Gilbert. 1988. *Language and Poverty: The Persistence of Scottish Gaelic in Eastern Canada.* St John's: Institute of Social and Economic Research, University of Newfoundland.

Gordon, Daniel V., and Gordon R. Munro. 1996. *Fisheries and Uncertainty.* Calgary: University of Calgary Press.

Greenspon, Edward. 1998. 'Ottawa Approves New Aid for Fishery', *Globe and Mail*, 12 June, A1, A4.

Hardin, Garrett. 1968. 'The Tragedy of the Commons', *Science* 162: 1243–8.

Hutchings, Jeffrey A., and Ransom A. Myers.

1994. 'What Can Be Learned from the Collapse of a Renewable Resource: Atlantic Cod, *Gadus morhua*, of Newfoundland and Labrador', *Canadian Journal of Fisheries and Aquatic Science* 51: 2126–46.

Isaac, Thomas. 2001. *Aboriginal and Treaty Rights in the Maritimes: The Marshall Decision and Beyond*. Saskatoon: Purich Publishing.

Jang, Brent. 1998. 'Hibernia Halves Output', *Globe and Mail*, 5 Mar., B1.

Keenan, Greg. 1998. 'AB Volvo to Close Halifax Plant', *Globe and Mail*, 10 Sept., B1, B4.

Kirby, J.L. 1982. *Navigating Troubled Waters: A New Policy for the Atlantic Fisheries: Report of the Task Force on Atlantic Fisheries*. Ottawa: Department of Fisheries and Oceans.

MacAfee, Michelle. 1998. 'Diversity in Fisheries Paying Off: Minister', *Globe and Mail*, 14 Aug., A3.

McCalla, Robert J. 1991. *The Maritime Provinces Atlas*. Halifax: Maritext.

McManus, Gary E., and Clifford H. Wood. 1991. *Atlas of Newfoundland and Labrador*. St John's: Breakwater.

MacKenzie, A.A. 1979. *The Irish in Cape Breton*. Antigonish, NS: Formac.

Macpherson, Alan G., ed. 1972. *The Atlantic Provinces: Studies in Canadian Geography*. Toronto: University of Toronto Press.

Marsh, James H., ed. 1988. *The Canadian Encyclopedia*, 2nd edn. Edmonton: Hurtig.

Matthews, Ralph. 1983. *The Creation of Regional Dependency*. Toronto: University of Toronto Press.

———. 1993. *Controlling Common Property: Regulating Canada's East Coast Fishery*. Toronto: University of Toronto Press.

Millward, Hugh, and Lorna Winsor. 1997. 'Twentieth-Century Retail Change in the Halifax Central Business District', *The Canadian Geographer* 41, 2: 194–201.

Natural Resources Canada. 2003a. Mineral Production of Canada, by Provinces and Territory [on-line database]. Searched 30 Aug. 2003: <http://mmsd1.mms.nrcan.gc.ca/mmsd/production/production_e.asp>.

———. 2003b. The State of Canada's Forests 2001/02 [on-line database]. Searched 31 Aug. 2003: <http://www.nrcan.gc.ca/cfs-scf/national/whatquoi/sof/sof02/profiles_e.html.>

Neis, Barbara, and Lawrence Felt, eds. 2000. *Finding Our Sea Legs: Linking Fishery People and Their Knowledge with Science and Management*. St John's: ISER Books.

Norcliffe, Glen, and Judy Bates. 1997. 'Implementing Lean Production in an Old Industrial Space: Restructuring at Corner Brook, Newfoundland, 1984–1994', *The Canadian Geographer* 41, 1: 41–60.

Palmer, Craig T., and Peter Sinclair. 1997. *When the Fish Are Gone: Ecological Disasters and Fishers in Northwest Newfoundland*. Halifax: Fernwood.

Phillips, David. 1993. *The Day Niagara Falls Ran Dry!* Toronto: Canadian Geographic and Key Porter Books.

Ponting, J. Rick. 1997. *First Nations in Canada: Perspectives on Opportunity, Empowerment, and Self-Determination*. Toronto: McGraw-Hill Ryerson.

Power, Thomas P., ed. 1991. *The Irish in Atlantic Canada, 1780–1900*. Fredericton: New Ireland Press.

Rumney, Thomas. 1998. *The Geography of Canada Bibliography Series: Vol. 3, Atlantic Canada*. Plattsburgh, NY: Plattsburgh State University, Center for the Study of Canada.

Samson, Colin. 2003. *A Way of Life That Does Not Exist: Canada and the Extinguishment of the Innu*. St John's: ISER Books.

Savoie, Donald. 2000. *Aboriginal Economic Development in New Brunswick*. Moncton: Canadian Institute for Research on Regional Development.

——— and Ralph Winter. 1993. *The Maritime Provinces: Looking to the Future*. The Canadian Institute for Research on Regional Development. Sackville, NB: Tribune Press.

Stanford, Quentin H., ed. 1998. *Canadian Oxford World Atlas*, 4th edn. Toronto: Oxford University Press.

Statistics Canada. 1982. *Census Metropolitan Areas and Census Agglomerations with Components*. Catalogue no. 95–903. Ottawa: Minister of Supply and Services Canada.

———. 1994. *Agricultural Economic Statistics*. Catalogue no. 21–603E. Ottawa: Statistics Canada.

———. 1995. *Annual Demographic Statistics, 1994*. Catalogue no. 91–213. Ottawa: Statistics Canada.

———. 1996. *Labour Force Annual Averages 1995*. Catalogue no. 71–220–XPB. Ottawa: Statistics Canada.

———. 1997a. *A National Overview: Population and Dwelling Counts*. Catalogue no. 93–357–XPB. Ottawa: Industry Canada.

———. 1997b. 1996 Census: National Tables— Population by Mother Tongue, Showing Age Groups, for Canada, Provinces and Territories, 1996 Census—20% Sample Data, 2 Dec. 1997 [on-line database], Ottawa. Searched 15 July 1998: <http://www.statcan.ca/english/census96 />.

———. 1997c. *The Daily*—1996 Census: Mother Tongue, Home Language and Knowledge of Languages, 2 Dec. 1997 [on-line database], Ottawa. Searched 14 July 1998: <http://www.statcan.ca/Daily/English/>.

———. 1998a. Canada Highlights: Census of Agriculture [on-line database], Ottawa. Searched 2 Sept.1998: <http://www.statcan.ca/english/ censuag>.

———. 1998b. *Canadian Economic Observer*. Catalogue no. 11–010–XPB. Ottawa: Statistics Canada.

———. 1998c. *The Daily*—1996 Census: Aboriginal Data, 13 Jan. 1998 [on-line database], Ottawa. Searched 14 July 1998: <http://www.statcan.ca/Daily/English/>.

———. 1998d. *The Daily*—1996 Census: Ethnic Origin, Visible Minorities, 17 Feb. 1998 [on-line database], Ottawa. Searched 16 July 1998: <http://www.statcan.ca/Daily/English/>.

———. 1998e. Table 2: Revised Statistics of the Mineral Production of Canada, by Province, 1996 [on-line database], Ottawa. Searched 3 Sept. 1998: <http://www.nrcan.gc.ca/mms/efab/mmsd/production>.

———. 1998f. 1981–1996 Census: Labour Force Activity [on-line database], Ottawa. Searched 2 Sept. 1998:<http://www.statcan.ca/english/census96/mar17/Labour/table6/>.

———. 2000a. Population [on-line database], Ottawa. Searched 25 Oct. 2000: <http://www.statcan.ca/english/Pgdb/People/Population/demo 02.htm>.

———. 2000b. Labour Force, Employed and Unemployed, Numbers and Rates [on-line database], Ottawa. Searched 8 Nov. 2000: <http://www.statcan.ca/english/Pgdb/People/Labour/labor07a.htm>.

———. 2000c. Gross Domestic Product [on-line database], Ottawa. Searched 25 Nov. 2000: <http://www.statcan.ca/english/Pgdb/Economy/Economic/econ15.htm>.

———. 2000d. Preliminary Estimates of the Mineral Production of Canada, by Province, 1999 [on-line database], Ottawa. Searched 7 Jan. 2001: <http://nrcan.gc.ca/mms/efab/mmad/production/1999/99p.pdf>.

———. 2002a. 2001 Census: Population and Dwelling Counts, for Census Metropolitan Areas and Census Agglomerations, 2001 and 1996 Censuses, 16 July 2002 [on-line database], Ottawa: <http://www.statcan.ca/english/IPS/Data/93F0050XCB2001013.htm>.

———. 2002b. Census of Canada 2001—Census Geography. Highlights and Analysis: Canada's 2001 Population: <http://www12.statcan.ca/English/census01/>.

———. 2003a Agriculture 2001 Census Farm Operations: Provincial/Regional Trends [on-line database], Ottawa. Searched 30 Aug. 2003: <http://www.statcan.ca/english/agcensus2001/first/regions/contents.htm>.

———. 2003b Canadian Statistics—Export of Goods on a Balance-of-Payments Basis [on-line database], Ottawa. Searched 2 Sept. 2003: <http://www.statcan.ca/english/Pgdb/gblec04.

htm>.

————. 2003c. Canadian Statistics: Distribution of Employed People, by Industry, by Province [on-line database], Ottawa. Search 20 May 2003: <http://www.statcan.ca/english/Pgdb/labor21b.htm>.

Stavely, Michael. 1987. 'Newfoundland: Economy and Society at the Margin', in L.D. McCann, ed., *Heartland and Hinterland: A Geography of Canada*, 2nd edn. Scarborough, Ont.: Prentice-Hall, 247–85.

Wynn, Graeme. 1987. 'The Maritimes: The Geography of Fragmentation and Underdevelopment', in L.D. McCann, ed., *Heartland and Hinterland: A Geography of Canada*, 2nd edn. Scarborough, Ont.: Prentice-Hall, 175–245.

Further Reading

Coates, Ken S. 2000. *The Marshall Decision and Native Rights*. Montréal and Kingston: McGill-Queen's University Press.

On 7 September 1999 the Supreme Court of Canada ruled in a case involving Donald Marshall Jr (who 16 years earlier had gained national attention when he was exonerated from a murder conviction after spending more than a decade in prison) that the Mi'kmaq could earn a 'modest income' from the fishery in the Maritimes. In one swoop, Atlantic Canada woke up to the Aboriginal desire and right to participate in resource development. While the Supreme Court limited its decision to the fishery, further legal action could see Aboriginal peoples gaining a share of other resources. Such events are taking place across Canada, as witnessed by the James Bay and Northern Québec Agreement, a series of comprehensive land-claim agreements in the North, and the Nisga'a Agreement in British Columbia. The Mi'kmaq, like other Indian, Inuit, and Métis peoples, want a share of Canada's natural resources, a place in the global market economy, and respect for their culture.

Coates places the *Marshall* decision, which changes the relationship between the Mi'kmaq and other Maritimers, within the larger Canadian context where a search for a new relationship between Aboriginal Canadians and other Canadians is underway. The tone and message of Coates's book is aptly captured by his call 'to restructure the relationship between First Nations and Other Canadians'. In Chapter 1, 'Of Eels, Judges, and Lobsters: The Marshall Challenge and the Supreme Court Decision', students will find an easy-to-read account of the complexity and urgency of revisiting historic treaties (in this case, a 1760–1 treaty) to ensure that they 'work' in our contemporary world.

Overview and Objectives

While the Territorial North is the largest geographic region in Canada, it has by far the smallest economy and population. Its cold climate and remote location greatly limit economic development and settlement. The region is a resource frontier for both Canada and the world, with energy and mineral products accounting for most of its output. While hunting remains an important source of food for Aboriginal peoples, trapping for commercial purposes has declined sharply. With the settlement of land claims and the formation of the new territory of Nunavut, more and more Aboriginal peoples are involved in the market economy. Because Aboriginal peoples constitute half of the region's population, their participation in economic development is crucial. Economic development in the Territorial North involves huge capital investments, so large-scale projects are common while social development calls for large transfer payments by Ottawa. The *Key Topic* in this chapter is megaprojects.

- Describe the Territorial North's physical and historical geography, including the birth of Nunavut and modern land claims.
- Present the dualistic nature of its population and economy.
- Examine the population and economy of the Territorial North's in the context of its physical setting and resource potential.
- Explore the region's changing economic and political position within Canada, North America, and the emerging global economy.
- Focus on the role of megaprojects in the North's development.

Chapter 10 The Territorial North

Introduction

Consisting of Yukon, the Northwest Territories, and Nunavut, the Territorial North is Canada's largest region (Figure 1.1). In Friedmann's regional scheme of the core/periphery model, the Territorial North would be described as a resource frontier (Table 1.5). Like other resource frontiers, the Territorial North has four principal characteristics:

- It is far from world markets.
- Resource development is dependent on external demand.
- Resource development is hampered by a cold environment.
- Its economy is sensitive to fluctuations in world prices for its resources.

The Territorial North is also a homeland for Aboriginal peoples. It is the only geographic region in Canada in which Aboriginal peoples constitute a majority of the total population (Figure 10.1). In 2001, just over half of the population of the Territorial North was classified as Aboriginal. Nunavut provides a more revealing example because 85 per cent of its population have declared themselves as Inuit. Nunavut, the most recently created territory, is a political expression of the homeland concept for the Inuit. (Nunavut is discussed later in this chapter.) The Territorial North, then, as both a frontier and a homeland, has a dual character. This duality is a theme that runs through this chapter. The *Key*

Topic explored is megaprojects in the Territorial North.

The Territorial North within Canada

Of the six geographic regions, the Territorial North has the smallest population. In 2001, it had 93,000 residents spread over nearly 4 million km². With a population density of only 2.5 people per 100 km², this region is one of

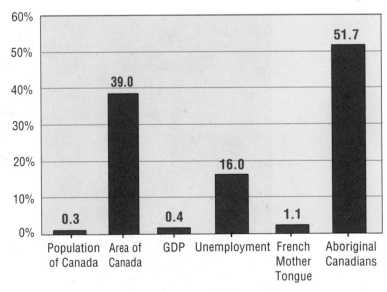

Figure 10.1 The Territorial North, 2001. Though the region is the largest in Canada, its population and economy are the smallest. Some French-speaking Canadians reside in the North but over half of the population is Aboriginal. The Territorial North suffers from high unemployment.

Notes: The unemployment figure is based on territorial data. Aboriginal unemployment is much higher, perhaps as high as 30 per cent.

Sources: Tables 1.1, 1.5, and 4.19.

the world's most sparsely populated areas. Most significantly, Aboriginal peoples form just over half of the total population of this northern region. The Territorial North has the smallest economic output of the six geographic regions in Canada (Figure 10.1).

The Territorial North's demographic features have been shaped by two factors. First, its small population is due to its weak economy, which can support relatively few people. Second, the Aboriginal population—especially the Inuit—has a very high birth rate that accounts for the population growth in the Territorial North. Third, the non-Aboriginal population tends to move to job opportunities so that when the North's economy stalls, non-Aboriginal residents are more likely to move to southern Canada than Aboriginal residents.

As a resource frontier, the Territorial North

has an export-based economy. Exploitation of non-renewable resources has spurred economic growth in the region, but this type of economic development lacks stability because it is subject to a boom-and-bust cycle. This cycle is caused by the finite nature of non-renewable resources and by fluctuations in world prices; downward price movements have caused temporary shutdowns of mining operations. Two recent shutdowns involved the Faro lead-zinc mine in Yukon and the Giant gold mine in the Northwest Territories.

Two Images

The Territorial North has two powerful and seemingly contradictory images—one is as a northern frontier, while the other is as a homeland. The traditional image of the northern

The South Nahanni is one of the world's great wild rivers. Located in boreal wilderness of Nahanni National Park Reserve in the southwestern part of the Northwest Territories, this untamed river is seen surging through the steep-walled First Canyon. Later, its waters rush past hot springs, plunge over a waterfall twice the height of Niagara, and cut through canyons more than one kilometre deep. (Stephen J. Krasemann/Valan Photos)

Vignette 10.1   Northern Lights

Northern lights (aurora borealis) are daz-zling displays of rapidly moving light that appear mostly as white-like flashes extend-ing across the sky. Sometimes these flashes display various colours. Northern lights occur most frequently in higher latitudes. At Yellowknife (62° 30' N), for example, north-ern lights occur, on average, nearly 250 nights a year; at Toronto (43° 42' N), north-ern lights may occur 20 nights a year. Since Japanese culture associates the aurora borealis with good luck for newly married couples, many Japanese couples travel to Yellowknife to see the northern lights. But what are northern lights? They are the visi-ble portions of the dissipation of solar ener-gy carried to the earth's magnetosphere by solar winds. Solar winds are the stream of electrically charged particles emitted by the sun. When these particles near the earth, they are drawn to the earth's two magnetic poles where they form a magnetic field known as the magnetosphere. While the aurora borealis causes spectacular displays of light in northern skies, northern lights are also associated with geomagnetic distur-bances that can disrupt communication and electrical systems.

frontier is one of great wealth just waiting to be discovered. For example, during the Klondike gold rush (1897–8), prospectors flooded the Yukon to pan for gold along the Klondike River. A more contemporary version of this image is one of large multinational cor-porations with their vast capital and advanced technology undertaking megaprojects—min-ing for gold, diamonds, lead, and zinc, and drilling for oil and gas. **Megaprojects** are large-scale resource developments financed and managed by multinational corporations designed to meet global needs for primary products. Such projects create an economic boom during the construction period, but in their operational phase fewer employment opportunities are available and economic spin-offs for local businesses are limited. Because of the risks associated with developing resources in a frontier—from overcoming physical barri-ers unique to the Territorial North to coping with downturns in world prices for resources—such projects are usually under-taken by large corporations. In return, these corporations reap large profits and supply the industrial cores of the world with raw materi-als and energy.

Northerners, however, particularly Aborig-inal peoples, see the North as a **homeland**. This perception is based on a special, deep

Offshore drilling in the Beaufort Sea. Megaprojects are common on resource frontiers, especially in the North, where adverse conditions require that developers have the investment power and experience to man-age risks. (CP/Bob Weber)

commitment to the North, which geographers often describe as regional consciousness. Local people have a strong appreciation for natural features, cultural traits, and the political and economic issues affecting their homeland. Regional consciousness, also known as a sense of place (discussed in Chapter 1), evokes a sense of belonging to a particular place.

The dual roles of homeland and frontier are not always compatible. For instance, the interests of multinational companies relate to the profitability of the northern frontier's untapped natural wealth. The interests of northerners, however, especially Aboriginal northerners, are best served when the long-term environmental and social well-being of their communities is considered. 'Frontier' economic activities often bring jobs and investment, but these activities, by focusing on short-term profitability, have little long-term or beneficial impact on Aboriginal communities.

In the Territorial North, the concepts of homeland and regional consciousness have resulted in the devolution of political power from the federal government to the territorial governments. Elected governments exist in the three territories and six land-claim agreements with First Nations have been concluded, with the most recent being the Tlicho Final Agreement (2003). Significantly, the Tlicho Final Agreement included provisions for self-government for the Dogrib. Future comprehensive land-claim agreements are likely to include a section on self-government. Aboriginal self-government received an enormous boost in 1992 when the provision for the territory of Nunavut was placed in the Nunavut Land Claims Agreement under the Nunavut Political Accord. Unlike Yukon and the Northwest Territories, Nunavut is an expression of 'ethnic' regional consciousness. The next challenge facing the Territorial North, but especially Nunavut, is to generate sufficient economic growth to break its economic dependency on Ottawa, while still preserving the homeland to which many northerners are committed.

The Territorial North's Physical Geography

The Territorial North includes the Yukon Territory, the Northwest Territories, and the new territory of Nunavut. The region extends over four of Canada's physiographic regions: the Canadian Shield, the Interior Plains, the Cordillera, and Arctic Lands (including the Arctic Archipelago) (Figure 2.1). As a result, the region encompasses a varied topography, from mountainous terrain and forest stands in the west, to barren plains, ice-covered islands, and Arctic seas further east and north. The vegetation in this region is also quite varied and includes a small portion of the boreal forest in the southwest, the tundra with its mosses and lichens further north, and a polar desert in the high reaches of the Territorial North (Figure 2.7). The region is also known for the several rivers that wind through it, the many lakes that dot its landscape, and the Arctic Ocean that supports a range of aquatic wildlife.

However, the physical geography of the Territorial North is governed not so much by physiography as by a cold environment. Cold persists throughout most of the year and, in many ways, it affects human activities in the Territorial North. The Territorial North's cold environment includes permafrost (Figure 2.9) and long winters with sub-zero temperatures. Except for the northern half of Yukon, the Territorial North was subjected to glaciation. For that reason the rich Klondike placer deposits were not affected by the scouring effects of the Cordillera ice sheets that affected southern Yukon. The region's main climate zones, the

Arctic and the Subarctic (Figure 2.6), are characterized by very short summers.[1] In the case of the Arctic climate, summer is limited to a few warm days interspersed with more autumn-like weather, including freezing temperatures and snow flurries. The Subarctic climate has a longer summer that lasts at least one month. During the short but warm summer, the daily maximum temperature often exceeds 20° C and sometimes reaches 30° C.

Arctic air masses dominate the weather patterns in the Territorial North. They are characterized by dry, cold weather and originate over the ice-covered Arctic Ocean, moving southward in the winter. The Arctic zone has an extremely cold and dry climate. Distinguished by long winters and a brief summer, the Arctic climate is normally associated with high latitudes and lower levels of solar energy. The ice-covered Arctic Ocean and continuous permafrost keep summer temperatures cool even though the sun remains above the horizon for most of the summer. These cool summer temperatures, which Köppen defined as an average mean of less than 10° C in the warmest month, prevent normal tree growth. For that reason, the Arctic climate region has tundra vegetation, which includes lichens, mosses, grasses, and low shrubs. In the very cold Arctic Archipelago, the ground is often bare, exposing the surface material. As there is little precipitation in the Arctic Archipelago (often less than 20 cm per year), this area's climate is sometimes described as a 'polar desert'.

Beyond 70° N, growing conditions become very limited. With lower temperatures and less precipitation than in the lower latitudes of this climatic zone, the tundra vegetation cannot sustain itself. In these high latitudes, much of the land is barren. The Arctic climate, however, does extend into lower latitudes in two areas: along the coasts of Hudson Bay and of the Labrador Sea. In both cases, these cold bodies of water chill the summer air along the adjacent land mass. In this way, the Arctic climate extends along the coasts of Ontario, Québec, and Labrador well below 60° N, sometimes extending as far south as 55° N.

The Arctic Ocean was called the 'Frozen Sea' by early explorers. This extensive ice cover, known as polar pack ice, drifts in a clockwise motion in the Beaufort Sea. Because of the extent and thickness of polar pack ice, few ships can navigate these waters without the assistance of icebreakers. The two mining operations in the High Arctic—the Nanisivik mine on northern Baffin Island and the Polaris mine on Little Cornwallis Island—must store their ore until the summer navigation season, when ships reinforced against ice, sometimes aided by icebreakers, transport the ore to European and other world markets.

Vignette 10.2   Resource Basins in the North

The Territorial North has many sedimentary basins, some of which contain petroleum deposits. The three main petroleum-containing basins are the Western Sedimentary Basin, the Mackenzie Basin, and the Canadian Arctic Basin. The Canadian Arctic Basin contains several smaller basins, including the Sverdrup Basin. Total reserves in the Mackenzie Delta and the Beaufort Sea are estimated at 12 trillion cubic feet of natural gas and 1.7 billion barrels of oil. Their market value at current prices (2001) is approximately $325 billion.

The geology of the Territorial North provides much of its wealth. For example, the vast sedimentary basins within the Interior Plains and Arctic Lands sometimes contain large deposits of oil and natural gas (Vignette 10.2). Similarly, the Cordillera and Canadian Shield have already yielded some of their mineral wealth to prospectors and geologists. These minerals include diamonds, gold, lead, uranium, and zinc.

Environmental Challenges

The Territorial North has a delicate environment. For that reason, industrial activities are more likely to cause long-lasting environmental damage than in more temperate lands. The exploitation of its resources has often scarred the landscape. The Yukon gold rush was no exception. Forests were ravaged for fuel while the river landscape was mutilated by hydraulic

mining that was employed as placer gold mining declined. To extract the gold, water and steam were forced by high-pressure hoses into the frozen sands and gravels of the valley terraces on the tributaries of the Yukon River. Gold was recovered after thawing and sorting these deposits, but in the process, trees were cut down for firewood and building materials, riverbanks were destabilized, and tailings (the huge piles of discarded sand and gravel) were scattered across the landscape. Environmentally, gold mining had a destructive impact on the Yukon landscape and those scars are still visible today.

In today's world, toxic wastes from mining operations pose a major threat to the environment. One of the worst examples is found near Yellowknife at the former Giant gold mine. A toxic time bomb of 270,000 tonnes of arsenic trioxide, a by-product of gold refining, sits in the Giant mine. Since the mining company

The Yukon River Valley near Dawson City. From Aboriginal peoples to fur traders and prospectors, this long and winding river has been an important transportation route in the history of the Territorial North. (Bill Terry/Take Stock Inc.)

(Royal Oak Mines) declared bankruptcy, the cost of the cleanup is left to the federal government at an estimated cost of a quarter of a billion dollars (Bone, 2003a: 160).

Resource conflicts also exist. Often, environmental issues are found at the heart of resource conflicts. Such a case has emerged near Dawson City in Yukon where placer gold mining has been a way of life since the Klondike gold rush of 1898. Here, some 280 family-operated placer gold-mining operations are threatened by new regulations designed to protect the fish habitat. Placer mining involves the use of water and sluices to separate gold particles and nuggets from gravel and sand. Such operations alter the water quality and disturb the fish habitat. In 2002, the Department of Fisheries and Oceans announced that, over a four-year period, new regulations will be introduced to restrict the amount of sediment released into the water. According to Tara Christie, president of the Klondike Placer Miners' Association, 'None of our existing operations will be able to meet the new standards consistently and 54 per cent of them will never be able to adapt to the changes, due to physical constraints. As far as we are concerned, DFO is out of control' (Canadian Press, 2003: FP10). By 2006, DFO plans to introduce site-specific reviews and thereby eliminate those mining operations that do not comply with the new guidelines.

The Territorial North's Historical Geography

At the time of contact with Europeans, seven Inuit groups and seven Indian groups belonging to the Athapaskan language family (also known as Dene) occupied the Territorial North. The Inuit stretched across the Arctic: the Mackenzie Delta Inuit lived in the west;

further east were the Copper Inuit, Netsilik Inuit, Iglulik Inuit, Baffinland Inuit, Caribou Inuit, and Sadlermiut Inuit. Inuit also lived in northern Québec and Labrador. By the early twentieth century, two groups (most of the Mackenzie Delta Inuit and all of the Sadlermiut Inuit) would succumb to diseases that European whalers brought to the Arctic. The Indian tribes that resided in the Subarctic were the Kutchin, Hare, Tutchone, Dogrib, Tahltan, Slavey, and Chipewyan.

These Inuit and Indians had developed hunting techniques that were well adapted to these two cold but different environments. The Inuit employed the kayak and harpoon to hunt seals, whales, and other marine mammals, which enabled them to occupy the Arctic coast from Yukon to Labrador. As a result, the Inuit depended extensively on marine mammals and fish. The Indians hunted and fished in the northern coniferous forest, where the birchbark canoe, the bow and arrow, and snowshoes enabled them to hunt in summer and winter. The Dene tribes relied heavily on big game like the caribou, the Chipewyans often following the caribou to their calving grounds in the northern barrens of the Arctic. Both the Inuit and the Dene moved across the land in a seasonal rhythm, following the migratory patterns of animals. Operating in small but highly mobile groups, these hunting societies depended on game for their survival. Cultural traits, such as the ethic of sharing, developed from this dependency on the land and sea for food.[2]

Though the Vikings were the first to make contact with northern Aboriginal peoples around 1000, little is known of those encounters.[3] About five centuries later, in 1576, Martin Frobisher, in searching for a Northwest Passage to the Far East, sailed to the northern Arctic. He reached Baffin Island where he encountered a group of Inuit, some on land, others in their kayaks. A skirmish between Frobisher's men and the Inuit took place in which Frobisher was wounded by an arrow. In the exchange, five of his men were lost and three of the Inuit were captured. The Inuit and one kayak were taken back to England as proof of Frobisher's discovery. All three of the captives soon succumbed to illness. Following Frobisher, various European explorers ventured into Arctic waters in search of the Northwest Passage, including John Franklin, whose famous last expedition ended in disaster (Vignette 10.3). However, cultural exchange between Europeans and the original inhabitants of these lands remained limited until the nineteenth century, when the trade in fur pelts and whaling peaked in North America.

Whaling and the Fur Trade

Whaling, which was the first commercial venture in the Arctic, began in the late sixteenth century in the waters off Baffin Island. During those early years of whaling, whalers had little opportunity or desire to make contact with the Inuit living along the Arctic coast. The Inuit probably felt the same, particularly those who could recall the nasty encounter with Frobisher's men. During early summer, whaling ships set sail from British, Dutch, and German ports to Baffin Bay, where they hunted whales for several months. By September, all ships would return home. In the early nineteenth century, the expeditions of John Ross (1817) and William Parry (1819) sailed farther north and west into Lancaster Sound. Their search for the Northwest Passage had limited success but opened virgin whaling grounds for whalers. These new grounds were of great interest as improved whaling technology had reduced the whale population in the eastern Arctic. In fact, the period from 1820 to 1840 is regarded as the peak of whaling activity in this area. At that time, up to 100 vessels were whaling in Davis Strait and Baffin Bay.

As whaling ships went further afield to find better whaling grounds, it became impossible

Vignette 10.3   The Franklin Search

In 1845, Sir John Franklin headed a British naval expedition to search for the elusive Northwest Passage through the Arctic waters of North America. He and his crew never returned. Their disappearance in the Canadian Arctic set off one of the world's greatest rescue operations, which was conducted both on land and by sea and stretched over a decade. The British Admiralty organized the first search party in 1848. Lady Franklin sent the last expedition to look for her husband in 1857. These expeditions accomplished three things: (1) they found evidence confirming the loss of Franklin's ships (the *Erebus* and *Terror*) and the death of his crew; (2) one rescue ship under the command of Robert McClure almost completed the Northwest Passage; and (3) the massive rescue effort resulted in a greater knowledge of the numerous islands and various routes in this part of the Arctic Ocean. The exact sequence of events that led to the Franklin disaster is not known. However, archaeological work, conducted in the early 1980s on the remains of members of the expedition, revealed that lead poisoning, caused by the tin cans in the ships' food supplies, probably contributed to the tragic demise of the Franklin expedition.

to return to their home ports within one season. By the 1850s, the practice of 'wintering over' (that is, allowing ships to freeze in sea ice along the coast) was adopted by English, Scottish, and American whalers. This allowed whalers to get an early start in the spring, providing for a long whaling season before the return trip home at the onset of the next winter. Wintering over took place along the indented coastlines of Baffin Island, Hudson Bay, and the northern shores of Québec and Yukon. Permanent shore stations were established at Kekerton and Blacklead Island in Cumberland Sound, at Cape Fullerton in Hudson Bay, and at Herschel Island in the Beaufort Sea. Life aboard whaling ships was dirty, rough, and dangerous, and many sailors died when their ships were caught in the ice and crushed.

The Inuit welcomed the whaling ships because of the opportunity for trade. The Inuit were attracted to shore stations and often worked for the whalers by securing game, sewing clothes, and piloting the whaling ship through difficult waters to promising sites for whale hunting. Some Inuit men signed on as boat crew and harpooners. In exchange for this work, the Inuit obtained useful goods, including knives, needles, and rifles, which made domestic life and hunting easier. While this relationship brought many advantages for the Inuit, there were also negative social and health aspects, including the rise in alcoholism and the spread of European diseases among the Inuit (Vignette 10.4). Perhaps the most devastating result of this trade relationship for the Inuit was the unexpected end of commercial whaling, which, for the Inuit, represented the loss of access to highly valued trade goods. Just as the twentieth century began, demand for products made from whales decreased sharply, halting the flow of whalers, and thus trade goods, that were sailing into Arctic waters. By now, the Inuit were dependent on trade goods for their hunting activities. Somehow, they had to find another means of obtaining these useful trade goods.

Fortunately for the Inuit, the fur trade had

Vignette 10.4 European Diseases

Whalers, fur traders, and missionaries introduced new diseases to the Arctic. As the Inuit had little immunity to measles, smallpox, and other communicable diseases such as tuberculosis, many of them died. In the late nineteenth century, the Sadlermiut and the Mackenzie Delta Inuit were exposed to these diseases. As a result, all the Sadlermiut died. The Mackenzie Delta Inuit, whose numbers were as high as 2,000, almost suffered the same fate but managed to survive.

The Mackenzie Delta Inuit occupied the northwestern Arctic coast, in present-day Yukon, the Northwest Territories, and part of Alaska. Herschel Island, lying just off the Yukon coast, was an important wintering station for American whaling ships. Whalers often traded their manufactured goods with the local Mackenzie Delta Inuit, who became involved with the commercial whaling operations. Through contact with the whalers, the Inuit were infected by European diseases. By 1910, only about 100 Mackenzie Delta Inuit were left. Gradually, Inupiat Inuit from nearby Alaska and white trappers who settled in the Mackenzie Delta area intermarried with the local Mackenzie Delta Inuit, which secured the survival of these people. Today, their descendants are called Inuvialuit.

been expanding northward into the Arctic, thereby providing a replacement for whaling. The fur trade had already been successfully operating in the Subarctic for some time—a relationship between European traders and the Indian tribes in this part of the Territorial North was established through the trade of fur pelts, especially beaver. However, by the beginning of the twentieth century, the fashion world in Europe had discovered the attractive features of the Arctic fox pelt. Demand for Arctic fox pelts rose, which led the Hudson's Bay Company to establish trading posts in the Arctic. Soon the Inuit were deeply involved in the fur trade. The working relationship between the Hudson's Bay Company and the Inuit was based on barter trade: white fox pelts could be traded for goods.

Until the 1950s, the fur trade dominated the Aboriginal land-based economy. It lasted for less than 100 years in the Arctic and for over three centuries in the Subarctic. Did the fur trade, as well as Arctic whaling, create a form of dependency whereby Indians and Inuit could not survive without trade goods? The answer is a qualified yes. At first, Aboriginal peoples had a form of partnership with European traders and whalers. Each side had power—for instance, the European traders needed the Indians to trap beaver and the Indians needed the traders to obtain European goods and technology. Gradually, however, the power relationship shifted in favour of the traders. By the nineteenth century, the fur companies controlled the fur economy. Fur-trading posts dotted the northern landscape. Indians, who had long ago integrated trade goods into their traditional way of life—including their hunting techniques and their migration patterns—were therefore heavily dependent on trade. In fact, when game was scarce, tribes relied on the fur trader for food.

Ironically, by securing game for the traders, Indians reduced the number of animals that would be available for their own sustenance. In the Territorial North, game became scarce around fur-trading posts not from natural causes but from overexploitation.

The problems of a growing dependency on European goods and a changing way of life for northern Aboriginal peoples were compounded after the arrival of Anglican and Catholic missionaries in the 1860s and the North West Mounted Police (NWMP) in the 1890s. The Indians and Métis were confronted with the full force of Western culture in the late nineteenth century and the Inuit in the early twentieth century. The Western ideas and rules introduced by the missionaries and police who now lived near the trading posts had a profound impact on Aboriginal culture. The NWMP (which added 'Royal' to its name in 1904 and, in 1920, was renamed the Royal Canadian Mounted Police) imposed Canada's system of law and order on Aboriginal peoples, while the missionaries challenged their spiritual values and encouraged the Inuit, Indians, and Métis to remain in the settlements. Also, both Anglican and Catholic missionaries placed young Aboriginal children in church-run residential schools, where they were taught in either English or French. In this attempted assimilation, most children learned to read and write in English or French, but they were inadequately prepared for northern life. As they lost the opportunity to learn from their parents about how to live on the land, they became trapped between the two very different worlds of their Aboriginal communities and the Euro-Canadians. Under these circumstances, many lost their indigenous language, animistic beliefs, and cultural customs. Fur traders opposed many of these induced Western cultural adaptations because they needed the Aboriginal peoples on the land to trap. Nevertheless, the influence of the churches, the power of the state, and the number of non-Aboriginal residents in the North increased in the twentieth century, placing Aboriginal cultures under siege and crippling their land-based economy. However, political changes were occurring at this time that would first lead to territorial governments and then to land-claim agreements.

Political and Territorial Evolution

When Canada was formed in 1867, the Territorial North remained a British possession. Britain had claimed British North America on the basis of settlement, trade, and exploration. In the Territorial North, the British declared their ownership of the Arctic islands, basing their claim on the British Navy's efforts to find the Northwest Passage, including the search for Franklin's missing ship. In the rest of the Territorial North, the Hudson's Bay Company had established a number of fur-trading posts in the forested lands of the Mackenzie Basin and the Yukon. By extending its fur-trading economy over this area, the British government claimed these Subarctic lands. Canada came into possession of the Territorial North with the purchase of Rupert's Land in 1869 and its transfer to Canada in 1870, and in 1880 Britain transferred the Arctic islands to Canada.

At first, these northern territories were governed from Ottawa. The first territorial government was established in Yukon after the population soared to 30,000. The demand for self-government came not from the Aboriginal population but from those who had been lured north by the Klondike gold rush. In 1898, Yukon became a separate territory—a territorial government was formed that consisted of a federally appointed commissioner and council located in Dawson City.[4] All appointed officials were residents of the newly formed territory. In 1899, the process towards electing members of council began. At that time, two were elected, and by 1908 all 10 members were elected. People began to leave the territory when the production of gold declined. This population loss led Ottawa to withdraw some of the territory's powers, including reducing the size of its elected council. After World War II, Yukon's population again increased, reaching 9,096 by 1951, so the number of elected council members increased from three to five. By 1996, Yukon's population was 30,766, permitting 17 elected members.

The evolution of the Northwest Territories was much different. In 1905, much of the Northwest Territories south of 60° N was assigned to two new provinces, Saskatchewan and Alberta. Another adjustment to the Northwest Territories border took place in 1912, when Manitoba's boundary was extended to 60° N (see Figures 3.4 to 3.7). The boundaries of the Northwest Territories underwent another change with the establishment of Nunavut in 1999. From 1905 until after World War II, the NWT was governed by an appointed commissioner and council, which was composed entirely of senior civil servants based in Ottawa. When its population reached 16,004 in 1951, elected members were gradually added to the previously appointed council until it became a fully elected body in 1975. Until 1963, the commissioner was a deputy minister in the federal department in charge of the administration of the Northwest Territories. In 1964, the first full-time commissioner was appointed to a separate territorial office. In 1967, the seat of territorial government was moved to Yellowknife and the commissioner was relocated there with the nucleus of what has since become a territorial public service. By 1996, the population had reached 64,402 and the council of the Northwest Territories consisted of 24 elected members.

The territories, including Nunavut (discussed later in the chapter), do not have all the powers given to the provinces through the Canadian Constitution. In fact, territorial powers do not stem from the Constitution but are assigned to the territories by the federal government. Territorial powers include education, social services, tax collection, highways, and community services. Given the importance of wildlife to Aboriginal peoples, the territorial governments are also responsible for wildlife. But the territories do not control tax revenues from their natural resources, which go to Ottawa. To offset this tax loss, the territories receive most of their revenue as transfer payments from the federal government. Without these transfer payments, the territorial governments would not be able to afford their existing programs.

Forgotten Frontier, Confederation to 1939

From the time the Territorial North in its entirety had been transferred to Canada in 1880 until World War II, the region was a forgotten part of Canada. The Territorial North was forgotten because it had little agricultural land and few commercial resources. For the federal government, the Territorial North was not a 'priority' region and thus received little attention from Ottawa. Aboriginal peoples, the vast majority of the population, continued to engage in hunting and trapping and, in times of need, turned to the fur trader for assistance. Ottawa adopted a laissez-faire policy towards the Territorial North to minimize federal

expenditures. At the same time it left the fur traders and missionaries to deal with the needs of a hunting society.

The federal presence in the Territorial North was through the Mounted Police, first assigned to Dawson City in Yukon during the Klondike rush. At numerous fur-trading posts across the North, the RCMP established detachments that not only enforced Canadian law but also provided a variety of other administrative services. During the winter, the local constable and his Native assistant would visit the various hunting camps by dog team to ascertain the well-being of the people at these camps and, if necessary, to deal with law-and-order matters and administer basic health care.

Before 1939, the only commercial interest expressed in this region of Canada came in the form of gold seekers. Individual prospectors of the Klondike gold rush in the late 1890s were followed by mining companies that invested in the region and established gold-mining operations. As a remote hinterland, the Territorial North had only a few mining operations, most of which were gold mines. Gold and other precious metals have such a high value per unit of weight that they can overcome the 'barrier of distance', that is, the cost of transporting them to external markets. The cost of shipping other minerals that could be mined in the region—such low-value minerals as copper, lead, nickel, and zinc—was simply too high at this time. Coal, however, was mined in Yukon because it supplied local heating needs.

Strategic Frontier and Resource Development

While the basic role of the Territorial North remains as a resource hinterland, a new role for the Territorial North emerged with the outbreak of World War II. As a strategic frontier, the North saw considerable investment in military bases and radar stations. While the nature of its strategic role changed over time, the Territorial North served as a buffer zone for over fifty years. This role ceased with the collapse of the Soviet Union and the end of the Cold War in 1991.

When World War II began, the United States military quickly recognized the strategic importance of Canada and its northern territories, an importance that would last until the collapse of the Soviet Union in 1991. For the Americans, Canada's North provided a secure transportation link to the European theatre of war and, in 1942, to Alaska where the Japanese threatened to attack the United States. The transportation links were named the Northwest Staging Route and Project Crimson. Each consisted of a series of northern landing strips that would enable American and Canadian warplanes to refuel and then continue their journey to either Europe or Alaska. In the Northeast, Project Crimson involved constructing landing fields at strategic intervals to allow Canadian and American airplanes to fly from Montréal to Frobisher Bay (now Iqaluit) and then to Greenland, Iceland, and, finally, England. In Canada's Northwest, American aircraft came to Edmonton and then flew along the Northwest Staging Route to Fairbanks, Alaska, where their major military base was located. The Alaska Highway, which was built at the same time, provided road access to the various landing fields and a truck convoy route to Alaska. The US Army command had decided that the oil needed by the American armed forces in Alaska must be made secure by increasing the oil production at Norman Wells in the NWT and sending the oil by pipeline across several mountain ranges to Whitehorse and then northward to the military facilities at Fairbanks. Known as the

Canol Project, the oil pipeline was completed in 1944, but with the disappearance of the Japanese threat it was closed within a year.

After World War II, the geopolitical importance of northern Canada changed. The Territorial North was now a buffer zone between the United States and the Soviet Union. The North's new strategic role was to warn of a surprise Soviet air attack. The defence against such an attack was a series of radar stations that would detect Soviet bombers and allow sufficient response time for American fighter planes and (later) American missiles to destroy the Soviet bombers. In the 1950s, 22 radar stations, called the Distant Early Warning line, were constructed in the Territorial North along 70° N. Before the end of the Cold War, these radar stations were abandoned and replaced with more sophisticated methods of detecting incoming Soviet planes. With the collapse of the Soviet Empire, Ottawa withdrew its military establishment at Inuvik, did not proceed with its plans for a military base at Nanisivik, and downsized its operation at Alert.

American military investment did expand the Norman Wells oil fields along with a pipeline to Whitehorse, but most of their investment went into improving the transportation system, such as the construction of the Alaska Highway. Private resource development moved along much more slowly as it responded to world demand. By the 1960s, multinational corporations had turned their attention to the region's mineral wealth, spending millions of dollars exploring to find profitable mineral deposits. By the early 1970s, three types of major resource development were underway: (1) major oil companies, such as Gulf Canada and Dome Petroleum, were drilling for oil and gas in the Mackenzie Delta and the Beaufort Sea; (2) Cominco, a large mining corporation, began extracting lead and zinc from a mine at Pine Point in the Northwest Territories (by 1988 this mine closed); (3) Cyprus Anvil, now a defunct mining company, began mining lead and zinc at Faro, Yukon.

Arctic and Subarctic Settlements

In addition to geopolitical and resource developments, the Territorial North underwent many social changes in the second half of the twentieth century—particularly among its Aboriginal peoples. The biggest change was the increase in the size of Arctic and Subarctic settlements.

At first, trading posts had few people except for the trader, the police officer, and the missionary. After the 1950s, however, trading posts evolved into tiny settlements due to the influx of Aboriginal families. Relocation is a controversial subject in northern history. For

some, relocation was a badly planned attempt at social engineering, while Ottawa saw it as a necessary step in the 'modernization' of northern Aboriginal peoples. The death by starvation of about 60 Caribou Inuit was the event that triggered relocation. In the early 1950s, the Caribou Inuit living in the Barren Lands of the central Arctic had difficulty finding game, particularly caribou. Reports of deprivation and even cases of death by starvation among the Caribou Inuit had reached Ottawa, but no action was taken. By 1958, Ottawa made the decision to relocate the Caribou Inuit to settlements, such as Baker Lake and Eskimo Point, but by then starvation had taken its toll—the Caribou Inuit population had dropped from 'about one hundred and twenty in 1950 to about sixty in 1959' (Williamson, 1974: 90). At the same time, Ottawa extended this relocation program to coastal Inuit and Indians in the Subarctic who also lived off the land. However, Ottawa was unprepared for the

The settlement of Pangnirtung on Baffin Island has undergone many of the stages of northern development: first the site of European whaling, then fur trading, today the community is known for the art its local Inuit inhabitants create. (Jim Fraser/Fraser Photos/Ivy Images)

economic, psychological, and social consequences of settlement life for hunting peoples.

Most Aboriginal peoples moved to trading posts. These settlements had certain attractions: ready access to food and other supplies at the Hudson's Bay store, medical services from the nursing station, and employment opportunities. Few jobs existed, however, because these settlements had no commercial purpose except fur trading. Most cash income came from transfer payments, such as child allowances and old-age pensions. A few traditionally oriented families remained on the land, but eventually they, too, came to live in settlements. Most adults spent some time hunting and trapping to escape from the stresses of settlement life and to enjoy once more the 'old ways' with family and friends. Within a generation of settlement living, however, many young, settlement-born Aboriginals were no longer interested in trapping.

Life in settlements had a number of social consequences. For example, access to food and medical services helped reduce the infant mortality rate, which caused the birth rate to rise. Since the 1940s, the Aboriginal birth rate has remained well above the national average, often more than twice as high. As a consequence, most northern settlements have tripled their population since 1951.[5] Even so, most are small by southern standards (under 1,000 people). For example, the largest centre in Nunavut is Iqaluit, with approximately 5,200 residents in 2001.

Life in settlements—affected as it was by Western ideas and goods—also meant that the pressure to obtain cash income rose sharply. As well, the cost of hunting and trapping increased because transportation to the hunting and trapping grounds was now by snowmobile rather than dog team.[6] The two principal sources of income became wages and various forms of government payments, including social assistance. However, employment opportunities in these former trading posts remained limited, while the Aboriginal population continued to grow rapidly. This dilemma is revealed in statistics for the Northwest Territories: from 1986 to 1996, the number of Aboriginal employees increased by 27 per cent, while the number of social assistance cases more than doubled (Government of the Northwest Territories, 1997: 11, 23).

Settlement life for northern Aboriginal peoples has been an established fact for nearly 50 years. During that time, Aboriginal society has become part of urban Canada and, in the process, its people have had to accept many of the cultural values and ways of the dominant society. Aboriginal children have learned English through the school system. Knowledge of English is necessary for employment in both private companies and public institutions. While some have succeeded in finding suitable employment in the community, many have not. The major employer is the government. Ironically, senior administrative positions and professional jobs are often held by southerners, who have the necessary education and job experience. Efforts to correct this situation through equity programs in the Northwest Territories have met with limited success. The Nunavut government has sought to break this pattern by hiring people who speak both Inuktitut and English. However, basic problems remain. For instance, most Aboriginal students are not completing their high school education. Also, most jobs are found only in capital cities (Whitehorse, Yellowknife, Iqaluit). Still, the idea of moving to a capital city where there are more job opportunities may be appealing, particularly for young Aboriginals who have graduated from high school and are attracted by urban amenities.

The Territorial North Today

The Territorial North consists of three territorial governments: Yukon, the Northwest Territories, and Nunavut. The capital cities of these three territories are Whitehorse, Yellowknife, and Iqaluit. Territorial governments have fewer powers than provincial governments and, in this sense, they are political hinterlands. For example, the federal government retains power over natural resources and collects a substantial amount of tax revenue from companies using natural resources, so this important source of revenue is not available to territorial governments. Instead, they depend on Ottawa for transfer payments. Without these transfer payments, the territorial governments could not offer the basic services available in southern Canada. For example, Nunavut receives 90 per cent of its revenue from Ottawa. The level of fiscal dependency is somewhat lower for Yukon and the Northwest Territories. The Northwest Territories is challenging Ottawa over the issue of resource royalties. Unlike provinces, royalties from natural resource developments in the territories flow to the federal government. With more diamond and natural gas projects on the drawing board, the Northwest Territories is demanding a share of resource revenues.

The Territorial North is changing, especially politically. One recent political change was the division of the Northwest Territories to create a new territory, Nunavut. The second political change is being triggered by the resolution of outstanding land claims between Aboriginal peoples and Ottawa through comprehensive land-claim agreements that transfer land, cash, and administrative powers to northern Aboriginal organizations. In a new development, the most recent comprehensive land-claims agreement—finalized in 2003 with the Dogrib First Nation and known as the Tlicho Final Agreement—included provisions for self-government.

The Territory of Nunavut

The new territory of Nunavut was made possible through a land settlement agreement between Canada and the Inuit of the Eastern Arctic in 1993. The terms of the agreement included the use of Crown lands for the Inuit to hunt, fish, and trap, and the transfer of part of the land to the Inuit, with a portion of this area involving rights to subsurface minerals. The same year the land agreement was reached, the federal government made the commitment to create a new territory by passing the Nunavut Act. This Act, which provided the legal basis for the creation of a distinct territory and territorial government, also allowed for a six-year transition period, giving the Inuit time to form their government, recruit their civil servants, and select a capital city. By means of a plebiscite, Iqaluit was selected as the capital of the new territory. Following an election in February 1999, the 19 members of the Nunavut Assembly took office on 1 April 1999.

The word 'Nunavut' means 'our land' in Inuktitut. By 2004, the vast majority of the approximately 28,000 people who reside in Nunavut are Inuit. A primary goal of the Nunavut government is to reflect the people's aims and aspirations. In addition to expanding economic development and increasing the number of jobs in the new public service for residents of the highly decentralized administration of Nunavut, the new government will strive to promote Inuit culture and the Inuktitut language. For many Inuit, Nunavut is a dream—one that ensures their cultural survival within Canadian society. The dream also brings with it the challenge of building a northern economy that can provide jobs for its growing labour force and thereby reduce its financial dependency on Ottawa. (Further resources: Student Web site, National Atlas section, Map 30. Web site instructions are found on p. xviii.)

Land Claims

Aboriginal land claims are based on the peoples' long-time use of the land for hunting and trapping. In Canada, Aboriginal land claims are settled by treaty: the Aboriginal tribe surrenders its claim to all the land in exchange for title to a smaller amount of land and a cash settlement. Other issues involved in land-claim negotiations are self-government, manage-

ment of wildlife resources, and preservation of language and culture. In simple terms, land claims are an attempt by Aboriginal peoples and the federal government to resolve the issue of Aboriginal rights. (See Chapter 3 for more information on these rights and the types of land claims.)

During the latter part of the twentieth century, Indian, Inuit, and Métis organizations in the Territorial North made land claims on behalf of their members. In the 1970s, three major land claims were made in the region: Yukon First Nations (Yukon), Dene/Métis (Denendeh), and Inuit (Nunavut). Over the next two decades, the Inuit land claim was split into the Western Arctic (Inuvialuit) and the Eastern Arctic (Inuit/Nunavut), and the Dene/Métis land claim for Denendeh was

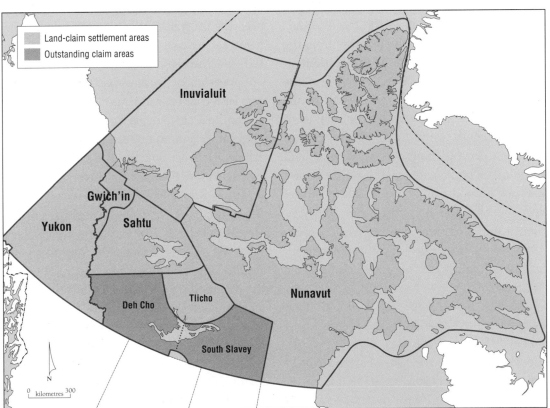

■ **Figure 10.2 Land-claim agreements in the Territorial North**. Six comprehensive land-claim agreements have been finalized while two land claims, the Deh Cho and South Slavey, remain outstanding.

divided into five separate land claims (Gwich'in, Sahtu/Métis, North Slavey (also referred to as Dogrib or Tlicho), South Slavey, and Deh Cho).[7] By 1984, the Inuvialuit reached a final agreement with Ottawa. They were followed by the Gwich'in in 1992, and, in the next year, the Sahtu/Métis, Inuit, and Yukon First Nations. In 2003, the Dogrib or Tlicho Agreement was completed, leaving only two outstanding claims—Deh Cho and South Slavey (Figure 10.2).

In 1984, the first comprehensive land-claim agreement was signed between the Inuvialuit and Canada. The Inuvialuit Final Agreement (IFA) was based on the Inuvialuit's traditional use and occupancy of lands in the Western Arctic, that is, the land that they and their ancestors used for hunting and trapping. The Inuvialuit's land claim was originally part of the Inuit land claim to the entire Arctic. However, the Inuvialuit broke away from the Inuit and, as a result, there are two **settlement areas** in the Arctic, the Inuvialuit and the Nunavut settlement areas.[8] The goals of the Inuvialuit were to preserve their cultural identity and values within a changing northern society; to enable themselves to be equal and meaningful participants in the northern and national economy and society; and to protect and preserve the Arctic's wildlife, environment, and biological productivity (Canada, 1985: 1). Canada's goal, on the other hand, was to extinguish the Inuvialuit claim to the Western Arctic.

As the first comprehensive land-claim agreement in Canada, the IFA served as a model for subsequent ones. (While the first modern treaty was the James Bay and Northern Québec Agreement in 1975, this agreement preceded the establishment of comprehensive land-claim negotiations and subsequent agreements.) Since 1984, as

shown in Table 10.1, five more comprehensive land-claim agreements were reached: Gwich'in (1992), Inuit (Nunavut) (1993), Sahtu/Métis (1993), Yukon (1993), and Dogrib (Tlicho) (2003). Each agreement is very similar to the IFA. One common feature is that each agreement has created economic and environmental administrative sectors. In the case of the IFA, the economic sector is the Inuvialuit Regional Corporation, which manages and invests the cash settlement received as part of the agreement.[9] The second sector, the Inuvialuit Game Council, is responsible for environmental issues that affect their hunting economy. The creation of these two sectors was the Inuvialuit's attempt to straddle two worlds; their old world was based on harvesting game from the land, while their new world is part of the global industrial economy. However, the IFA was silent on one important element so necessary for Aboriginal peoples—self-government. Over time, the issue of self-government became part of comprehensive land-claim agreements. The Nunavut and Tlicho final agreements do spell out the specific nature and unique structures of self-government for the Inuit and the Dogrib.

The Yukon First Nation agreement, signed in 1993, was different in that it only established the basic elements of final agreements for each of the 14 Yukon First Nations. Each First Nation was responsible for negotiating the local aspects of their agreement. Known as the Umbrella Final Agreement, this 1993 arrangement provided the basic framework within which each of the 14 Yukon First Nations (Carcross/Tagish; Champagne and Aishihik; Dawson; Kluane; Kwanlin Dun; Liard; Little Salmon/Carmacks; Nacho Nyak Dun; Ross River Dena; Selkirk; Ta'an Kwäch'än Council; Teslin Tlingit Council; Vuntut Gwitchin; and White River) can conclude a

Table 10.1	Comprehensive Land-Claim Agreements in the Territorial North		
Aboriginal Peoples	**Date of Agreement**	**Cash Values***	**Land (km²)**
Inuvialuit	1984	$45 million (1977)	90,650
Gwich'in	1992	$75 million (1990)	22,378
Sahtu/Métis	1993	$75 million (1990)	41,000
Inuit (Nunavut)	1993	$580 million (1989)	350,000
Yukon First Nations	1993	$243 million (1989)	41,440
Dogrib (Tlicho)	2003	$152 million (1997)	39,000

*By dollar for the stated year.
Sources: Canada (1985: 6, 31; 1991: 3; 1993a: 3; 1993b: 81, 215; 2004); *Globe and Mail* (1993: A1).

final claim settlement agreement. So far, five First Nations have achieved a final agreement: Teslin Tlingit Council (1993), Selkirk First Nation (1997), Little Salmon/Carmacks (1997), Ta'an Kwäch'än Council (2002), and the Kluane First Nation (2003).

For the Inuvialuit, the agreement was an important step towards defining their place within Canadian society. In exchange for cash and land, the Inuvialuit gave up their Aboriginal rights, including their claim to the vast lands of the Western Arctic. In exchange, they received legal title to about 90,650 km² of land, which is slightly less than 20 per cent of the settlement area (Canada, 1985: 6). The land to which the Inuvialuit gained title lies within the Inuvialuit settlement area, which extends from the Arctic mainland into the Arctic Ocean (Figure 10.2). A number of islands, including Banks Island, Prince Patrick Island, and the western parts of Melville and Victoria islands, are situated here. Of the total settlement land, 12,950 km² include surface and subsurface mining rights for the Inuvialuit. The Inuvialuit enjoy exclusive rights to hunting, trapping, and fishing over the remaining 77,700 km².

Population

The population of the Territorial North is small and concentrated in settlements. In 2001, the Territorial North had nearly 93,000 people. At that time, the Northwest Territories had 37,360 residents, Nunavut 26,745, and Yukon 28,674. Overall, Aboriginal peoples make up close to 52 per cent of the northern population. However, the percentage of Aboriginal peoples varies widely between the three territories. Nunavut has the highest percentage of Aboriginal peoples at 85 per cent, followed by the Northwest Territories at 50 per cent and Yukon at 23 per cent (Table 10.2).

Because of the nature of the environment, almost everyone in the Territorial North lives in a settlement, town, or city. Most urban centres are very small (Figure 10.3). In 2003, for example, approximately three-quarters of these urban centres had populations under 1,000, and more than 40 per cent of the Territorial North's population was in three cities: Whitehorse (19,058), Yellowknife (16,541), and Iqaluit (5,236). By function, urban centres in this region fall into three categories: Native settlements, resource towns, and regional service centres. Most Aboriginal peoples reside in

Table 10.2	Population and Aboriginal Population, Territorial North, 2001			
Territory	Total Population	% Change, 1996–2001	Aboriginal Population	Aboriginal Population as % of Total Population
Yukon	28,674	(6.8)	6,540	22.8
Northwest Territories	37,360	(5.8)	18,725	50.1
Nunavut	26,745	8.1	22,720	85.0
Territorial North	92,779	(2.5)	47,985	51.7

Source: Bone (2003a: 86).

Native settlements, where they form more than half the population. A growing number, however, live in regional centres, particularly the capital cities. This population shift has occurred because there are more job opportunities, particularly in the public sector, in these larger cities (especially the capitals). Also, a small number of Aboriginal families have relocated to cities in southern Canada for similar reasons.

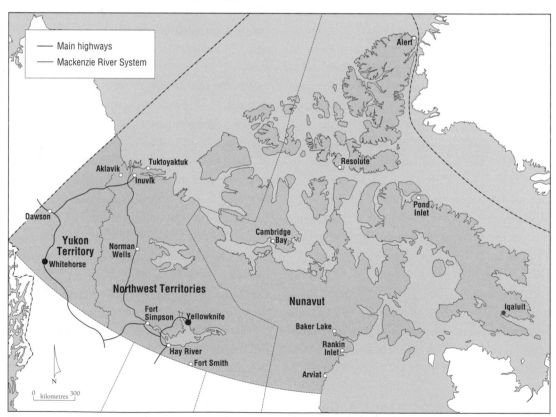

Figure 10.3 Major urban centres in the Territorial North. The major cities are the territorial capitals, Whitehorse, Yellowknife, and Iqaluit. Smaller, but more common settlements have populations under 1,000.

Table 10.3	Components of Population Growth for the Territories, 1980 and 2000–1						
Demographic Event	Canada 1980	Canada 2000–1	Yukon 1980	Yukon 2000–1	NWT* 1980	NWT 2000–1	Nunavut 2000–1
Births/1,000 persons	15.1	11.0	19.4	12.0	28.2	16.6	25.4
Deaths/1,000 persons	7.0	7.6	5.2	4.8	5.2	3.8	5.3
Natural rate of increase (%)	0.8	0.3	1.4	0.7	2.3	1.3	2.0
Net migrants			400	(846)	(900)	(606)	220

*Includes Nunavut.
Source: Bone (2003a: 91).

In 2000–1, the rate of natural increase in the Territorial North was approximately 2 per cent. In comparison, the national figure was only 0.3 per cent. The reason for this significant difference is the much higher birth rate in the Territorial North (17 births per 1,000 compared to the national figure of 11 births per 1,000) and a lower death rate, the result of a younger population, in the Territorial North (3.5 deaths per 1,000 compared to the national figure of 7.6 deaths per 1,000). Within the Territorial North, the highest birth rates and lowest death rates are among Aboriginal peoples, especially among the Inuit in Nunavut (Table 10.3).

The population of the Territorial North is also affected by migration. Migration to the North normally occurs when economic expansion creates jobs. Since World War II, many southerners have moved north to take jobs in the mining industry, the public service, and the business sector. However, since many newcomers only remain for a few years, a rapid turnover in the non-Aboriginal population exists. Since the end of the resource boom in the early 1980s, there has been a net flow of people moving to southern Canada. From 1996 to 2001, there was a net loss of just over 6,200 migrants (Table 10.4). There are three basic reasons for this out-migration. First, these migrants have an education and job skills that allow them to find employment in many areas of Canada, so they are economically mobile. Second, with friends and rela-

Table 10.4	Net Number of Migrants and Migration Rates, 1986–2001			
Territory	1986–91	1991–6	1996–2001	1996–2001 rate (%)
Yukon	790	665	(2,760)	(9.5)
Northwest Territories	(1,700)	(400)	(3,170)	(8.6)
Nunavut			(330)	(1.4)
Territorial North	(911)	265	(6,260)	(6.9)

Sources: Statistics Canada (1998d, 2002b).

tives living in southern Canada, they consider southern Canada their 'homeland'. Third, families tend to move south when their children reach school age.

Two Societies

A distinctive feature of the Territorial North's population is that it can be divided into two societies. These two societies have evolved from two different cultures. The first is a product of Western culture and early French and English presence in North America. The second has its roots in Aboriginal cultures and the fur trade.

At the time of Confederation, many officials believed that Aboriginal peoples would assimilate into Canadian society, while others believed that diseases would destroy them. Neither has happened. In the Territorial North, there are obvious signs of Indian, Inuit, and Métis culture, particularly in Native settlements and hunting camps. Aboriginal culture, which is founded on a land-based economy, is prominent in this region of Canada because of the high percentage of Aboriginal peoples. Aboriginal presence is most prominent in the smaller communities, where residents have a close relationship with the land. Here they often form 90 per cent of the population.

In sharp contrast, cities such as Whitehorse and Yellowknife not only house the majority of the non-Aboriginal population but are similar in function and structure to cities in southern Canada. The two cityscapes reveal the dominant position of the English language and Western popular culture, i.e., street signs and business advertising. These are visible signs of a Western industrialized society. The transportation and communications networks spread across the North not only illustrate the importance of connections between many small centres across an extensive land mass,

but they also show the impact that Western society has had on the landscape.

Over time, the cultural differences between the two societies of the Territorial North have been narrowing. Whether or not this narrowing process is the result of cultural integration or assimilation is a moot point. The struggle over the language (Inuktituk/English) used on the streets of Iqaluit represents the clash of languages (cultures). Without some form of self-government, Aboriginal languages and cultures have little protection and losses are likely to continue. Four factors reduce the cultural differences between Aboriginal and non-Aboriginal peoples: (1) English is the common language used by all people in the Territorial North (in Nunavut, the government is making every effort to keep Inuktitut the main language, but it is a difficult struggle because few non-Inuit speak Inuktitut (Bone, 2003a: 222); (2) satellite television brings predominantly English-language programming and Western values into most homes and even the smallest communities, and, similarly, modern telecommunications, such as the Internet, are dominated by the English language; (3) Aboriginal and non-Aboriginal children attend the same schools and are exposed to similar knowledge and ideas; and (4) Aboriginal men and women are entering the wage economy in greater numbers. The fear of cultural loss is based on the principle that language is the core of a culture. Nevertheless, at this point in time, each society retains key characteristics. Three factors that illustrate a continuing difference between the two societies are:

- Aboriginal peoples are beneficiaries of comprehensive land-claim agreements, which have given them more control over the environment and wildlife in their settlement areas—control that is critical to their land-based economy.

- Aboriginal peoples remain culturally attached to the land. For instance, they harvest and eat game in much greater quantities than non-Aboriginal peoples.
- Aboriginal peoples, by taking greater pride in their culture, are reclaiming their identity through education and the renaming of geographic places. For example, in 1996 Fort Norman was changed to Tulita (meaning, where the waters meet).

The Aboriginal Economy

The Aboriginal economy has been and continues to be based on the harvesting of land and sea resources, but this economy is not static. In fact, the Aboriginal economy is a dynamic one that has constantly incorporated new technologies and ideas to make use of natural resources. The Aboriginal economy has passed through two stages—the hunting stage and the trapping stage—and it is now entering the commercial stage. In all three stages, this economy has focused on the utilization of land and sea resources, but whereas the first two stages were based on subsistence and trade, the current, commercial stage involves a greater interest in deriving 'cash' for resources and labour.

The commercialization of the Aboriginal economy was a natural outcome of the integration of the Territorial North into the global economy and the relocation of Aboriginal peoples into settlements, where they became more reliant on consumer goods. Aboriginal commercial enterprises include the sale of fish and meat, such as the marketing of muskox meat, earnings generated from eco-tourism and from fishing and hunting operations, and revenue from the sale of artworks by Aboriginal artists, especially sculptors and printmakers.

Potentially, the Aboriginal economy could provide a sustainable use of northern resources primarily involving Aboriginal workers and owners. Yet, a fundamental weakness is its relatively marginal economic nature: the income generated by these activities is limited, resulting in low wages for Aboriginal workers and low profits for Aboriginal entrepreneurs. For example, the value of Aboriginal commercial enterprises is insignificant compared to the value of mineral production. These commercial enterprises do not generate the necessary cash to support Aboriginal families who live in settlements, so these families must supplement their income either by working for wages (in the non-Aboriginal economy) or through welfare payments.

Wages can provide a partial answer but the number of Aboriginal workers far outstrips the number of jobs. Two other problems exist: geographic location and education. Most jobs are found in regional centres and capital cities, while most Aboriginal workers live in small communities, and educational levels among Aboriginal workers are much lower than those in the national labour force. Given this situation, the strengthening of the commercial land-based economy seems the only logical solution (another solution would be the relocation of Aboriginal peoples to larger centres—like the controversial Newfoundland relocation program that lasted from 1954 to 1975). Neither solution has gained favour with government (investing in commercial land-based enterprises) and Aboriginal peoples ('forced' relocation).

Some land-based activities have experienced great instability over the years, due to changes in the commercial economy. For instance, since the 1950s, declining prices for furs and efforts by animal rights groups to ban the sale of seal pelts in Europe made the situation for Inuit seal hunters very difficult. As shown in Figure 10.4, the value of fur production in the Northwest Territories dropped drastically from the mid-1980s to 2002. In 1992–3, for example, the value of fur sales dropped below $1 million for the first time. In the subsequent six years, trapping exceeded $1 million in sales only once (1996–7). By 2000–1, the value of fur production in the Northwest Territories was $477,000. With low prices and rising costs, trapping does not generate enough cash income, and individual trappers must supplement their revenue from trapping with other income to make ends meet.

Trapping, hunting, and other land-based activities persist among Aboriginal peoples in the North not because of their commercial value but largely because of their cultural importance. For instance, hunting produces food for the table and country food remains a core cultural feature among northern Aboriginal families.

Country Food

Though cultural change has occurred among Aboriginal peoples, such as the growth of the wage economy, some core cultural elements have remained. For Aboriginal peoples, the core elements are a strong attachment to the land, to country food, and to the ethic of sharing. **Country food** is food obtained from the land, a preferred source of meat and fish. As equivalent store-bought foods are expensive, most Aboriginal northerners keep their food costs low by consuming country food. While it is true that there are substantial costs expended in harvesting country food, to some degree this cost is offset by the pleasure and spiritual rewards of being on the land and participating in hunting and fishing. Sharing also remains an important component in the harvesting and distribution of country food among family members, relatives, and close friends. However, this ethic does not apply to

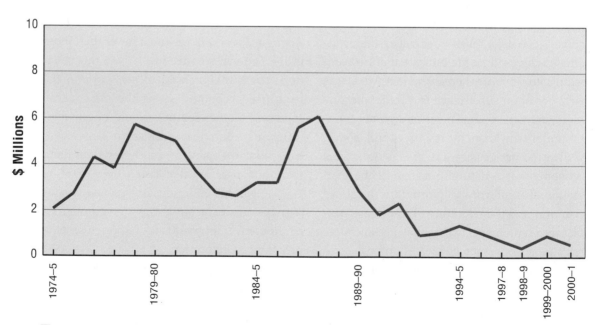

Figure 10.4 Value of fur production in the Northwest Territories. Since peaking in the late 1980s, the value of fur production has been steadily declining, signalling the shrinking importance of the commercial trapping economy in the North.
Source: Government of the Northwest Territories (2003, 2004).

the wage economy—wages, for instance, are not shared.

Figures on the cost, size, and value of harvesting country food are not available. However, estimates suggest that the 'substitute value' of country food (the value of equivalent store-bought food) was about $40–50 million per year in the mid-1980s (Usher and Wenzel, 1989). More recent estimates are not available, but a visit to an Aboriginal community quickly reveals the importance of wildlife harvesting to the daily diet. While some store-bought foods can be found on the tables of Aboriginal families, most of the meat and fish they consume is from the land and waters.

Country food is so important to Aboriginal northerners that they have made wildlife an essential issue in land-claim negotiations. For instance, in northern Québec, both the Cree and Inuit were able to obtain financial support for their hunters and trappers in the James Bay and Northern Québec Agreement. Though such financial support was not included in the comprehensive land-claim agreements of the Territorial North, they did assign harvesting of wildlife exclusively to the Aboriginal claimants in their respective settlement areas. The agreements also establish co-management committees that administer the environment and wildlife. For example, these committees approve (or reject) proposals for new industrial projects and determine the total allowable harvest of wildlife based on biological principles and **traditional ecological knowledge**. In the case of the Inuvialuit Final Agreement, the members of the co-management committees are named by the Inuvialuit Game Council and the three governments (Canada, Northwest Territories, and Yukon). The members of the Inuvialuit Game Council are selected by the Hunters and Trappers Associations found in each of the six Inuvialuit communities. Under the direction of these committees, professional scientists record and monitor the state of the environment and wildlife. As a result of comprehensive land-claim agreements, power to control the use of the environment and wildlife now rests, in part, with most Aboriginal peoples in the Territorial North.

Industrial Structure

As a northern frontier, the Territorial North's economy depends heavily on its primary industries. Energy and mining are the princi-

Country food among Aboriginal peoples in the Territorial North is not only an alternative to costly store-bought food but also an intrinsic feature of their culture. (John Eastcott/Ya Momatiuk/Valan Photos)

pal elements of this economy. In terms of employment, the primary sector in the Territorial North is much larger than the same sector in other geographic regions. For example, approximately 15 per cent of the workers fall into the primary sector in the Territorial North compared to only 2.7 per cent in Ontario. Similarly, the tertiary sector is much larger than in the other geographic regions. The Territorial North has approximately 83 per cent of its workforce allocated to the service sector compared to about 73 per cent for Ontario. A large tertiary sector is characteristic of resource frontiers in modern industrial nation-states. In Canada, the large transfer payments from Ottawa to the three territorial governments play a major role in their budgets, allowing them to have similar levels of education, health, and social services to those found in the provinces. As a result, the contemporary economic structure of the Territorial North mirrors these two forces—resource extraction and transfer payments—making the tertiary and primary sectors of greatest significance. The secondary sector, manufacturing, is almost non-existent in the Territorial North (Table 10.5).

The resource industries, comprised of energy and mineral developments, export their production to world markets. The north-

ern transportation system is geared for that role. The value of production from the primary sector accounts for nearly 90 per cent of all production in the Territorial North. While the value of mineral output varies from year to year, it hovered around $3 billion for 2002. However, this highly efficient and capital-intensive industry employs relatively few people and generates few indirect benefits for the Territorial North because companies purchase their equipment and supplies from firms in southern Canada or in foreign countries.

Another drawback to the resource industry is that its labour force often redirects money to other parts of the country. For instance, mineral deposits are often found in remote places, where the mining industry has sometimes built resource towns to house its labour force. But commuting by air is a common alternative to living in an isolated resource town. Several mining companies fly their workers to the mine site and then transport them back to their home community, often in more southern parts of the country. The diamond mines just east of Great Bear Lake transport workers by air from Edmonton and Yellowknife to the mine site once every three weeks. For the Territorial North, such air-commuting systems that terminate in southern cities are a drawback for the region's economy because: (1)

Table 10.5	Estimated Employment by Industrial Sector in the Territorial North, 2003		
Industrial Sector	North Workers (per cent)	Ontario Workers (per cent)	Difference (percentage points)
Primary	15	2.7	12.3
Secondary	2	24.6	(22.6)
Tertiary	83	72.7	10.3
Total	100.0	100.0	

Source: Author's estimate.

workers spend their wages in their home community in southern Canada, thereby stimulating provincial, not territorial, economies; and (2) workers who reside in a province but work in the territories pay personal income tax to provincial rather than territorial governments, thereby depriving territorial governments of valuable personal income tax.

In 2001, the Territorial North employed about 50,000 workers, but only 15 per cent of those employees worked in the primary sector. Of the more than 80 per cent of the labour force in the tertiary or service sector, most are employed by one of three governments: federal, territorial, or local. Local governments are settlement councils, band councils, and other organizations funded by a higher level of government. The reason the tertiary sector is so large in the Territorial North is that geography demands such an investment of people and capital to ensure the delivery of public services. Territorial governments must spend more money per resident than do provincial governments to provide basic services. Much of the cost differential is attributed to overcoming distance in the Territorial North and hiring staff for small communities. The drawback is that economies of scale are difficult to achieve in small communities where teachers and nurses often have a relatively small number of students and patients compared to those working in larger centres.

However, the social importance of the public service sector goes beyond the number of employees. The wide geographic distribution of public jobs across the North is a major social benefit to those in small communities. As a result, employment opportunities, while concentrated in the three capital cities, exist in every community. In small, remote communities, where unemployment rates are often as high as 30 per cent, virtually all jobs are associated with one or more levels of government. A second social benefit is the governments' ability to implement social policies in their hiring practices. For example, Nunavut's government seeks to employ mainly Inuit in its civil service.

Resource Development in the Territorial North

Since the sixteenth century, explorers—now replaced by multinational companies—sought to exploit the North's natural wealth, particularly its minerals. The search for precious metals began soon after explorers reached North America. In Mexico and South America, the Spanish and Portuguese were particularly successful in obtaining vast quantities of gold and silver. In North America, English and French explorers searched for similar wealth. Northern Indians trading at the Hudson's Bay posts often wore copper ornaments and had copper tools. Richard Norton, the factor or chief trader at Fort Prince of Wales, ordered one of his men, Samuel Hearne, to find the source of this copper and assess its commercial worth to the company. In one of the great overland journeys of all time, Hearne travelled with Chipewyan leader Matonabbee's tribe from Fort Prince of Wales on the west coast of Hudson Bay (present-day Churchill, Manitoba) to the mouth of the Coppermine River at Coronation Gulf on the Arctic Ocean. In 1771, Hearne found the source of copper, but this natural copper only occurred at the surface in small amounts. Others seeking mineral wealth in the North were more fortunate. In 1896, George Carmack, an American prospector, and his Indian brothers-in-law, Skookum Jim and Tagish Charley, found gold nuggets in a small tributary (later renamed Bonanza Creek) of the Yukon River. When word of their find reached the outside world, the Klondike gold rush began.

The Klondike Gold Rush

The discovery of gold in Yukon sparked the greatest gold rush of all time. It caught the imagination of the world and gave credence to the myth of a northern Eldorado. While the discovery of gold took place in 1896, the Klondike gold rush did not begin until the following year. Dawson City was the centre of this frenzy. Thousands of prospectors flooded the area to scour the sandbars of the rivers and pan for gold. Within two summers, the more accessible placer gold was gone. When word of a new gold find in Alaska reached Dawson City, most prospectors moved down the Yukon River into Alaska.

By 1900, gold mining in the Klondike had changed from an individual pursuit to a large-scale capital-intensive operation. With the accessible placer gold effectively gone, the only remaining gold was deep in the frozen **terraces** found along the sides of the Klondike River. Permafrost is widespread at latitudes above 60° N. Mining in permafrost was beyond the abilities of a prospector, so a second phase of mining began, organized by companies with both capital and technology.

Mining companies soon purchased the leases to large sections of the land surrounding the rivers. Gold mining became highly organized, capital-intensive, and technically advanced. To expose the gold locked in the frozen terraces, hydraulic and steam-mining techniques were used to thaw the ground and force the sand, gravel, and fine gold particles into separating devices commonly known as sluice boxes (Vignette 10.5). The emergence of commercial, large-scale gold mining marked the birth of the mining industry in the Territorial North.

Environmental and Social Impacts

In the late nineteenth century, Western society regarded the North as a source of wealth. Little regard was given to its natural resources and Aboriginal peoples. The Klondike gold rush exemplifies this attitude. The North, popularized by the writings of Jack London and Robert Service, and later by Charlie Chaplin's portrayal of a prospector in the motion picture *The Gold Rush*, was seen as a frontier, a place of adventure and excitement. In those days, Yukon's environment was at the mercy of

---

Vignette 10.5　Gold Rushes

Placer gold was found in commercial quantities mainly in the western Cordillera region from California to Alaska where a series of gold rushes occurred in the mid- to late nineteenth century. The discovery of placer gold in some remote places triggered a sudden influx of prospectors, traders, gamblers, merchants, bankers, prostitutes, and various hangers-on. All sought to increase their wealth either by finding gold or by 'extracting' gold from the miners.

Placer gold can be worked cheaply by amateurs. It involves the separation of gold nuggets and particles from sand and gravel. The common practice is to pan for gold, that is, to jiggle the sand, gravel, and gold in a pan, causing the heavier gold to separate from the other material. Sluicing is another more elaborate method of separating the gold from other loose material obtained from the stream bed. This requires sluice boxes—channels or troughs fitted with grooves to separate gold from gravel.

prospectors and mining companies. Wood was in great demand for power by steamboats and the hydraulic operations in the goldfields, and for heat by companies, government offices, and individuals. As a result, the forests adjacent to the Yukon River and its tributaries were logged, leaving the riverbanks and adjacent lands bare. Wildlife habitat suffered and so did the Aboriginal peoples who depended on wildlife for their food supply. In the post-placer gold-mining period, the gold was washed out of the frozen terraces, leaving behind a scarred landscaped (see the earlier section on Environmental Challenges).

The Klondike gold rush also had social impacts on the Territorial North, but these were more subtle. The influx of miners, administrators, and, eventually, the North West Mounted Police transformed southern Yukon into an organized territory within Canada. The impact on Aboriginal tribes in Yukon was overwhelming—now they were not only a minority in their own country but were also living within a Western culture with values often very different from their own. Aboriginal peoples found themselves on the margins in this new world.

The disregard for the environmental and

BHP Diamonds' Ekati diamond mine. The value of mineral production from this mine far exceeds that of all other mines in the Territorial North. (Tim Atherton)

social impacts of resource development in the North continued until after World War II. Following the Berger Report (1977) on the effects of building a gas pipeline in the North, the public began demanding that Ottawa take action to deal with the hidden costs of resource development.[10] The federal government responded by enacting legislation that requires an environmental and social assessment of development projects before they can be approved. The federal agency responsible for enforcing these regulations is the Canadian Environmental Assessment Agency. By their very nature, industrial projects alter the natural and social environments. This federal agency's task is to review the proposals of planned developments, to assess the degree of their impact on the natural and social environments, and, if necessary, to recommend measures that will mitigate any identified impacts.

However, this process does not always manage to reconcile the interests of all the parties affected by developments. For instance, in 1994 the NWT Diamond Project was submitted to Ottawa for review under the Federal Environmental Assessment and Review Process. BHP Diamonds Inc. and the Blackwater Group proposed to construct open-pit and underground diamond mines in a remote area about 300 km northeast of Yellowknife near Lac de Gras. The ensuing review process did not satisfy everyone. Aboriginal people complained that they did not have enough funding and time to examine and understand the project. They also complained that the public review process did not hold enough meetings in their communities. (Public hearings were held in eight Native communities but most took place in Yellowknife.) The Northern Environmental Coalition expressed concern that the panel did not pay enough attention to the cumulative effects of the project on the

environment (O'Reilly, 1996). Nevertheless, in 1996, the federal environmental assessment panel filed its report, recommending approval subject to a series of restrictions and qualifications (MacLachlan, 1996: 1).

Though resource developers often promise long-term economic development, non-renewable resource projects frequently have a short lifespan and therefore fail to generate lasting economic benefits. Because of their boom-and-bust nature, mining operations cause social dislocations when they cease to operate. The question of the long-term economic and social value of resource developments is further complicated by the North's need to attract investment and create jobs. As a result, the region is extremely vulnerable. This vulnerability is very evident when developers request public subsidies for their projects. For instance, in the case of the Izok Lake lead-zinc deposit located about 250 km south of Kugluktuk (formerly called Coppermine), a multinational company (Metall Mining Corporation) proposed to develop a lead-zinc mine, but the company wanted considerable public financial support. Metall argued that if the government built a highway to an ocean port in this remote area, Metall would open a mine and other minerals could be developed. In the *Slave River Journal*, two opinions, one in favour of development and the other opposed, capture the essence of the debate surrounding such development projects (Vignette 10.6).

Development in Remote Locations

In the Territorial North, the development of known mineral deposits in remote areas is often prevented by high transportation costs. The pattern of mining developments in the North clearly indicates the importance of

Vignette 10.6 Izok Lake Mining Proposal: Pro and Con Views

Pro:

The plan for the Izok Lake mining project is the centrepiece of northern Canada's economic strategy. The proposed lead-zinc development near Coppermine would secure a prosperous future for the Western Arctic and Nunavut. The minister of economic development for the Northwest Territories, John Todd, knows all too well the importance of Izok Lake and now that Metall Mining Corporation is apparently having second thoughts, he is courting the mining company with financial assistance to convince them to go ahead with the project. Financial support for needed infrastructure, such as an all-weather road north from Yellowknife and a deep sea port at Coppermine, would make the project economically viable for Metall and create thousands of good jobs in the NWT. In addition to the Izok Lake development on the Arctic Coast, many new mineral deposits along the road will become viable, both because of the road access and because the new port will provide access for large shipping vessels, providing a cost-effective method of transporting raw materials. A development of this scale will shore up the economy of the central Arctic, providing tremendous opportunities for business people and jobs for the locals, both during the project construction phase and while the ore is being extracted. The next two decades will be looked after. It will be a major boost for the sluggish Yellowknife economy, especially as it suffers from the loss of jobs due to the division of the NWT. And it would ultimately greatly reduce the NWT's financial dependence on Ottawa. Everyone will win with this one.

Con:

Why invest hundreds of millions of dollars in an operation that is marginal and will likely fail? Remember the pain in shutting down Pine Point? The proposed lead-zinc development at Izok Lake near Coppermine is teetering and the NWT government wants to prop it up—supporting a multinational corporation with our tax dollars. Lead and zinc are low in value and even if the prices rise, they will likely fall again. Look at how long pundits have been predicting the resurgence of the gold price. What then, a bailout? The developers will never stop coming back once they get a taste of free government money. And the more marginal the operation, the more environmental protection laws will be overlooked. Didn't we just finish cleaning up a bunch of old industrial sites in the north at a cost of millions of dollars to Canadian taxpayers? It is admirable to want to create jobs, but why put everything we have into a megaproject with a limited life span? There are other northern roads half finished that will be forgotten and small, long-term community-based economic projects that will be lost in the stampede to make quick, big bucks. Twenty years is not a long time and once the ore has been high-graded and the mine is being mothballed, what will the workers do? What will the cost of social problems in the northern communities be then? Are all these costs being considered in that equation? It seems John Todd, the minister of economic development for the NWT, has a vision that lines the pockets of the shareholders of Metall Mining Corporation and leaves (most) Northerners out in the cold.

Source: 'Opinion: Pick a Side', *Slave River Journal*, 9 Mar. 1994, 4. Reprinted from the *Slave River Journal*, Fort Smith, NWT, Canada.

transportation, especially water transportation. In 1935, the first major gold discovery in the Northwest Territories took place near Yellowknife. The Con-Rycon gold mine was a viable proposition because the mine was located on the shores of Great Slave Lake and therefore had excellent access to the Mackenzie River transportation system. Mines located far from water transportation, such as the Lupin gold mine (about 250 km northeast of Yellowknife), required the construction of a winter road. The need for roads in mining is not always so much for transporting minerals to external markets but for bringing equipment and supplies to the mine site. Mines

along the Arctic coast can take advantage of ocean transportation. Three examples are the former nickel mine at Rankin Inlet on the shores of Hudson Bay, the lead-zinc mine at Nanisivik near the northern tip of Baffin Island, and the Polaris lead-zinc mine on Little Cornwallis Island in the Arctic Archipelago (see Figure 10.5).

In assessing the cost of transportation, the value of the mineral is taken into account. For example, copper, lead, nickel, and zinc are all low-grade ores (much of the ore has no commercial value). Even after the first-stage separation of some of the waste material from the valuable mineral, the enriched ore still remains

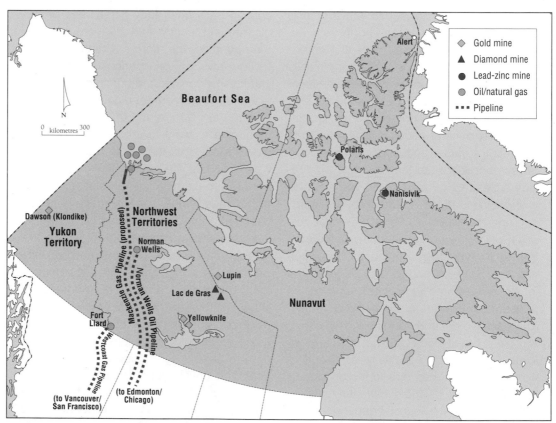

Figure 10.5 Resource development in the Territorial North. The mineral wealth of the Territorial North lies mainly in the western Northwest Territories. Diamonds, gold, natural gas, and oil drive this territory's resource economy. In contrast, the resource economies of Yukon and Nunavut are much smaller and are based on gold and lead-zinc mines. (Further resources: Student Web site, National Atlas section, Map 29. Web site instructions are found on p. xxi.)

a bulky product, with significant waste material remaining that can only be removed through smelting. Shipping such a low-value commodity is very expensive. Ideally, such ore is transported to a smelter by ship. Railways are the second most effective transportation carrier for low-grade ore. The critical nature of transportation for such mines is clear in the example of the lead-zinc deposit at Pine Point in the Northwest Territories. Discovered in 1898 by prospectors heading overland to the Klondike gold rush, the Pine Point deposit was not developed until 1965, when a railway was extended to the mine site. With a means of transporting the ore to a smelter, the company, Cominco, could begin sending massive amounts of ore by rail to its smelter at Trail in British Columbia. Unfortunately, the mine closed in 1983, transforming Pine Point into a ghost town.

Few Aboriginal workers are employed in the mining industry. The Rankin Inlet nickel-mining operation proved to be an exception to this. The majority of the employees at this mine were from the local community. The local Inuit had a chance at these jobs because it was difficult to recruit outside workers to work in such a remote place, the mining operation was small, and mine officials were willing to hire entire Inuit crews. This meant that most underground communications were in Inuktitut.

Key Topic: Megaprojects

The Territorial North has entered a new phase of resource development characterized by megaprojects controlled by multinational companies. Megaprojects usually cost more than $1 billion and require several years to complete. This phase of development has integrated the Territorial North's economy into the global economy, thereby firmly locking the North into a resource hinterland role in the world economic system. The most recent megaprojects in the Territorial North have involved diamond mining. The first diamond mine to come into production was the Ekati mine. As part of the NWT Diamonds Project (Canada, 1996), preliminary construction of buildings, roads, and mines began in the early 1990s, but the Ekati mine was not completed until 1998. Another mine, Diavik, is now in production while the Snap Lake, Jericho, and Kennedy Lake projects are at an advanced stage of planning and all are located in the Slave geological region that straddles the border between the Northwest Territories and Nunavut.

Proponents of resource development describe megaprojects as the economic engine of northern development, though others challenge this assumption, claiming that they offer few benefits to the region (Vignette 10.6; Bone, 1992: 143–5). These large-scale ventures are designed for the export market. By injecting massive capital investment into the construction of giant engineering projects, megaprojects create a short-term economic boom. However, most construction monies are spent outside hinterlands because the manufactured equipment and supplies are produced not in hinterlands but in core industrial areas. This reduces the benefits of megaprojects on the hinterland economy and virtually eliminates any opportunity for economic diversification. As well, since all megaprojects in the Territorial North are based on non-renewable resources, these developments have a limited lifespan. At the end of these projects, the local economy suffers a collapse. Three examples are the closure of mines at Faro, Yukon; Pine Point, Northwest Territories; and Rankin Inlet, Nunavut.

However, megaprojects can more readily inject much-needed capital and create development in the region. Megaprojects in resource hinterlands are high-risk ventures. Multinational companies can reduce their risks in three ways. First, they can create a consortium of companies and thereby spread the investment risk among several firms. Second, they can arrange for long-term sales of the product at a fixed price before proceeding with construction. Third, they can obtain government assistance, which often takes the form of low-interest loans, cash subsidies, and tax concessions. Three megaprojects are discussed in the following sections: the Mackenzie Valley Pipeline Project: 1970s and 2000s, the Norman Wells Oil Expansion and Pipeline Project, and the NWT Diamonds Project. Except for the 2000 version of the Mackenzie Valley Pipeline Project, each project was approved after an environmental impact assessment. The Norman Wells Project received its approval in 1980, while the NWT Diamonds Project was approved in 1996.

The Mackenzie Valley Pipeline Project: 1970s

By 1970, the plans formulated by oil and pipeline companies to bring natural gas from Prudhoe Bay, Alaska, to major markets in the United States, primarily in the Midwest, were submitted to Ottawa.[11] Economic conditions at the time included a high demand for natural gas in the continental United States and the fact that natural gas was a by-product of oil production at Prudhoe Bay. One plan was to build a pipeline, known as the Mackenzie Valley Pipeline Project, which would transport natural gas from the Arctic coast of Alaska to the Mackenzie Delta and then south along the Mackenzie Valley, eventually reaching Alberta and then the United States. Shortly after these plans were announced, world oil prices

increased sharply as a result of joint action to limit supply by the Organization of Petroleum Exporting Countries (OPEC). OPEC's control of world oil production changed the economic landscape for energy and added another economic reason for the Mackenzie Valley Pipeline Project. At these higher prices, the Mackenzie Delta deposits were no longer too expensive to develop, providing its natural gas could be added to the proposed pipeline by means of a short spur pipeline.

In Canada, this project was subjected to the Mackenzie Valley Pipeline Inquiry (also known as the Berger Inquiry). The Berger Inquiry began in 1974 and was concluded in 1976. Led by British Columbia Justice Thomas Berger, who a few years earlier had been chief counsel for the Nisga'a in their landmark land-claim case (*Calder*), this federal investigation examined the potential environmental, social, and economic impacts of the proposed gigantic construction project. The Berger Inquiry established a new precedent by holding community hearings in numerous remote Native settlements, and its final report, *Northern Frontier, Northern Homeland*, issued in 1977, recommended that a natural gas pipeline from the Mackenzie Delta to Alberta was feasible but should only proceed after further study and after settlement of the Dene land claims. To complete these two tasks, Berger recommended a 10-year delay in the construction of such a pipeline. At the same time, he rejected the construction of a pipeline from Prudhoe Bay to the Mackenzie Delta because of the danger of harming the delicate Arctic environment and wildlife found along the North Slope of Alaska and Yukon.

Without the link to Prudhoe Bay and faced with rising construction costs, the proposal lost its economic attraction and the pipeline was never built. With new supplies of Mexican and

Albertan natural gas reaching the United States in the 1980s, prices declined in the North American market. However, by 1998, this surplus had disappeared and prices began to rise. From 1998 to 2003, natural gas prices more than quadrupled. Consequently, deposits in the Territorial North are again commanding the attention of the major oil and gas companies servicing the North American market. In 2000, for instance, natural gas from northeastern British Columbia (just south of the Northwest Territories) began to flow through the Alliance Pipeline to the Chicago market, underlining the continental system of energy production and marketing in North America.

The Mackenzie Valley Pipeline Project: Revisited

The possible development of the natural gas deposits in the Mackenzie Delta has brought about a revisiting of the old Mackenzie Valley Pipeline Project. Oil companies—Imperial Oil, Gulf Canada Resources, Shell Canada, and Exxon Mobil—are taking a second look at the Mackenzie Delta, where they hold considerable natural gas reserves. These companies have already produced new plans for developing the gas fields in the Mackenzie Delta and then shipping this gas by pipeline to southern markets.[12] Such plans must be submitted to Ottawa for review and approval. In the past, strong opposition from the Aboriginal peoples posed a major stumbling block for two sets of reasons. First, the Dene insisted that their land claims be settled before the construction of a Mackenzie Valley pipeline. Second, the Dene had concerns about the impact of this construction project on their lands and people. More specifically, they feared that wildlife would become less plentiful, causing great harm to their hunting/trapping life. Other concerns also were identified, such as the Dene fear that their communities would not gain a fair share of jobs and business opportunities, and that the expected increase in non-Aboriginal population would generate negative impacts on Dene culture and social well-being.

By 2003, these concerns were no longer seen as major obstacles. Several land-claim agreements have been concluded, and the Dene workforce and business community are more comfortable and better prepared to take advantage of such a mega-construction project. The Aboriginal Pipeline Group has obtained a one-third interest in the pipeline. The cost of such a pipeline is approximately $4 billion. Nellie Cournoyea, chairwoman and chief exec-

---

Vignette 10.7 Will Natural Gas Take Centre Stage?

Oil production from Norman Wells in the Northwest Territories continues to fuel its economy. A small amount of natural gas is produced at Fort Liard (NWT), Pointed Mountain (NWT), and Kotaneelee (Yukon Territory) near the 60th parallel. Before 2010, however, the value of natural gas is expected to replace that of oil as the leading form of energy for three reasons. First, rapidly rising prices of natural gas are expected to continue to outpace anticipated increases in the price of oil. Second, oil production at Norman Wells may decrease before 2010. Third, the Mackenzie Delta natural gas field is expected to come into production by 2009. For these three reasons, the value of natural gas output in 2010, estimated at $2 billion, will exceed the value of oil.

utive officer of the Inuvialuit Regional Corporation, believes that 'it signals the recognition we are building the capacity to participate in the economy of the twenty-first century.'

The attention focused on the Mackenzie Delta gas fields is clearly related to the rising price of natural gas, which in turn is a result of the growing demand for natural gas in the United States. By 2015, demand for natural gas is projected to soar to 31.3 trillion cubic feet a year in North America, well above the current production of 24 trillion cubic feet, possibly pushing prices well over current rates. By then, existing production in southern Canada and the United States is expected to diminish—perhaps to 20 trillion cubic feet—leaving a possible gap of 11 trillion cubic feet. One of the two proposed Arctic pipelines should be completed by that time, possibly making up as much as half of this expected shortfall.

The race to build the first Arctic pipeline to North American markets pits Prudhoe Bay against the Mackenzie Delta. At the moment, the Mackenzie Delta is favoured because of two main advantages: (1) a much lower construction cost because of a shorter distance to market and because the diameter of its pipe is much smaller (and therefore less costly) than that proposed for the Alaskan pipeline; and (2) a much shorter construction time (six years) compared to the proposed Alaskan pipeline (10–12 years). However, the US government may subsidize the construction of the much more expensive Alaskan pipeline (estimated at some $20 billion compared to $4 billion for the all-Canadian pipeline). The stakes are high, causing much lobbying in Washington. Alaska has two strong points: (1) Alaska's reserves are much larger than those in the Mackenzie Delta (Alaska holds known reserves of 35 trillion cubic feet and an estimate of an additional 100 trillion cubic feet,

compared to about 9 trillion cubic feet of known reserves at Mackenzie Delta, plus estimated reserves similar in size to those in Alaska; and (2) American security concerns call for the development of American petroleum deposits, though the Prudhoe Bay pipeline would follow the Alaska Highway and therefore must pass through Canadian territory (Yukon, British Columbia, and Alberta) to reach the main markets in the United States.

The Norman Wells Project, 1982–5

The Norman Wells Project, which cost almost $1 billion in 1996 dollars, was hailed by industry as a model megaproject for the North. Both the federal government and Esso Resources Canada believed that the project would not harm the natural and social environments in the construction impact zone, which stretched from Norman Wells to Fort Simpson and then to the Alberta border. Aboriginal organizations, such as the Dene Nation, and environment groups, such as the Canadian Arctic Resources Committee, did not agree.

Plans for this major oil and pipeline construction effort by Esso Resources Canada began in the late 1970s, when the price of oil was rising sharply. Once the proposal was approved by the federal government, construction began, running from 1982 to 1985. The petroleum pipeline route begins at the resource town of Norman Wells in the Northwest Territories and ends at Zama in northern Alberta, the northern terminus for the national oil pipeline system. Much of the terrain north of 60° N lies in the zone of discontinuous permafrost. As a result, special construction measures were necessary. Limiting construction to the winter months prevented excessive damage to the terrain and vegetation

A drilling rig near Norman Wells (second island from the left). The Norman Wells project allowed for a significant increase in oil production and supply to southern markets. (Terry A. Parker/Viewpoints West)

by trucks and other heavy equipment, and efforts to re-establish vegetation along the pipeline route following construction were designed to minimize soil erosion and the warming of ground temperatures by solar radiation. The danger from higher ground temperatures in the summer is that this could accelerate the melting of ground ice, leading to the subsidence (sinking) of the ground and possible rupturing of the buried pipeline.

The local oil fields were first discovered in 1920. By 1982, geological investigations had revealed the quantity of proven reserves to be as high as 100 million m³ (Bone, 1992: 145).

The purpose of the Norman Wells Project was to increase the oil produced and shipped to southern markets by means of a pipeline. Prior to 1985, the output from the Norman Wells oil fields served only the local Mackenzie Valley market. At that time, annual output was less than 180,000 m³. With the completion of the pipeline and the expansion of oil production by a factor of 10, the Norman Wells oil fields had the capacity to send vast quantities of oil to southern customers. Since 1985, annual output has exceeded 1 million m³ and almost reached 2 million m³ several times.

The Norman Wells Project represents a

successful and profitable operation. For nearly 20 years, Esso has produced and transported crude oil to markets in southern Canada and the United States. During that time, while some subsidence of the ground along the pipeline route has occurred, the pipeline has not been damaged and there have been no oil spills. Equally important, the operation of the Norman Wells Project has not triggered social problems in the communities along the pipeline route. From the oil industry's perspective, the Norman Wells Project demonstrated that pipelines can be built in areas with permafrost. However, from the Dene Nation's perspective, land claims should have been settled before the project was approved. Under such circumstances, the Dene would have become business partners in the construction and ensuing operation.

The NWT Diamonds Project

Canada is now the third largest producer of diamonds in the world, behind Botswana and Russia, and accounts for 15 per cent of the world supply, thanks to the two diamond mines in the Northwest Territories. In 2002, '[d]iamond mining accounted for just over one-fifth of the territories' gross domestic product' (*Toronto Star*, 2004: C1). How did this remarkable development come about? In 1991, two prospectors, Charles Fipke and Stewart Blusson, defied conventional opinion that the Canadian Shield was not a geological structure where diamond could be formed. In their search, they discovered diamond-bearing kimberlite near Lac de Gras in the Northwest Territories. Following extensive drilling, their find proved to have commercial viability, and BHP Diamonds Inc. decided to develop a mine. The company applied to Ottawa to secure its approval for its NWT Diamond Project. In

1996, the federal government approved the NWT Diamonds Project, which entails the operation of the Ekati diamond mine in the Lac de Gras area of the Northwest Territories for about 25 years. Five diamond-bearing **kimberlite pipes** are being mined, four of them located within a few kilometres of each other. All five kimberlite pipes lay under lakes that had to be drained to facilitate mining operations.

During the operations of the Ekati diamond mine, between 35 and 40 million tonnes of waste rock are removed from the kimberlite pipes each year. Since only a small proportion of that rock contains diamonds, the rest of the rock is placed in piles near the mines. Recovery of diamonds from the ore takes place in a processing plant where the ore is crushed and diamonds are separated. Final sorting is done using X-rays to separate the diamonds from the remaining waste material. BHP then sells rough-cut diamonds at international diamond markets. The mining operation employs about 800 workers with an estimated annual wage bill of $40 million. BHP uses an air-commuting system to bring workers from Yellowknife to its mine site on a two-weeks-in and two-weeks-out rotation.

Diamond mining has quickly become the backbone of the mining industry in the Territorial North. By value of production, number of employees, and spinoff effects, BHP Diamonds is the leading mining company in the North. In 2002, the two mines, Ekati and Diavik, produced nearly $1 billion worth of diamonds (Table 10.6). Their effect on the economy of the Northwest Territories has been profound. One spinoff was the designation of Yellowknife as the pickup point for miners who commute by air on a rotational scheme to the mine site. This reverses the usual trend of cash and tax flowing out of the territories. A second spinoff was the location of three diamond-cutting and -polish-

Table 10.6	Mineral Production in the Territorial North, 2002 ($ millions)			
Mineral Product	Yukon	Northwest Territories	Nunavut	Territorial North
Metals	31.4	52.7	268.8	352.9
Non-metals	3.6	811.2	0	814.8
Fuels	43.9	397.2	0	441.1
Total	78.9	1,261.1	268.8	1,608.8

Source: Natural Resources Canada (2003).

ing businesses in Yellowknife.

Another diamond mine, Snap Lake, is likely to begin production in 2006. Two other projects, Jericho and Kennedy Lake, are at an advanced stage of planning, and promising finds recently took place on Melville Peninsula, which may translate into Nunavut's first diamond mine.

How important is the diamond industry to the Northwest Territories? The figures are dazzling—diamond production has gone from zero in 1997 to over $800 million in 2002, and the total value of the diamond reserves at the three mines is around $23.3 billion (Statistics Canada, 2004). According to Randy Turner, president of Winspear Resources Ltd, the Ekati diamond field contains about 71 million carats with an average value of $133 a carat or a total value of $9.4 billion; Diavik field has about 101 million carats with an average value of $80 a carat or a total value of $8 billion; and Snap Lake has 42 million carats with an average value of $140 a carat or a total value of $5.9 billion (Belford, 2000: C9).

Megaprojects: Achilles Heel

Megaprojects in the Territorial North are based on non-renewable resources, which consist of petroleum deposits and mineral bodies. Forests, water, and wildlife are classified as renewable resources. The value of non-renewable resources far outweighs the economic value derived from renewable resources. In 2002, mineral production in the Territorial North was valued at $1.6 billion, which was almost 2.5 per cent of Canada's mineral output (Table 10.6). That same year, approximately 66 per cent of the Territorial North's production came from the Northwest Territories. The leading minerals by value were diamonds, gold, zinc, oil, and lead. The major mines were located at Ekati (diamonds), Nanisivik (lead-zinc), Polaris (lead-zinc), Con (gold), and Fort Liard (natural gas). Placer gold, valued at $50 million per year, is obtained from the valleys of the Yukon River and its tributaries. Norman Wells is the only oil-producing site in the Territorial North, but there are much larger oil and gas deposits in the Sverdrup Basin and the Beaufort Sea (Vignette 10.8).

Non-renewable resource development, because of the limited lifespan of such projects, subjects the Territorial North to a boom-and-bust economic cycle. Since mineral and petroleum deposits are in finite quantities, they do not offer long-term economic stability for the Territorial North. This economic cycle is the Achilles heel of northern development. In addition, because resource development is based on world demand, production fluctuates with this demand, creating great instabil-

Vignette 10.8   Beaufort Sea

The Beaufort Sea is part of the Arctic Ocean. Named after the British hydrographer, Sir Francis Beaufort, this 'frozen' sea occupies an area of approximately 450,000 km². Most of its surface is permanently covered by the polar pack ice, which is often over 5 m thick. Only large icebreakers can penetrate this formidable ice barrier. In August 1994, two icebreakers, the American *Polar Sea* and the Canadian *Louis S. St Laurent*, ploughed through this ice on a scientific voyage to the North Pole. During the short summer, the shore ice melts, leaving a narrow stretch of open water between the shore and the polar pack ice. Small ships take advantage of this open water to bring supplies to communities located along the coast of the Beaufort Sea. Most of these supplies are transported by barge northward along the Mackenzie River.

The shallow waters of the Beaufort Sea are often less than 50 m, so offshore drilling for oil and gas is possible. Sparked by high oil prices in the 1970s, oil companies drilled hundreds of wells in the Beaufort Sea over the subsequent 20 years. Significant deposits of oil and gas were discovered, but the high cost of production and transportation to southern markets has prevented further offshore oil and gas development.

ity in resource communities. Fluctuations in mineral production provide a measure of this instability. For instance, over the past 20 years, the value of mineral production in the Territorial North reached a high of $1.7 billion in 1989 and a low of $350 million in 1976. In 1996 it was $1.2 billion, but with the closure of the Faro mine it dropped to $1 billion.

The negative economic and social effects of such a drop are deeply felt in resource towns. Mines even suspend their operations during periods of low prices. In 1997, for example, the Faro mine was closed for the fourth time in its short history. Again, low metal prices made the zinc, lead, and silver-mining operation unprofitable. The Faro mine had been the principal mining activity for 30 years in Yukon. Each time the mine closed, the value of mineral production in Yukon dropped drastically. For example, in 1992 Yukon accounted for nearly half a billion dollars worth of minerals. After the Faro mine closed in 1993, production dropped to $81 million in 1994.

Megaprojects have made important contributions to the northern economy. These contributions include: generating the bulk of the North's GDP, expanding the northern transportation infrastructure, and providing high wages for employees. The diamond industry holds out some hope of even greater spinoffs such as the processing of diamond gems in Yellowknife. Megaprojects, however, though touted as the engine of northern development in the 1970s, have so far failed to transform the Territorial North's economy. While generating profits for the corporations, they have failed to diversify the northern economy, have not solved the massive unemployment problem, and have increased the region's economic vulnerability due to sudden changes in world demand for its primary products. Worse yet, all northern megaprojects are based on non-renewable resources, which, having a fixed lifespan, cannot provide the basis for a stable, long-term economy. From this perspective, megaprojects are not the engine, but rather are the Achilles heel of the northern economy.

The Territorial North's Future

Old ways and structures are changing in the Territorial North. The people of the North's two cultures are building new worlds within the region. To that end, the political landscape has changed dramatically. Power has shifted in two directions. Land-claim agreements have resulted in a devolution of economic and environmental powers from Ottawa to the organizations representing Aboriginal claimants in the North. A second shift took place in 1999, when Nunavut became a territory and political powers were transferred from Yellowknife to Iqaluit.

The economic landscape also is changing. Megaprojects will draw the Arctic, as they have the Subarctic, into a resource hinterland serving the global economy. A sign of this transformation is evident in the diamond mines along the southern edge of the Arctic— the Territorial North has become one of the major diamond producers in the world. However, resource development, particularly non-renewable development, has not yet triggered sufficient economic growth to create a stable economy or to reduce the Territorial North's financial dependency on Ottawa.

Nevertheless, in the twenty-first century, multinational corporations will participate in more and more resource developments. Natural gas prices have risen so much that the high cost of developing the natural gas fields in the Mackenzie Delta is now feasible. Given the vast reserves of natural gas in the Mackenzie Delta and the offshore fields in the Beaufort Sea, the Northwest Territories may become the Alberta of the North.

More economic changes are taking place through land-claim agreements in Yukon, much of the Northwest Territories, and Nunavut. These agreements, which so far involve over three-quarters of the Aboriginal peoples in the Territorial North, provide capital and create an Aboriginal business corporation to manage and invest this capital on behalf of its members. But there are risks in such ventures and economic success is not assured. The economic place of Aboriginal peoples in the Territorial North will become better defined in this century. Much depends on the extent to which land-claim agreements can protect wildlife and the environment. If successful, then harvesting of country and commercial food will continue and eco-tourism may well become an important 'sustainable' industry. Business corporations owned and operated by the Inuvialuit, Gwich'in, Inuit, Sahtu/Métis, Yukon, and Tlicho First Nations may provide the vehicle for entering the market economy and still allow them control over decision-making, including hiring Aboriginals to run these businesses. Aboriginal peoples who have agreements are controlling their future. Unfortunately, the future of those Aboriginal groups who have not yet negotiated a land-claim settlement is less clear.

The social landscape in the Territorial North is changing more slowly. While a rural-to-urban migration has taken place, most settlements have no economic purpose. A second migration, driven by economic considerations, is underway, drawing Aboriginal residents to regional centres and capital cities within the Territorial North or to metropolitan centres in southern Canada. Until there is a better geographic match between economic opportunities and Aboriginal peoples, social problems, including high unemployment rates, will continue to plague the Territorial North. This population imbalance is fuelled even further by the exceptionally high birth rates among Aboriginal peoples, especially among the Inuit. Another obstacle to northern development is the limited schooling achieved by Aboriginal workers, which either blocks their chances to enter the labour market or relegates them to 'low-end' jobs. Can the social landscape improve? The political and economic changes in the Territorial North are an important first step in that direction.

Summary

The Territorial North is both a resource frontier and a homeland for Aboriginal peoples. It is the only geographic region in Canada where Aboriginal peoples form a substantial proportion of the total population. Nunavut is a political expression of this homeland concept for the Inuit of the Eastern Arctic. Comprehensive land-claim agreements in six geographic areas have equipped Aboriginal peoples with capital, land, and control over wildlife to chart a new future. Although they are involved in the market economy, Aboriginal peoples retain a strong attachment to the land, country food, and the ethic of sharing. Such cultural traits strengthen their recent political and economic advances.

As a resource hinterland, the Territorial North has three principal characteristics: it is far from world markets; resource development is dependent on external demand; and its economy is sensitive to fluctuations in world prices for resources. The economy of the Territorial North has two serious flaws, namely, that it is based on non-renewable resources and that economic benefits from resource development accrue mainly to firms and workers in southern Canada and foreign countries. Combined, these two flaws lead to a state of underdevelopment known as a staple trap.

In recent years, megaprojects have drawn the Territorial North more closely into the global economy as a resource frontier. The implication of such resource development is that stable economic growth and economic diversification remain elusive goals. Without a more diversified economy, the Territorial North will continue to suffer from a boom-and-bust economic cycle and its three governments will remain financially dependent on Ottawa.

Notes

1. There is a third climatic zone in the Territorial North—the Cordillera. However, the Cordillera climate, often described as a mountain climate, is affected by elevation (as elevation increases, temperature drops). North of 60° N, the Cordillera climate is also affected by latitude so that boreal natural vegetation is found at lower elevations and tundra natural vegetation at high elevations.

2. Sharing food was essential for small hunting groups living together harmoniously. By sharing in all hunters' successes, they could adjust for the vagaries of individual luck and reduce the threat of starvation. Today, the sharing of food remains a pivotal component of Aboriginal culture and the Native economy.

3. The Vikings made contact with the ancestors of the Inuit, the Thule (c. AD 1000 to 1600). The Thule originated in Alaska where they hunted bowhead whales and other large sea mammals. They quickly spread their whaling technology across the Arctic, travelling in skin boats and dogsleds. With the onset of the Little Ice Age in the fifteenth century, climate conditions affected the distribution

of animals and the Thule who were dependent on them. An increased amount of sea ice blocked the large whales from their former feeding grounds, resulting in the collapse of the Thule whale hunt. With the loss of their main source of food, the Thule had to rely more and more on locally available foods, usually some combination of seal, caribou, and fish. By the eighteenth century, the Thule culture had disappeared and was replaced by the Inuit hunting culture.

4. Territorial governments, unlike provinces, are governed under delegated powers from the federal government. In that sense, they are dependent on Ottawa for their political powers. Under the Canadian Constitution, Ottawa has the power to govern its territories. In the past, responsible government occurred when a part of these territories had a sufficiently large population and tax base to warrant having an elected council. Until that time, the governing members were appointed by the federal government.

5. Social programs may have encouraged large families. For example, the Family Allowance program begun in 1945 provided a payment for each child. For a family of five, the annual cash derived from a family allowance was often greater than the cash a father might earn from the sale of fur pelts.

6. Today, capital and operating costs are still serious problems for hunters. Transportation from a settlement to hunting/trapping areas is a major expenditure. Snowmobiles, for example, may cost as much as $20,000 in a northern retail store. (In the 1990s, they cost $5,000 to $10,000.) With a continuing rise in the prices of manufactured goods, the cost of a snowmobile 10 years from now may double again. Since a snowmobile used for long-distance travel has a short lifespan, the hunter/trapper must replace it about every three or four years.

7. The first comprehensive land-claim submission came from the Dene/Métis in 1974. The land claimed by the Dene/Métis extended over most of the Mackenzie Basin north of 60° N. This land was called Denendeh. In 1988, an agreement-in-principle was signed by the two negotiating parties (the federal government and Dene Nation). This agreement called for a cash payment of $500 million over 15 years plus title to 181,230 km^2 of land. Nearly 6 per cent of this land (10,000 km^2) would include subsurface rights for the Dene/Métis. As well, the Dene/Métis would obtain a share of federal resource royalties, including those generated by the Norman Wells oil field. Chiefs and elders from the Great Slave Lake area refused to approve this agreement for two main reasons—the agreement-in-principle contained no reference to self-government and it called for the surrendering of Aboriginal rights. With this rejection, the Gwich'in and Sahtu/Métis subsequently negotiated their own comprehensive land-claim agreements with Ottawa.

8. The Inuvialuit broke away from the other Inuit who were seeking a common land-claim settlement that would have stretched from the Arctic Coast of the Yukon to Baffin Island. The Inuvialuit wanted to advance their own land claim before the huge oil and gas deposits in the Beaufort Sea were developed. The Inuvialuit, with their separate agreement, hoped to obtain land with oil and gas deposits and achieve a taxation arrangement similar to that of the Arctic Slope Regional Corporation of the Inupiat in Alaska near Prudhoe Bay. This municipality received substantial tax revenues from the oil companies.

9. Under the IFA, Canada agreed to transfer $45 million in 1977 dollars. By 1984, these funds were valued at $152 million. The payments to the Inuvialuit Regional Council, the business corporation of the IFA, began on 31 December 1984 with $12 million. By the year 1997, there were three annual pay-

ments of $1 million beginning 31 December 1985; five annual payments of $5 million beginning 31 December 1988; four annual payments of $20 million beginning 31 December 1993; and a final payment on 31 December 1997 of $32 million (Canada, 1985: 107).

10. Until the 1970s, Canadians did not question the merit of northern industrial projects. The Mackenzie Valley Pipeline Inquiry headed by Thomas Berger made Canadians aware of the hidden costs of building a gas pipeline from Prudhoe Bay along the Arctic coast to the Mackenzie River and then southwards along the Mackenzie Valley to markets in southern Canada and the United States. The hidden costs identified by Berger were: the potential threat to wildlife, especially the Porcupine caribou herd; the ground subsidence that could result from a buried pipeline; and the feared social impacts on Aboriginal peoples living in the construction zone. For more on this subject, see Bone (2003a: 169–70).

11. This pipeline would be constructed by

Canadian Arctic Gas Pipeline Ltd, which represented 27 Canadian and American oil producers, such as Exxon, Gulf, and Shell, and Trans-Canada Pipelines. Later, a rival bid was made by Foothills Pipeline Ltd, which consisted of Alberta Gas Trunk Line and Westcoast Transmission. Prudhoe Bay is located along the North Slope of Alaska. Alaskan oil accounts for 20 per cent of US oil production. An oil pipeline, known as the Trans-Alaska Pipeline System (TAPS), transports Alaskan oil from Prudhoe Bay to Valdez and then by tanker to markets along the west coast of the United States.

12. In 1977, a rival proposal by Foothills Pipeline called for a natural gas pipeline to follow the route of the Alaska Highway. In 1982, Ottawa approved the Alaska Highway Gas Pipeline Project. However, this proposed project was never built because of declining natural gas prices in the 1980s. Now that gas prices have soared to historically high levels, the Yukon government is calling for the construction of the Alaska Highway Gas Pipeline.

Key Terms

country food
As hunters, Indians, Inuit, and Métis fished and hunted for food. Now settlement dwellers, they still fish and hunt for cultural and economic reasons. Such food is called game or country food.

frontier
The perception of the Territorial North as a place of great mineral wealth that awaits development by outsiders.

homeland
People who live in a region develop a strong attachment to that place; a sense of place.

kimberlite pipe
An intrusion of igneous rocks in the earth's crust that takes a funnel-like shape. Diamonds are sometimes found in these rocks.

megaprojects
Large-scale construction projects that exceed $1 billion and take more than two years to complete.

settlement area
Each comprehensive land claim has a geographic extent that is known as a settlement area. While less than 25 per cent of this land is allocated to the Aboriginal beneficiaries as a collective (not individual) landholding, the entire area is subject to the environmental and wildlife regulations exercised by the settlement area's co-management boards.

terraces
Streams and rivers form terraces when they cut downward into the flood plain and form a new and lower flood plain. The old flood plain

(now a terrace) is found along the sides of the stream or river.

traditional ecological knowledge
Aboriginal peoples are very familiar with their natural surroundings. This knowledge of the environment and wildlife, which has accumulated over the centuries, is known as traditional ecological knowledge or TEK.

Bibliography

Anderson, Robert B., and Robert M. Bone. 1995. 'First Nations Economic Development: A Contingency Perspective', *The Canadian Geographer* 39, 2: 120–30.

——— and Robert M. Bone. 2003. *Natural Resources and Aboriginal People in Canada: Readings, Cases and Commentary.* Concord, Ont.: Captus Press.

Belford, Terrence. 2000. 'Moiling for diamonds in Canada's North', *National Post*, 8 May, C9.

Berger, Thomas R. 1977. *Northern Frontier, Northern Homeland: The Report of the Mackenzie Valley Pipeline Inquiry*, 2 vols. Ottawa: Minister of Supply and Services.

Bockstoce, John R. 1994. *Whales, Ice and Men: The History of Whaling in the Western Arctic.* Seattle: University of Washington Press.

Boden, Jürgen F., and Elke Boden, eds. 1991. *Canada North of Sixty.* Toronto/Osteinbek: McClelland & Stewart/Alouette Verlag.

Bone, Robert M. 1992. *The Geography of the Canadian North: Issues and Challenges.* Toronto: Oxford University Press.

———. 1998. 'Resource Towns in the Mackenzie Basin', *Cahiers de Géographie du Québec* 42, 116: 249–59.

———. 2003a. *The Geography of the Canadian North: Issues and Challenges*, 2nd edn. Toronto: Oxford University Press.

———. 2003b. 'Power Shifts in the Canadian North: A Case Study of the Inuvialuit Final Agreement', in Anderson and Bone (2003: 382–93).

——— and Shane Long. 1995. 'Population Change in Aboriginal Settlements in the Mackenzie Basin North of 60°', *Proceedings of the 1995 Symposium of the Federation of Canadian Demographers.* Ottawa: St Paul University.

Canada. 1985. *The Western Arctic Claim: The Inuvialuit Final Agreement.* Ottawa: Department of Indian Affairs and Northern Development.

———. 1991. 'Comprehensive Land Claim Agreement Initialled with Gwich'in of the Mackenzie Delta in the Northwest Territories', Communiqué 1–9171. Ottawa: Department of Indian Affairs and Northern Development.

———. 1993a. 'Formal Signing of Tungavik Federation of Nunavut Final Agreement', Communiqué 1–9324. Ottawa: Department of Indian Affairs and Northern Development.

———. 1993b. *Umbrella Final Agreement between the Government of Canada, Council for Yukon Indians and the Government of the Yukon.* Ottawa: Department of Indian Affairs and Northern Development.

———. 1996. *NWT Diamonds Project.* Report of the Environmental Assessment Panel. Canadian Environmental Assessment Agency. Ottawa: Minister of Supply and Services.

———. 2004. *Agreements.* Ottawa: Department of Indian and Northern Affairs [on-line database], Ottawa. Searched 29 Jan. 2004: <http://www.ainc-inac.gc.ca/pr/agr/index_e.html#Comprehensive%20Claims%20Agreements>.

Canadian Press. 2003. 'Ottawa to phase out Yukon placer mining', *National Post*, 28 Jan., FP10.

Cattaneo, Claudia. 2000. 'Inuvialuit press PM for pipeline support', *National Post*, 24 Nov., C5.

Coates, Ken, and Bill Morrison. 1992. *The Forgotten North: A History of Canada's Provincial Norths.* Toronto: Copp Clark.

Crowe, Keith J. 1991. *A History of the Original Peoples of Northern Canada*, 2nd edn. Montréal and Kingston: McGill-Queen's University

Press.

Dahl, Jens, Jack Hicks, and Peter Jull, eds. 2000. *Nunavut: Inuit Regain Control of Their Lands and Their Lives*. Copenhagen: International Work Group for Indigenous Affairs.

Dickason, Olive Patricia. 2002. *Canada's First Nations: A History of Founding Peoples from Earliest Times*, 3rd edn. Toronto: Oxford University Press.

Difrancesco, R.J. 2000. 'A Diamond in the Rough? An Examination of the Issues Surrounding the Development of the Northwest Territories', *The Canadian Geographer* 44, 2: 114–34.

Elias, Peter Douglas. 1995. *Northern Aboriginal Communities: Economies and Development*. North York, Ont.: Captus Press.

French, Hugh M., and Olav Slaymaker. 1993. *Canada's Cold Environments*. Montréal and Kingston: McGill-Queen's University Press.

Frideres, James. 1993. *Native Peoples in Canada*. Scarborough, Ont.: Prentice-Hall.

Globe and Mail. 1993. 'Mackenzie Land Claim Settled', 7 Sept., A1.

Government of the Northwest Territories, Bureau of Statistics. 2003. *Statistics Quarterly* 25, 2 [online database], Yellowknife. Searched 12 Sept. 2003: <http://www.stats.gov.nt.ca/Statinfo/Generalstats/statsquarterly/_statq.html>.

———. 2004. Northwest Territories—2003 . . . by the numbers [on-line database], Yellowknife. Searched 29 Jan. 2004: <http://www.stats.gov.nt.ca/Statinfo/Generalstats/bythenumbers/BTNhome(dvo).html>.

Grant, Shelagh D. 1988. *Sovereignty or Security? Government Policy in the Canadian North, 1936–1950*. Vancouver: University of British Columbia Press.

Hamelin, Louis-Edmond. 1978. *Canadian Nordicity: It's Your North, Too*, trans. William Barr. Montréal: Harvest House.

———. 1982. 'Originalité culturelle et régionalisation politique: Le project Nunavut des Territoires-du-Nord-Ouest (Canada)', *Recherches Amérindiennes au Québec* 12, 4: 251–62.

Hamilton, John David. 1994. *Arctic Revolution: Social Change in the Northwest Territories, 1935–1994*. Toronto: Dundern Press.

Hesketh, Bob, ed. 1996. *Three Northern Wartime Projects: The Alaska Highway, the Northwest Staging Route, and Canol*. Edmonton: Canadian Circumpolar Institute.

Ironside, R.G. 2000. 'Canadian Northern Settlements: Top-down and Bottom-up Influences', *Geografiska Annaler* 82 B, 2: 103–14.

Johnston, Margaret E. 1994. *Geographic Perspectives on the Provincial Norths*. Northern and Regional Studies Series, vol. 3. Centre for Northern Studies, Lakehead University. Mississauga, Ont.: Copp Clark Longman.

Keith, Robbie. 1998. 'Arctic Contaminants: An Unfinished Agenda', *Northern Perspectives* 25, 2: 1–3.

Kenney, Gerard. 1994. *Arctic Smoke and Mirrors*. Prescott, Ont.: Voyageur Publishing.

Légaré, André. 1993. 'Le project Nunavut: Bilan des revendications des Inuit des Territoires-du-Nord-Ouest', *Études/Inuit/Studies* 17, 2: 29–62.

———. 1996. 'Le gouvernement du Territoire du Nunavut', *Études/Inuit/Studies* 20, 1: 7–43.

McGhee, Robert. 1996. *Ancient People of the Arctic*. Vancouver: University of British Columbia Press.

MacLachlan, Letha. 1996. *NWT Diamonds Project: Report of the Environmental Assessment Panel*. Ottawa: Canadian Environmental Assessment Agency.

Marcus, Alan R. 1995. *Relocating Eden: The Image and Politics of Inuit Exile in the Canadian Arctic*. Hanover, NH: University Press of New England.

Marsh, James H., ed. 1988. *The Canadian Encyclopedia*, 2nd edn. Edmonton: Hurtig.

Natural Resources Canada. 2003. Mineral Production of Canada, by Provinces and Territory [on-line database]. Searched 30 Aug. 2003: <http://mmsd1.mms.nrcan.gc.ca/mmsd/production/production_e.asp>.

Notzke, Claudia. 1994. *Aboriginal Peoples and*

Natural Resources in Canada. Toronto: Captus Press.

O'Reilly, Kevin. 1996. 'Diamond Mining and the Demise of Environmental Assessment in the North', *Northern Perspectives* 24, 1–4: 1–6.

Page, Robert. 1986. *Northern Development: The Canadian Dilemma.* Toronto: McClelland & Stewart.

Poelzer, Greg. 2001. 'Aboriginal Peoples and Environmental Policy in Canada: No Longer at the Margins', in Debora L. VanNijnatten and Robert Boardman, eds, *Canadian Environmental Policy: Context and Cases*, 2nd edn. Toronto: Oxford University Press, 87–106.

Purich, Donald. 1992. *The Inuit and Their Land: The Story of Nunavut.* Toronto: Dormer.

Ray, Arthur J. 1990. *The Canadian Fur Trade in the Industrial Age.* Toronto: University of Toronto Press.

Rowley, Graham W. 1996. *Cold Comfort: My Love Affair with the Arctic.* Montréal and Kingston: McGill-Queen's University Press.

Royal Commission on Aboriginal Peoples. 1996. *Report of the Royal Commission on Aboriginal Peoples*, 5 vols. Ottawa: Minister of Supply and Services.

Saku, James A., and Robert M. Bone. 2000. 'Looking for Solutions in the Canadian North: Modern Treaties as a New Strategy', *The Canadian Geographer* 44, 3: 259–70.

———, ———, and Gérard Duhaime. 1998. 'Towards an Institutional Understanding of Comprehensive Land Claim Agreements in Canada', *Etudes/Inuit/Studies* 22, 1: 109–21.

Shidelar, Janet. 2000. *The Geography of Canada Bibliography Series: Vol. 6, The North.* Plattsburgh, NY: Plattsburgh State University, Center for the Study of Canada.

Slave River Journal. 1994. 'Opinion: Pick a Side', 9 Mar., 4.

Stanford, Quentin H., ed. 1998. *Canadian Oxford World Atlas*, 4th edn. Toronto: Oxford University Press.

Statistics Canada. 1997a. *A National Overview: Population and Dwelling Counts.* Catalogue no.

93–357–XPB. Ottawa: Industry Canada.

———. 1997b. 1996 Census: Nation Tables—Population by Mother Tongue, Showing Age Groups, for Canada, Provinces and Territories, 1996 Census—20% Sample Data, 2 Dec. 1997 [on-line database], Ottawa. Searched 15 July 1998: <http://www.statcan.ca/english/census96/>.

———. 1997c. *The Daily*—1996 Census: Mother Tongue, Home Language and Knowledge of Languages, 2 Dec. 1997 [on-line database], Ottawa. Searched 14 July 1998: <http://www.statcan.ca/Daily/English/>.

———. 1998a. *Canadian Economic Observer.* Catalogue no. 11–010–XPB. Ottawa: Statistics Canada.

———. 1998b. *The Daily*—1996 Census: Aboriginal Data, 13 Jan. 1998 [on-line database], Ottawa. Searched 14 July 1998: <http://www.statcan.ca/Daily/English/>.

———. 1998c. *The Daily*—1996 Census: Ethnic Origin, Visible Minorities, 17 Feb. 1998 [on-line database], Ottawa. Searched 16 July 1998: <http://www.statcan.Daily/English/>.

———. 1998d. *The Daily*—1996 Census: Education, Mobility and Migration, 14 Apr. 1998 [on-line database], Ottawa. Searched 12 Sept. 2003: <http://www.statcan.Daily/English/>.

———. 2002a. 2001 Census: Population and Dwelling Counts, for Census Metropolitan Areas and Census Agglomerations, 2001 and 1996 Censuses. 16 July 2002 [on-line database], Ottawa: <http://www.statcan.ca/english/IPS/Data/93F0050XCB2001013.htm>.

———. 2002b. *The Daily*—2001 Census: Language, Mobility and Migration, 10 Dec. 2002 [on-line database], Ottawa. Searched 12 Sept. 2003: <http://www.statcan.Daily/English/>.

———. 2003a. Canadian Statistics, International Trade. Exports of Goods on a Balance-of-Payments Basis [on-line database], Ottawa. Searched 3 May 2003: <http://www.statcan.ca/english/Pgdb/gblec04.htm>.

———. 2003b. Canadian Statistics: Population of Census Metropolitan Areas [on-line database], Ottawa. Searched 12 May 2003: <http://www.statcan.ca/english/Pgdb/demo05.htm>.

———. 2003c. Census of Canada 2001—Labour, Employment and Unemployment, by Industry, by Province [on-line database], Ottawa. Searched 3 May 2003: <http://www.statca.ca/english/Pgdb/labor21b.htm>.

———. 2003d. Census of Canada 2001—Canada's Ethnocultural Portrait: The Changing Mosaic. Analytical Series 96F0030XIE 2001008, 21 Jan. 2003 [on-line database], Ottawa. Searched 28 Jan. 2003: <http://www12.statcan.ca/english/census01/products/analytic/companion/etoimm/subprovs.cfm>.

———. 2003e. Canadian Statistics: Distribution of Employed People, by Industry, by Province [on-line database], Ottawa. Searched 20 May 2003: <http://www.statcan.ca/english/Pgdb/labor21b.htm>.

———. 2004. *The Daily*—Study: Diamonds are adding lustre to the Canadian economy [on-line database], Ottawa. Searched 29 Jan. 2004: <http://www.statcan.ca/Daily/English/040113/d040113a.htm>.

Stevenson, Marc, Culley Schweger, Valerie Raey, Elaine L. Maloney, and Susan Macara. 1996. *Environmental and Economic Issues in Fur Trapping*. Edmonton: Canadian Circumpolar Institute.

Toronto Star. 2004. 'Canada climbs to third in world diamond output', 14 Jan., C1.

Usher, Peter J. 1982. 'The North: Metropolitan Frontier, Native Homeland?', in L.D. McCann, ed., *Heartland and Hinterland: A Geography of Canada*. Scarborough, Ont.: Prentice-Hall, 411–56.

——— and George Wenzel. 1987. 'Native Harvest Surveys and Statistics: A Critique of Their Construction and Use', *Arctic* 40, 2: 145–60.

——— and ———. 1989. *A Strategy for Supporting the Domestic Economy of the Northwest Territories*. Report for the Legislative Assembly's Special Committee on the Northern Economy. Ottawa: P.J. Usher Consulting Services.

——— and ———. 1989. 'Socio-Economic Aspects of Harvesting', in Randy Ames, Don Axford, Peter Usher, Ed Weick, and George Wenzel, *Keeping on the Land: A Study of the Feasibility of a Comprehensive Wildlife Harvest Support Program in the Northwest Territories*. Ottawa: Canadian Arctic Resources Committee, ch. 1.

Wenzel, George. 1991. *Aboriginal Rights, Animal Rights: Ecology, Economy and Ideology in the Canadian Arctic*. Toronto: University of Toronto Press.

Williamson, Robert G. 1974. *Eskimo Underground: Socio-Cultural Change in the Canadian Central Arctic*. Occasional Papers II. Uppsala, Sweden: Almqvist & Wiksell.

Further Reading

Morrison, William R. 1998. *True North: The Yukon and Northwest Territories*. Toronto: Oxford University Press.

This is a finely illustrated history of Canada's North. Some 200 photographs and five maps accompany the lively narrative that describes the main elements of the North's history. From the coming of the First Peoples to our contemporary period, Professor Morrison has etched out an entertaining story of the dual nature of the Canadian North—both a northern frontier and a northern homeland. In Chapter Eight, 'Invasion: War and the Militarization of the North', construction of the Alaska Highway is described both in words and photographic images. Black American troops cleared the right-of-way for the highway and horses hauled supplies to the work camps in the first year of this mega-construction project. The difficulty of constructing a highway across the rugged northern terrain is well illustrated by a photograph of the highway turning into a quagmire after a heavy rain. The completed military highway was not suited for faint-of-heart drivers—places like 'Suicide Hill' are discussed and illustrated. The final chapter, 'Search for a Future: the Modern North, 1970–1995', depicts the North as a place where 'its future seems to be in a state of permanent uncertainty', but Morrison goes on to identify one constant, the people, particularly the Native people. Geography students will find this illustrated book not only easy to read but also very informative.

Overview and Objectives

Canada is a country composed of six major geographic regions. This final chapter summarizes the complex regional differences caused by Canada's physical and historical geography, its core/periphery economic structure, and the tensions existing in its social and political structures. The principal tensions or faultlines are between centralists and decentralists, Aboriginal and non-Aboriginal Canadians, French-speaking and English-speaking Canadians, and new immigrants and those born in this country. The future, while full of uncertainty, contains one truth: Canada will remain a country of cultural and regional diversity. Though problems have arisen because of this diversity in the past, and will continue to arise in the future, Canadians have recognized that differences cannot be obliterated. Instead, compromises must and can be found.

- Demonstrate that Canada is and will remain a country of regions.
- Explain why it will remain so.
- Re-examine the core/periphery model in the light of globalization.
- Re-evaluate Canada's four faultlines not as centrifugal forces but as centripetal ones.
- Stress that Canada's future lies not in division but in the 'process' of seeking compromises.

Chapter 11 Canada: A Country of Regions

■ Introduction

'Canada is big—preposterously so', wrote geographer Kenneth Hare (1968: 31). The sheer size of Canada has meant that the country spans a diverse physical geography, varying climate conditions, and a wide range of vegetation. These geographic variations, combined with regional histories, have transformed the country into a country of regions. Geography and history have woven Canada's regions into a federated nation and, in so doing, have shaped and then reshaped Canada's identity and the identities of its regions.

During the historical evolution of the country, tensions between certain groups have arisen. These tensions, known as faultlines, are manifested in four different contexts: between centralists and decentralists, between French-speaking and English-speaking Canadians, between Aboriginal Canadians and non-Aboriginal Canadians, and between new immigrants and those born in Canada. However, these cracks, rather than fragmenting the country, have forced Canadians to build bridges—between regions and governments and between different peoples. Canada, by incorporating cultural diversity and regional differences within its socio-political structure, provides an alternative to monolithic nation-states, where the majority settles differences. Canada, on the other hand, has accepted minority positions, whether same-sex marriages or self-government for Aboriginal groups. In doing so, Canada is prepared to live

with contradictions—and these contradictions make Canada an exciting and different place to live. In this concluding chapter, five issues will be addressed: (1) Canada's regional character; (2) Canada's core/periphery spatial structure; (3) Canada's faultlines; (4) Canada's past; and (5) Canada's future.

Regional Character

The geographic complexity of Canada begins with its six geographic regions—Ontario, Québec, British Columbia, Western Canada, Atlantic Canada, and the Territorial North. These regions, though primarily created by geography, were shaped further by Canada's historical development. For instance, southern Ontario presented some of the most favourable geographic conditions for economic growth and industrialization. It had fertile soils and a moderate climate, which supported agriculture and a growing population, as well as geographic proximity to water transportation routes and to US markets. Complementing these favourable natural features, historically Ontario was supported by economic conditions that allowed it to prosper, such as the Reciprocity Treaty and then protective tariffs—policies that ensured a stable market for the region's industries.

All of Canada's regions have been similarly shaped by a combination of geography and history. For instance, Québec has been shaped by the St Lawrence River and by its French-speaking inhabitants. British Columbia has been shaped by the Cordillera, which separates it from the rest of Canada, and by the railway that first connected it to the country. Western Canada has been shaped by its fertile soils and by a sense of alienation from the country's centre of power. Atlantic Canada has been shaped by its coastal geography and by economic downturns in its resource-oriented industries. The Territorial North has been shaped by its cold and harsh climate and by Aboriginal land-claim agreements. This is to name but a few of the geographic and historical features that have defined Canada's regions.

From each region's history and geography, a sense of place has developed. Regional geography is very much determined by a sense of place, the power of which has been elegantly expressed by Canadian writer John Ralston Saul (1997: 69):

Because, in spite of intellectual claims to the contrary, not religion, not language, not race but place is the dominant feature of civilizations. It decides what people can do and how they will live.

Within each region of Canada, the collective experiences of people, both past and present, evoke a sense of belonging to a place. Like other geographic concepts, this psychological sentiment operates at several geographic levels—national, regional, local. In this book, attention has been directed to regions and the common experiences within those regions that have fuelled a sense of place—for instance, grain farming in Western Canada.

Within a federation of regions, sense of place can take on political connotations as divisions between regions feed geographic 'separatism'. Though Québec is identified with this sense of separation, almost every region experiences it to some degree. For instance, historian Margaret Ormsby (1958: 257) claimed that 'British Columbia was in, but not of, Canada', meaning that union with Canada for British Columbia was not based on sentiment but on material advantage, such as the promise of a railway and the absorption of BC's debt by Ottawa. In British Columbia, history created a sense of independence while geography formulated a sense of isolation from the rest of Canada. That Canada 'ends' at the Rocky Mountains is part of this region's sense of identity. Another part of this identity, or sense of place, comes from British Columbians' pride in their natural surroundings—this feeling is expressed in the provincial licence-plate motto, 'Beautiful British Columbia'.[1]

If Canada is so clearly made up of very different regions that are separate from the whole, how did it ever overcome its geographic and historical differences to evolve from regional clusters into a modern nation-state? Why is Canada more than a collection of economic regions dominated by self-interest? Canada was able to overcome its regional differences through a willingness to share power and thereby create a balance between central interests and those of regions and groups. In fact, the sharing of power was embedded in the Canadian Constitution right from the start. With the creation of a federation, the BNA Act divided governing powers between the provinces and the federal government. This initial sharing of powers set the stage for a trend to develop in which different Canadian regions and peoples would share power and economic wealth. In fact, Canadian history illustrates this trend. A number of struggles between those holding power and those wanting a share of that power have occurred. Often, the result has been compromise, with power being shared between parties. Equalization payments (1957), official bilingualism (1969), multiculturalism (1971), and modern land-claim agreements (the first in 1975) are examples of how the federation has been able to share wealth and cultural recognition between various regions and peoples. Spatial and cultural diversity did not create barriers to national unity that needed to be obliterated, but rather, served as the very building blocks necessary to construct a humanitarian society.

Canada's strength—you might even say what makes it interesting—is its regional diversity. Physical geography laid the groundwork for Canada's regional structure while Canada's history unfolded within these natural settings. Economic, political, and social change has occurred—some regions have grown swiftly while others have lagged behind and in the process lost some of their best people—but Canada's regions have proven remarkably stable. Furthermore, compromises have been reached to reconcile the inequities between regions as well as the peoples across these regions. Equalization payments play a key role in keeping regional economic differences from getting out of balance. Calculating equalization payments is a complex process and some provinces are demanding changes. Change is possible but the federal government must pass legislation that sets out the new rules. Each region of the country sees this review as an opportunity to correct regional injustices. Atlantic Canada, for instance, would like relief from the link between oil royalties and equalization payments. The Territorial North but especially the Northwest Territories would like to obtain control over natural resources and thereby receive the royalties from resource developments. British Columbia has gone through a rough economic patch,

with the forest fires of 2003 costing over half a billion dollars, and is just happy that it is now receiving equalization payments.

The Core/Periphery Model

Canada's regional geography has significantly affected the country's national and regional economies. Physiography, climate, permafrost, and other natural elements have in many ways determined what economic activities could be pursued in each region and, therefore, how regional economies would develop. A way to examine this phenomenon is through the core/periphery model—cores represent the more prosperous and industrialized regions, which draw on the peripheries, the more resource-oriented, less economically diversified regions. Using Friedmann's version of this model (see Chapter 1), Canada's six geographic regions fall into: core regions (Ontario and Québec); upward transitional regions (British Columbia and, to a lesser degree, Western Canada); a downward transitional region (Atlantic Canada); and a resource frontier (the Territorial North). Furthermore, within each of Canada's geographic regions, natural variations have created internal core/periphery structures. For instance, within Ontario, southern Ontario represents an industrial core, while northern Ontario is a resource periphery.

For years, the core/periphery structure of Canada remained relatively stable. Central Canada benefited from geography and favourable federal policies that allowed its industrial sector to grow by guaranteeing a domestic market for its goods in other regions. The rest of Canada's regions acted as peripheries, supplying raw materials and purchasing the core's manufactured items. However, during the latter part of the twentieth century a new global economic order emerged that is affecting and will continue to affect Canada's core/periphery structure.

Since World War II, Canada has taken several steps to a new economic goal of integration into larger economies. First, Canada became a participant in the GATT (more recently, WTO) regime in order to be involved more fully in the global post-war economic order. Then came the Auto Pact, an agreement that integrated the North American automobile industry. Finally, the FTA and NAFTA were signed in order to achieve uninhibited access to the American economy. The net effect of trade liberalization has been to strengthen Canada's economic ties with the United States. For example, from 1992 to 2002 the percentage of Canada's exports to the United States increased from 75.5 per cent to nearly 85 per cent (Statistics Canada, 2003).

The effects of trade liberalization on Canada's regional economies have been mixed. All the regions have increased their service sectors but, except for Ontario, exports are predominantly primary products, with most destined for the American market. Ontario, thanks to the Auto Pact, saw its automobile industry thrive, almost doubling its share of the North American market. But following the Free Trade Agreement in 1989, the rest of the manufacturing industry in Central Canada was either subject to fierce competition from American firms or to closure of American branch plants. As a result, the manufacturing sectors of both Ontario and Québec went through difficult times.

For Canada's peripheries, the process of globalization accomplished two, long-sought-after goals—more local processing of primary products and access to foreign manufactured goods. New trade links with the United States and, to a lesser degree, with Japan and Europe quickly took shape. British Columbia, for example, benefited from the rapid economic expansion of Japan and other Pacific Rim countries. In the early 1990s, BC exported around 37 per cent of its goods to Pacific Rim countries, with Japan accounting for 25 per cent of this trade (IDAC, 1998: 6). However, by 1997, the economies of Pacific Rim countries faltered, chilling BC's export-dependent economy. By 2003, British Columbia was receiving equalization payments and had therefore slid into the position of a have-not province. Other hinterlands gained greater access to particular parts of the American market through liberalization. For instance, Atlantic Canada has access to the New England market. McCain Foods and the various Irving enterprises have taken advantage of the American market, but smaller producers were less able to achieve economies of scale and therefore have had more difficulty selling their products in the American market. New Brunswick's Irving family is building a plant at Moncton to specialize in diaper production for the North American market. Certainly, trade liberalization has allowed various regions of Canada to tap into the much larger American market and high production levels allow economies of scale to take place, keeping cost per unit low. In this way, Canadian firms can compete in the North American marketplace. Greater trade with the United States has brought economic benefits, but it has made Canada more dependent on access to American markets—and when that access is disrupted, Canadian firms and workers suffer.

In becoming a member of NAFTA and the GATT/WTO, Canada surrendered certain controls over its economy. No longer can Ottawa or the provinces impose policies reflecting economic nationalism, such as having one set of rules for foreign companies and another set for

Canadian companies, or maintaining regional development programs that lend support to businesses competing with foreign firms. For instance, subsidies to sustain regional industries, such as the Cape Breton Development Corporation (Devco), or to reduce grain transportation costs for Prairie farmers by means of the Crow Benefit are no longer an option for the federal government. In fact, Canada, being the smaller partner, has had to bend on sensitive trade issues that affect American producers. One testy example is the Canada–US softwood lumber dispute in which Washington is determined to limit the amount of lumber Canada can export to the United States.

Given the changes that have already resulted from trade liberalization, what future transformations await Canada in these times of global uncertainty? Four questions arise, all having implications for Canada's core/periphery structure:

- Will the economic regional alignment of Canada shift from an east–west alignment to a north–south one?
- Will the national core/periphery structure be supplanted by a continental or global core/periphery structure?
- As the powers of the federal government wane, is more and more of Canada's economic future in the hands of external market forces led by multinational corporations?
- Will Washington continue to stress internal security over trade with Canada?

The north–south 'grooves' of geography naturally encourage north–south regional trade flows. For instance, trade moves more naturally between the provinces of Western Canada and the American states to their south than between these three provinces and the more distant provinces to the east. With the erosion of trade barriers, north–south trade has increased rapidly. All along the US–Canada border, ever-increasing quantities of products move between Canadian regions and American markets. Ontario trades with New York and the Midwest, Québec with New England and New York, British Columbia with the US Pacific Northwest, Atlantic Canada with New England, and Western Canada with the American Great Plains. The Territorial North is an exception. While its oil and gas flows southward to American markets, mineral production has a more widespread distribution, including Europe and the Far East.

However, east–west trade, while growing more slowly, remains an essential element of Canada's economic structure. Recent economic studies by McCallum (1995), Helliwell (1996), and Anderson and Smith (1999) support the notion that the flow of trade between Canadian regions and provinces remains much more powerful than north–south trade between Canada and the United States, perhaps by a factor of 10 to 20 times. The explanation lies in the 'national border effect'. This effect is likely caused by a number of factors: the east–west Canadian transportation and communications system that encourages interprovincial trade; trade inertia based on pre-liberalization trading patterns between Canadian firms that remained strong in spite of slightly lower US prices; surviving trade barriers such as the milk marketing boards that shut out American milk imports and that, through the assignment of a large milk quota to Québec, ensure a high level of Québec milk sales to Ontario customers; and finally, Canadian taste preferences, such as Canadians consuming more domestic beer rather than lower-priced American beer.

Still, integration into the North American

economy has realigned Canada's industrial core and resource peripheries from within a national structure to within continental and global structures. The Canadian core serves both the North American market and the domestic one while the Canadian peripheries are receiving more of their manufactured goods from adjacent parts of the United States and from other industrial cores, and are shipping more of their primary products to the United States and world markets.

However, the core/periphery structure of Canada's regions has so far remained relatively intact. For instance, Central Canada is still Canada's core. The resilience of the Canadian core/periphery spatial structure under trade liberalization is remarkable, given the noticeable increase in north–south trade flows. This persistency represents a paradox: as regional economic growth occurs, the regional economic structure remains relatively stable. Geography, by limiting the economic opportunities available to each region, has promoted a particular type of economic growth in each region, even while foreign countries have provided new markets. At the same time, regional economies have undergone major modifications in response to trade liberalization through finding new markets and processing more primary products. For instance, livestock production and alternative crops in Western Canada have expanded because of increased trade with the United States and Japan. But economic integration of Western Canada with the outside world makes this region's prosperity dependent on continued exports to these countries. Trade dependency brings with it a measure of vulnerability to unexpected events. In May 2003, Canadian beef exports were banned by other countries, including the United States, because one cow was found to have mad cow disease (BSE). With about one-third of

beef production exported, the impact on ranchers and the beef industry has been immense, and countries are unlikely to remove their ban on Canadian beef for some time. Will this mean that the livestock industry must reduce its size by one-third?

In re-examining the core/periphery model in the light of trade liberalization, much change has taken place and more is expected. Canada's core now forms a subset of the American industrial core and, as a result, competes for customers in all regions of North America. Canadian peripheries are changing, too. With new trade opportunities, especially in the United States, regional peripheries are realigning their economic ties along continental and global lines. Among the peripheries, however, there are significant differences. Those peripheries with a rich resource base and a highly skilled labour force are accelerating their economic diversification and are moving towards a more independent economic status. Those with a weak economy, high unemployment rates, and an outflow of young people are becoming more and more dependent on Ottawa and on world market prices for their resources.

The global economy has taken Canada and other countries into unexplored territory. Economic structures, such as trade agreements, are in some ways overcoming political structures, such as national borders. Furthermore, multinational corporations, growing ever larger and more powerful, hold more and more of the world's investment capital and therefore have a significant impact on global economic structures. Capital, now the key element in the new market economy, is so mobile and multinational firms are so powerful that a new political order will have to emerge to regulate this world economy. Of particular concern is the need to ensure global environmental standards and labour practices, as well as to maintain employment levels in Western economies, while bringing greater development to global peripheries. Varying opinions exist on what form a new political order should take, but most agree that political globalization, which can match the economic globalization already underway, is greatly needed (Wallerstein, 1997: 141, 158; Dicken, 1998: 462–3).

What will be the consequences for Canada if an efficient political global order is not established? Stripped of more and more of its economic powers, Canada will, at best, retain control over its domestic affairs. Even domestically, however, international market forces, unless kept in check by a new political order, can exert exceptional pressure on governments. Take Canada's tax system, for instance. Ottawa is under pressure to lower Canadian taxes to the level of US taxes. While such a policy would keep more money in the pockets of Canadians, it could lead to the underfunding of our public health-care system, the slashing of social programs, and the widening of the income gap between Canada's regions. Consider the unexpected impact of international terrorism on world trade. For Canada, exports to the United States are critical for our economic well-being and any delays at the border are costly.

Canada's Faultlines

Canada is a complex and dynamic country. Part of its complexity is rooted in the four faultlines that represent the main tensions within the country. The first faultline, between non-Aboriginals and Aboriginal peoples, has existed since Europeans first arrived on what is now Canadian soil: for centuries, Aboriginal Canadians have struggled with non-Aboriginals over such issues as land, rights, and the

environment. The second faultline, between English-speaking and French-speaking Canada, represents the relationship between the two founding European nations that first settled Canada. The third faultline, between centralists and decentralists, developed as governing and economic structures began to create 'have' and 'have-not' regions. The last faultline—between 'newcomers' and 'old-timers'—reflects a gap in values and expectations between recent immigrants to the country and those who were born and raised in Canada. While Canada's early history is characterized by imposed solutions to these faultlines, the country's wisest leaders, acknowledging the strength of Canada's complexity, sought compromises instead. Compromise usually took the form of shared power rather than power enforced by the dominant party. Sharing power allowed Canada to evolve into a pluralistic society where both individual and collective differences are accepted and respected.

The sharing of power, however, has not been easy. For instance, in the past, Aboriginal peoples were marginalized by such 'imposed solutions' as reserves and residential schools. Today the place of Aboriginal peoples within Canada has changed. A shift in power is embodied in modern land-claim agreements (Sparke, 1998: 463–95). The Nisga'a, for example, reached a compromise with British Columbia and the federal government that secured them territory, guaranteed them a share of resources, and established a locally designed form of self-government. The Nisga'a gained full title to some of their lands in northern BC and then began the process of re-establishing their place within Canadian society. This journey of reconciliation between non-Aboriginals and Aboriginal peoples through land-claim agreements began in the 1970s and will continue well into the twenty-first century. For some, like the Métis and non-status Indians, the process may be more difficult.[2]

Canada's French/English faultline is perhaps the one that is the most deeply felt. Since Confederation, these two linguistic groups have struggled for power. Over time, French Canadians tried to win compromises from the more dominant power of English Canada. Despite the cultural divide between the two groups, the fact that both societies have contributed to Canada's unique diversity is recognized. However, the rise of Québec nationalism and a series of failed attempts at constitutional reform have made the need for further reconciliation even more urgent. Hollow gestures, such as the Calgary Declaration following the 1995 Québec referendum, need to be backed up by a greater sharing of power and compromise between Canada's two major linguistic groups and visions. Political reconciliation would mute the separatist voices calling for Québec's separation from the rest of the country.

Canada's third faultline, between centralists and decentralists, stems from one of Canada's most prominent features—regional diversity. Regional diversity has led to regional disparity—some regions are endowed with both political and economic power while other regions lack both. This disparity has generated a tension between those regions having power, as well as the federal government, and those regions struggling for more power. Though federalism already affords a sharing of power between the central government and regional (provincial and territorial) governments, some argue that power is not shared equally. Because of this sentiment as well as other issues, such as the unilateral cuts in transfer payments to the provinces by Ottawa in the 1990s, provincial governments have been demanding a greater share of power, including more taxing

power. With trade liberalization and the federal government's reduced ability to inject money into depressed regions of the country, centralist/decentralist tensions will only intensify. Compromises will have to be reached if provinces are to meet their socio-economic responsibilities, while still maintaining a level of power for the central government that ensures the preservation of national standards among the services available to Canadians in all regions. As Canada entered the twenty-first century, western alienation loomed as a major political problem. The National Atlas map of the voting pattern in the 2000 federal election presents a spatial version of western alienation (see Student Web site, National Atlas section, Map 31. Web site intructions are found on p. xxi).

Canada's last faultline, between newcomers and old-timers, has always been a factor in Canada's history. In the last decade, immigration has not only added to Canada's labour force but it has also kept Canada's population increasing. Immigrants come from all corners of the world, but the vast majority have settled in the three largest cities—Toronto, Vancouver, and Montréal. As a result, two Canadas have been created—one with a high proportion of newcomers and the other with relatively few newcomers. Ottawa has toyed with the idea of requiring new immigrants to locate outside of the three main cities, partly to ensure a source of labour and partly to disperse the newcomers more broadly across Canada.

Canada's faultlines continually challenge the country and its regions, but these dynamic struggles also strengthen us. One of Canada's leading thinkers, John Ralston Saul, sees these faultlines as pillars. In his book *Reflections of a Siamese Twin: Canada at the End of the Twentieth Century*, Saul argues that the Aboriginal, the francophone, and the anglophone are dependent on each other, forming a triangular reality. 'Each of their independent beings has been interwoven with the other two for over 450 years of continuous existence on the northern margins of the continent' (1997: 81). He further argues that this interdependence has, over time, allowed Canada to avoid the rigidity of nation-states that compel their citizens to conform. In a sense, then, Canada's faultlines, rather than dividing the country, have managed to unite its different regions and peoples.

The Last Century: An Overview

Since 1867, Canada's society and economy have undergone a remarkable transformation. At Confederation, Canada was a small rural society of some 3 million people clustered into four regions of Canada (New Brunswick, Nova Scotia, southern Ontario, and southern Québec). Canada's population was composed of Aboriginal peoples and settlers from two main ethnic groups—English and French. Eventually, the country's territory expanded and new provinces joined or came into being. The country was rich in resources but small in population and therefore seemed destined to produce primary goods for export markets. Ottawa, recognizing the problem of an economy entirely devoted to primary production, set in motion an economic policy that imposed high tariffs on imported manufactured goods. That policy stimulated industrial growth in southern Ontario and southern Québec and encouraged foreign firms to build branch plants in the emerging industrial core of Canada. The same policy forced the rest of Canada to buy Central Canada's manufactured goods but to sell their primary goods on world markets. The settling of Western Canada added a

third dimension to Canada's population—people from Northern, Central, and Eastern Europe added to the diversity of Canada's population and culture. As Canada grew more independent from Britain, it had to face the harsh realities of federalism, including Aboriginal issues, the concept of two founding peoples, and regional self-interests.

Canada has evolved into an independent nation consisting of 10 provinces and three territories. Technological advances have transformed Canada. Natural barriers like the Canadian Shield and the Rocky Mountains have been overcome by innovations in transportation; advances in plant and seed genetics have shortened the growing time for crops and provided more crop alternatives in Western Canada; new techniques such as insulated buildings, winter roads, and utilidors have helped overcome the cold environment found in the Territorial North.

Political decisions have also transformed Canada. The liberalization of immigration and trade has had a profound impact on Canada's society and economy. Immigration from a broader range of countries is creating a more cosmopolitan society with strong international links. Free trade has moved Canada into a North American regional economy and strengthened its economic links with other countries of the world. Canada is now a more urbane society composed of Canadians whose origins are from all corners of the globe. While progress has been made in building compromise among Canada's four faultlines, more work remains. In fact, Canada's faultlines will continue to demand attention because absolute solutions are impossible within such a diverse but democratic country—but then that combination of democracy and diversity has become Canada's true strength.

The Twenty-First Century

First and foremost, the feature that characterizes Canada and makes it strong—diversity—has evolved through compromise. But compromise is never fully achieved: it takes ongoing and committed efforts to maintain. In the twenty-first century, Canadians will face several challenges. The most daunting issue remains the French/English faultline, one that could literally split the country. In the aftermath of the 1995 referendum, the sharper edges of conflict were blunted, and in 2003 the separatist government was pushed from political power in Québec. Except for hardcore separatists, few Canadians can contemplate their country without Québec and some fear that Québec separation would lead to Canada's disintegration. The potential outcomes of Québec separating are sobering to consider. Many argue that once split, the durability of 'the rest of Canada' is uncertain. For instance, history and geography have shown that discontiguous states have short lifespans.[3]

Canada's regionalism, if not fostered with ongoing compromise, also has the potential to fragment the country. Some, like writer Joel Garreau, argue that Canada's regions could be 'picked off', one by one, and absorbed into the United States. Robert Kaplan, an American writer, argues that economic integration with the United States will inevitably lead to regional self-interest fracturing Canada into new political amalgamations. For instance, British Columbia, Kaplan speculates, could become part of the Pacific Northwest, which might be renamed 'Cascadia' (Kaplan, 1998: 322–4). Economists Courchene and Telmer (1998: 1–3) support this economic integration thesis by declaring that Ontario is no longer Canada's heartland but is now a North American region

state. They also base their thesis on 'regional self-interest'. But does this thesis of regional self-interest apply to conservation? The answer lies in the fact that Ontario, in 2003, gave Ottawa a vast area of the Lake Superior lakebed to create the world's largest freshwater conservation area. Ontario has ceded one million hectares of lakebed and 6,000 hectares of islands and shoreline in the Nipigon area to create the Lake Superior National Marine Conservation Area.

However, geographic regions and their senses of place are enduring. In the next century, the six major regions of Canada will continue to exist. But are Canada's regions likely to become part of a new spatial arrangement (i.e., North America), as predicted by some experts? Or do Canada and its regions comprise an idea driven by more than just economic concerns? Clearly, to ensure that Canada is enduring, internal tensions will have to be addressed. This demands a political will and a clear recognition that Canada is not a homogeneous nation-state but a pluralistic country where regional differences not only exist but are cherished by those living in these regions. Canada's survival will depend on our ability to accept the country's geographic structure and, as a result, to recognize that Canada's identity is a regional matter (Frye, 1971: i–iii). In the twenty-first century, Canada will need to build cultural bridges between its regions and, in so doing, strengthen national identity and unity.

Accepting Canada's cultural diversity, then, is the linchpin holding Canada and its six regions together. The future, while full of uncertainty, contains two certainties: Canada will remain a heterogeneous country and, as such, will continue to experience regional differences. However, by engaging in open and vigorous debate, compromises—not solutions—will follow. Compromise strikes at the heart of the exercise of democratic power in a diverse society by demanding a balance between individual and collective rights, between regional and national interests, and between those with power and those in need of it. The ability to strike this balance is critical to Canada's well-being and to regional harmony.

Notes

1. Geographers view cultural landscapes, particularly cityscapes, as creating a human imprint that evokes a sense of place. However, this human imprint has been affected in recent years by a sense of urban placelessness wherein all cities seem alike due to a spatial standardization by global-based firms like McDonald's and Shell. Uniformity is erasing the uniqueness of place (Relph, 1976: 79–80). While the impact of standardization of cultural landscapes is most apparent in urban areas, it does exist elsewhere, especially in international tourist centres. Some cities have opposed such imposed architectural regularity in an effort to conserve their unique cultural landscapes. For instance, Québec City forced McDonald's to construct a fast-food restaurant that matched the architectural style of surrounding buildings in old Québec City.

2. The ability to negotiate modern land-claim agreements varies among Aboriginal peoples. Those Aboriginal peoples who did not sign treaties are best positioned to negotiate comprehensive land-claim agreements, which generally include rights to land, resources, and self-government. First Nations in British Columbia, the Territorial North, and Labrador fall into this category. Aboriginal peoples who have signed treaties but have not received all the benefits entitled to them under their treaties can seek redress under specific land-claim agreements. Aboriginals who are classified as non-status Indians by the federal government have lost their legal right to land claims.

3. In today's world, discontiguous states are rare. Of the more than 200 nations in the world, less than 10 states are divided. Most are small enclaves, such as Kaliningrad, which is separated from Russia by Lithuania and Belarus. Others have evolved into separate states, such as Pakistan, which existed as two parts for 25 years but saw its eastern lands form a new state, Bangladesh, in 1972. If Québec separated, Atlantic Canada would be cut off from the rest of the country, much like Alaska is from the rest of the United States. Alaska, like Atlantic Canada, is a large, remote region that relies on a resource economy. However, several differences exist between Alaska and Atlantic Canada. Alaska has less than 1 per cent of the US population while Atlantic Canada represents 8 per cent of Canada's population. Furthermore, unlike Alaska, Atlantic Canada relies more heavily on road transportation and road links to keep its economy functioning. Finally, while both regions receive large subsidies, the cost of supporting an isolated Atlantic Canada would place a much greater financial burden on the rest of Canada than Alaska now places on the United States.

Bibliography

Anderson, Michael A., and Stephen L.S. Smith. 1999. 'Canadian Provinces in World Trade: Engagement and Detachment', *Canadian Journal of Economics* 32, 1: 22–38.

Barnes, Trevor J. 1993. 'A Geographical Appreciation of Harold A. Innis', *The Canadian Geographer* 37, 4: 352–64.

Britton, John N.H., ed. 1996. *Canada and the Global Economy: The Geography of Structural and Technological Change*. Montréal and Kingston: McGill-Queen's University Press.

Clark, Joe. 1994. *A Nation Too Good to Lose: Renewing the Purpose of Canada*. Toronto: Key Porter Books.

Courchene, Thomas J., and Colin R. Telmer. 1998. *From Heartland to North American Region State: The Social, Fiscal and Federal Evolution of Ontario*. Monograph Series on Public Policy, Centre for Public Management. Toronto: University of Toronto Press.

Dicken, Peter. 1998. *Global Shift: Transforming the World Economy*. London: Guilford Press.

Frye, Northrop. 1971. *The Bush Garden: Essays on the Canadian Imagination*. Toronto: Anansi Press.

Garreau, Joel. 1981. *The Nine Nations of North America*. Boston: Houghton Mifflin.

Hare, F. Kenneth. 1968. 'Canada', in John Warkentin, ed., *Canada: A Geographical Interpretation*. Toronto: Methuen.

Haynes, David M., ed. 1997. *Can Canada Survive? Under What Terms and Conditions?* Toronto: University of Toronto Press.

Helliwell, John F. 1996. 'Do National Boundaries Matter for Quebec's Trade?', *Canadian Journal of Economics* 29, 3: 507–22.

Innis, Harold A. 1930. *The Fur Trade in Canada: An Introduction to Canadian Economic History*. New Haven: Yale University Press.

Investment Dealers Association of Canada (IDAC). 1998. *British Columbia: Economic & Fiscal Outlook*. Vancouver: Investment Dealers Association of Canada.

Joyce, Greg. 1999. 'Treaty with the Nisga'a', *Saskatoon Star-Phoenix*, 13 Apr., B8.

Kaplan, Robert D. 1998. *An Empire Wilderness: Travels into America's Future*. New York: Random House.

McCallum, John. 1995. 'National Borders Matter: Canada–US Regional Trade Patterns', *American Economic Review* 85, 3: 615–23.

Ormsby, Margaret A. 1958. *British Columbia: A History*. Toronto: Macmillan.

Pally, Thomas. 1998. 'Building Prosperity from the Bottom Up', *Challenge* 51, 5: 59–71.

Penrose, Jan. 1997. 'Construction, De(con)struction and Reconstruction: The Impact of Globalization and Fragmentation on the Canadian Nation-State', *International Journal of Canadian Studies* 16: 15–50.

Relph, Edward. 1976. *Place and Placelessness*. London: Pion.

Sparke, Matthew. 1998. 'A Map that Roared and an Original Atlas: Canada, Cartography, and the Narration of Nation', *Annals of the Association of American Geographers* 88, 3: 463–95.

Statistics Canada. 1997. *The Daily*—Provincial GDP and Interprovincial Trade 1996, 16 May 1997 [on-line database], Ottawa. Searched 14 Oct. 1998: <http://www.statcan.ca/Daily/English/ 970516/d970516.htm>.

———. 2003. Imports and Exports of Goods on a Balance-of-Payments Basis [on-line database], Ottawa. Searched 14 Sept. 2003: <http://www.statcan.ca/english/Pgdb/gblec02a.htm>.

Wallerstein, Immanuel. 1998. 'Contemporary Capitalist Dilemmas, the Social Sciences, and the Geopolitics of the Twenty-first Century', *Canadian Journal of Sociology* 23, 2–3: 141–58.

Further Reading

Saul, John Ralston. 1997. *Reflections of a Siamese Twin: Canada at the End of the Twentieth Century*. Toronto: Viking.

John Ralston Saul is one of Canada's outstanding thinkers. In *Reflections of a Siamese Twin* the reader is offered an intellectual challenge, for knowledge of Canada's past is necessary to interpret Saul's concept of complexity and his many metaphors. Canada's complexity is linked by Saul to his concept of '*un grand pays mou*' (a big soft country). 'Soft' does not mean weak but flexible enough to seek compromise rather than confrontation, and the idea is related to the diversity of our society—Canada as 'a country of minorities'. Saul's vision of Canada hinges on relations between anglophone, francophone, and Aboriginal Canadians, and these three pillars of Canada form the basis of what the author calls our 'triangular reality'. The regional faultline (the centralist/decentralist argument) discussed in this textbook adds another dimension to Saul's triangle and reinforces the importance of Canada's regional nature.

Appendix

1   Exports of goods on a balance-of-payments basis

	1998	1999	2000	2001	2002
	$ millions				
Exports	**327,161.5**	**367,170.9**	**425,587.2**	**414,638.2**	**410,686.5**
Agricultural and fishing products	**25,039.7**	**25,572.2**	**27,501.4**	**30,883.4**	**30,541.6**
Wheat	3,642.3	3,356.2	3,609.0	3,807.2	3,057.6
Other agricultural and fishing products	21,397.4	22,216.0	23,892.4	27,076.2	27,484.0
Energy products	**23,812.4**	**29,821.1**	**53,159.1**	**54,743.1**	**50,419.3**
Crude petroleum	7,829.8	11,017.1	19,165.9	15,370.2	18,925.2
Natural gas	8,967.1	10,951.4	20,536.8	25,595.1	19,162.5
Other energy products	7,015.5	7,852.6	13,456.4	13,777.8	12,331.6
Forestry products	**35,440.5**	**39,744.3**	**42,163.7**	**39,309.2**	**36,650.3**
Lumber and sawmill products	16,760.7	19,996.9	18,682.1	17,762.6	17,514.5
Wood pulp and other wood products	6,004.6	6,703.8	8,920.9	6,711.0	6,238.1
Newsprint and other paper and paperboard products	12,675.2	13,043.6	14,560.7	14,835.6	12,897.7
Industrial goods and materials	**59,169.7**	**59,412.5**	**67,245.2**	**66,797.4**	**69,434.7**
Metals and metal ores	5,370.9	5,078.4	5,821.6	5,541.9	5,747.3
Chemicals, plastics, and fertilizers	18,351.7	19,492.0	22,804.8	23,428.5	23,918.8
Metals and alloys	19,937.4	18,355.9	20,649.4	20,220.9	22,158.9
Other industrial goods and materials	15,509.7	16,486.2	17,969.4	17,606.1	17,609.7
Machinery and equipment	**80,704.0**	**87,920.7**	**107,798.7**	**99,732.1**	**94,718.0**
Industrial and agricultural machinery	17,491.4	17,058.5	18,790.2	19,230.8	19,517.5
Aircraft and other transportation equipment	16,619.8	18,104.7	20,030.8	23,899.9	22,144.7
Other machinery and equipment	46,592.8	52,757.5	68,977.7	56,601.4	53,055.8
Automotive products	**78,461.5**	**97,291.7**	**98,112.2**	**92,860.9**	**97,081.1**
Passenger autos and chassis	41,840.0	51,059.2	51,501.8	48,525.3	49,824.2
Trucks and other motor vehicles	14,018.8	19,399.9	18,174.1	17,336.4	17,968.3
Motor vehicle parts	22,602.7	26,832.6	28,436.3	26,999.2	29,288.6
Other consumer goods	**12,565.6**	**13,690.6**	**14,898.6**	**15,972.8**	**17,341.0**
Special transactions trade	5,563.4	7,348.2	7,980.1	8,118.5	7,909.8
Unallocated adjustments	**6,405.3**	**6,369.9**	**6,728.3**	**6,220.8**	**6,590.8**

Source: Statistics Canada, CANSIM II, tables 228–0001, 228–0002, and 228–0003.

| | 2 | Average earnings for all earners and those working full-year, full-time, Canada, provinces, and territories, 1980, 1990, 2000 | | | | | | | |

	All earners				Earners working full-year, full-time[1]			
	1980	1990	2000	Per cent change 1990–2000	1980	1990	2000	Per cent change 1990–2000
Canada	$29,229	$29,596	$31,757	7.3	$40,943	$41,013	$43,231	5.4
Newfoundland and Labrador	$23,530	$22,017	$24,165	9.8	$37,082	$37,703	$37,806	0.3
Prince Edward Island	$20,210	$21,546	$22,303	3.5	$32,575	$34,812	$33,381	-4.1
Nova Scotia	$24,422	$25,587	$26,632	4.1	$35,892	$37,518	$37,800	0.8
New Brunswick	$23,501	$24,173	$24,971	3.3	$35,705	$36,828	$35,982	-2.3
Québec	$29,285	$28,516	$29,385	3.0	$39,726	$38,569	$39,150	1.5
Ontario	$29,360	$32,181	$35,185	9.3	$41,103	$43,831	$47,247	7.8
Manitoba	$25,988	$25,859	$27,178	5.1	$36,888	$36,017	$36,549	1.5
Saskatchewan	$27,460	$24,159	$25,691	6.3	$38,901	$33,901	$35,252	4.0
Alberta	$31,857	$29,241	$32,603	11.5	$44,659	$40,540	$44,080	8.7
British Columbia	$31,950	$30,170	$31,544	4.6	$45,389	$42,439	$44,231	4.2
Yukon	$33,554	$31,402	$31,526	0.4	$48,672	$45,359	$44,605	-1.7
Northwest Territories	**	**	$36,645	***	***	**	$51,823	***
Nunavut	*	*	$28,215	***	*	*	$48,017	

1. Full-year, full-time earners are those who worked 49 to 52 weeks during the year for 30 or more hours per week.
* Not available for any reference period
** Not available for a specific reference period.
*** Not applicable.

Source: Statistics Canada: <http://www12.statcan.ca/english/census01/Products/Analytic/companion/earn/canada.cfm>.

| 3 | Employment, employment growth, and employment rate, Canada, provinces, and territories, 1991 and 2001 | | | | |

	Employment		Growth in employment 1991–2001	Employment rate	
	1991	2001		1991	2001
	Number		%	%	
Canada	13,005,505	14,695,135	13.0	61.0	61.5
Newfoundland and Labrador	192,890	188,820	-2.1	44.2	45.1
Prince Edward Island	59,070	63,935	8.2	59.8	59.9
Nova Scotia	390,785	402,295	2.9	55.3	54.9
New Brunswick	300,965	325,335	8.1	53.2	55.2
Québec	3,110,795	3,434,270	10.4	57.3	58.9
Ontario	5,041,935	5,713,900	13.3	63.6	63.2
Manitoba	521,490	549,990	5.5	62.1	63.3
Saskatchewan	470,475	479,735	2.0	63.7	63.5
Alberta	1,308,800	1,608,835	22.9	68.2	69.3
British Columbia	1,568,780	1,883,975	20.1	60.7	59.6
Yukon	15,040	15,860	5.5	72.1	70.6
Northwest Territories	17,805	18,810	5.6	69.3	69.8
Nunavut	6,670	9,380	40.6	51.8	56.2

Source: Statistics Canada: <http://www12.statcan.ca/english/census01/Products/Analytic/companion/earn/canada.cfm>.

| 4 | Average earnings of all earners and proportion of low earners, high earners, and full-year, full-time earners, Canada and census metropolitan areas, 2000 |

	All earners				
	Number	Average earnings ($)	Percentage of low earners[1]	Percentage of high earners[2]	Percentage of full-year, full-time earners[3]
Canada	16,415,790	31,757	40.6	2.7	52.2
St John's	93,755	28,872	43.1	2.2	54.0
Halifax	204,715	30,614	41.0	2.5	55.7
Saint John	64,355	28,817	44.1	1.9	52.5
Chicoutimi-Jonquière	75,835	29,681	41.1	1.3	49.4
Québec	377,595	29,789	39.4	1.1	53.3
Sherbrooke	82,245	26,866	44.0	1.4	50.0
Trois-Rivières	68,390	27,950	44.8	1.4	48.9
Montréal	1,840,335	31,730	39.0	2.7	53.6
Ottawa-Hull	614,090	38,011	33.7	4.3	57.7
Kingston	79,895	30,497	43.2	2.4	51.8
Oshawa	165,750	36,290	33.1	3.0	57.8
Toronto	2,659,225	38,598	34.0	4.8	56.0
Hamilton	358,240	35,360	35.8	3.3	55.8
St Catharines-Niagara	200,310	30,384	41.9	1.9	52.8
Kitchener	241,515	33,985	35.4	3.0	58.0
London	239,330	32,393	39.2	2.6	54.1
Windsor	166,160	37,655	33.7	4.2	56.1
Sudbury	80,740	31,043	42.1	1.7	47.3
Thunder Bay	65,245	31,498	39.4	2.0	48.9
Winnipeg	383,080	29,359	41.4	2.0	56.3
Regina	111,535	30,127	41.1	1.9	54.5
Saskatoon	128,940	28,174	45.4	1.9	51.1
Calgary	594,040	36,851	36.8	4.9	54.6
Edmonton	556,115	31,999	40.1	2.7	52.2
Abbotsford	77,575	28,567	43.8	1.6	47.5
Vancouver	1,108,050	34,007	37.8	3.2	50.1
Victoria	173,680	30,529	39.8	1.8	49.7

1. Earners making less than $20,000.
2. Earners making $100,000 or more.
3. Earners who worked 49 to 52 weeks during the year for 30 or more hours per week.

Source: Statistics Canada: <http://www12.statcan.ca/english/census01/Products/Analytic/companion/earn/canada.cfm>.

5	Most common occupations of women earning $100,000 or more who were working full-year, full-time, Canada, 1990 and 2000			
Occupations	2000		1990	
	Number	Per cent	Number	Per cent
1. Lawyers and Québec notaries	3,030	5.6	1,195	6.0
2. General practitioners and family physicians	2,825	5.2	1,565	7.9
3. Sales, marketing, and advertising managers	2,745	5.1	565	2.9
4. Senior managers: financial, communications carriers, and other business services	2,600	4.8	400	2.0
5. Computer and information systems occupations	1,885	3.5	135	0.7
6. Financial managers	1,730	3.2	450	2.3
7. Financial auditors and accountants	1,540	2.8	270	1.4
8. Specialist physicians	1,520	2.8	1,085	5.5
9. Retail trade managers	1,500	2.8	765	3.9
10. Real estate agents and salespersons	1,355	2.5	1,280	6.5
11. Information systems and data processing managers	1,325	2.5	90	0.5
12. Human resources managers	1,210	2.2	200	1.0
13. Senior managers: trade, broadcasting, and other services	1,205	2.2	270	1.4
14. Banking, credit, and other investment managers	1,110	2.0	120	0.6
15. Senior managers: goods production, utilities, transportation, and construction	1,075	2.0	340	1.7
Women earning $100,000 or more who were working full-year, full-time in most common occupations	26,655	49.2	8,730	44.2
Women earning $100,000 or more who were working full-year, full-time in any occupation	54,185	100.0	19,755	100.0

Source: Statistics Canada: <http://www12.statcan.ca/english/census01/Products/Analytic/companion/earn/canada.cfm>.

6	Comparison of male and female earnings		
($ constant 2000) Year	Women	Men	Earnings ratio
1991	20,757	33,871	61.3
1992	21,459	33,665	63.7
1993	21,235	33,120	64.1
1994	21,432	34,555	62.0
1995	21,895	33,783	64.8
1996	21,678	33,633	64.5
1997	21,857	34,544	63.3
1998	22,832	35,649	64.0
1999	23,101	36,076	64.0
2000	23,796	37,210	64.0

Source: Statistics Canada, CANSIM II, table 202–0102.

7	Employment by industry and sex, 2002		
	Both sexes	**Men**	**Women**
	(thousands)		
All industries	**15,411.8**	**8,262.0**	**7,149.8**
Goods-producing sector	**3,942.6**	**3,013.3**	**929.3**
Agriculture	330.0	232.1	97.9
Forestry, fishing, mining, oil, and gas	272.0	229.6	42.4
Utilities	131.5	99.0	32.5
Construction	882.8	794.7	88.0
Manufacturing	2,326.2	1,657.8	668.4
Services-producing sector	**11,469.3**	**5,248.8**	**6,220.5**
Trade	2,430.0	1,246.5	1,183.5
Transportation and warehousing	756.2	588.1	168.1
Finance, insurance, real estate, and leasing	895.6	380.8	514.8
Professional, scientific, and technical services	993.3	568.8	424.5
Management of companies and administrative and other support services	591.4	314.3	277.1
Educational services	1,015.9	348.4	667.4
Health care and social assistance	1,607.0	286.3	1,320.7
Information, culture, and recreation	704.8	368.5	336.3
Accommodation and food services	1,003.9	398.9	605.0
Other services	693.2	333.8	359.3
Public administration	778.0	414.4	363.6
All industries	**15,411.8**	**8,262.0**	**7,149.8**

Source: Statistics Canada, CANSIM II, table 282–0008.

| 8 | Net migrants[1] and net migration rates, provinces and territories,[2] 1976 to 2001 |

Province or territory	1976–81 Number	Rate (%)	1981–6 Number	Rate (%)	1986–91 Number	Rate (%)	1991–6 Number	Rate (%)	1996–2001 Number	Rate (%)
Newfoundland and Labrador	-19,860	-3.7	-16,550	-3.1	-13,945	-2.6	-23,240	-4.3	-31,055	-6.1
Prince Edward Island	-15	0.0	1,540	1.4	-850	-0.7	1,455	1.2	135	0.1
Nova Scotia	-8,420	-1.1	6,275	0.8	-4,885	-0.6	-6,450	-0.8	-1,275	-0.2
New Brunswick	-8,505	-1.3	-1,370	-0.2	-6,060	-0.9	-1,950	-0.3	-8,425	-1.2
Québec	-141,725	-2.4	-63,295	-1.1	-25,560	-0.4	-37,430	-0.6	-57,315	-0.9
Ontario	-78,070	-1.0	99,355	1.2	46,965	0.5	-47,025	-0.5	51,905	0.5
Manitoba	-43,600	-4.6	-1,555	-0.2	-35,260	-3.5	-19,390	-1.9	-18,560	-1.8
Saskatchewan	-5,820	-0.7	-2,830	-0.3	-60,365	-6.4	-19,780	-2.1	-24,940	-2.7
Alberta	197,645	11.3	-27,675	-1.3	-25,005	-1.1	3,575	0.1	119,420	4.7
British Columbia	110,930	4.8	9,515	0.4	125,870	4.6	149,935	4.5	-23,630	-0.7
Yukon	-545	-2.6	-2,655	-11.4	790	3.4	685	2.5	-2,760	-9.5
Northwest Territories	-2,015	-5.0	-755	-1.6	-1,695	-3.4	-465	-1.3	-3,170	-8.6
Nunavut							80	0.4	-330	-1.4

1. Difference between the number of incoming and outgoing migrants.

2. These numbers are for internal migration only. They do not include the number of people who were outside Canada in 1996 and entered Canada between 1996 and 2001.

Source: Statistics Canada, Profile of the Canadian population by mobility status: Canada, a nation on the move: <http://www12.statcan.ca/english/census01/products/analytic/companion/mob/contents.cfm>.

9	Earnings of recent immigrants[1] relative to earnings of the Canadian-born, by sex, Canada, 1980, 1990, 2000					
	Men			Women		
Number of years since arrival	1980	1990	2000	1980	1990	2000
	Earnings as a percentage of earnings of the Canadian-born					
1 year	71.6	63.4	63.1	64.7	70.0	60.5
2 years	86.9	73.3	71.4	79.3	79.8	68.4
3 years	93.4	77.0	75.5	84.4	84.4	71.7
4 years	88.8	77.1	77.3	87.8	82.0	74.3
5 years	92.7	78.5	77.1	91.7	83.8	77.4
6 years	93.5	81.5	76.5	94.9	83.3	77.8
7 years	95.1	84.5	76.6	97.9	87.3	76.8
8 years	89.9	97.5	75.2	96.3	94.6	80.2
9 years	97.3	97.2	78.3	103.1	93.7	82.2
10 years	100.4	90.1	79.8	103.1	93.3	87.3

1. Those who arrived in Canada from 1990 to 1999.

Source: Statistics Canada: <http://www12.statcan.ca/english/census01/Products/Analytic/companion/earn/canada.cfm>.

10 Population reporting Aboriginal identity, Canada, provinces, and territories, 2001	Number	%
Canada	976,310	100.0
Newfoundland and Labrador	18,780	1.9
Prince Edward Island	1,345	0.1
Nova Scotia	17,015	1.7
New Brunswick	16,990	1.7
Québec	79,400	8.1
Ontario	188,315	19.3
Manitoba	150,040	15.4
Saskatchewan	130,190	13.3
Alberta	156,220	16.0
British Columbia	170,025	17.4
Yukon Territory	6,540	0.7
Northwest Territories	18,725	1.9
Nunavut	22,720	2.3

Source: Statistics Canada: <http://www12.statcan.ca/english/census01/products/analytic/companion/abor/contents.cfm>.

| 11 | Population and growth rate of language groups by mother tongue, Canada, provinces, territories, and Canada less Québec, 1991, 1996, and 2001 |

	Anglophone[1]							
	1991		1996		2001		Growth rate	
	Number	%	Number	%	Number	%	1991–1996	1996–2001
Canada	16,311,210	60.4	17,072,435	59.8	17,521,880	59.1	4.7	2.6
Newfoundland and Labrador	555,920	98.6	539,045	98.5	500,075	98.4	-3.0	-7.2
Prince Edward Island	120,770	94.3	125,020	94.1	125,390	94.0	3.5	0.3
Nova Scotia	831,580	93.3	838,280	93.1	834,780	93.0	0.8	-0.4
New Brunswick	462,875	64.6	476,395	65.3	468,090	65.0	3.4	-1.7
Québec	626,200	9.2	621,860	8.8	591,365	8.3	-0.7	-4.9
Ontario	7,443,535	74.6	7,777,735	73.1	8,041,995	71.3	4.5	3.4
Manitoba	793,330	73.5	822,225	74.7	831,820	75.4	3.6	1.2
Saskatchewan	812,600	83.3	823,745	84.3	822,640	85.4	1.4	-0.1
Alberta	2,045,905	81.2	2,175,750	81.5	2,395,765	81.5	6.3	10.1
British Columbia	2,562,240	78.9	2,809,395	76.1	2,849,185	73.6	9.6	1.4
Yukon Territory	24,550	88.8	26,615	86.8	24,770	86.8	8.4	-6.9
Northwest Territories			30,250	76.7	28,860	77.8		-4.6
Nunavut			6,080	24.6	7,175	26.9		18.0
Canada less Québec	15,685,005	77.7	16,450,570	76.6	16,930,515	75.2	4.9	2.9

(continued . . .)

11	(continued)							
	Francophone[2]							
	1991		**1996**		**2001**		**Growth rate**	
	Number	**%**	**Number**	**%**	**Number**	**%**	**1991–1996**	**1996–2001**
Canada	6,562,060	24.3	6,711,630	23.5	6,782,320	22.9	2.3	1.1
Newfoundland and Labrador	2,855	0.5	2,430	0.4	2,360	0.5	-14.8	-3.1
Prince Edward Island	5,750	4.5	5,715	4.3	5,890	4.4	-0.6	3.1
Nova Scotia	37,525	4.2	36,310	4.0	35,380	3.9	-3.2	-2.6
New Brunswick	243,690	34.0	242,410	33.2	239,400	33.2	-0.8	-1.3
Québec	5,585,650	82.0	5,741,430	81.5	5,802,020	81.4	2.8	1.1
Ontario	503,345	5.0	499,687	4.7	509,265	4.5	-0.7	1.9
Manitoba	50,780	4.7	49,110	4.5	45,920	4.2	-3.3	-6.5
Saskatchewan	21,800	2.2	19,895	2.0	18,645	1.9	-8.7	-6.3
Alberta	56,730	2.3	55,290	2.1	62,240	2.1	-2.5	12.6
British Columbia	51,590	1.6	56,755	1.5	58,891	1.5	10.0	3.8
Yukon Territory	905	3.3	1,170	3.8	945	3.3	29.6	-19.5
Northwest Territories			1,005	2.5	1,000	2.7		-0.5
Nunavut			415	1.7	400	1.5		-2.8
Canada less Québec	976,415	4.8	970,200	4.5	980,270	4.4	-0.6	1.0

(continued . . .)

562 The Regional Geography of Canada

11 (continued)

| | Allophone[3] | | | | | | Growth rate | |
| | 1991 | | 1996 | | 2001 | | 1991–1996 | 1996–2001 |
	Number	%	Number	%	Number	%		
Canada	4,120,770	15.3	4,744,060	16.6	5,334,770	18.0	15.1	12.5
Newfoundland and Labrador	5,145	0.9	5,665	1.0	5,645	1.1	10.3	-0.6
Prince Edward Island	1,580	1.2	2,135	1.6	2,120	1.6	34.5	-0.2
Nova Scotia	21,855	2.5	25,380	2.8	27,405	3.1	16.1	8.0
New Brunswick	9,355	1.4	10,825	1.5	12,315	1.7	6.5	13.7
Québec	598,450	8.8	681,785	9.7	732,160	10.3	13.9	7.4
Ontario	2,030,170	20.3	2,365,370	22.2	2,734,280	24.2	16.5	15.6
Manitoba	235,290	21.8	228,940	20.8	225,950	20.5	-2.7	-1.3
Saskatchewan	141,640	14.5	132,970	13.6	121,875	12.7	-6.1	-8.3
Alberta	416,550	16.5	438,150	16.4	483,125	16.4	5.2	10.3
British Columbia	633,670	19.5	823,605	22.3	960,780	24.8	30.0	16.7
Yukon Territory	2,200	8.0	2,870	9.4	2,835	9.9	30.2	-1.2
Northwest Territories			8,210	20.8	7,250	19.5		-11.7
Nunavut			18,170	73.7	19,070	71.6		5.0
Canada less Québec	3,522,320	17.5	4,062,275	18.9	4,602,675	20.4	15.3	13.3

1. Anglophone: the population with English as mother tongue.
2. Francophone: the population with French as mother tongue.
3. Allophone: the population with a non-official language as mother tongue.

Source: Statistics Canada, *The Daily*—Census of Population: Language, Mobility and Migration, 10 Dec. 2002 [on-line database], Ottawa: <http://www.statcan.ca/Daily/English/02120/d021210a.htm>.

| 12 | Population by language spoken most often at home, Canada, provinces, territories, and Canada less Québec, 1991, 1996, and 2001 |

	English					
	1991		**1996**		**2001**	
	Number	**%**	**Number**	**%**	**Number**	**%**
Canada	18,440,535	68.3	19,294,835	67.6	20,011,535	67.5
Newfoundland and Labrador	559,505	99.2	542,630	99.2	503,985	99.2
Prince Edward Island	124,620	97.3	129,190	97.2	129,950	97.4
Nova Scotia	858,130	96.3	866,260	96.3	863,730	96.2
New Brunswick	488,560	68.2	502,530	68.9	496,680	69.0
Québec	761,810	11.2	762,455	10.8	746,895	10.5
Ontario	8,499,520	85.2	8,900,845	83.6	9,337,615	82.7
Manitoba	947,085	87.7	971,610	88.3	983,270	89.1
Saskatchewan	921,085	94.4	923,445	94.6	916,790	95.2
Alberta	2,305,200	91.5	2,432,680	91.1	2,681,525	91.2
British Columbia	2,909,930	89.6	3,189,880	86.5	3,279,345	84.8
Yukon Territory	26,740	96.7	29,240	95.4	27,310	95.8
Northwest Territories			34,970	88.6	33,370	89.9
Nunavut			9,160	37.1	11,060	41.5
Canada less Québec	17,678,730	87.6	18,532,415	86.3	19,264,640	85.6

(continued . . .)

12	(continued)					
	French					
	1991		**1996**		**2001**	
	Number	%	Number	%	Number	%
Canada	6,288,425	23.3	6,448,615	22.6	6,531,375	22.0
Newfoundland and Labrador	1,340	0.2	1,015	0.2	990	0.2
Prince Edward Island	3,045	2.4	3,045	2.3	2,820	2.1
Nova Scotia	22,260	2.5	20,710	2.3	19,700	2.2
New Brunswick	223,270	31.2	222,450	30.1	217,775	30.3
Québec	5,651,790	83.0	5,830,0805	82.8	5,918,310	2.7
Ontario	318,695	3.2	306,790	2.9	307,295	82.7
Manitoba	25,040	2.3	23,140	2.1	20,895	1.9
Saskatchewan	7,155	0.7	5,830	0.6	4,810	0.5
Alberta	20,180	0.8	17,815	0.7	20,670	0.7
British Columbia	14,555	0.4	16,585	0.4	16,905	0.4
Yukon Territory	395	1.4	540	1.8	430	1.5
Northwest Territories			370	0.9	385	1.1
Nunavut			230	0.9	225	0.8
Canada less Québec	636,635	3.2	618,535	2.9	612,9850	2.7

Source: Statistics Canada, *The Daily*—Census of Population: Language, Mobility and Migration, 10 Dec. 2002 [on-line-database], Ottawa: <http://www.statcan.ca/Daily/English/02120/d021210a.htm>.

Index

Abbotsford, BC, 375
Aboriginal peoples, 88–95, 112, 139, 151, 444; Aboriginal identity, 121, 210, 490, 502, 559; Aboriginal title, 123–8, 159, 341–2; culture regions of, 94–5; in BC, 341–2, 348–9, 350; political power and, 159, 160, 209; population of, 25, 168, 206–10, 211; relocation of, 500–1, 509; resource development and, 318–19, 363–4, 425, 437, 471–2, 465–6; rights of, 104, 122–7; self-government and, 327; in Territorial North, 487–90, 493–8, 500–11, 515, 519, 521–2, 527–8; trade and, 91–2, 128, 495–7; treaties and, 102, 104, 122–3, 128–9, 341–2, 396; in Western Canada, 395–7, 408; and White Paper (1969), 124, 159; *see also* First Nations; Indians; Inuit; land claims; Métis; specific groups
Aboriginal/Non-Aboriginal faultline, 121–7, 471–2, 543–4
Aboriginal Pipeline Group, 521–2
Acadians, 109, 128, 442, 444, 445
acid rain, 76
Act of Union (1841), 137, 292
aerospace industry, 300–1, 306–7
age: dependency, 177–8; dependency ratio, 177, 219; median, 176; structure, 176–8, 209
agriculture: agricultural duties, 415, 416; in Atlantic Canada, 441, 449, 472–4; communal, 132–3; dairy, 304–5, 314; 'fringe', 409–10, 413–15, 431; livestock, 393, 408–10, 412, 415, 417–18, 419–20; oilseed crop, 410, 411, 418–19; specialty crop, 248–9, 410–13, 417–19, 473–4; in Western Canada, 408–15; *see also* grain industry
Air Canada, 307
Aklavik, NWT, 56
Alaska, 101, 104
Alaska Highway, 499
albedo effect, 77, 81
Alberta, 103, 107–8, 398; Badlands, 52; identity in, 187; Kyoto and, 403; NEP and, 117–18; Plateau, 53; population in, 131, 216; *see also* Western Canada
Alcan Aluminum, 312, 340, 362, 363–4
Algoma Steel, 271
Alvarez, Pierre, 403
Annapolis Valley, 442, 473

Appalachian Uplands, 56, 440–1
apparel industry, 300, 306
Arctic: Archipelago, 55, 58, 102; Basin, 74–5; Coastal Plain, 55; Lands, 54–6, 92–3; Ocean, 491; Platform, 55
arêtes, 51, 81
Asia, 351; trade with, 338; *see also* Japan
Atlantic Basin, 73
Atlantic Canada, xvii, 1, 30–22, 436–85; cities, 474–7; future of, 477–80; historical geography, 444–6; Opportunities Agency, 455–6; physical geography, 439–43; population, 446, 450–7
Atlantic Development Board, 455
Atlantic Groundfish Strategy (TAGS), 456, 463, 479
Atlantic Ocean, 437, 440
aurora borealis, 489
automobile industry, 192, 198, 239–40, 245, 249, 250–6, 300, 306
Auto Pact, *xix*, 17, 239–40, 249, 250–1

baby boom effect, 172
Bagnall, James, 269
Barkerville, BC, 347
Barnes, Trevor J., 198, 379
'barrier of distance', 498
barriers, trade, 193–4, 240, 241, 416, 453–4, 541–2
basins, 52–3, 81, 491, 492; drainage, 72–5, 81
Batoche, SK, 130, 139, 395–6
BC Hydro, 362, 364
Beaufort Sea, 525
Bennett, W.A.C., 352
Berger Inquiry/Report, 125, 516, 520
Beringia, 88, 89, 90
Bernier, Jacques, 212
biculturalism, 141–2
Big Bear, 139, 396
Big Three, 239, 253, 255–6
bilingualism, 109, 212, 213
biomass, 461, 481
biotechnology, 198–9, 300
birth rate, 290, 501, 507; *see also* crude birth rate; fertility rate
bitumen, 423, 431
Blackbourn, Anthony, 408, 468
Blood people, 122
Bombardier, 300, 306–7
Borden Commission, 117
borders: fishing and, 460; internal, 106–8; provincial, 236, 293–4, 347;

US–Canada, 4–5, 18, 101, 104–5, 194
boreal: barrens, 259; forest, 259, 261
Bouchard, Lucien, 145
Boundary Waters Treaty, 233
Bourassa, Henri, 141
Bourassa, Robert, 191, 214
Bourne, Larry S., 3, 205
Britain, 95–9, 444–6; immigration from, 98, 138
British Columbia, xvii, 12, 28–9, 102, 112, 118, 336–85, 538, 539; alienation in, 8–9; cities, 373–7; economy, 338–40, 350–73, 378–9; future of, 378–9; historical geography, 346–51; land claims in, 124–5, 126, 159, 341–2; physical geography, 343–6; population, 338–9, 348–50, 352–3
British North America (BNA) Act, 101–2, 141, 189, 395
Britton, John N.H., 248
BSE (bovine spongiform encephalopathy), 415, 542
Bull Arm complex, 457, 471
Busch, Marc, 248

Cabot, John, 90, 93, 444
CAE Inc., 300–1
Calder case, 124, 159
Calgary, AB, 170, 171, 426, 427; Declaration, 146
call centres, 437–8
Cambridge, ON, 244, 249–50
Canada: as nation-state, 546; territorial evolution of, 101–8
Canada East, 292
Canada Health and Social Transfer Act, 120
Canada Pension Plan, 119
Canadian Environmental Assessment Agency, 516
Canadian National Railway, 18, 158, 160, 193
Canadian Pacific Railway, 104; in BC, 58, 112, 347, 348, 349; in Western Canada, 99, 100, 129, 397, 398, 399, 404
Canadian Shield, 47–50; in Atlantic Canada, 440–1; in Ontario, 237, 256–8; in Québec, 286; in Territorial North, 490; in Western Canada, 389–90
Canadian Wheat Board, 272, 415
Canol Project, 499
Cape Breton Development Corporation

(Devco), 468–9, 479
Cardinal, Harold, 159
Cartier, George-Étienne, 293
Cartier, Jacques, 90, 93, 94, 287, 290
Cascade Mountains, 343
Cascadia, 337–8, 548
Cassidy, Frank, 342
Cenozoic Era, 47
census metropolitan areas (CMAs), 169, 170–1, 219
Central Canada, 110, 111–16, 231
centralist/decentralist faultline, 109–20, 337, 352–3, 402, 404, 543, 544
Champlain, Samuel de, 94, 98, 128, 287, 290
Chan, Raymond, 117
Charlottetown, PEI, 476
Charlottetown Accord, 145, 297
Charter of Rights and Freedoms, 15, 145, 214, 297
Château Clique, 136, 151, 291
chernozemic soil, 81
Chilliwack, BC, 375
Chrétien, Jean, 145, 404
Churchill, MB, 54, 388
Churchill Falls, NL, 73
circulation systems, 61, 62–3
cirques, 51, 81
cities, 221, 168–71; in Atlantic Canada, 474–7, 478; in BC, 373–7; capital, 167; immigration and, 205–6; in Ontario, 261–72; population of, 168–71; in Québec, 320, 321–5; in Territorial North, 505, 508; in Western Canada, 407
Clay Belt, 312–14
Clayoquot Sound, 369
clear-cutting, 340–1, 367, 465
climate, 58–65, 81; Arctic, 55–6, 439, 491; in Atlantic Canada, 439–40, 449; BC, 340, 343–5; change, 77; controls on, 61; Cordillera, 340, 343–5; Köppen classification, 60; in Ontario, 230–1, 246–7; in Québec, 284–5, 304; Pacific, 340, 342–5; Prairie, 390–2; soils and vegetation and, 64–5; Subarctic, 66–7, 68, 70, 439, 491; in Territorial North, 490–1; in Western Canada, 390–2; zones of, 65–9, 81; see also global warming
'clusters', high-tech, 248
coal mining, 353, 361, 389, 421, 441, 447, 466–9
Coalition for Fair Lumber Imports, 371–3
Coast Forest, 366–8
coastline, 58
Coast Mountains, 343

Coderre, Denis, 206
cod fishery, xix, 437, 443, 458; moratorium, 456, 461–3
Cohen, Stewart J., 77
Columbia Forest, 367–8
Columbia Mountains, 343
Columbia River, 361–2
Confederation, 100–1; Atlantic Canada and, 447–8; BC and, 347–9; Québec and, 293–95; Western Canada and, 402
Confederation Bridge, 449
constitution, 538; patriation of, 296–7
Constitution Act (1982), 121, 145, 159
Constitutional Act (1791), 136–7
contact, Aboriginal–European, 93–4
continental air masses, 63–4, 81
continental divide, 72
continental drift, 46, 81
continental effect, 66–7
continentalism, 240; see also integration, economic; North America
continental shelves, 46
Cook, James, 346
Cordillera, 50–2, 343, 345–6, 388–9, 490
Cordillera ice sheet, 48, 88
core, 19, 37; regions, 22
core/periphery model, 3, 18–21, 34, 37, 43, 88, 110, 112–13; Canada and, 21–4; globalization and, 539–43; population and, 163–4, 215–17
country food, 510–11, 530
Courchene, Thomas, 229
Cournoyea, Nellie, 521–2
Cree, 94–5, 121–2, 146; Hydro-Québec and, 286, 309–10, 318–19; in Western Canada, 396–7
Crow Benefit, 272, 404, 417–18
Crown: corporations, 294–5, 307, 326–7; lands, 465–6
crude birth rate (CBR), 171, 173–4, 175, 219
crude death rate (CDR), 171, 174, 175, 219
cryosolic soil, 81
cultural regions, 37
culture, 185, 219; Aboriginal, 91, 94–5, 496–7, 508–9; Québec, 325–6, 327
culture areas, 91, 151
culture regions, 94–5
Cypress Hills subregion, 53

DaimlerChrysler, 251, 253, 256
death rate, 507; see also crude death rate; mortality rate
De Blij, J.H., 5
debris, glacial, 48

debt, federal, 120, 191
Delgamuukw case, 159
de Gaulle, Charles, 144
deindustrialization, 467–9
De Löe, R., 59
demographic transition theory, 158, 175–6, 219
demography, 158, 161, 219; see also population
Denbigh people, 90, 92
Dene, 493, 520, 521, 524
Denevan, William M., 207
Denison Mines, 361
Department of Fisheries and Oceans (DFO), 358, 493
Department of Regional Economic Expansion (DREE), 23, 455
dependency theory, 20, 37
depopulation, rural, 216
deposition, 48–9, 81
desert, 'polar', 56, 491
determinism, environmental, 20, 36
development, stages of, 160
diamond mining, 516, 519, 524–6
Dickason, Olive, 207
Dicken, Peter, 250
Diefenbaker, John, 112, 117
Dion, Léon, 297
disagreements, regional, 87, 108–9
diseases: European, 494, 495; pollution and, 232–3, 287, 443
disparities, regional, 23
distance, friction of, 58, 81
distinct society, 145, 146
diversification, economic, 24, 400–1, 428, 457
diversity: cultural, 546–7; regional, 539
divides, drainage, 72
Dixon, Paul, 319
Dogrib (Tlicho), 126
dollar, Canadian, 252
Dominion Lands Act, 99, 133, 397
Dominion Steel and Coal Co. (Dosco), 467–8
Donnacona, 94
Dorset people, 90, 92
'double contract', 141
Doukhobors, 130–4, 398
downward transitional regions, 22
draggers, 462
drainage basins, 72–5, 81
drought, 7, 65, 399, 414–15
drumlins, 49, 81
Dry Belt, 392, 409–10, 412–14, 431
'dual-earner family', 202
duality: cultural, 281; linguistic, 212
Duplessis, Maurice, 143
Durham, Lord, 137, 291–2

Dust Bowl, 7, 390
duties: agricultural, 415, 416; counter-vailing, 372–3

earnings, 200–2, 203, 551, 553; immi-grants', 558; male and female, 555
earthquakes, 50–1
Eastern Townships (Estrie), 303–4
economic diversification, 24, 400–1, 428, 457
economic integration, 16–18, 240, 540–3, 546
economic transition, 416–20
economies: 'of association', 451, 481; 'of scale', 24, 37
economy, 191–203; continental, 158, 160, 192–4, 195, 240–2, 300; private sector, 192; structure of, 194–6; trans-formation of, 160
ecumene, 12, 37, 164
Edmonton, AB, 170, 426, 427
education, 501, 509; income and, 201, 202; religion and, 189–90
El Niño effect, 358
employment, 552, 554, 556; in Atlantic Canada, 454, 455, 463; in auto indus-try, 253; in BC, 353–6; levels of, 191; in Ontario, 242–5; in Québec, 301; restructuring and, 195; in Territorial North, 501, 512–13; in Western Cana-da, 407, 408; see also labour force
Employment Insurance, 119
energy, solar, 61, 62
English/French, see French/English
environmental assessment, 516
environmental challenges, xx, 44, 57–8, 75–6; in Atlantic Canada, 443; in BC, 340–1, 363–4; in Ontario, 232–4; in Québec, 287; in Territorial North, 492–3, 504, 514–16; in Western Canada, 392–3, 420
environmental determinism, 20, 36
equator, 59
equity, pay, 203
erosion, glacial, 48, 82
Erskine, David, 448
eskers, 49, 50, 81
Established Programs Financing Act, 119–20
Estrie (Eastern Townships), 303–4
ethnic group, 185, 219–20
ethnicity, 132, 184, 185–7
Eureka, NWT, 12
Europe: fishing and, 462–3; immigration from, 100–1, 131, 179–80, 181–2
European Union, 251
evapotranspiration, 391
exploitation, regional, 20, 21

explorations, 93–4
exports, 550; after FTA, 160; see also spe-cific industries

fallow, summer, 399, 412, 431
Family Compact, 136, 151
faulting, 45–6, 81
faultlines, xvii–xviii, xix, 3, 87–8, 108–48, 537, 543–5, 545; demo-graphic, 203–18
federal–provincial relations, 87, 88, 108–9, 111, 113, 119–20
Fenians, 101
Fertile Belt, 392, 409–12, 431
fertility rate, 173–4, 175, 209, 220; see also birth rate; crude birth rate
fires, forest, 338, 340
First Nations: names of, 121–2; self-gov-ernment and, 113; see also Aboriginal peoples; Indians; Inuit; Métis; specific groups
fish: contaminated, 296; farms, 357; stocks, 358–9, 460–3, 480
fishing industry: in Atlantic Canada, 437, 441, 444, 446, 457–63, 480; in BC, 357–9; whaling, 494–5; modernization in, 458, 462–3
Flin Flon, MB, 422–3, 426
floods, 58–9, 285
folding, 45–6, 81
Folsom culture, 91
Ford Motor Co., 249, 251, 253
forest, commercial, 370, 380
forest industry, xix, 112; annual allowable cut, 369, 380, 465; in Atlantic Canada, 446, 465–6; in BC, 337, 338, 339–41, 351, 353–4, 364–73, 378–9; free trade and, 370–3; leases and, 259, 261, 369–70; management of, 369–70; 'myth', 369–70; in Ontario, 242, 259–61, 268–9; in Québec, 284, 314–15; in Western Canada, 400–1, 423–5
forest subregions, 367–8
Fort McMurray, AB, 403, 423, 426
'founding peoples', 141–3, 281, 190
France: Atlantic Canada and, 444–5; col-onization by, 95–6
Franklin, John, 494
Fraser, Simon, 347
Fraser Canyon, 343
Free Trade Agreement (FTA), 17, 18, 23–4, 160, 195, 240–2, 274; effect of, 250; see also North American Free Trade Agreement
French, H.M., 66
French/English balance, 210–15
French/English faultline, 109, 134–47,

296–7, 298, 543, 544, 546
French/English relations, 98–9, 141–2
friction of distance, 58, 81
Friedmann, John, 21
Frobisher, Martin, 93–4, 493–4
Front de libération du Québec (FLQ), 144
Frontenac Axis, 57
frontier, 487, 488–90, 530; strategic, 498–500
fur trade, 287–8, 322, 346–7, 495–7

gas: natural, 359, 389, 422, 521; off-shore, 437, 457, 466, 469; see also oil; petroleum industry
Gaspé Peninsula, 286, 303
Gatineau, QC, 266–9, 321
G-8 countries, 191
General Agreement on Tariffs and Trade (GATT), 11, 17, 240
general circulation models (GCM), 77
General Motors, 249, 251, 253, 254–5, 269, 300, 306
geography: defined, 6; as discipline, 5–9; historical, 86–155; human, 156–227; physical, 43–59; regional, 2–41, 536–49
geology, 47
geomorphic processes, 45
Georges Bank, 457–9, 460
Gérin-Lajoie, Antoine, 292
Gilbert, Anne, 211
glacial lakes, 53
glacial spillways, 53, 82
glacial striations, 49, 82
glaciation, 47–8
glacial troughs, 51, 82
glaciers, alpine, 48, 55–6
global circulation system, 62–3, 82
globalization, xix; core/periphery and, 539–43
global system, core/periphery, 21, 22
global warming, 48, 76–9, 403
gold, placer, 491, 492, 493, 514
Golden Horseshoe, 263–6
gold rushes, 341, 347, 348, 489, 492, 498, 513–15
government: Atlantic Canada and, 442–3, 447, 448, 455–7; Atlantic provincial, 453–5; elected, 291; federal system of, 87; fishery and, 357–9, 443, 460–3; forestry and, 369–70; in Québec, 294–5; territorial, 497–8, 502–3, 513; in US and Canada, 34–5; see also federal–provincial relations
grain industry, xix, 112, 130, 398–400, 404–5, 408–15; subsidies and, 417–20; see also agriculture

Grand Banks, 457–9, 462
Granville Island (Vancouver), 377
grassland, 392, 395
Great American Desert, 392, 394
Great Harbour Deep, 452, 453
Great Lakes, 233–4
Great Lakes-St Lawrence Lowlands, 56–7, 166–7, 237, 245
Great Lakes Water Quality Agreement (GLWQA), 234
Great Whale River project, 318, 319
greenhouse: effect, 77, 82; gases, 18, 108, 393
gross domestic product (GDP), 14–15, 37
groundfish, 458, 481
Gwich'in, 125, 126

habitants, 135, 288–90, 330
Haggett, Peter, 6
Halifax, NS, 63, 170, 444–5, 474, 476
Hamelin, Louis-Edmond, 13
Hamilton, ON, 244, 245, 248–9, 264
Hamm, John, 478
Hare, Kenneth, 537
Harris, Mike, 232–3
Hartshorne, Richard, 6
Hayter, Roger, 378–9
Hearne, Samuel, 513
heartland, 19, 37
heartland/hinterland model, 19
Herbert, Edward N., 203–4
Hibernia oil field, 441, 456–7, 469, 470
Hiebert, Daniel, 204
high-technology industry, 426, 437, 480; in BC, 351, 353, 354; 'clusters', 248; in Ottawa, 269; in Québec, 300–1, 306
Hind, Henry, 395
hinterland, 19, 37; resource, 16, 111–12
Holocene Epoch, 47, 82
homeland, 487, 488–90, 530
home language, 212
homesteaders, 99, 129, 394, 397–8
House of Commons, 113–14, 295, 351
Hudson Bay Basin, 73
Hudson Bay Lowland, 54, 237, 256, 257, 286–7
Hudson Bay Railway, 112, 388
Hudson's Bay Co., 497; settlement and, 99, 129, 395; territory of, 104–6, 347, 393–4, 496
hunting, 90–1, 92, 493, 501, 509–10
Huronia, 93, 128
Hutchison, Bruce, 8
hydroelectricity projects, xix-xx, 73–5; BC, 340, 361–4; consequences of, 75–6; and industry, 310–12; in

Ontario, 237, 257; in Québec, 286, 299, 307–12, 316–19; in Western Canada, 389–90, 400
Hydro-Québec, 107, 143, 295, 307–12

ice advance, late Wisconsin, 48
ice ages, 47–9, 78, 82, 88, 90
icebergs, 55–6
ice storms, 285
identity: Aboriginal, 121, 210, 490, 502, 559; regional/national, 186
immigration: xix; British, 98–9; BC and, 352–3; 'early', 179; European, 179–80, 181–2; faultline, 128–34; importance of, 182–5; language and, 213, 214; non-British, 100–1; non-European, 181–2, 184–5; in Ottawa, 268; policy, 100–1, 178–9, 181, 182; population and, 164, 172, 178–85; Québec and, 295–6; recent, 180–2; tensions and, 203–6; in Toronto, 265–6; in Western Canada, 397–8; from US, 96–8
Impact and Benefits Agreement, 472
Inco Ltd, 470–2
income gap, 200; *see also* earnings
Indians, 91–2, 121; *see also* Aboriginal peoples; First Nations; specific groups
Indian Act, 121, 124, 159
industrialization, 175–6
industrial stage, 160, 220
industrial structure: in Atlantic Canada, 455–7; in BC, 353–6; in Ontario, 242–5; in Québec, 301–2; in Territorial North, 511–12; in Western Canada, 408–15
industry: hydroelectricity and, 310–12
information technology, 198, 199, 437–8
Innis, Harold, 23
Innu, 472
Innuitian Mountain Complex, 55
integration, economic, 16–18, 240, 540–3, 546
Intercolonial Railway, 447
Intergovernmental Panel on Climatic Change (IPCC), 77
Interior Forest, 366–8
Interior Plains, 52–3, 389, 392, 490
Interior Plateau, 343–5
International Joint Commission (IJC), 233–4
Inuit, 92, 94, 121, 151–2; homeland of, 487–8, 493–7; Hydro-Québec and, 286, 309–10, 318–19; Labrador, 124, 126, 127; self-government and, 327, 502–5
Inuvialuit, 125, 126, 495
Inuvialuit Final Agreement, 504–5
Inuvialuit Game Council, 504, 511

Inuvik, NWT, 56
Iqaluit, 501, 502, 505
Ireland, 98
iron and steel industry, 467–9; *see also* steel industry
iron ore, 315–16
Iron Ore Company of Canada, 316
Iroquois, 94, 444
irrigation, agricultural, 412–13
isostatic rebound, 54, 82
Izok Lake mine, 516, 517

James Bay and Northern Quebec Agreement, 124, 125–6, 309–10, 318–19, 330
James Bay Project, 74, 309–10, 316–19
Japan: auto industry and, 251, 253–4; BC and, 353, 368
Japan Current, 62
Jesuits' Estates Act, 140
Jones, David, 414–15
just-in-time principle, 253, 274

Kahnawake, 122
Kalbach, W.E., 184
Kaplan, Robert, 546
Kelowna, BC, 340, 375
Kemano, BC, 75, 340, 363–4
Kilger, Bob, 195
kimberlite pipes, 524, 530
Kitimat, BC, 75, 340, 363
Klondike gold rush, 489, 492, 498, 513–15
Köppen classification of climatic types, 60
Kyoto Protocol, 18, 87–8, 108–9, 403

labour force: age dependency and, 177–8; earnings of, 200–2, 203; restructuring and, 195; structure of, 198–200; women in, 202–3; *see also* employment
Labrador, 107, 437, 439, 442; *see also* Atlantic Canada
Labrador City, 474, 476–7
Labrador Current, 63, 67, 439, 440
La Grande Rivière, 74; La Grande project, 286, 317
Lake Agassiz, 53
Lake Erie, 234
lakes, glacial, 53
land claims, 472; in BC, 124–5, 126, 159, 341–2; comprehensive agreements, 124, 151, 502–5; modern, 123–8, 318–19, 543–4; specific agreements, 124, 152; in Territorial North, 490, 502, 503–5, 511, 520, 521, 524, 527

landforms, 44–6
language(s), 98–9, 109, 560–4; Aboriginal, 96, 121, 188; culture and, 15–16, 24–5; ethnicity and, 187–8; immigrants and, 138; in French/English balance, 210–15; home language, 212; in Quebec, 281, 283, 292, 295–6, 298, 303, 325–6, 327; laws on, 101, 144–5, 214–15, 295–6, 327; mother tongue, 212, 281, 283; non-official, 211–12; official, 144–5, 187–8; rights of, 140–1; in Territorial North, 496, 501, 508
La Salle, René-Robert Cavelier de, 288
latitude, 58–9, 82
Laurentide ice sheet, 48, 53, 54, 88, 92
Laurentides, 50, 320
Leduc, AB, 422
Lesage, Jean, 143, 294
Lévesque, René, 113, 144
Ley, David, 204
Li, Peter, 203–4
liberalization, trade, xvii, 540–3; *see also* globalization
Little Ice Age, 78
location, geographic, 58–9, 66
London, ON, 269–70
longitude, 58–9, 82
Low, A.P., 316
Lower Canada, 136–7, 290–3
Loyalists, 96–8, 135–6, 137, 152, 234, 446
lumber, 259–61; *see also* forest industry

McCain, Michael, 420
Macdonald, John A., 16, 101, 103, 110
Mackenzie, Alexander, 346
Mackenzie, William Lyon, 137
Mackenzie Valley Pipeline Project, 125, 520–22
MacLachlan, Ian, 242
MacMillan Bloedel, 378
Macpherson, Alan G., 439
McVey, Wayne W., 184
'mad cow disease', 415, 542
Magna International, 254–5, 269
Makivik Corporation, 327
Maliseet people, 444
Manicouagan River, 308
Manitoba, 102, 106–7, 395; immigration to, 128–30; *see also* Western Canada
Manitoba Act, 129, 139
Manitoba Lowland, 53
Manitoba Schools Question, 140–1
manufacturing industry, 192–3, 197–8; in Atlantic Canada, 437, 447, 448, 455; in BC, 355, 356, 368; in Ontario, 236–7, 239–42, 248–56, 269–70,

272–3; in Québec, 284, 299–302, 306, 322, 325, 326; in Western Canada, 408
'maritime nation', 58
Maritimes, 437, 439, 442; Confederation and, 101, 111; *see also* Atlantic Canada; individual provinces
Maritime Union, 454–5
Marquis wheat, 130, 399
Martineau, Daniel, 287
Meech Lake Accord, 145, 297
megaprojects, *xix-xx*, 469–71, 489, 519–27, 530
megaregion, 10, 38
mesoregion, 10, 38
Mesozoic Era, 47
metallurgical industry, 310–12
Métis, 98, 102, 121, 129–31, 152, 394; rebellion of, 138–40, 395–7
microregion, 10, 38
migration, 171–2, 557; French-Canadian, 290–1, 325; interprovincial, 215–17, 447; net, 171, 220; rural-to-urban, 216, 407, 426–7, 451, 453; Territorial North and, 507–8, 527–8; *see also* outmigration
Mi'kmaq, 444, 445, 465–6
military bases, 499
Millar-Western pulp mill, 425
mining industry, 50; Aboriginal workers in, 519; in Atlantic Canada, 441, 466–72; in BC, 359–61; closures in, 258–9, 488, 519, 526; impact of, 514–16; in Ontario, 258–9, 272; in Québec, 315–16, 303; in Territorial North, 492–3, 511–12, 516–19, 524–7; in Western Canada, 389–90, 421–3
minorities, visible, 205
modernization: in fishing industry, 458, 462–3; in Québec, 294–5; theory, 20–1, 38
Moffat, Ben, 129
Mohawk, 122
Montréal, 282, 320, 321–3; high tech in, 248, 301, 306; immigration to, 205–6; population of, 170, 171
Mooney, James, 207
Moosonee, ON, 54
mortality rate, 174, 220
mother tongue, 212, 281, 283
Mount Barbeau, 55
Mount Jacques Cartier, 56
Mount Logan, 51–2
Mulroney, Brian, 145, 297
multiculturalism, 186–7, 190–1; in Québec, 191
Multiculturalism Act, 190

Multilateral Agreement on Investment, 17
Murphy, Alexander B., 5
muskeg, 54, 70, 82

Nanaimo, BC, 375
Nass River, 341, 342
National Atlas of Canada, xx-xxi
'national border effect', 541
National Energy Program (NEP), 117–18, 402, 423
nationalism: Canadian, 296; Québec, 143–8, 295, 326; *see also* separatism
National Policy, 16, 110, 111, 237
nation-state, Canada as, 546
Nechako Forest, 367
Nechako River, 361–2, 363–4
Nelson River, 74
neo-Marxism, 20, 21
New Brunswick, 101, 109, 445, 447; ice storm in, 8; language in, 213; *see also* Atlantic Canada
newcomers/old timers faultline, xix, 204–6, 543, 544; *see also* immigration
Newfoundland, 103, 107, 437, 439, 442, 448; education in, 189; outports in, 451–3; Québec and, 293–4; settlement of, 444; *see also* Atlantic Canada
New France, 95–6, 287–90
Newhouse, David R., 188
newsprint, 365, 366, 380
'new world economic order', 240–2
'Next Year Country', 390, 414–15
Niagara Falls, 74
Niagara fruit belt, 247
Nisga'a, 124, 126, 159, 341, 342, 543–4
Nootka Convention, 346
Noranda Inc., 311–12
nordicity, 12, 13
Norman Wells, NWT, 499, 521, 522–4
Norsask Forest Products, 425
Norsk Hydro, 310–11
Nortel Networks, 269
North (Extreme, Far, Middle, Near), 13–14
North America: integration of, 16–18, 539–43, 545; security perimeter of, 4–5, 18, 194; trade patterns in, 12, 541–2
North American Air Defence Agreement (NORAD), 4
North American Free Trade Agreement (NAFTA), 4, 11, 17, 2, 19, 24, 240–2, 274, 540–1; forest industry and, 370–3
Northeast Coal Project, 353, 361
Northern Forest subregion, 368
northern lights, 489

'northernness', 12; *see also* nordicity
'northern vision', 112
North Pole, 12, 525
North West Company, 346–7, 393–4
North West Mounted Police (NWMP), 496
Northwest Passage, 493–4, 497
North-West Rebellion, 139–40
Northwest Staging Route, 499
Northwest Territories, 114, 487, 497–8, 502; *see also* Territorial North
North-West Territory, 102
Nova Scotia, 101, 447; *see also* Atlantic Canada
nunatuks, 53, 82
Nunavik, 327
Nunavut, 25, 114, 501; creation of, 107–8, 487, 497–8; as modern treaty, 124, 126, 160, 502–3; *see also* Territorial North
NWT Diamond Project, 516, 524–6

ocean: currents in, 62–3; distance from, 61
oil: dual market for, 117–18; industry, 52–3, 108–9, 387, 389, 400, 401, 402, 405–6, 422, 423, 428, 441; offshore, 437, 456–7, 466, 469; royalties from, 477–8; sands, 403; in Territorial North, 491, 492, 520–4, 525, 526, 527
Okanagan Valley, 340
old-age dependency ratio, 177–8, 220
Olfert, M.R., 427
Oliver, Frank, 133
Ontario, *xvii*, 12, 25–71, 107, 114, 115, 228–79, 301–2; environmental challenges in, 232–4; future of, 272–3; historical geography, 234–7; industry in, 236–45; language in, 213; northern, 229, 231, 244, 256–61, 270–2; physical geography of, 230–1; population in, 229, 230; southwestern, 244, 269–70
Orange Order, 140
Oregon Territory, 105, 347
Organization of Petroleum Exporting Countries (OPEC), 118
Ormsby, Margaret, 538
Ottawa, 248, 266–9
out-migration, 325, 437, 448, 450–1, 478, 507–8
outports, 451–3
outsourcing, 253, 274–5

Pacific Basin, 75
Pacific Ocean, 343
Pacific Rim, 338, 351

Paleo-Eskimos, 90, 92
Paleo-Indians, 90–1
Paleozoic Era, 47
Palliser, John, 392, 394; Palliser's Triangle, 390, 392, 395, 431
Papineau, Louis-Joseph, 137, 292
Parizeau, Jacques, 113, 145, 146, 297, 298
parkland, 391–2, 395
parks, national, 388–9
Parry, William, 494
partitionists, 146
patriation, 296–7, 330
patterned ground, 55, 82
pay equity, 203
payments: equalization, 438, 477–8, 539; transfer, 119–20, 498, 501, 502, 512
Peace River, 74, 361–2; country, 398, 409–10, 413–14
peripheral regions, 22
periphery, 19, 38
permafrost, 55, 69, 71–2, 77, 82, 491; alpine, 71, 81; construction in, 522–4; continuous, 71, 81; discontinuous, 71–2, 81; mining in, 514; sporadic, 72, 83
Petro-Canada, 403
petroleum industry, 108–9, 117–18; *see also* gas; oil
pH factor, 76
Phillips, David, 440
physiography, 10, 38, 43–59
physiological density, 164, 220
pingos, 55, 82
pipelines, 406, 499, 520–4
Plains Indians, 395–7
Plano culture, 91
plate tectonics, 45–6, 50
Pleistocene Epoch, 47, 48, 83
pluralism, 178, 190
podzolic soil, 83, 345
political parties, 87, 112, 143, 144, 159, 297, 402
pollution, 75–6; air, 232; water, 232–3, 286, 287; *see also* environmental challenges
population, 98, 100, 161–5; Aboriginal, 93, 96, 206–10, 211; age and sex structure of, 176–8; Canada and US, 43; change in, 171–8; density of, 163; distribution of, 164–5, 220, 258; foreign-born, 170; growth, 171–8, 183, 220; increase, 171–4, 183, 208–9, 220; in New France, 96; by province, 138; regional, 14, 16, 24, 157, 165–8; representation and, 114; strength, 170, 220; trends in, 178–91; in Upper and Lower Canada, 136, 137–8; urban,

168–71; zones, of, 165, 166–8
postglacial uplift, 54, 83
post-industrial stage, 160, 220
Poundmaker, 139
power: imbalances in, 22–3; sharing of, 538–9, 543–4, 546–7; struggles, 112–18
'Prairie psyche', 401–2, 404
Precambrian Era, 47
precipitation: annual, 63; in BC, 340, 344–5; convectional, 64, 65; frontal, 64, 65, 81–2; orographic, 64, 65, 82; in Western Canada, 390–1; *see also* snowfall
pre-industrial stage, 160, 220
primary sector, 194, 197; activities, 160, 161, 220; prices, 406, 431, 479; products, 23, 38
prime meridian, 59
Prince Edward Island, 56, 102, 114, 449, 473–4; *see also* Atlantic Canada
producer services, 355–6, 380–1
profits, corporate, 192, 193, 244
Project Crimson, 499
Province of Canada, 137, 292
provinces: equal, 142–3; language rights and, 210; powers of, 87, 88, 113, 142, 544
Prudhoe Bay, 520, 522
pulp: types of, 365, 380, 381
pulp and paper industry, 259–61, 314–15, 365–6, 424–5, 465; *see also* forest industry
Putnam, Robert G., 408, 468

quaternary activities (sector), 160, 161, 194, 197, 199, 220
Quaternary Period, 47, 83
Québec, xvii, 9, 12, 27–8, 101, 107, 109, 114, 115, 280–335, 546; culture in, 141–2; economy, 283–4, 299–302; education in, 189–90; environment and, 287; future of, 325–8; historical geography of, 287–95; identity in, 187; language in, 212–115; modernization in, 294–5; multiculturalism in, 191; nationalism in, 143–4; northern, 312–21; physical geography of, 284–7; population, 283; separatism in, 144–8; southern, 302–7
Québec Act, 135
Québec Cartier Mining Company, 316
Québec City, 96, 170, 320, 323–4
Quiet Revolution, 143, 152, 294, 296
quotas: fish, 463; grain, 415

racism, 205
radioactive waste, 392–3

railways: in BC, 347–8, 349, 351; orientation of, 18, 110, 158, 160, 193; resource development and, 257, 316; in Western Canada, 112, 388, 404, 447
rain shadow effect, 65, 83
rebellions (1837), 137, 291, 292
rebound, isostatic, 54
Reciprocity Treaty, 237
Red River: colony, 98, 394, 395; flooding, 59–60; Rebellion, 102, 129, 138–9, 395
referendums: in BC, 126, 341; in Québec, 144, 145–6, 297–9
reforestation, 261
refugees, 182
Regan, Gerald, 454
Regina, SK, 427
regional consciousness, 490
regional core, 388, 431
regional geography, 2–41
regionalism, 109–10, 546–7
regional service centres, 505–6, 168, 220
regional system, 21, 22
regions, 38; Canadian, xvii, 11, 12–15; changing, 15–16; characteristics of, 24–5; cultural, 6–7, 10; culture, 94–5; federation of, 537–9; natural, 10, 38; nature of, 9–10; physical, 6–7; physiographic, 46–57, 82
religion: culture and, 188–90; education and, 140–1, 189–90
Remedial Action Plans (RAPs), 234
Rennie, John, 195
replacement fertility rate, 175
representation, political, 113–15; see also House of Commons; Senate
reserves, Aboriginal, 122–3, 152
residual uplift, 54, 83
Resnick, Philip, 9
resource development, 112, 117, 480, 502
resource economy: Atlantic Canada, 339–40, 350–1, 353–7, 378–9, 463–74; Québec, 299, 326; Territorial North, 487, 488, 499–500, 512, 513–19; Western Canada, 387–8, 420, 421–5
resource frontier, 22
resource towns, 50, 168, 220, 422–3, 426, 474, 505–6
restrained rebound, 54, 83
restructuring, 239, 240, 275, 325; agricultural, 428; economic, 195, 197–8; effects of, 244–5
revolution, social, 158–9
Reynolds Aluminum, 310
Richards, Howard, 402

Riel, Louis, 129, 138–40, 395–6
roads, 516, 517, 518
rock: igneous, 45, 82; metamorphic, 45, 82; sedimentary, 45, 83
Rocky Mountains, 8, 50–1, 58
Roman Catholic Church, 135, 136, 140, 188, 189, 287, 291, 292, 314
Rose, Damaris, 3, 205
Ross, John, 494
Royal Canadian Mounted Police (RCMP), 496, 498
Royal Commission on Aboriginal Peoples, 127
Royal Commission on Bilingualism and Biculturalism, 142
Royal Commission on Energy, 117
Royal Proclamation, 104, 123
Rupert's Land, 138, 393, 497
Russia, 101, 104–5, 346; immigration from, 100–1, 131–2; see also Soviet Union
Ryan, Claude, 191

Sable Island, 469
Sahtu people, 124, 126
Saint John, NB, 476
Saint John River Valley, 473
St John's, 440, 476
St Lawrence Lowlands, 57, 58, 285–6, 288, 304–7
St Lawrence River, 9, 27, 281, 282, 287, 288, 289, 322
St Lawrence Seaway, 238, 282, 300
Sainte-Thérèse, QC, 254, 300, 306
SARS (Severe Acute Respiratory Syndrome), 265
Saskatchewan, 103, 107–8, 118, 398; Doukhobors in, 132–3; land claims in, 124; population, 131; rebellion in, 139–40; see also Western Canada
Saskatchewan Plain, 53
Saskatchewan Wheat Pool, 7
Saskatoon, SK, 426
Saul, John Ralston, 3, 326, 538, 544–5
Sault Ste Marie, ON, 271
scale: regions and, 10
Scandinavia, 100–1
schools: public/religious, 189–90; residential, 496; see also education
sea, inland, 52
secondary activities, 160, 161, 220
secondary sector, 195, 197–9
section, land, 99, 395
seigneurial system, 135, 288–90
seigneurs, 288–90, 330
Selkirk, Lord, 394
semiperipheral regions, 22
Senate, 114–15

'sense of belonging', 3
'sense of place', 7–9, 38, 490, 538
separatism, 109, 144–8, 160, 284, 295, 296–9, 326
service industry, 199–200, 355
settlement: physiography and, 57–8; in Territorial North, 500–1, 505–9; in Western Canada, 99–101, 395–8
settlement areas, 502–3, 530
sex: ratio, 177, 220; structure, 175–8
Sherbrooke, QC, 303–4
Shields, Douglas, 195
shipbuilding, 446
ships, 'post-Panamax', 474, 476
Sifton, Clifford, 100, 101, 397
Simpson, Jeffrey, 87, 134
Six Nations, 122
Slaymaker, O., 66
Smallwood, Joey, 454
Smith, Peter, 118
smog, 232
snowfall, 67, 233, 440
Social Credit, 402
social programs, 119–20
softwood lumber, 192, 315, 338, 364, 371–3, 381; see also forest industry
Softwood Lumber Agreement, 259, 373
soil(s): in BC, 345; climate and, 64–5, 67–8; order, 65, 83; in Western Canada, 390–2
sovereignty-association, 144, 152
Soviet Union, 499; see also Russia
spawning biomass, 461, 481
spillways, glacial, 53, 82
Springhill Mine, 467
Stabler, Jack C., 427
staples: production, 368, 381; thesis, 23
steel industry, 264, 271, 447, 467–9; see also iron and steel industry
Stelco, 245, 264
Stephens, Tom, 378
Steyn, Mark, 4
Stikine Plateau, 343
strata, 45, 83
striations, glacial, 49, 82
strip farming, 412, 431
stumpage fees, 369–70
subregions, 10, 13–14
subsidence, 83
subsidies: agricultural, 406, 408–9, 416; government, 541; regional, 455–6, 468–9, 479; transportation, 399, 404, 408, 416–18
Sudbury, ON, 270–1, 272
superport, 474, 476
Supreme Court, 159, 214, 298
surveying, land, 99, 129, 395
Sydney Steel Co. (Sysco), 443, 468

Sydney Tar Ponds, 443

Taiwan Sugar Corporation, 393
Talon, Jean, 288
tar ponds, 443
tar sands, 523
tariffs, 16, 24, 111, 237, 251, 447
taxation, 115–16, 372, 513, 543
Taylor, Charles, 144
technology: in forest industry, 260–1; hydroelectric, 308; new, 198–9; *see also* biotechnology; high technology
Teck Corporation, 361
tectonics, plate, 45–6, 50
Tellier, Paul, 160
temperature, 61, 62, 64, 76–9
tensions, regional, 110–11; *see also* fault-lines
terraces, 514, 530
Terra Nova project, 457, 469
Territorial Formula Financing, 119
Territorial North, *xvii*, 12, 112, 32–3, 115, 486–535; economy, 509–13; environment and, 492–3, 514–16; future of, 527–8; historical geography, 493–501; land claims and, 503–5; megaprojects in, 519–27; physical geography, 490–2; population, 488–9, 497–8, 501, 505–9; resource development, 513–19
terrorism, 3, 4–5, 17–18
tertiary activities (sector), 160, 161, 194, 197, 199–200220
textile industry, 300, 306
'thermokarst topography', 77
Thompson, MB, 422–3, 426
Thorpe, Jacqueline, 191
Thule people, 78, 90, 92–3
Thunder Bay, ON, 271–2
till, 48–9, 83
Timmins, ON, 270–1
Tlicho Final Agreement, 490
Torngat Mountains, 49
Toronto, 263–6, 322–3; high tech in, 248; immigration to, 205–6; population of, 170, 171
tourism, 272, 286, 319–21, 352, 355, 449, 480, 527
Tracie, Carl, 133–4
trade: Aboriginal peoples and, 91–2, 128, 495–7; agreements over, 192–4; in Atlantic Canada, 446–7; barriers to, 193–4; 240, 241, 416, 453–4, 541–2; in BC, 351; direction of, 541–2; disputes over, 192, 193–4, 241; global, *xvii*, 240–2, 539–43; Ontario and,

239–42; US and, 3–5, 10–12, 17–18, 110, 362
traditional ecological knowledge, 511, 530
'tragedy of the commons', 358, 381, 443, 461
transition, economic, 416–20
transportation, 112, 282, 322, 447–8; BC, 349, 350–1, 360–1, 376–7; integration of, 193; in Ontario, 256–7, 260–1, 271–2; technology, 199; in Territorial North, 491, 498, 499, 512–13, 517, 518–19; in Western Canada, 387–8, 399, 411, 417–18, 428
trapping, 509–10
treaties: Aboriginal peoples and, 102, 104, 122–3, 128–9, 341–2, 396; *see also* land claims
Treaty Negotiations Referendum, 341
troughs, glacial, 51, 82
Trudeau, Pierre, 4, 117, 159
Tuktoyaktuk, NWT, 56
Tumbler Ridge, BC, 353, 361
tundra, 70, 491
Turner Valley, AB, 422
Tyrrell Sea, 54

Umbrella Final Agreement, 504–5
unemployment, 191, 192, 200, 201, 245, 284, 437, 479
unemployment insurance, 455
unilingualism, 214–15
United Grain Growers, 402
United States: borders with, 104–6; cultural impact of, 187; disputes with, 311, 370–3, 415, 460; geography of, 4, 43; integration with, 160, 192–4, 540–3, 546; Manifest Destiny and, 101; Ontario and, 229, 237, 239–42, 244–5, 250–6, 272–3; Pacific Coast and, 347–8, 358–9; Quebec and, 299–300, 316; Territorial North and, 499, 522; trade with, 3–5, 10–12, 17–18, 110, 160, 362
Université Laval, 140
Upper Canada, 136–7, 234–5
upward transitional regions, 22
uranium, 389–90, 400, 421–2
urban population areas, 221
urbanization, 168–71

value added, 251, 253, 275; production and, 368, 381
Vancouver, George, 346
Vancouver, BC, 12, 64, 337, 345, 354–5, 373–7; immigration to, 205–6; popu-

lation of, 170, 171, 349
Vancouver Island, 58, 106
vegetation, natural: climate and, 64–5, 67–8, 70; BC, 345; in Western Canada, 390–2
Victoria, BC, 63, 345, 349, 354–5, 374, 375
Vikings, 90
Villeneuve, Paul, 147
visible minorities, 205
Voisey's Bay nickel mine, 127, 441, 466, 469–72
volcanoes, 50

Walkerton, ON, 232–3
Wallaceburg, ON., 244
Wallerstein, Immanuel, 21
War of 1812, 234–5
water: 'deficit', 391; fresh, 58; tainted, 232–3
waterways, inland, 58
weather: events, 8, 59; extreme, 284–5, 440; *see also* climate
weathering, 45
Welland Canal, 238
western alienation, 116–17, 401–2, 404, 544
Western Canada, xvii, 12, 29–30, 386–455; cities, 425–8; economy of, 387–8, 398–401, 404–6, 408–28; historical geography of, 393–404; population, 406–8; settlement of, 99–101, 395–8; unrest in, 111–12
Western Sedimentary Basin, 52, 389, 422, 423, 431
Westray Mine, 467
Weyerhaeuser Canada, 368–9, 378
wheat: Marquis, 130, 399; Red Fife, 130
Whitehorse, 505
White Paper on Aboriginal policy (1969), 124, 159
White Rose project, 457, 469
winds, 62
Winnipeg, MB, 59–60, 64, 426, 170, 427
Wisconsin ice advance, late, 48
women: labour force and, 202–3; occupations of, 554
Wood Buffalo, AB, 403
World Trade Organization (WTO), 11, 17, 192, 240, 251, 305, 540–1

Yellowknife, 505
Yukon, 487, 497, 502; *see also* Territorial North
Yukon First Nations, 124, 126, 504